The electrical resistivity of metals and alloys

Cambridge Solid State Science Series

PAUL L. ROSSITER

Department of Materials Engineering, Monash University
Clayton, Victoria, Australia

The electrical resistivity of metals and alloys

CAMBRIDGE UNIVERSITY PRESS

Cambridge
New York Port Chester
Melbourne Sydney

Published by the Press Syndicate of the University of Cambridge
The Pitt Building, Trumpington Street, Cambridge CB2 1RP
40 West 20th Street, New York, NY 10011–4211
10 Stamford Road, Oakleigh, Melbourne 3166, Australia

First published 1987
First paperback edition (with corrections) 1991

British Library cataloguing in publication data

Rossiter, Paul L.
The electrical resistivity of metals and alloys.
1. Metals—Electric properties
I. Title
620.1′697 TN690

Library of Congress cataloguing in publication data

Rossiter, Paul L.
The electrical resistivity of metals and alloys.
Bibliography:
Includes index.
1. Free electron theory of metals. 2. Metals—Electric properties. 3. Alloys—Electric properties. 4. Electric resistance. I. Title.
QC176.8.E4R67 1987 620.1′697 86-9599

ISBN 0 521 24947 3 hardback
ISBN 0 521 40872 5 paperback

Transferred to digital printing 2003

To my Mother and Father

Contents

	Preface	xiii
1	**Basic concepts**	1
1.1	Introduction	1
1.2	Conduction electron scattering in solids	1
1.3	Scattering anisotropy	7
1.4	Effects of the scale of microstructure	7
1.5	Matthiessen's rule	9
1.6	Simple and non-simple metals	10
1.7	Elastic and inelastic scattering	15
	1.7.1 Electron and phonon energies	15
	1.7.2 Conservation of momentum	15
	1.7.3 Magnetic scattering	16
1.8	The Boltzmann equation and relaxation time	17
	1.8.1 Wavepackets	17
	1.8.2 The linearised Boltzmann equation	21
	1.8.3 The relaxation time approximation	22
	1.8.4 Calculation of the resistivity in the relaxation time approximation	22
	1.8.5 Other solutions of the Boltzmann equation – anisotropic relaxation times	24
	1.8.6 Other formalisms	28
2	**Atomic configuration of an alloy**	30
2.1	Dilute and concentrated alloys	30
2.2	Correlation parameters in crystalline materials	30
2.3	Composition waves	39
2.4	Reciprocal space representation	40
2.5	Short range atomic configurations	41
	2.5.1 Mode of decomposition	41
	2.5.2 Phase separation	43
	(a) Clustering	43
	(b) Precipitation	45
	(c) Spinodal decomposition	46
	2.5.3 Atomic ordering	49
	(a) Type I homogeneous (statistical) SRO	49

(b) Type II(a) heterogeneous SRO (microdomain model) 50
(c) Type II(b) heterogeneous SRO (antiphase domain model) 50
2.6 Long range atomic correlations 50
2.6.1 Long range ordering 51
2.6.2 Two-phase mixtures 54
2.6.3 Some general comments 54
2.7 Atomic displacement effects 57
2.7.1 Atomic size effects 58
2.7.2 Dynamic atomic displacements 61
(a) Einstein model 61
(b) Debye model 63
2.7.3 Displacive phase transitions 67
2.8 Amorphous alloys 69
2.8.1 Static atomic structure 71
2.8.2 Dynamic fluctuations 75

3 The structure of magnetic materials 76
3.1 Collective electron and localised moment models 76
3.2 Magnetic configuration 80
3.2.1 Isolated moments 80
3.2.2 Spin glasses 84
3.2.3 Magnetic clusters 87
3.2.4 Long range magnetic order, $T < T_c$ 89
3.2.5 Short range magnetic order, $T > T_c$ 94
3.2.6 Magnons 95
3.3 Nearly magnetic metals – spin fluctuations 98
3.4 Effects of atomic rearrangements 102
3.4.1 Long range effects 103
3.4.2 Short range effects 104

4 Electrons in simple metals and alloys 107
4.1 Scattering potentials and electron wavefunctions 107
4.2 Pseudo- and model potentials 115
4.3 Electron–electron interactions 119
4.3.1 Screening in metals 120
4.3.2 Exchange and correlation 122
4.4 Nearly free electron theory 124
4.5 The scattering matrix 127
4.5.1 The first and second Born approximations 128
4.5.2 Factorisation of the matrix elements 129

4.5.3 The pseudopotential in alloys 130
4.5.4 The pseudopotential in a deformed lattice 134

5 Electrical resistivity of simple metals and alloys 137
5.1 A general resistivity expression 138
5.2 The resistivity of alloys with short range atomic correlations 139
5.2.1 Homogeneous atomic correlations 141
5.2.2 Inhomogeneous atomic correlations 143
(a) Small zone limit 145
(b) Intermediate zone size 147
(c) Large zone limit 150
5.3 Homogeneous long range atomic ordering 160
5.3.1 Conduction electron scattering effects 160
(a) Bragg–Williams model 162
(b) Coexisting long and short range ordering 162
5.3.2 Electron band structure effects 164
5.4 Inhomogeneous long range ordering 167
5.5 Long range phase separation 169
5.5.1 Scale of phase separation $\gg \Lambda$ 170
5.5.2 Scale of phase separation $\sim \Lambda$ 184
5.6 Atomic displacement effects 195
5.6.1 Point defects and displacements 195
(a) Vacancy 195
(b) Substitutional impurity 197
(c) Self-interstitials 197
(d) Impurity interstitial 198
5.6.2 Thermally induced displacements 198
5.6.3 Static atomic displacements in a concentrated alloy 203
5.6.4 Displacive transitions 207
5.6.5 Combined effects 208
5.7 Some applications 212
5.7.1 Phonon scattering 212
(a) Alkali metals 213
(b) Noble metals 218
5.7.2 Residual resistivity of disordered random solid solutions 220
(a) Dilute alloys 221
(b) Concentrated alloys 236
5.7.3 Homogeneous short range atomic correlations 237
5.7.4 Long range ordering 253

	5.7.5 Precipitation	257
	5.7.6 Long range phase separation	261
	(a) Scale of phase separation $\gg \Lambda$	261
	(b) Scale of phase separation $\leqslant \Lambda$	264
	5.7.7 Displacive transitions	271
6	**Non-simple, non-magnetic metals and alloys**	272
6.1	Band structure and the electrical resistivity	273
6.2	Models and pseudopotentials in non-simple metals	274
6.3	The phase shift method	279
6.4	The *T*-matrix	284
6.5	Advanced phase shift methods: the KKR–Green's function method	287
6.6	Some applications	289
	6.6.1 Pure noble and transition metals	289
	6.6.2 Dilute alloys: bound and virtual bound states	291
6.7.	Concentrated alloys	300
	6.7.1 First-order theories: the virtual crystal and rigid band approximations	300
	6.7.2 Advanced theories: the average *t*-matrix approximation (ATA) and coherent potential approximation (CPA)	306
7	**Magnetic and nearly magnetic alloys**	318
7.1	Magnetic materials with long range magnetic order	318
	7.1.1 Overview	318
	7.1.2 Two-sub-band model	323
7.2	Local environment effects and magnetic clusters	334
7.3	Nearly magnetic systems: local spin fluctuations	339
	7.3.1 Kondo alloys	339
	7.3.2 Exchange-enhanced alloys	341
	7.3.3 Composition dependence	345
	7.3.4 Nearly magnetic pure metals and concentrated alloys	348
7.4	Spin glasses	351
8	**Other phenomena**	356
8.1	Resistivity at the critical point	356
	8.1.1 Some general comments	356
	8.1.2 The electrical resistivity near T_c	358
	(a) Ferromagnets	363
	(b) Antiferromagnets	365

(c) Atomic order–disorder 366
(d) Miscibility gap 368
8.1.3 Related phenomena 370
8.2. Highly resistive materials 372
8.2.1 Some general observations 372
8.2.2 $\Lambda > a_0$ 375
(a) Diffraction models and the Debye–Waller factor 375
(b) CPA, interband and other band-based calculations 377
8.2.3 $\Lambda \sim a_0$ 378
8.2.4 Some general comments 379
8.3 Amorphous metals 380
8.3.1 General observations 380
8.3.2 Resistivity in non-magnetic glasses 382
(a) $T \geqslant \Theta_D$ 382
(b) $\Theta_D > T > 0$ 386
8.3.3 Resistivity of metallic glasses containing magnetic components 388
(a) Ferromagnetic behaviour 388
(b) Spin glasses 389
8.3.4 Resistivity minima 391

Appendices
A Units 393
B Integrations over d**k**, dS, dE and dΩ 394
C The average $\langle \exp(\mathrm{i}x) \rangle$ 396
D High and low temperature limits of $\rho_p(T)$ 397
E Determination of $2k_F R_i$ in a nearly free electron solid 398

References 399

Index 421

Preface

The electrical resistance or resistivity of a conducting solid can be experimentally determined without much difficulty and for many years it has been used as a research tool to investigate various microstructural and physical phenomena. However, unlike conventional diffraction methods which are capable of mapping out scattered intensities in two- or three-dimensional space, an electrical resistivity measurement gives only a single value (at any fixed temperature and structural state) representing an average over all directions of conduction electron scattering. As there is no means of performing the back-transform from this single point, the analysis of resistivity data in terms of microstructure must incorporate calculations of conduction electron scattering based on some model of the structure or microstructure concerned. With the refinement and greater availability of more direct methods, particularly X-ray, neutron or electron diffraction, transmission and analytical electron microscopy (especially atom-probe field ion microscopy which allows an atom-by-atom picture of a material to be established) there has been a declining utilisation of such an indirect method in microstructural investigations. Nevertheless, while a resistivity study may require support from some other technique to allow an unambiguous interpretation of the results, there are many cases where such studies still have particular value, either by virtue of their simplicity or lack of alternative techniques. These include studies of defects, pre-precipitation processes, short and long range ordering or phase separation (particularly with respect to transformation kinetics) and determination of critical transformation compositions and/or temperatures. These studies make use of the sensitivity of the conduction electron scattering process to microstructural details right down to the atomic scale, and the fact that it provides a convenient average over the volume of a specimen.

The electrical properties of metals and alloys are also of great practical importance, especially in applications involving heating, temperature measurement, signal and power transmission, precision resistances, switching devices, semiconducting and thin film devices or simply specification of purity. Development of new methods and materials for such applications will be assisted by a knowledge of the physical processes which determine those properties. On a more fundamental level there is the basic need to understand the process of scattering of conduction electrons in solids, particularly in inhomogeneous (i.e. real) solids. In this regard, determination of the resistivity also provides a useful test of some of the elegant electronic band structure determinations that have been carried out over the last few years.

There have been many good texts devoted to the basic aspects of electrons in solids and reviews of specific topics such as electrical properties of pure metals, galvanomagnetic properties of pure metals and deviation from Matthiessen's rule. However, to the author's knowledge there has been no other text devoted to the problem of understanding the electrical properties of concentrated and often inhomogeneous solids. It is hoped that this text will help fill the gap.

This book is thus unashamedly devoted to understanding the electrical properties of real metals and alloys. Because of the complexity of the structures concerned, this often means that some aspects of the work lack the elegance of the more profound theoretical works on ideal materials. Nevertheless, it is hoped that the text will be of use to those interested in such properties and indicate where more research effort is required. Rather than devote space to a formal derivation of basic equations concerning electron states in solids (which is available in many other texts), we will assume that these are known and concentrate more on the aspects of electrical conduction, particularly its dependence upon composition and atomic or magnetic structure.

The general problem is introduced in Chapter 1 in terms of simple diffraction concepts, modified slightly to take band structure effects into account. This has the advantage that those readers who have expertise in the allied fields of X-ray, neutron or electron diffraction but who are less confident in matters concerning electron states will nevertheless be readily able to gain a feeling for the problem. However, it would be wrong to pretend that the electrical properties of alloys which may have complicated electronic, atomic and magnetic structures could be understood quantitively on the basis of a simple theory. In such cases a full understanding of the problem requires facility with complicated and often highly specialised techniques. While it is beyond the scope of this book to give a full tutorial in such techniques, the general models and formalisms will be introduced and related to the problems at hand. Thus, while a reader may not be familiar with ensemble averages or Green's functions, it is hoped that he or she will be able to gain an understanding of the direction taken by modern theoretical approaches, of the problems that have been addressed and to what extent a satisfactory solution has been found. It may then be possible to judge whether efforts should be made to become more familiar with the particular techniques concerned. However, it is necessary to have a good grounding in quantum mechanics in order to achieve mastery of many of the advanced techniques and readers are so warned. The first chapter also introduces the problems associated with anisotropy of scattering over the Fermi surface and finite conduction electron mean free path, and a working definition of 'short' and 'long' range effects is given. The similarity between replacive disorder (i.e. that to do with atom type) and displacive disorder (i.e. that to do with atomic position) is also briefly discussed and Matthiessen's rule is introduced. The foundation of the Boltzmann equation is also considered as are some of the methods of solution and alternative approaches.

The second chapter is entirely concerned with a discussion of microstructure and definition of parameters which are required for the description of atomic positions and correlations in crystalline or amorphous solids. These concepts are extended to magnetic structures in Chapter 3 and some of the dynamic aspects of isolated spins and spin systems are also discussed. Nearly free electron theory and the pseudopotential approximations are discussed in Chapter 4.

The concepts developed in Chapters 1, 2 and 4 are brought to bear on the determination of the electrical resistivity in Chapter 5, allowing formulation of equations relating this to a variety of structures containing short and long range atomic correlations. The effects of static and dynamic atomic displacements are also considered. Some of the methods appropriate to non-simple metals and alloys are introduced in Chapter 6. The resistivity of the magnetic and nearly magnetic structures discussed in Chapter 3 is considered in Chapter 7. However, the situation here is much less satisfactory as it appears that the assumption of independent electrons made explicit in the earlier chapters is not adequate to determine the magnetic spin–spin correlations. There is still much work to be done in deriving a realistic first principles calculation of the resistivity of such materials. The particular problems associated with the critical point, high resistivity materials and amorphous metals are considered in the final chapter.

In order to allow for an uninterrupted development of the theory, particularly in Chapters 5 and 6, examples and applications of the concepts and equations derived are generally given in separate sections following those devoted to the presentation and development of that theory. Thus, if a reader is interested in a specific problem, he or she should be able to follow almost from first principles and without too much interruption the development of the theory relating to that problem, but will need to turn to the appropriate later section to find the examples and applications.

The symbols used throughout this book are generally the same as widely employed in the literature. However, as the range of topics covered is quite broad, this often means that the same symbol is used in a different context in relation to different problems. For example, the symbol α is initially used as an atomic correlation parameter but is also used to indicate the ratio of sub-band resistivities in a two-band model of conduction as well as a critical exponent. This problem could be avoided by the invention of a new set of symbols but only at the risk of greater confusion. Where practical a distinction is made with the aid of sub- and superscripts and in all cases the parameters are redefined when they take on a new meaning. Similarly, this work draws on the results of many different fields and, despite efforts to promote the acceptance of SI units, many of these fields have evolved their own 'preferred' units such as Rydbergs or electron-volts for energies associated with electron states and still the micro-ohm cm as the unit of resistivity. Thus, while the formulas derived in this text are correct within the SI system of units, again in order to avoid confusion at the interface between this work and the majority of other published results, the input data and results will generally be given in terms of these preferred units. The conversion factors necessary to obtain SI units are given in Appendix A.

Finally, it is with much pleasure that I acknowledge the valuable discussions that I have had with many colleagues throughout the world and which have been invaluable in moulding the contents of this book. Rather than offend anybody by my forgetfulness which could lead to unintentional omissions from a list of names, I would simply like to thank them all for their interest and helpfulness. With regard to the actual production of this text the situation is much more straightforward as most of the work has fallen on comparatively few shoulders. In this regard I would especially like to thank my wife Kathy for word-processing my scratchy handwriting and for putting up with a rather obsessed author for the past twelve months, Mrs L. Lyons for producing the bulk of the artwork and Ms J. Fraser for photographic assistance. Much of the work was completed while I was on study leave from the Department of Materials Engineering and I would like to thank that Department and Monash University for the opportunity of taking the leave and the members of the Physics Department at Monash for their hospitality during this period.

P. L. Rossiter
1986

1 Basic concepts

1.1 Introduction

Understanding the physical processes that determine the electrical resistivity of a concentrated metallic alloy is a daunting task because of the large number of possible contributions that could be involved. In addition to conduction electron scattering from thermally induced atomic displacements (which may depend upon concentration and degree of atomic and magnetic order) there will be other direct contributions from atomic and magnetic disorder, strain and band structure effects. The magnitude of such effects will be influenced by the homogeneity of the microstructure and will depend specifically upon whether the spatial extent or 'scale' of the inhomogeneity is greater or less than the conduction electron mean free path length.

The purpose of this first chapter is to introduce in a general way the relationship between the electrical resistivity and conduction electron scattering and band structure effects. It will be assumed that the reader is familiar with the fundamental concepts of electron waves in solids which have been very adequately considered in a variety of other texts (Ashcroft & Mermin 1976; Coles & Caplin 1976; Harrison 1970; Kittel 1976; Mott & Jones 1936; Blatt 1968; Ziman 1960, 1969, 1972). Other topics which are not specifically considered in detail in this text but which have been considered elsewhere include the electrical properties of pure metals (Meaden 1966; Wiser 1982; Pawlek & Rogalla 1966; Bass 1984; van Vucht *et al.* 1985), galvanomagnetic effects (Hurd 1974; Jan 1957), deviations from Matthiessen's rule (Bass 1972) and the electrical properties of intermetallic compounds (Gratz & Zuckermann 1982; Gratz 1983; Schreiner *et al.* 1982; Dugdale 1977, p. 279). A compilation of experimental data relating to the electrical resistivity of binary metallic alloys and rare-earth intermetallic compounds has recently been published by Schröder (1983).

1.2 Conduction electron scattering in solids

The electrical resistivity of a solid can be determined by passing a current i through the specimen of cross-section area a and measuring the resultant voltage drop v over a distance l. The electrical resistivity ρ is then given by

$$\rho = \frac{va}{il}$$
$$= \frac{ra}{l}, \tag{1.1}$$

where r is the resistance of the specimen between the potential contacts. Despite the general acceptance of SI units, the resistivities of metals and alloys are usually given in units of $\mu\Omega$ cm (units are discussed in more detail in Appendix A). Under the influence of an applied field the conduction electrons drift through an ionic array, the resistivity being determined by the rate at which they are scattered from some initial state $\Phi_{\mathbf{k}}$ into a final state $\Psi_{\mathbf{k}'}$. This may be represented in **k**-space as shown in Figure 1.1. As evident from the Fermi–Dirac distribution of electron energies (discussed later in relation to equation (1.23)), only electrons within an energy range $\sim k_B T$ about the Fermi surface can increase their energy by some small amount under the influence of the external field. However, since the Fermi energy $E_F \gg k_B T$ over the normal range of temperatures of interest, the vectors **k** and **k**′ must terminate on the sharply defined Fermi surface. Note also that in the case of a spherical Fermi surface the maximum amplitude of the scattering wave vector is equal to $2k_F$. This scattering rate will be determined by the strength of the scattering potential $V(\mathbf{r})$ and, in non-simple metals, the availability of states into which the electrons can be scattered. In terms of Fermi's

Fig. 1.1. Schematic representation of the scattering of a conduction electron from an initial state **k** to a final state **k**′.

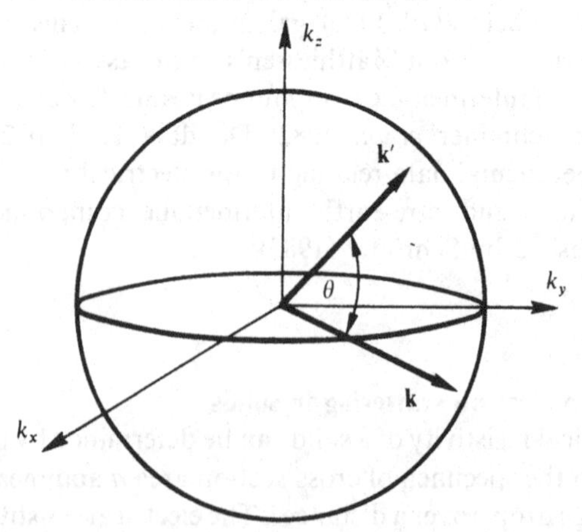

'golden rule' the scattering probability can be written as

$$P_{\mathbf{kk}'} = \frac{2\pi}{\hbar} |\langle \Psi_{\mathbf{k}'} | V(\mathbf{r}) | \Phi_{\mathbf{k}} \rangle|^2 N(E_F), \tag{1.2}$$

where $\langle \Psi_{\mathbf{k}'} | V(\mathbf{r}) | \Phi_{\mathbf{k}} \rangle$ is the scattering amplitude for transitions between an initial state $\Phi_{\mathbf{k}}$ and a final state $\Psi_{\mathbf{k}'}$ (i.e. the matrix element of the scattering potential $V(\mathbf{r})$ between the states $\Phi_{\mathbf{k}}$ and $\Psi_{\mathbf{k}'}$) and $N(E_F)$ is the density of states at the Fermi energy E_F into which the electrons can be scattered. This latter term arises because of the necessity to have vacant states ready to accept the scattered electrons. Readers who may be unfamiliar with the notation in equation (1.2) should not despair. Examples will be given in the following chapters which show that many of the quantities required are available in the literature and that, in certain simplified cases, calculations can be performed on a programmable calculator or personal computer.

This scattering rate may be approximately described in terms of a relaxation time τ averaged over the Fermi surface. In the simple case of a spherical Fermi surface (i.e. free conduction electrons) $|\mathbf{k}'| = |\mathbf{k}| = k_F$ and the scattering probability $P_{\mathbf{kk}'}$ will depend only upon the angle θ between $\mathbf{k}$ and $\mathbf{k}'$. The relaxation time averaged over the Fermi surface can then be written as (see Chapter 5)

$$\frac{1}{\tau} \propto \int P(\theta)(1 - \cos\theta)\, d\mathbf{S}$$

or (1.3)

$$\frac{1}{\tau} \propto \int P(\theta)(1 - \cos\theta) \sin\theta\, d\theta,$$

where $P(\theta)$ is now simply the probability of scattering through an angle θ into the element of area $d\mathbf{S}$ on the Fermi surface and the integration variables are discussed in Appendix B. The term $(1 - \cos\theta)$ essentially arises because we are only interested in the total change in momentum resolved in the direction of the electric field. For example, it we take the x direction to be in the direction of the applied field, the change in the contribution to the current from an electron will depend only upon the change in the x component of its velocity. Since the wavevector $\mathbf{k}$ of a free electron will be in the same direction as its velocity, the change in the contribution to the current will be proportional to $(k_x - k'_x)/k_x$. (A more rigorous derivation will be given in Section 5.1.) The geometry of the problem in three dimensions is shown in Figure 1.2 and leads to the following relationships:

$$\left.\begin{aligned} k_x &= k_F \cos\alpha \\ k'_x &= k_F(\cos\alpha \cos\theta + \sin\alpha \sin\theta \cos\phi) \end{aligned}\right\}. \tag{1.4}$$

Averaging over ϕ then gives the required result $(k_x - k'_x)/k_x = 1 - \cos\theta$. This term also has an important physical implication: it indicates that scattering through large angles is more important in determining the resistivity than small angle scattering. The significance of this fact will be emphasised in later chapters.

The final step in determining the resistivity is given by the Drude formula

$$\rho = \frac{m}{ne^2\tau}, \tag{1.5}$$

where n is the number per unit volume of electrons of mass m and charge e. It should be emphasised that this simple derivation has been given mainly to illustrate the link between the scattering process and the electrical resistivity. The assumptions of free electrons and a uniform scattering rate over a spherical Fermi surface are clearly severe restrictions on its applicability. One might expect them to be reasonable in the case of monovalent metals where the Fermi surface lies entirely within the first Brillouin zone, but generally the distortion of the Fermi surface and its intersection with the Brillouin zone boundaries might be expected to produce deviations from such simple behaviour in most metals or alloys. Nevertheless, in many concentrated alloys of interest (particularly those that do not contain transition metals) it appears that the Fermi surface is still roughly spherical, at least as far as the majority charge carriers are concerned, so that a 'nearly' free electron calculation can proceed. In other cases one clearly must take into account the effects of stronger scattering on the band structure of the alloy. These problems are discussed in more detail in relation to specific alloy systems in the chapters that follow. At this stage let us continue to use this simple model to illustrate some more features of the scattering process.

Fig. 1.2. Scattering geometry in three dimensions: ϕ is the angle between the planes defined by **k**, **k**′ and **k**, x axis and α is the angle between **k** and the x axis.

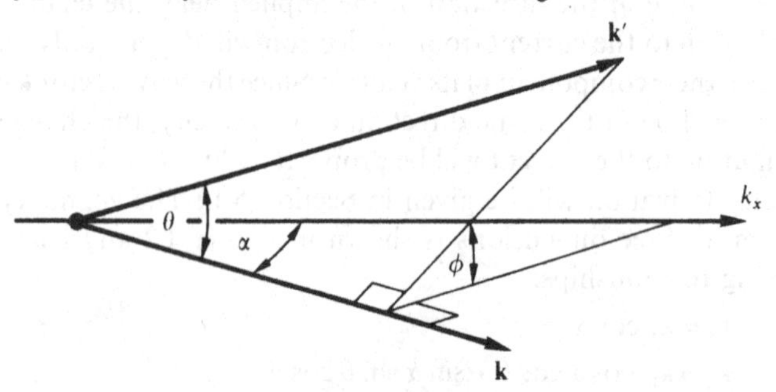

The degree of periodicity of the ionic array determines the amount of electron scattering and hence the electrical resistivity. For example, in a perfectly periodic array (implying an infinite array of identical ions each at rest on a periodic lattice site) the electrons will suffer only Bragg scattering. (We treat here the general case where the electron wavelength is small enough to allow diffraction by the lattice, i.e. the Fermi surface intersects one or more Brillouin zone boundaries. If this is not the case there will be no Bragg scattering to worry about.) While the probability of scattering is then very large at the Bragg wavevectors, the scattering probability averaged over the Fermi surface is zero. This is because the Bragg scattering is very sharp and localised to a vanishingly small fraction of the Fermi surface, the width of the Bragg peaks being inversely proportional to the number of ions in the array, N. This behaviour is illustrated in Figure 1.3 which shows schematically the

Fig. 1.3. Bragg scattering: (*a*) in an ideal diffraction experiment and (*b*) superimposed on the Fermi surface. The Brillouin zone boundary is shown as a dashed line.

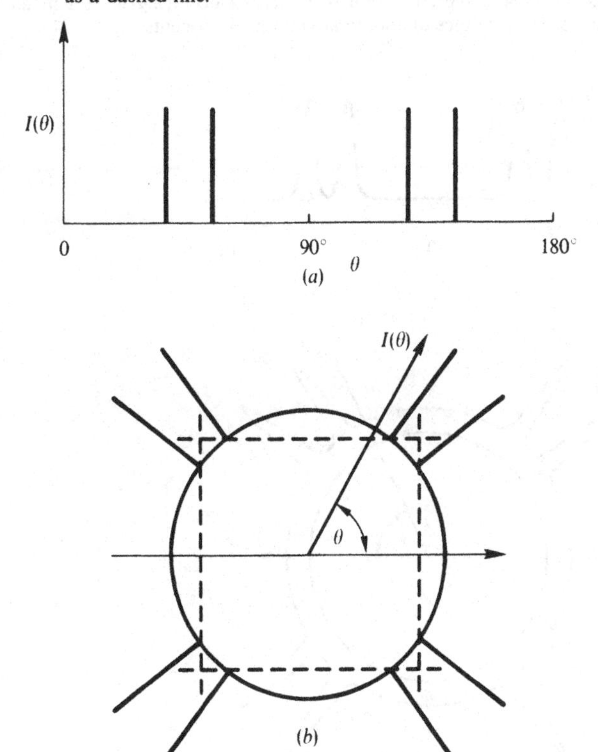

Bragg scattering as it would be measured in a conventional diffraction experiment (corrected for all other sources of scattering and experimental broadening) and shows how such scattering would appear when plotted in a two-dimensional polar form in **k**-space. This form of plotting allows visualisation of the scattering superimposed on the Fermi surface.

The first Brillouin zone is also shown to emphasise the fact that Bragg diffraction occurs where the Fermi surface intersects the Brillouin zone boundary, this simply being a geometrical restatement of Bragg's law. In real solids the lack of perfect periodicity will give rise to additional diffuse scattering contributions, in direct analogy to the additional scattering (e.g. size or strain induced X-ray line broadening, short range order diffuse scattering and thermal diffuse scattering) observed in a diffraction experiment, as shown in Figure 1.4. This additional scattering will lead to a non-zero scattering probability when averaged over the Fermi surface and hence result in a finite value of the electrical resistivity.

Fig. 1.4. Schematic representation of scattering occurring in real materials showing the appearance of additional diffuse components.

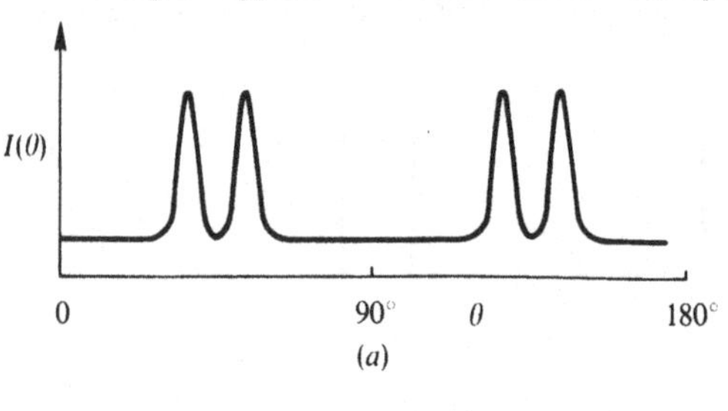

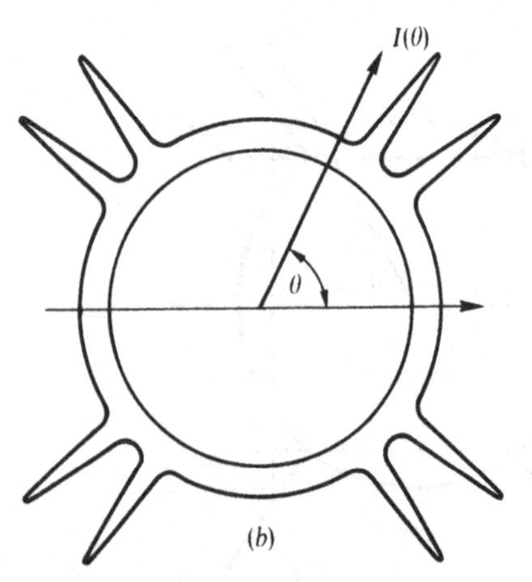

1.3 Scattering anisotropy

Figure 1.4 also illustrates another important point. A purely random, free electron alloy would produce isotropic scattering with the scattering probability (and hence the relaxation time τ) having the same value at all points on the Fermi surface. However, as the deviation from randomness increases, the diffuse scattering will become unevenly distributed over the Fermi surface, leading to an anisotropic relaxation time $\tau(\mathbf{k})$ that varies over the Fermi surface. Similar behaviour will result if the electron wavefunctions are not spherically symmetric since the electrons at different places on the non-spherical Fermi surface will have wavefunctions of different s-, p- and d-like character. Applying equation (1.3) would then lead to errors since it gives only an average reciprocal relaxation time $\langle 1/\tau(\mathbf{k}) \rangle$ rather than properly taking into account the parallel contribution of all electrons to the total current flow. We return to this important problem later on. In the meantime it is important to note that an anisotropic relaxation time does not imply a direction dependence of the resistivity in real space. This is because the resistivity is determined from an average over all scattering directions (see, e.g. equation (1.3)) and will thus be isotropic in a cubic system in real space. In fact it is this very averaging process that makes interpretation of resistivity data so difficult. Unlike conventional diffraction techniques, which can map out scattered intensities in two- or three-dimensional reciprocal space, an electrical resistivity measurement gives only a single value at any particular temperature and state of disorder. As there is no means of performing the back-transform from this single point, analysis of resistivity data must rely on calculation of electron scattering based on some model of the structure or microstructure concerned.

1.4 Effects of the scale of microstructure

It is also important to consider the spatial extent or 'scale' of the atomic or magnetic correlations. This is because there is a characteristic conduction electron mean free path Λ defined by

$$\Lambda = v_F \tau, \tag{1.6}$$

where v_F is the velocity of the electron at the Fermi level. (Note that this suggests that τ may be regarded as a 'mean free time'.) This quantity is based on an average scattering probability and should not be simply interpreted as the real distance between scattering centres (see e.g. Sondheimer 1952). However, it may be taken as a measure of the coherence length of the electron wave. Within this context, the conduction electrons will only be scattered coherently by deviations from perfect periodicity which occur within a volume $\sim\Lambda^3$. This is because scattering centres separated by a distance greater than Λ do not produce coherent scattering, the electron having effectively 'forgotten' its phase once it has travelled a distance $\sim\Lambda$. This increases the complexity of the problem considerably: we need to consider deviations from perfect periodicity on a scale Λ, but Λ is in turn determined by electron scattering effects, and hence the degree of imperfection! While the ways around this problem will not be discussed until Chapter 5, such considerations do allow for the definition of 'short' and 'long' range atomic or magnetic correlations, depending upon whether their spatial extent (or correlation length) is less than or greater than Λ respectively. In fact this raises a point which is often overlooked. Whether a correlation is described as short or long range depends entirely upon the coherence length (or time) of the probe being used. For example, 'long range order' in a resistivity study typically implies strong correlations over volumes larger than 1000 Å^3, whereas in a Mössbauer experiment 'long range' effects (e.g. quadrupole splitting) will result from correlations over much smaller distances, and may even be due to near neighbour effects representative of volumes of the order 10 Å^3. In a dynamic problem such as spin fluctuation, similar considerations apply. For example, the lifetime of an excited state in a Mössbauer experiment (typically $\sim 10^{-8}$ s) is very much longer than mean free time of a conduction electron (typically $\sim 10^{-14}$ s) or a neutron ($\sim 10^{-8}$–10^{-12} s). Thus, while spins in a material may appear to be strongly correlated (in time) in a Mössbauer experiment, a neutron diffraction or electrical resistivity study may lead to a quite different conclusion. Within the context of this book we will use the terms 'short range' and 'long range' as they apply to electrical resistivity studies, i.e. according to whether the correlation length is less than or greater than Λ, where Λ is typically $\sim$10–50 Å in concentrated binary alloys. These considerations are particularly important in the critical region since the atomic or magnetic correlation length is then rapidly changing with temperature and will pass from the short range to the long range regime as the temperature approaches the critical transition temperature from above. Problems associated with this behaviour are considered in Chapter 8.

One final point that relates to the significance of Λ: if a solid is so strongly disordered that the mean free path becomes comparable with the conduction electron wavelength ($\sim$3–4 Å) then one might question the use of a diffraction model to determine the resistivity. This will occur when the residual resistivity reaches values over $\sim 100\mu\Omega$ cm. However, a diffraction theory of the resistivity of liquid metals (Ziman 1969) produces results in reasonable agreement with experiment (even though the mean free path may be only roughly double the mean interatomic spacing) and so there does seem to be some justification for its use. The particular problems associated with very high resistivity solid alloys and the adiabatic approximation are also considered in Chapter 8.

1.5 Matthiessen's rule

In the above discussion, no distinction has been made between disorder that is frozen into the lattice by quenching or equilibrium disorder, as would occur in an ordering alloy at high temperatures and at compositions away from stoichiometry, for example, and disorder that results from thermal excitation of the lattice. In dilute alloys (see Chapter 2) the scattering from impurity atoms is nearly independent of temperature and so the total resistivity $\rho_{tot}(T)$ at any measuring temperature T can be written as the sum of two components

$$\rho_{tot}(T)=\rho_0+\rho_h(T), \tag{1.7}$$

where ρ_0 is the temperature independent residual (i.e. impurity) resistivity and $\rho_h(T)$ is the resistivity of the pure host material at that temperature. This is known as Matthiessen's rule and will only be valid if the impurity and phonon scattering are independent and if the relaxation time is isotopic. These assumptions are only partly true in most systems and there is a large body of work devoted to studying 'deviations from Matthiessen's rule', DMR (see e.g. Bass 1972).

Matthiesen's rule is often applied to the case of concentrated alloys. This rather unfair extrapolation of Matthiessen's original work is then written as

$$\rho_{tot}(T)=\rho_0+\rho_p(T) \tag{1.8}$$

and collects all contributions from scattering due to atomic disorder (excluding that due to thermally induced lattice displacements) into a residual resistivity ρ_0, leaving separate the phonon scattering of the alloy lattice $\rho_p(T)$. The first of these is usually described (with some confusion) as being 'temperature independent'. By this it is meant that, while ρ_0 may of course depend indirectly upon temperature if the degree of atomic disorder changes with temperature, it does not have the *intrinsic* temperature dependence of $\rho_p(T)$. For example, an ordering alloy may be quenched to a temperature low enough to prevent atomic diffusion and a small change in measuring temperature T_m would then produce a change in $\rho_p(T)$ but not ρ_0. Here one needs to be very careful about the specification of temperature and make a clear distinction between the measuring temperature T_m and the temperature that characterises the degree of disorder (the quench temperature T_q, for example). At measuring temperatures high enough to allow significant atomic diffusion, equilibrium may be attained, in which case these two temperatures will coincide. In magnetic alloys the dynamics of the spin system are usually such that the spins remain in thermal equilibrium down to very low temperatures, in which case it is more usual to express $\rho_{tot}(T)$ as

$$\rho_{tot}(T) = \rho_0 + \rho_p(T) + \rho_m(T), \tag{1.9}$$

where $\rho_m(T)$ is the temperature-dependent magnetic contribution resulting from spin–disorder.

However, in both the concentrated alloy and magnetic cases there are reasons to expect strong deviations from simple additivity. This is because the phonon spectrum is likely to depend upon both the concentration and degree of order, as are the electronic band structure and relaxation time anisotropy. Equations (1.8) and (1.9) should then properly include an additional term $\Delta(T)$ to allow for these interactions. Nevertheless, it is often useful to identify the different contributions contained in an experimental result. Some aspects of this complicated problem are discussed in more detail in the following chapters.

1.6 Simple and non-simple metals

We move now to the definition of 'simple' and 'non-simple' metals. In order to make such a distinction we need to consider formally the ideas of electronic band structures, Fermi surfaces and density of states. In metals, all of these are determined by the valence or outer electron states. The band structure is obtained by solving the Schrödinger equation to obtain the energy as a function of wavenumber $E(\mathbf{k})$ (see Chapter 4). The solutions are usually obtained along various

Fig. 1.5. Symmetry lines and points in the Brillouin zone for the fcc (*a*) and bcc (*b*) structures.

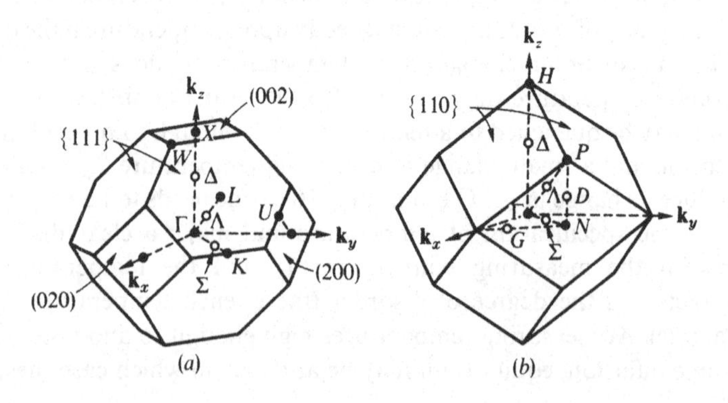

directions of symmetry in **k**-space, as defined in Figure 1.5. As an example, the band structure of Al is shown in Figure 1.6. The highest filled level is the Fermi level which is also shown in Figure 1.6. For each partly filled band there will be a surface in **k**-space separating the occupied from the unoccupied levels. The set of all such surfaces defines the Fermi surface, the shape of which is directly related to the band structure. Experimental determination of the Fermi surface (see e.g. Ashcroft & Mermin 1976, ch. 14; Schoenberg 1969) provides a test for calculated band structures although there are other techniques such as angular resolved photoemission spectroscopy which allow a more direct mapping of the band structure.

For a free electron solid the Fermi surface is a sphere of radius k_F, but Fermi surfaces are usually plotted in a reduced zone scheme. This is illustrated for a two-dimensional square lattice in Figure 1.7 and shows that while all parts of the Fermi surface are constructed from segments of a circle it is easy to lose sight of the fact that the surface is indeed circular. A similar construction is shown in three dimensions in Figure 1.8 where the Fermi sphere just projects into the third Brillouin zone of an fcc monatomic Bravais lattice. The Fermi surface in the reduced zone scheme is now constructed from portions of spheres. Illustrations of the free electron Fermi surfaces for fcc, bcc and hcp structures of various valencies are given in Harrison (1966). The conduction electron energy is given by the familiar expression (see e.g. Kittel 1976)

$$E = \frac{\hbar^2 k^2}{2m} \tag{1.10}$$

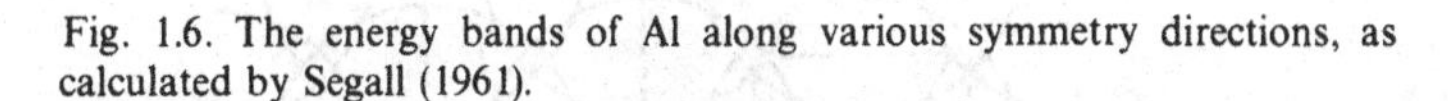

Fig. 1.6. The energy bands of Al along various symmetry directions, as calculated by Segall (1961).

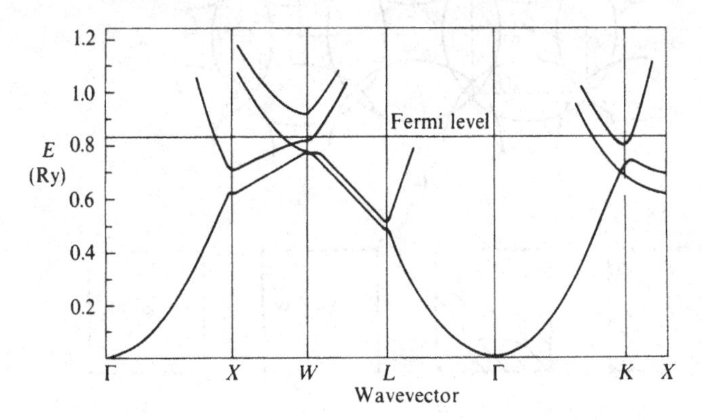

If all the N electrons in a solid of volume Ω were s-like, the maximum (Fermi) energy of states occupied would be

$$E_F = \frac{\hbar^2}{2m}\left(\frac{3\pi^2 N}{\Omega}\right)^{2/3} \tag{1.11}$$

which depends only upon the number of electrons per unit volume N/Ω.

The density of states is the number of states per unit energy at any given energy level. For any particular band this is given by

$$N(E) = \frac{1}{4\pi^3}\int \frac{\mathrm{d}S}{|\nabla E(\mathbf{k})|}, \tag{1.12}$$

where the integration is over the appropriate part of the Fermi surface, and so it depends upon the curvature of the Fermi surface. The density of states of a free electron metal is thus given by

$$N(E) = \frac{mk}{\pi^2 \hbar^2}. \tag{1.13}$$

In metals having a more complex band structure there will be points on the Fermi surface at which $\nabla E(\mathbf{k})$ vanishes. While this leads to a divergence in the integrand in equation (1.12), such singularities are integrable giving finite values for $N(E)$, although they do produce discontinuities in the slope $\mathrm{d}N(E)/\mathrm{d}E$ which are known as van Hove singularities.

Fig. 1.7. Fermi surface in a two-dimensional square array: (*a*) circular Fermi surfaces in the extended zone scheme, (*b*) Fermi surfaces in the first four bands redrawn in the reduced zone scheme centred on either point *P* or *P*′.

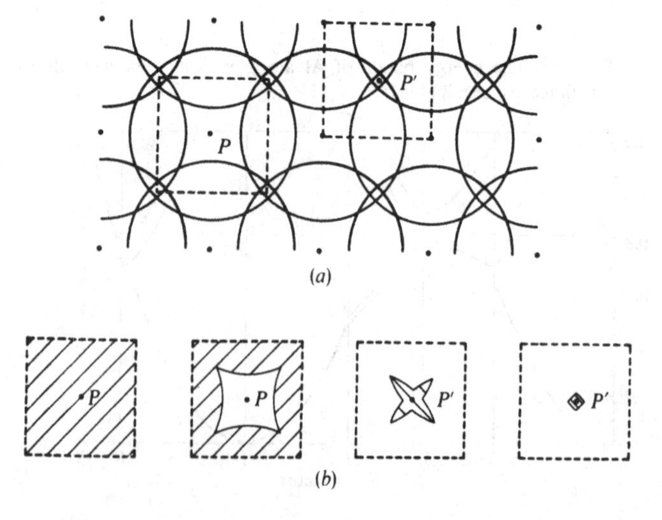

By the term 'simple metal' it is meant that virtually all electrons at the Fermi surface are s-like. This implies that the Fermi surface is spherical and the conduction electrons essentially 'free'. The wavefunctions of these electrons are plane waves

$$\psi_{\mathbf{k}} = \Omega^{-1/2} \exp(-i\mathbf{k} \cdot \mathbf{r}). \quad (1.14)$$

This definition implies that the Fermi level occurs at an energy far removed from any complicated band structure effects, as indicated by E_{F1} in Figure 1.9. As we are only concerned with characteristics of electrons at the Fermi level, equations (1.10) and (1.13) are then of some relevance. Equation (1.11) would of course be in error because of the number of electrons residing in the filled bands at lower energies.

Non-simple metals are those in which there is a significant density of states at the Fermi surface that do not have an s-like character, as in the

Fig. 1.8. (*a*) First Brillouin zone for an fcc crystal. (*b*) Second Brillouin zone for an fcc crystal. (*c*) Fermi sphere just projecting beyond the second Brillouin zone. (*d*) and (*e*) Fermi surface of (c) redrawn in the reduced zone scheme. (*e*) may be redrawn to appear as (*f*) (after Harrison 1966).

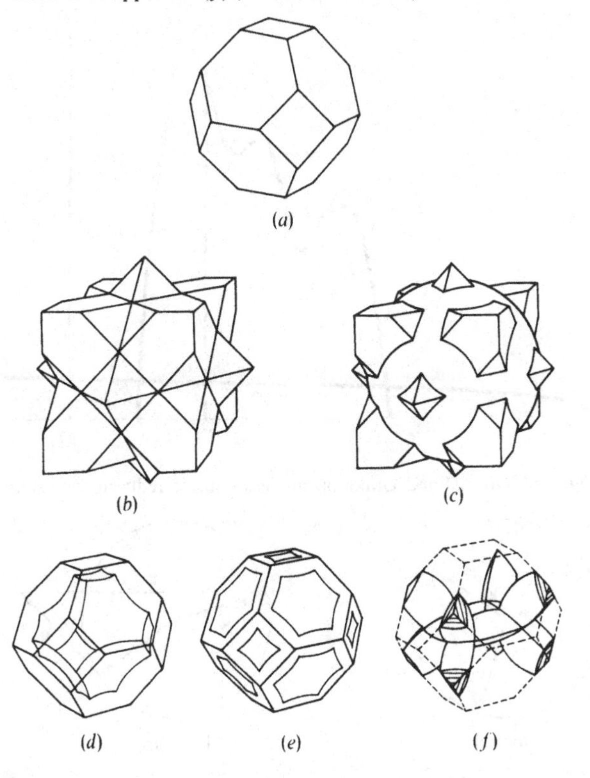

case of a material with the Fermi energy E_{F2} shown in Figure 1.9. In such a case the motion of both s and d electrons in the applied field should be considered, as should both the density of states $N_s(E_F)$ and $N_d(E_F)$ in the s- and d-bands respectively. These factors add considerable complexity even to the determination of the electrical resistivity of an ideal solid. In fact, in such cases the weak scattering approach outlined above is quite inappropriate and one must resort to an entirely different approach based on the determination of phase shifts or a current–current correlation function, for example. Such techniques are discussed in Chapter 6. The further complications that arise in inhomogeneous solids extend current theoretical approaches to the limit. However, if the strength of scattering is weak (or if there is simply no suitable alternative) a so-called nearly free electron calculation is quite tractable, although the results should always be interpreted with caution. This approach will be discussed in greater detail in Chapter 3.

Fig. 1.9. Density of states of a hypothetical solid containing both s and d electrons. The Fermi level occurs at E_{F1} in a simple metal and E_{F2} in a non-simple metal.

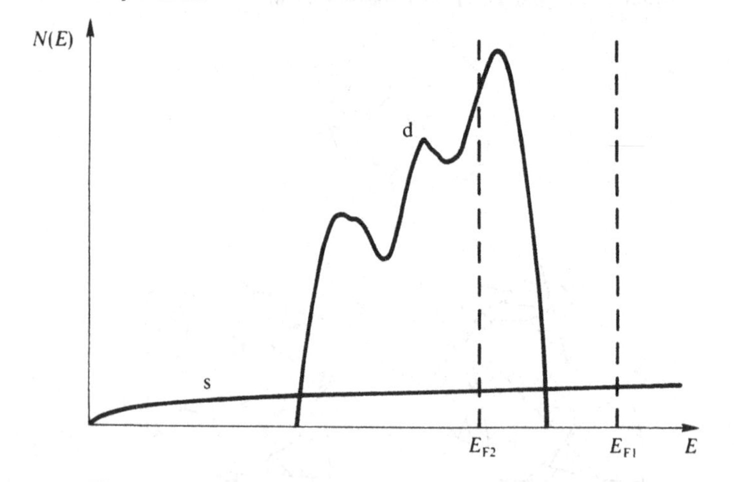

Fig. 1.10. Normal and Umklapp processes in the reduced zone scheme.

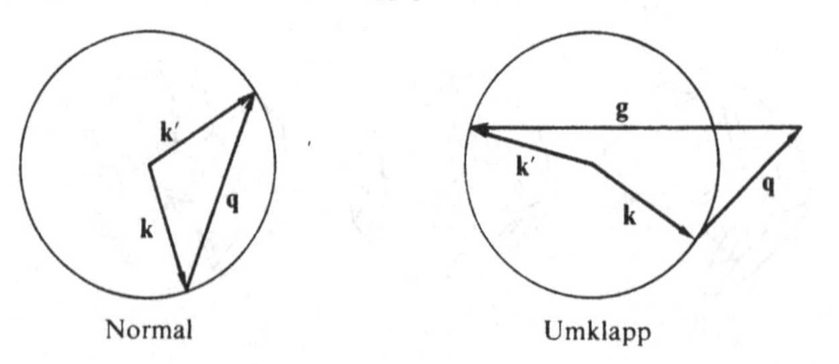

1.7 Elastic and inelastic scattering

So far we have assumed that the initial and final states **k** and **k**′ lie on the Fermi surface, i.e. the scattering is elastic since there is no change in energy. We now want to consider this point in more detail. There are two constraints that must be applied to the scattering process: one relates to the fact that the vibrational energy levels of the lattice are restricted by the discrete nature of the lattice and the other is that the momentum of the system is conserved.

1.7.1 *Electron and phonon energies*

If there were no loss of energy from the electrons to the lattice, the electrons would steadily gain energy by being accelerated in the applied field and their effective temperature would rise indefinitely above that of the lattice. Since this does not occur, the scattering cannot be purely elastic. However, if the collision between an electron and an ion is treated classically the fraction of the kinetic energy lost by the electron to the ion would be proportional to the ratio of their masses. For a typical metal this ratio is $\sim 10^{-5}$–10^{-6} and is small enough in relation to the Fermi energy to be neglected in the scattering process. It is quite large enough, however, to ensure that the electrons remain in thermal equilibrium with the lattice (see e.g. Dugdale 1977, p. 116).

This problem can also be considered in terms of the elementary excitations of the lattice. In the Debye model of a solid the upper limit of phonon energy is by definition $k_B\Theta_D$, where Θ_D is the Debye temperature which typically lies in the range 100–500 K. Thus, at any temperature the maximum energy that can be transferred in a single phonon process (the most likely event) is $\sim 200k_B$–$400k_B$. As this is much less than typical Fermi energies ($\sim 10^5 k_B$) the scattering can again be considered as essentially elastic, although there is still the small but necessary energy transfer to the lattice. Note that this assumption is only true as far as resistivity is concerned. Other properties such as the thermopower and thermal conductivity by electrons are sensitive to the magnitude of energy transfer in relation to k_BT rather than E_F and so one cannot neglect the inelastic scattering in such cases. The slight inelasticity in the scattering has also been invoked as a possible explanation for DMR in some dilute alloys (Bass 1972).

1.7.2 *Conservation of momentum*

In a simple metal the conduction electrons are described by a wavefunction of the form given in equation (1.14) and carry a crystal momentum $\hbar\mathbf{k}$ (see Section 1.8). In a non-simple metal the momentum is no longer constant but varies from place to place throughout the lattice, in which case $\hbar\mathbf{k}$ can be taken as a measure of the average momentum. When scattered for a state $\mathbf{k}$ to a state $\mathbf{k}'$ the momentum change $\hbar(\mathbf{k}'-\mathbf{k})$ is transmitted ultimately to the centre of mass of the solid. The condition of conservation of momentum requires that

$$\mathbf{k}'-\mathbf{k}=\pm\mathbf{q}+\mathbf{g}, \tag{1.15}$$

where $\mathbf{g}$ is a reciprocal lattice vector (or zero) and $\pm\mathbf{q}$ corresponds to the creation $(-)$ a destruction $(+)$ of a phonon of wavevector $\mathbf{q}$ with a consequential change in electron state. If $\mathbf{g}=0$, $\mathbf{k}'-\mathbf{k}=\pm\mathbf{q}$ corresponds to the simple absorption or emission of a phonon. However, in the Debye model there is an upper limit to $|\mathbf{q}|$, this being the Debye wave number $|\mathbf{q}_D|$ (Ziman 1972), and so the scattering angle in such 'normal' processes is limited. The condition $\mathbf{k}'-\mathbf{k}=\mathbf{g}$ (i.e. $\mathbf{q}=0$) corresponds to Bragg diffraction of the electrons from the set of lattice planes described by $\mathbf{g}$. Equation (1.15) with non-zero $\mathbf{g}$ thus corresponds to Bragg diffraction and simultaneous creation or annihilation of a phonon. This is called an Umklapp-process. These two scattering processes ($\mathbf{g}=0$ and $\mathbf{g}\neq 0$) are often referred to simply as N-processes or U-processes and are illustrated in a reduced zone scheme in Figure 1.10.

The general importance of both of these scattering processes has been discussed in many texts (see e.g. Ziman 1972; Dugdale 1977) and so will not be repeated here. The transfer of momentum to the phonons can cause them to drift in real space along with (N-processes) or against (U-processes) the electrons. This is called phonon drag and has a significant effect on the thermoelectric properties (and possibly deviations from Matthiessen's rule) of dilute alloys. However, in concentrated alloys there will usually be sufficient impurity scattering of the phonons to prevent such a phenomenon.

1.7.3 *Magnetic scattering*

There is one final case where it is important to distinguish between elastic and inelastic scattering. Many alloy systems of interest may contain one or more magnetic phases and disorder within the itinerant or localised spin systems will also cause scattering of conduction electrons. This is due to the spin state of the conduction electrons $\boldsymbol{\sigma}$ which can adopt only one of two values, ± 1 (spin 'up' and 'down'), and which interacts with other spins $\mathbf{S}$ through the effects of symmetry and the Pauli exclusion principle. The energy associated with this exchange interaction may be written as

$$E_{ex}=-J(\mathbf{r})\boldsymbol{\sigma}\cdot\mathbf{S}, \tag{1.16}$$

where the exchange parameter $J(\mathbf{r})$ falls off rapidly with distance $\mathbf{r}$. Below some critical temperature the itinerant or localised spins may condense into an ordered array (ferromagnetic or antiferromagnetic order). As in the case of atomic scattering, any disorder within this array will cause electron scattering. This scattering may be elastic, in which case there is no change in energy or spin–flip, or it may be inelastic, in which case the spin state of the electron changes. In both situations the scattering rate will be different for the spin-up and spin-down conduction electrons and so one usually has to consider a two sub-band model. The resistivity of materials that contain magnetic elements will be considered in more detail in Chapters 3 and 7.

1.8 The Boltzmann equation and relaxation time

Finally we come to an examination of the concept of a relaxation time.

1.8.1 *Wavepackets*

When an electric field is applied to a conductor all of the electrons are displaced at a uniform rate in **k**-space provided that they are not in a completely filled band (Kittel 1976, p. 213). As we noted in the previous section, random scattering effects tend to restore the electrons to their equilibrium distribution. The nature of this distribution (i.e. the population of the electron states) is thus determined by a dynamical balance between the acceleration in the field and the scattering by the lattice. We can approach this problem mathematically using the method introduced by Boltzmann in consideration of the properties of a classical gas. The gas particles were treated as hard atomic spheres, the position and momentum of which were well defined at any instant of time. As such, the state of the system could be specified by the distribution of particles over the individual states. However, we are now dealing with a strongly interacting quantum-mechanical system and it is by no means clear that such a distribution function defined in terms of single electron states adequately describes the state of the system (cf. the Einstein and Debye models of atomic excitation: the former describes the single particle excitation but the latter takes into account the *collective* excitations). However, if we regard the particles as wavepackets, this objection becomes less significant (Lewis 1958). For example, we can construct a simple Gaussian packet for the electron wavefunction centred on a state $\mathbf{k}_0$ (see e.g. Harrison 1970, p. 65):

$$\psi=\sum_{\mathbf{k}} u_{\mathbf{k}} \exp\left(i\mathbf{k}\cdot\left[\mathbf{r}-\frac{1}{\hbar}\frac{\mathrm{d}E(\mathbf{k})}{\mathrm{d}\mathbf{k}}t\right]\right)\exp(-\alpha[\mathbf{k}-\mathbf{k}_0])^2. \tag{1.17}$$

The group velocity of this wavepacket is

$$\mathbf{v}(\mathbf{k})=\frac{1}{\hbar}\frac{\mathrm{d}E(\mathbf{k})}{\mathrm{d}\mathbf{k}}, \tag{1.18}$$

and gives the velocity associated with the electron state **k** in the crystal. This is analogous to the classical result if $\hbar\mathbf{k}$ is regarded as the momentum. However, $\hbar\mathbf{k}$ is not a true momentum as is evident from the motion of the electron in an applied field. For example, if a force **F** is applied:

$$\frac{\mathrm{d}E}{\mathrm{d}t}=\mathbf{F}\cdot\mathbf{v}(\mathbf{k})=\frac{1}{\hbar}\mathbf{F}\cdot\frac{\mathrm{d}E(\mathbf{k})}{\mathrm{d}\mathbf{k}}. \tag{1.19}$$

But $\mathrm{d}E(\mathbf{k})/\mathrm{d}t=\partial E(\mathbf{k})/\partial\mathbf{k}\cdot\partial\mathbf{k}/\partial t$, and so

$$\hbar\partial\mathbf{k}/\partial t=\mathbf{F}.$$

Thus, while the rate of change of the momentum of the electron is given by the total force (including the crystal forces) the rate of change of $\hbar\mathbf{k}$ is determined only by the external forces. For example, if an electric field **E** is applied,

$$\hbar\partial\mathbf{k}/\partial t=e\mathbf{E},$$

and $\hbar\mathbf{k}$ changes at a constant rate in **k**-space in a direction parallel to **E**. However, the velocity of the electron in real space may behave in a complicated fashion as **k** runs over the bumps and valleys of the energy bands and will even be zero at the Brillouin zone boundary. Furthermore, since $\mathbf{v}(\mathbf{k})$ is normal to the constant energy surface in **k**-space, it will not in general be parallel to $\hbar\mathbf{k}$. For this reason (although somewhat misleadingly) $\hbar\mathbf{k}$ is called the crystal momentum of the electron. If the wavepacket contains a spread $\Delta\mathbf{k}$ in wavenumber, it must extend over a region $\sim 1/\Delta\mathbf{k}$ in real space. It is required that $\Delta\mathbf{k}$ be small compared with the Brillouin zone to obtain the necessary energy resolution and so the wavepacket must extend over many unit cells. This means that any fields and temperature gradients must be essentially constant over such regions. Note that this represents the practical classical limit and that the periodic potential of the ions which occurs on a finer scale must be treated within a quantum-mechanical framework. It is also clear that the model will fail if the mean free path Λ becomes much shorter than the wavelength of the packet λ, i.e. we require $\Lambda>\lambda$. Provided that these conditions are satisfied the distribution function $f(\mathbf{k},\mathbf{r},t)$ is then a function of position **r**, wavenumber **k** and time t and defines the location of an electron state in six-dimensional $(\mathbf{r},\mathbf{k})$ phase space. It will be defined such that

$$(4\pi^3)^{-1}f(\mathbf{k},\mathbf{r},t)\,\mathrm{d}\mathbf{k}\,\mathrm{d}\mathbf{r} \tag{1.22}$$

is the number of electrons which lie in an element $\mathrm{d}\mathbf{r}$ of real space and in $\mathrm{d}\mathbf{k}$ of wavenumber space (or in an element $\mathrm{d}\mathbf{r}$, $\mathrm{d}\mathbf{k}$ of phase space) at time t, such that $f=1$ if all states are occupied and $f=0$ if all are empty. If the electron gas is at equilibrium at temperature T in a uniform solid, f is then the familiar Fermi–Dirac distribution function:

$$f_0(\mathbf{k},\mathbf{r})=\frac{1}{\exp\left(\left[\dfrac{E(\mathbf{k})-E_{\mathrm{F}}(\mathbf{k})}{k_{\mathrm{B}}T(\mathbf{r})}\right]+1\right)}. \tag{1.23}$$

This is shown plotted as a function of energy together with its derivative $\partial f_0/\partial k$ in Figure 1.11. In this expression the Fermi energy $E(\mathbf{k})$ should strictly be interpreted as a chemical Fermi *potential*, though, for most metals, the two are essentially the same up to at least room temperature (see e.g. Ashcroft & Mermin 1976, p. 43).

Returning now to the Boltzmann equation and following the discussion above, we assume a semiclassical description of the electron gas whereby the 'particles' are wavepackets of well-defined position $\mathbf{r}$ and crystal momentum $\hbar\mathbf{k}$, but allow also for the Pauli exclusion principle. In the presence of applied fields, but without scattering, any particular state will move through phase space according to the semiclassical equations (see Ashcroft & Mermin 1976, p. 319). If we follow the path of some

Fig. 1.11. The Fermi–Dirac distribution f_0 and its derivative $\partial f_0/\partial E$ as a function of energy for $E_{\mathrm{F}}/k_{\mathrm{B}}T \sim 40$. (For Ag this would correspond to a temperature near the melting point.)

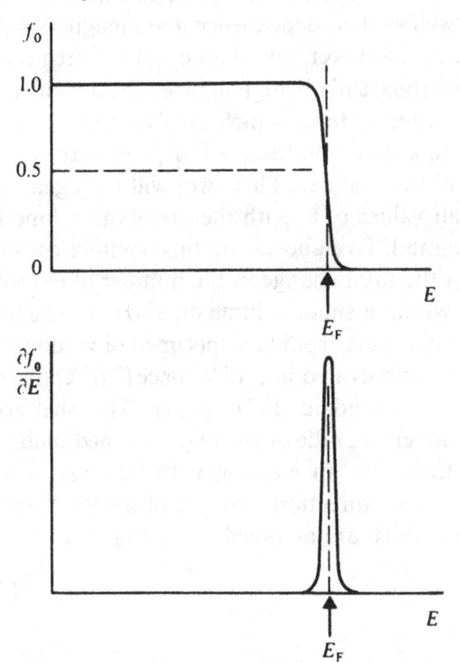

particular state, its occupation will not change with time and the total derivative of f with respect to time must be zero. But if we now admit the possibility of scattering, an electron can discontinuously change its momentum (i.e. in the terminology of equation (1.2) it will be scattered from some initial state $\Phi_{\mathbf{k}}$ into a final state $\Psi_{\mathbf{k}'}$). The transport equation can then be written as

$$\frac{\mathrm{d}f}{\mathrm{d}t}=\left.\frac{\partial f}{\partial t}\right|_{\mathrm{scatt}}, \tag{1.24}$$

where the last term is the change in the distribution function due to collisions or scattering. This equation can be rewritten as

$$\frac{\partial f}{\partial t}+\frac{\partial f}{\partial \mathbf{r}}\frac{\mathrm{d}\mathbf{r}}{\mathrm{d}t}+\frac{\partial f}{\partial \mathbf{k}}\frac{\mathrm{d}\mathbf{k}}{\partial t}=\left.\frac{\partial f}{\partial t}\right|_{\mathrm{scatt}}. \tag{1.25}$$

If we apply both an electric field $\mathbf{E}$ and a magnetic field $\mathbf{H}$, equations (1.20) and (1.25) give

$$\frac{\partial f}{\partial t}+\mathbf{v}(\mathbf{k})\frac{\partial f}{\partial \mathbf{r}}+\frac{e}{\hbar}\left[\mathbf{E}+\mathbf{v}(\mathbf{k})\times\mathbf{H}\right]\frac{\partial f}{\partial \mathbf{k}}=\left.\frac{\partial f}{\partial t}\right|_{\mathrm{scatt}}, \tag{1.26}$$

which is one form of the celebrated Boltzmann equation discussed at length in many texts.

A solution of this equation for some particular system will give the distribution function in the presence of an external field and/or temperature gradient and is sufficient to determine the electrical and thermal conductivities as well as their dependence upon magnetic fields and all thermoelectric effects. However, one should not be lured into a false sense of security by its apparently benign nature. The difficulty lies in the complexity of the scattering term, which involves the transition rates from all other states to $\mathbf{k}'$, say, and these will depend in turn upon the occupation numbers of those states. This term will thus generally involve an integral over all values of $\mathbf{k}'$ with the distribution function itself appearing in the integrand. To elaborate on this point we consider the effects of scattering on the total change in the number of electrons with wavevectors that lie within a small volume $\mathrm{d}\mathbf{k}$ about $\mathbf{k}$. The total number of such states of a particular spin in a specimen of volume Ω is $\mathrm{d}\mathbf{k}\Omega/(2\pi)^3$, since each state occupies a volume of $\mathbf{k}$-space $(2\pi)^3/\Omega$ (see e.g. Mott & Jones 1936, p. 56; Dugdale 1977, p. 27). The scattering probability can be written in terms of the quantity $Q_{\mathbf{k}\mathbf{k}'}$ defined such that the probability per unit time that an electron with wavevector $\mathbf{k}$ is scattered without change of spin direction into one of the levels in the volume $\mathrm{d}\mathbf{k}'$ about $\mathbf{k}'$, all of which are assumed to be empty, is

$$Q_{\mathbf{k}\mathbf{k}'}\,\mathrm{d}\mathbf{k}'\frac{\Omega}{(2\pi)^3}. \tag{1.27}$$

(It should be noted that some texts define this probability simply as $Q_{\mathbf{kk'}}\,\mathrm{d}\mathbf{k'}$ and so the additional factor $\Omega/(2\pi)^3$ must be taken into account when $Q_{\mathbf{kk'}}$ is determined.) A scattering process can cause a change in the number of such electrons either by a scattering out of these states to a new state in volume $\mathrm{d}\mathbf{k'}$ about $\mathbf{k'}$ or by the scattering into them from some other states in a volume $\mathrm{d}\mathbf{k'}$ about $\mathbf{k'}$. The number of electrons that can be scattered in a small time interval $\mathrm{d}t$ out of the volume $\mathrm{d}\mathbf{k}$ about $\mathbf{k}$ into some new set of levels in $\mathrm{d}\mathbf{k'}$ about $\mathbf{k'}$ is given by the product of the number of electrons available for the scattering process $f_{\mathbf{k}}\,\mathrm{d}\mathbf{k}\Omega/(2\pi)^3$, the fraction of these that will be scattered into $\mathrm{d}\mathbf{k'}$ about $\mathbf{k'}$ in the time interval $\mathrm{d}t$ if all such levels are available $Q_{\mathbf{kk'}}\,\mathrm{d}t\,\mathrm{d}\mathbf{k'}\Omega/(2\pi)^3$ and the probability that such levels are available $1-f_{\mathbf{k'}}$. The total change in the population of the volume $\mathrm{d}\mathbf{k}$ about $\mathbf{k}$ by scattering out of the volume is then given by integrating over all volume elements $\mathrm{d}\mathbf{k'}$. However, this change in population can also be written in terms of the change in the distribution $(\mathrm{d}f_{\mathbf{k}}/\mathrm{d}t)_{\mathrm{out}}\,\mathrm{d}t$ and the number of levels $\mathrm{d}\mathbf{k}\Omega/(2\pi)^3$ such that

$$\left(\frac{\mathrm{d}f_{\mathbf{k}}}{\mathrm{d}t}\right)_{\mathrm{out}} \mathrm{d}t\,\mathrm{d}\mathbf{k}\frac{\Omega}{(2\pi)^3} = f_{\mathbf{k}}\,\mathrm{d}\mathbf{k}\frac{\Omega}{(2\pi)^3}\,\mathrm{d}t\int Q_{\mathbf{kk'}}\,\mathrm{d}\mathbf{k'}\frac{\Omega}{(2\pi)^3}(1-f_{\mathbf{k'}}). \quad (1.28)$$

Similarly, the total number of electrons which are scattered into levels $\mathrm{d}\mathbf{k}$ about $\mathbf{k}$ from other levels $\mathrm{d}\mathbf{k'}$ about $\mathbf{k'}$ may be written as

$$\left(\frac{\mathrm{d}f_{\mathbf{k}}}{\mathrm{d}t}\right)_{\mathrm{in}} \mathrm{d}t\,\mathrm{d}\mathbf{k}\frac{\Omega}{(2\pi)^3} = (1-f_{\mathbf{k}})\,\mathrm{d}\mathbf{k}\frac{\Omega}{(2\pi)^3}\,\mathrm{d}t\int Q_{\mathbf{k'k}}\,\mathrm{d}\mathbf{k'}\frac{\Omega}{(2\pi)^3}f_{\mathbf{k'}}. \quad (1.29)$$

The total rate of change of the distribution function due to scattering is given by $(\mathrm{d}f_{\mathbf{k}}/\mathrm{d}t)_{\mathrm{in}}-(\mathrm{d}f_{\mathbf{k}}/\mathrm{d}t)_{\mathrm{out}}$ and is thus

$$\left.\frac{\partial f_{\mathbf{k}}}{\partial t}\right|_{\mathrm{scatt}} = \frac{\Omega}{(2\pi)^3}\int [Q_{\mathbf{k'k}}f_{\mathbf{k'}}(1-f_{\mathbf{k}})-Q_{\mathbf{kk'}}f_{\mathbf{k}}(1-f_{\mathbf{k'}})]\,\mathrm{d}\mathbf{k'}. \quad (1.30)$$

The Boltzmann equation is thus a non-linear integro-differential equation in its most general form leading to computational difficulties.

1.8.2 *The linearised Boltzmann equation*

The deviation from equilibrium is usually small in the steady state and so, if one writes the distribution function f in terms of its deviation from the equilibrium distribution f_0:

$$f(\mathbf{k},\mathbf{r},t)=f_0(\mathbf{k},\mathbf{r})+g(\mathbf{k},\mathbf{r},t), \quad (1.31)$$

one may substitute f_0 for f on the left-hand side of equation (1.26) and keep only the lowest power of $f-f_0$ which does not vanish for the scattering term. This leads to a linearised Boltzmann equation which is easier to solve. In particular, (1.30) becomes

$$\left.\frac{\partial f_{\mathbf{k}}}{\partial t}\right|_{\text{scatt}} = \frac{\Omega}{(2\pi)^3} \int (g_{\mathbf{k}'} - g_{\mathbf{k}}) Q_{\mathbf{k}\mathbf{k}'} \, d\mathbf{k}', \tag{1.32}$$

where we have used the principle of microscopic reversibility (or detailed balancing) which requires $Q_{\mathbf{k}'\mathbf{k}} = Q_{\mathbf{k}\mathbf{k}'}$ (see Messiah 1961, ch. xv; Rodberg & Thaler 1967, p. 274).

The Boltzmann equation (1.26) then becomes (neglecting henceforth the magnetic terms)

$$\mathbf{v}(\mathbf{k}) \frac{\partial f_0}{\partial T} \nabla_{\mathbf{r}} T + \mathbf{v}(\mathbf{k}) \cdot e \frac{\partial f_0}{\partial E_{\mathbf{k}}} \mathbf{E} = \frac{\Omega}{(2\pi)^3} \int (g_{\mathbf{k}'} - g_{\mathbf{k}}) Q_{\mathbf{k}\mathbf{k}'} \, d\mathbf{k}', \tag{1.33}$$

since

$$\partial f_0 / \partial \mathbf{k} = \partial f_0 / \partial E_{\mathbf{k}} \cdot \partial E_{\mathbf{k}} / \partial \mathbf{k} = \partial f_0 / \partial E_{\mathbf{k}} \hbar \mathbf{v}(\mathbf{k}), \tag{1.34}$$

and

$$\mathbf{v}(\mathbf{k}) \frac{\partial f}{\partial \mathbf{r}} \approx \mathbf{v}(\mathbf{k}) \frac{\partial f_0}{\partial T} \nabla_{\mathbf{r}} T. \tag{1.35}$$

Such a linearised form is in accordance with the observed linear response of the electric and thermal fluxes to electric and thermal gradients and so appears to be a reasonable approximation. Some theoretical justification for it and comments about the range of conditions which ensure its validity are given later in Section 1.8.6.

1.8.3 *The relaxation time approximation*

Solution of equation (1.33) is much simplified if we assume that the distribution relaxes exponentially in time to the equilibrium form when the field is switched off, in which case

$$\left.\frac{\partial f_{\mathbf{k}}}{\partial t}\right|_{\text{scatt}} = -\frac{f_{\mathbf{k}} - f_0}{\tau_{\mathbf{k}}} = -\frac{g_{\mathbf{k}}}{\tau_{\mathbf{k}}} \tag{1.36}$$

where $\tau_{\mathbf{k}}$ is the relaxation time which may vary with the magnitude of $\mathbf{k}$ but not its direction. In fact this relaxation time approximation can be shown to be quite rigorous provided that the scattering is elastic and the Fermi surface is spherical (Ashcroft & Mermin 1976, p. 325).

1.8.4 *Calculation of the resistivity in the relaxation time approximation*

The current density $\mathbf{J}(\mathbf{r}, t)$ at position $\mathbf{r}$ and time t is obtained by summing the contribution $e\mathbf{v}(\mathbf{k})$ for all electrons. This is achieved by integrating over **k** and weighting each contribution by the distribution function

$$\mathbf{J}(\mathbf{r},t)=\frac{e}{4\pi^3}\int v(\mathbf{k})f(\mathbf{k},\mathbf{r},t)\,\mathbf{dk}, \tag{1.37}$$

since there are $(1/4\pi^3)f(\mathbf{k})\,\mathbf{dk}$ electrons per unit volume, allowing for both spin directions. There is no current flow if the electrons are in equilibrium ($\int \mathbf{v}(\mathbf{k})f_0\,\mathbf{dk}=0$) and so equation (1.37) becomes

$$\mathbf{J}(\mathbf{r},t)=\frac{e}{4\pi^3}\int \mathbf{v}(\mathbf{k})g_{\mathbf{k}}\,\mathbf{dk}. \tag{1.38}$$

From equations (1.33) and (1.36) (assuming now that $\nabla_{\mathbf{r}}T=0$):

$$g_{\mathbf{k}}=e\tau_{\mathbf{k}}\left(-\frac{\partial f_0}{\partial E_{\mathbf{k}}}\right)\mathbf{v}(\mathbf{k})\cdot\mathbf{E}, \tag{1.39}$$

and so

$$\mathbf{J}(\mathbf{r},t)=\frac{e^2}{4\pi^3}\iint \tau_{\mathbf{k}}\mathbf{v}(\mathbf{k})[\mathbf{v}(\mathbf{k})\cdot\mathbf{E}]\left(-\frac{\partial f_0}{\partial E_{\mathbf{k}}}\right)\frac{\mathbf{dS}}{\hbar|\mathbf{v}(\mathbf{k})|}\,\mathrm{d}E_{\mathbf{k}}, \tag{1.40}$$

where the integration over a volume **dk** of **k**-space has been transformed into an integration over constant energy surfaces (see Appendix B). In a metal at normal temperatures $-(\partial f_0/\partial E_{\mathbf{k}})$ behaves like a δ-function, as shown in Figure 1.11, and so after integration over energy equation (1.40) becomes

$$\mathbf{J}(\mathbf{r},t)=\frac{e^2}{4\pi^3\hbar}\int \tau_{\mathbf{k}}\mathbf{v}(\mathbf{k})[\mathbf{v}(\mathbf{k})\cdot\mathbf{E}]\frac{\mathbf{dS}}{|\mathbf{v}(\mathbf{k})|}, \tag{1.41}$$

and is correct for any band structure. The relationship between **J** and **E** is usually written as Ohm's law

$$\mathbf{J}=\boldsymbol{\sigma}\cdot\mathbf{E}, \tag{1.42}$$

where $\boldsymbol{\sigma}$ is the conductivity tensor, i.e.

$$J_i=\sigma_{ij}E_j, \tag{1.43}$$

and the components of the conductivity tensor σ_{ij} are given by

$$\sigma_{ij}=\frac{e^2}{4\pi^3\hbar}\int\frac{\tau_{\mathbf{k}}v_i(\mathbf{k})v_j(\mathbf{k})\,\mathbf{dS}}{|\mathbf{v}(\mathbf{k})|}. \tag{1.44}$$

The tensor is symmetric and can therefore be reduced to diagonal form leaving only three principal conductivity coefficients to be determined, σ_1, σ_2 and σ_3 which describe the conductivity parallel to the principal axes of the crystal, and the associated resistivities $\rho_i=1/\sigma_i$ ($i=1,2,3$). The observed resistivity of a rod cut at angles $\alpha_1,\alpha_2,\alpha_3$ to the principal axes will then be

$$\rho=\rho_1\cos^2\alpha_1+\rho_2\cos^2\alpha_2+\rho_3\cos^2\alpha_3 \tag{1.45}$$

(Meaden 1966 gives some further discussion of the resistivity of non-cubic materials). In cubic crystals the conductivity tensor becomes a scalar

$$\sigma = \frac{e^2}{12\pi^3\hbar}\int \tau_{\mathbf{k}} v(\mathbf{k})\,\mathrm{d}\mathbf{S} \tag{1.46}$$

since, if **E** and **J** are both in the x direction, for example, $(\mathbf{v}(\mathbf{k})[\mathbf{v}(\mathbf{k})\cdot\mathbf{E}])_x = v_x^2 E = \frac{1}{3}v^2 E$. In the case of free electrons, $mv = \hbar k_F$ and the number of electrons n per unit volume within the Fermi sphere is $\frac{4}{3}\pi k_F^3/(4\pi^3)$ and so (1.46) reduces to the simple Drude formula, equation (1.5).

1.8.5 *Other solutions of the Boltzmann equation – anisotropic relaxation times*

There are situations (particularly in non-simple metals) where the restrictions of isotropic elastic scattering over a spherical Fermi surface are unacceptable. Furthermore, it is clear that, in contrast to the relaxation time approximation, the general probability per unit time that an electron will suffer a collision (also written as $\tau^{-1}(\mathbf{k})$) must depend upon the form of the non-equilibrium distribution function since this is, by definition,

$$\frac{1}{\tau(\mathbf{k})} = \frac{1}{(2\pi)^3}\int \mathrm{d}\mathbf{k}' Q_{\mathbf{k}\mathbf{k}'}(1 - f_{\mathbf{k}'}). \tag{1.47}$$

It is thus important to pursue other methods of solution of the Boltzmann equation. One alternative is based on the variational principle of Köhler (1948) and involves defining a function $\Phi_{\mathbf{k}}$ (not to be confused with the wavefunction introduced in Section 1.2) such that

$$f_{\mathbf{k}} = f_0 - \Phi_{\mathbf{k}}\frac{\partial f_0}{\partial E_{\mathbf{k}}}. \tag{1.48}$$

Some suitable parametric trial function is chosen for $\Phi_{\mathbf{k}}$ and the variational principle used to optimise the parameters so that the trial function is then the nearest approximation (within the limits of its basic form) to the true solution (Ziman 1960, p. 275; Blatt, 1968, p. 130). The trial function is usually constructed from a linear combination of standard functions with variable coefficients

$$\Phi_{\mathbf{k}} = \sum_i \eta_i \phi_i(\mathbf{k}). \tag{1.49}$$

By using the identity

$$-\frac{\partial f_0}{\partial E} = \frac{f_0(1 - f_0)}{k_B T}, \tag{1.50}$$

equation (1.33) may then be written as

$$\mathbf{v}(\mathbf{k})\frac{\partial f_0}{\partial T}\nabla_r T+\mathbf{v}(\mathbf{k})\cdot e\frac{\partial f_0}{\partial E_{\mathbf{k}}}\mathbf{E}$$
$$=-\frac{\Omega}{(2\pi)^3 k_B T}\int(\Phi_{\mathbf{k}}-\Phi_{\mathbf{k}'})f_0(1-f_0)Q_{\mathbf{k}\mathbf{k}'}\,d\mathbf{k}'$$
$$=-\frac{\Omega}{(2\pi)^3 k_B T}\int(\Phi_{\mathbf{k}}-\Phi_{\mathbf{k}'})P_{\mathbf{k}\mathbf{k}'}\,d\mathbf{k}', \qquad (1.51)$$

where we have written the actual transition probabilities $P_{\mathbf{k}\mathbf{k}'}\,d\mathbf{k}\Omega/(2\pi)^3$ in terms of the intrinsic transition probabilities

$$P_{\mathbf{k}\mathbf{k}'}\,d\mathbf{k}'\frac{\Omega}{(2\pi)^3}=f_{\mathbf{k}}(1-f_{\mathbf{k}'})Q_{\mathbf{k}\mathbf{k}'}\,d\mathbf{k}'\frac{\Omega}{(2\pi)^3}. \qquad (1.52)$$

If we now assume that there is no temperature gradient $\nabla_r T$, and if the scattering is elastic this leads to an expression for the resistivity (Ziman 1960, p. 283) given by a minimum of the ratio:

$$\rho=\frac{\dfrac{\Omega}{(2\pi)^3 2k_B T}\displaystyle\iint(\Phi_{\mathbf{k}}-\Phi_{\mathbf{k}'})^2 P_{\mathbf{k}\mathbf{k}'}\,d\mathbf{k}\,d\mathbf{k}'}{e^2\left[\displaystyle\int\mathbf{v}(\mathbf{k})\Phi_{\mathbf{k}}\frac{\partial f_0}{\partial E_{\mathbf{k}}}\,d\mathbf{k}\right]^2}. \qquad (1.53)$$

For more information about the variational solution the reader is referred to Ziman & Blatt.

Returning now to the linearised Boltzmann equation (1.33), the most general solution will be of the form

$$g_{\mathbf{k}}=-\frac{\partial f_{\mathbf{k}}}{\partial E_{\mathbf{k}}}e\mathbf{E}\cdot\mathbf{\Lambda}(\mathbf{k}), \qquad (1.54)$$

where $\mathbf{\Lambda}(\mathbf{k})$ is a vector defined at each point $\mathbf{k}$ of the Fermi surface. The Boltzmann equation can then be written as (Taylor 1963)

$$\mathbf{v}(\mathbf{k})\cdot\mathbf{E}=\frac{\Omega}{(2\pi)^3}\int Q_{\mathbf{k}\mathbf{k}'}[\mathbf{\Lambda}(\mathbf{k})-\mathbf{\Lambda}(\mathbf{k}')]\cdot\mathbf{E}\,d\mathbf{k}. \qquad (1.55)$$

Sondheimer (1962) has shown that if $Q_{\mathbf{k}\mathbf{k}'}$ can be written in the form

$$Q_{\mathbf{k}\mathbf{k}'}=\sum_{i,j}p_i(\mathbf{k})q_j(\mathbf{k}'), \qquad (1.56)$$

then a solution of equation (1.55) can be obtained in closed form. In the relaxation time approximation $\mathbf{\Lambda}(\mathbf{k})=\tau_{\mathbf{k}}\mathbf{v}(\mathbf{k})$ and so $\mathbf{\Lambda}(\mathbf{k})$ can be interpreted as a vector mean free path (Ziman 1972, p. 219). Beyond this approximation there has been some confusion as to the definition of an anisotropic relaxation time $\tau(\mathbf{k})$, which may vary with both the magnitude and direction of **k**. Ziman has argued against a definition such as

$$\mathbf{\Lambda}(\mathbf{k}) = \tau(\mathbf{k})\mathbf{v}(\mathbf{k}), \tag{1.57}$$

suggesting that $\mathbf{\Lambda}(\mathbf{k})$ is the only well-defined quantity. However, this objection is based on the concept of a universal relaxation time which is independent of $g_{\mathbf{k}}$. Sorbello (1974*a*), on the other hand, takes as a definition of a relaxation time

$$\left.\frac{\partial f_{\mathbf{k}}}{\partial t}\right|_{\text{scatt}} = -\frac{g_{\mathbf{k}}}{\tau(\mathbf{k})}. \tag{1.58}$$

In contrast to the universal relaxation times of Ziman and Springford (1971) this relaxation time is regarded as a functional of $f_{\mathbf{k}}$. As shown by Blatt (1968, p. 123) it has a meaning when the energy change per collision is small compared with $k_B T$ (i.e. nearly elastic scattering). However, as noted by Chambers (1968), it is not universal and may vary from problem to problem depending upon the particular physical behaviour being investigated. With this definition equation (1.54) becomes

$$g_{\mathbf{k}} = \tau_z(\mathbf{k}) v_z(\mathbf{k}) e E_z \frac{\partial f_{\mathbf{k}}}{\partial E_{\mathbf{k}}}, \tag{1.59}$$

where the subscript z indicates that the relaxation time depends upon the field direction z, and $v_z(\mathbf{k})$ is the z component of the velocity of an electron of state **k**. This leads to a new form of the Boltzmann equation:

$$\frac{1}{\tau_z(\mathbf{k})} = \frac{\Omega}{(2\pi)^3} \int \left[1 - \frac{v_z(\mathbf{k}')\tau_z(\mathbf{k}')}{v_z(\mathbf{k})\tau_z(\mathbf{k})}\right] Q_{\mathbf{k}\mathbf{k}'}\, d\mathbf{k}', \tag{1.60}$$

where **k** and **k**′ are confined to the Fermi surface. Once this has been solved for $\tau_z(\mathbf{k})$ the zz component of the conductivity tensor is given by (Sorbello 1974*a*):

$$\sigma_{zz} = \frac{e^2}{4\pi^3 \hbar} \int \frac{\tau_z(\mathbf{k}) v_z(\mathbf{k})^2}{|\mathbf{v}(\mathbf{k})|}\, dS. \tag{1.61}$$

As noted before, in a cubic system $\sigma_{xx} = \sigma_{yy} = \sigma_{zz}$ and all off-diagonal components are zero, but in non-cubic systems one needs to solve equation (1.61) for each direction. With this definition of $\tau(\mathbf{k})$, equation (1.57) should be written in terms of the i vector components:

$$\Lambda_i(\mathbf{k}) = v_i(\mathbf{k})\tau_i(\mathbf{k}). \tag{1.62}$$

A different relaxation time was defined by Taylor (1963) and Coleridge (1972):

$$\tau(\mathbf{k}) = \frac{\mathbf{v}(\mathbf{k}) \cdot \mathbf{\Lambda}(\mathbf{k})}{\mathbf{v}(\mathbf{k}) \cdot \mathbf{v}(\mathbf{k})}, \tag{1.63}$$

which leads to a conductivity expression

$$\sigma = \frac{e^2}{12\pi^3\hbar}\int \frac{\mathbf{v}(\mathbf{k})\cdot\boldsymbol{\Lambda}(\mathbf{k})}{|\mathbf{v}(\mathbf{k})|}\,\mathrm{d}\mathbf{S}. \tag{1.64}$$

The relation between this relaxation time and $\tau_i(\mathbf{k})$ is (Sorbello 1974*a*)

$$\tau(\mathbf{k}) = \sum_{i=1}^{3} \frac{\tau_i(\mathbf{k})v_i(\mathbf{k})}{\mathbf{v}(\mathbf{k})\cdot\mathbf{v}(\mathbf{k})}. \tag{1.65}$$

With this definition the vector mean free path can be calculated by iteration of equation (1.55), written in the form (Coleridge 1972)

$$\boldsymbol{\Lambda}(\mathbf{k}) = \tau_0(\mathbf{k})\left[\mathbf{v}'(\mathbf{k}) + \sum_{\mathbf{k}'} Q_{\mathbf{k}\mathbf{k}'}\boldsymbol{\Lambda}(\mathbf{k}')\right], \tag{1.66}$$

where

$$\tau_0(\mathbf{k}) = \left(\sum_{\mathbf{k}'} Q_{\mathbf{k}\mathbf{k}'}\right)^{-1}. \tag{1.67}$$

Using $\boldsymbol{\Lambda}(\mathbf{k}) = \mathbf{v}(\mathbf{k})\tau_0(\mathbf{k})$ as a starting value, Coleridge (1972) found that for impurities in copper, convergence was achieved within a few iterations and that $\boldsymbol{\Lambda}(\mathbf{k})$ and $\mathbf{v}(\mathbf{k})$ were approximately parallel. Ziman (1961*a*,*b*) has suggested that, to a good approximation, the effects of anisotropy may be incorporated by assuming that the vector mean free path is indeed in the same direction as the velocity and writing

$$\frac{1}{\tau(\mathbf{k})} = \frac{2\pi}{\hbar}\sum_{\mathbf{k}'} P_{\mathbf{k}\mathbf{k}'}[1 - \cos(\mathbf{v}_{\mathbf{k}}, \mathbf{v}_{\mathbf{k}'})], \tag{1.68}$$

where $(\mathbf{v}_{\mathbf{k}}, \mathbf{v}_{\mathbf{k}'})$ is taken to be the angle between the electron velocities $\mathbf{v}_{\mathbf{k}}$ and $\mathbf{v}_{\mathbf{k}'}$. We will consider the accuracy of this approximation in relation to some specific systems in Sections 5.7.2 and 6.4. Further comments upon the above and other relaxation times are to be found in Sorbello (1974*a*). One should note that, if the relaxation time is anisotropic, special care is required in comparing relaxation times obtained by different techniques because of the different averaging processes involved. For example, the Dingle temperature or cyclotron resonance line-width are determined by an average over the Fermi surface $\langle\tau_0(\mathbf{k})\rangle$ of the true mean collision time $\tau_0(\mathbf{k})$ (see equation (1.67) whereas the resistivity involves an average of the 'conductivity' relaxation time $\tau(\mathbf{k})$ which incorporates a $1-\cos\theta$ weighting factor and is thus of the form $\langle(1-\cos\theta)/\tau_0(\mathbf{k})\rangle$. Similarly the thermopower involves an average $\langle\tau(\mathbf{k})\rangle$ which is not the same as $\langle\tau(\mathbf{k})^{-1}\rangle^{-1}$ if $\tau(\mathbf{k})$ is anisotropic. We will always take the relaxation time τ or $\tau(\mathbf{k})$ to be the conductivity relaxation time. We can also see why Matthiessen's rule fails in the relaxation time approximation if the scattering is anisotropic. For example, equation (1.7) requires

$$\frac{1}{\langle\tau_{\text{tot}}\rangle}=\frac{1}{\langle\tau_0\rangle}+\frac{1}{\langle\tau_{\text{h}}\rangle}, \tag{1.69}$$

whereas equations of the form (1.3) give only

$$\left\langle\frac{1}{\tau_{\text{tot}}}\right\rangle=\left\langle\frac{1}{\tau_0}\right\rangle+\left\langle\frac{1}{\tau_{\text{h}}}\right\rangle. \tag{1.70}$$

These two equations will be equal only if τ_0 and τ_{h} are isotropic. In general this will not be true and their anisotropies may even be temperature dependent. Furthermore we have shown that the scattering rate depends upon the distribution of the other electrons and this can be strongly affected by the presence of two scattering mechanisms unless the distribution function in the presence of each scattering mechanism separately is the same (which is unlikely). In fact, Ziman (1960, p. 286) has shown that beyond the relaxation time approximation Matthiessen's rule holds as an inequality

$$\rho_{\text{tot}}\geqslant\rho_0+\rho_{\text{h}}. \tag{1.71}$$

Allen (1978), Engquist (1980) and Julianus & de Chatel (1984) have proposed other methods of solution which go beyond the lowest order variational solution and which are effectively based on series expansions of $\Phi_{\mathbf{k}}$. Such expansions can take into account any energy dependence of τ or $\Phi_{\mathbf{k}}$ (as well as anisotropy effects) which also lead to deviations from Matthiessen's rule (see also Kus 1978; Rice & Bunce 1970; Leavens 1977). Lekner (1970) has considered a solution which includes inelastic scattering processes, although it is not certain that the Boltzmann equation is valid for such a case.

1.8.6 *Other formalisms*

Use of the Boltzmann equation and its associated problems may be avoided by employing the Kubo formalism (Kubo 1957; Greenwood 1958) in which the conductivity tensor derived from linear response theory is given exactly in terms of a time correlation between the current component $j_\nu(0)$ at time 0 and $j_\mu(t)$ at time t:

$$\sigma=\frac{1}{k_{\text{B}}T}\int\langle j_\mu(t)j_\nu(0)\rangle\,\text{d}t. \tag{1.72}$$

The evaluation of such an expression is not a simple matter although the correlation functions are directly related to two-particle Green's functions (see e.g. Elliott *et al.* 1974) and so this formalism is usually employed in Green's function based calculations, as discussed in

Chapter 4. Fortunately the results obtained generally confirm the use of the linearised Boltzmann equation (Mahan 1984; Schotte 1978), at least for the case of elastic scattering if the scattering is weak or if the concentration of impurities is small. A similar conclusion is also reached from density matrix studies (Kohn & Luttinger 1957; Luttinger & Kohn 1958). There has been interest in another expression for the resistivity derived also from linear response theory in terms of force–force correlation functions (Edwards 1965; Jones 1974; Ballentine & Heaney 1974; Christoph *et al.* 1974; Röpke & Christoph 1975; Marsch 1976):

$$\rho = \frac{1}{n^2 e^2} \int \mathrm{d}t \int \mathrm{d}\lambda \langle F_x(0) F_x(t + \mathrm{i}\lambda) \rangle. \tag{1.73}$$

Sorbello (1981) and others have shown that this equation can also be derived either from the Boltzmann equation or the Kubo formula, but it is generally only valid if the relaxation time is not anisotropic. We might also note in passing that the optical theorem (see Rodberg & Thaler 1967) has been used to determine the total transition probability (Sorbello 1974*b*, 1977; Coleridge 1972):

$$\frac{1}{\tau_0(\mathbf{k})} = -\frac{2}{\hbar}\,\mathrm{Im}(T_{\mathbf{kk}'}), \tag{1.74}$$

where the *T*-matrix is one form of the scattering matrix (see Section 6.4), although Braspenning & Lodder (1981) have pointed out that its use may lead to considerable quantitative errors if the *T*-matrix is not exact.

A more-detailed account of these various approaches is beyond the scope of this book and the interested reader should consult the references for more information. For the purposes of the discussions which follow we will generally accept the validity of the relaxation time approximation for the simple metals and alloys discussed in Chapter 5, while the studies of non-simple metals and alloys considered mainly in Chapters 6 and 7 will generally employ some of the more exact methods described above.

2 Atomic configuration of an alloy

In order to proceed with calculation of the electrical resistivity of a concentrated alloy it is necessary to find some suitable description of the atomic configuration.

2.1 Dilute and concentrated alloys

We start by defining a dilute alloy as one in which the solute atoms are in sufficiently small concentration that they have no effect on the electronic structure or phonon spectrum of the bulk alloy, although localised perturbations may result. In this regime the impurity atoms will act as independent scattering centres for the conduction electrons so that the effect of N impurity atoms is just N times that of a single impurity. This will only be possible if the solute atoms are widely separated (and hence randomly distributed), implying no direct or indirect interaction between them. In real alloys this composition range is usually restricted to less than 1 or 2% solute concentration.

At higher solute concentrations the electronic structure or phonon spectrum of the bulk alloy will start to suffer perturbations, and the solute atoms may no longer be randomly distributed. Such alloys will be described as concentrated alloys and are the main concern of this text.

2.2 Correlation parameters in crystalline materials

The configuration of a concentrated alloy may be described in terms of some suitable set of site occupation parameters defined by

$$\left.\begin{aligned} \sigma_i^x &= 1 \quad \text{if a particular 'defect' } x \text{ is at site } \mathbf{r}_i \\ &= 0 \quad \text{if not} \end{aligned}\right\}. \tag{2.1}$$

For example, if the atomic distribution of an alloy is under consideration, one would define

$$\left.\begin{aligned} \sigma_i^{\mathrm{A}} &= 1 \quad \text{if an A atom is at site } \mathbf{r}_i \\ &= 0 \quad \text{if not} \end{aligned}\right\}. \tag{2.2}$$

(Note that this is not the only way that such parameters could be defined, and one could take ± 1, for example, or some other set of values such as $+1, 0, -1$ in a ternary alloy). Similar parameters could be defined to describe a structural transformation in which case they would refer to the presence or absence of a particular atomic displacement (or defect).

Table 2.1. *Motifs and corresponding averages*

Motif	Average
e.g. single site: ○	average σ_i^x over all sites: $\langle\sigma_i^x\rangle$
two-site cluster: ○—○	average σ_i^x over all (two-site) pairs: $\langle\sigma_i^x\sigma_j^x\rangle$
three-site cluster: (triangle of three sites)	average σ_i^x over all (three-site) triplets: $\langle\sigma_i^x\sigma_j^x\sigma_k^x\rangle$

For example, one could define

$$\left.\begin{aligned}\sigma_i^\beta &= 1 \quad \text{if a displacement (or defect)}\\ &\qquad \text{of type } \beta \text{ is at site } \mathbf{r}_i\\ &= 0 \quad \text{if not}\end{aligned}\right. \tag{2.3}$$

Similarly, in a magnetic system the parameters would be chosen to refer to the spin direction at a particular site (Sanchez *et al.* 1982):

$$\left.\begin{aligned}\sigma_i^\uparrow &= 1 \quad \text{if a} \uparrow \text{spin is at site } \mathbf{r}_i\\ &= 0 \quad \text{if not}\end{aligned}\right\}. \tag{2.4}$$

A complete set of these parameters $\{\sigma_i^x\}$, one for each lattice site, would give the exact configuration of a system. However, it is not possible (or even useful) to specify such a set and some more suitable average values must be found. In order to proceed, we consider the averages over various motifs, as indicated in Table 2.1.

In the general case of a motif containing n lattice sites, the n-site ensemble average is written as $\langle\sigma_i^x\sigma_j^x\cdots\sigma_n^x\rangle$, where here and above the angled brackets $\langle\ \rangle$ indicate an ensemble average over the products of n-terms $(\sigma_i^x\times\sigma_j^x\times\cdots\times\sigma_n^x)$ that are obtained from all possible locations of the motif within the lattice. We can now see that the parameters defined in equations (2.1)–(2.4) are actually projection operators as they only allow particular configurations (e.g. an A atom at site i, etc.) to contribute to the average.

Let us consider now the information that is available from these averages in the case of a substitutional binary alloy. The single-site average $\langle\sigma_i^A\rangle$ is just the probability of finding an A atom at any site and is simply given by the atomic fraction of A atoms c_A. This quantity will clearly be important since we know that the resistance of an alloy is composition dependent.

By itself such a parameter contains no information about the distribution of atoms, although if a long range ordered solid is decomposed into sublattices it can give some description of the state of order. For example, the B_2 (bcc) ordered structure may be regarded as two interpenetrating sublattices, as shown in Figure 2.1. In the fully ordered state, one sublattice (α) will be fully occupied by A atoms and the other (β) by B atoms. The probability of finding an A atom in the α sublattice (i.e. concentration of A atoms on the α sublattice) or B atoms on the β sublattice then gives some measure of the degree of long range order (LRO). Figure 2.2 shows the appropriate sublattices for the fcc $L1_0$ and $L1_2$ ordered structures. In such a single-site approximation (also known as the point approximation or mean-field approximation) the familiar Bragg–Williams LRO parameter S for a stoichiometric alloy is defined by

$$S = \frac{\langle \sigma^{A} \rangle^{\alpha} - c_{A}}{1 - y^{\alpha}} = \frac{\langle \sigma^{B} \rangle^{\beta} - c_{B}}{1 - y^{\beta}} \tag{2.5}$$

where y^{α}, y^{β} are the fraction of α, β sites and the α, β superscripts on the averages $\langle\ \rangle$ indicate that the average is to be taken over all sites on the appropriate sublattice. This parameter is normalised so that $S = 1$ for complete LRO (all A atoms on the α sublattice) and $S = 0$ for complete randomness (random probability of finding an A or B atom on any site). For example, for complete order,

$$\langle \sigma^{A} \rangle^{\alpha} = 1,$$

and so

$$S = \frac{1 - c_{A}}{1 - y^{\alpha}}.$$

For a bcc B_2 alloy with 50 at.%A, $c_A = \frac{1}{2}$, $y^{\alpha} = \frac{1}{2}$, giving $S = 1$.

Fig. 2.1. The B_2 ordered lattice decomposed into appropriate α and β sublattices.

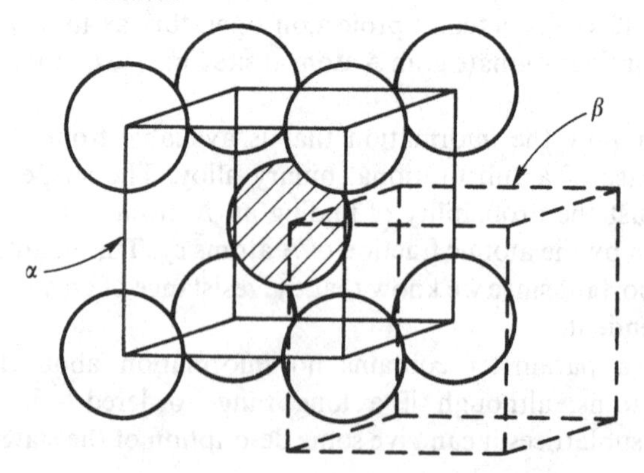

For complete randomness,

$$\langle\sigma^{A}\rangle^{\alpha}=c_{A}.$$

Therefore

$$S=\frac{c_{A}-c_{A}}{1-y^{\alpha}}$$

$$=0.$$

The temperature dependence of S is shown schematically for systems that undergo first- or second-order phase transition in Figure 2.3.

There are other definitions of an LRO parameter which could be used, although if used consistently within this single-site approximation they must of course produce the same final result. For example, a weighted average over the sublattices is more convenient for a non-stoichiometric alloy:

$$S'=\sum_{n}y^{n}\left[\frac{\langle\sigma^{A}\rangle^{n}-c_{A}}{1-c_{A}}\right]+\sum_{m}y^{m}\left[\frac{\langle\sigma^{B}\rangle^{m}-c_{B}}{1-c_{B}}\right]. \tag{2.6}$$

Fig. 2.2. (*a*) The ($L1_0$) ordered lattice decomposed into appropriate α and β sublattices. (*b*) The $L1_2$ ordered lattice decomposed into appropriate α and β sublattices.

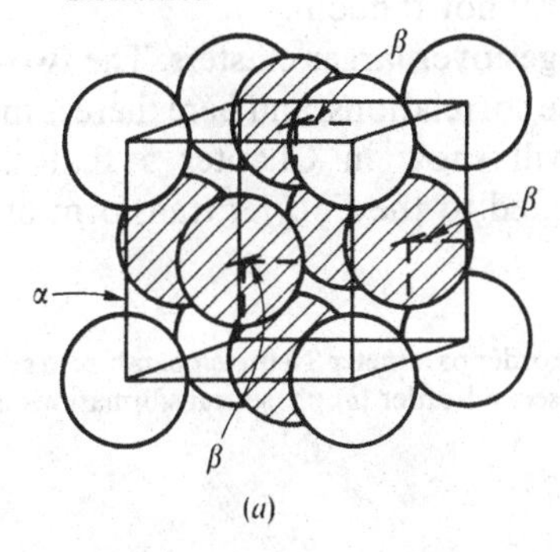

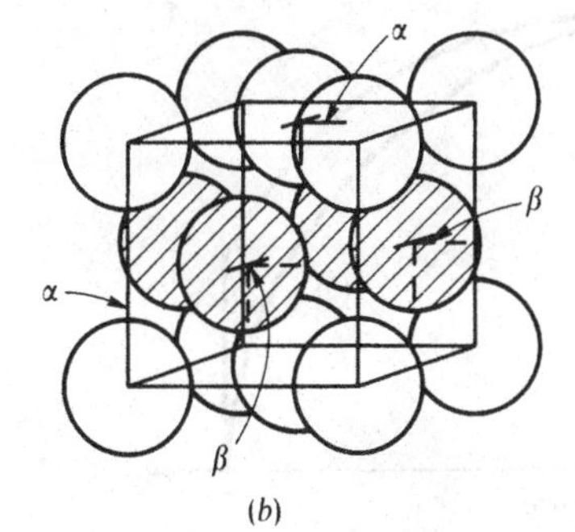

where it is assumed that the n sublattices contain only A atoms and the m sublattices contain only B atoms in the fully ordered state. At stoichiometric compositions, the two LRO parameters S and S' are identical. Conversely these single-site averages can be expressed as simple functions of S or S'. For example, in the case of fcc $L1_0$ structures:

$$\begin{aligned}\langle\sigma^{A}\rangle^{\alpha}&=\tfrac{4}{3}c_{A}c_{B}S'+c_{A}\\ \langle\sigma^{B}\rangle^{\beta}&=4c_{A}c_{B}S'+c_{B}\end{aligned}\quad. \tag{2.7}$$

For bcc B_2 structures:

$$\begin{aligned}\langle\sigma^{A}\rangle^{\alpha}&=2c_{A}c_{B}S'+c_{A}\\ \langle\sigma^{B}\rangle^{\beta}&=2c_{A}c_{B}S'+c_{B}\end{aligned}\quad. \tag{2.8}$$

Note that these parameters are simply averages taken over the sublattice single-site motifs and give only the concentration of A and B atoms on the α, β sublattices. Once again they can give no information about the detailed distribution of atoms over the sublattices (i.e. the local atomic arrangements) as they are still based on an average over single sites. As such they do not adequately describe the complete atomic arrangement. For example, introduction of an antiphase boundary through the centre of an otherwise perfectly ordered body would give $S=0$ from equation (2.5) even though the alloy is clearly not random.

We thus need to consider averages over larger clusters. The two-site average takes into account pairwise correlations, and here there is much information to be gained (we will show in Chapter 5 that these parameters are in fact directly related to the Fourier transform of the

Fig. 2.3. Variation of long range order parameter S with temperature in systems that undergo first-order (a) or second-order (b) phase transformations at the critical temperature T_c.

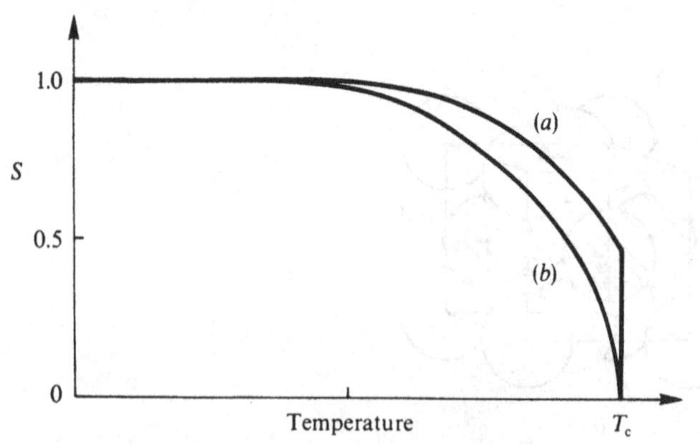

scattering intensity of the atomic array). Returning to the definition of the site occupation parameters (equation (2.1)), the average probability of finding an A–B pair at sites i, j is, by definition,

$$P_{ij}^{AB} = \langle \sigma_i^A \sigma_j^B \rangle. \tag{2.9}$$

However, the parameters σ_i^A, and σ_j^B must be interconnected since the total number of A and B atoms is constant and so they could be replaced by a single parameter which expresses the deviation from the mean occupation:

$$\begin{aligned} \sigma_i &= \sigma_i^A - \langle \sigma_i^A \rangle \\ &= \sigma_i^A - c_A \\ (&= c_B - \sigma_j^B) \end{aligned} \tag{2.10}$$

If an A atom is at site i, this gives

$$\begin{aligned} \sigma_i &= 1 - c_A \\ &= c_B. \end{aligned} \tag{2.11a}$$

If a B atom is at site i,

$$\begin{aligned} \sigma_i &= 0 - c_A \\ &= -c_A. \end{aligned} \tag{2.11b}$$

Such a parameter is sometimes referred to as a Flinn occupation parameter (or a spin–deviation parameter in the magnetic case to be considered in Chapter 3). The conditional probability P_{ij}^{AB} of finding an A atom at site i and B atom at site j can now be written as

$$P_{ij}^{AB} = \langle \sigma_i^A \sigma_j^B \rangle = \langle (\sigma_i + c_A)(c_B - \sigma_i) \rangle,$$

$$P_{ij}^{AB} = c_A c_B - \langle \sigma_i \sigma_j \rangle,$$

and similarly

$$\begin{aligned} P_{ij}^{BB} &= c_B^2 + \langle \sigma_i \sigma_j \rangle, \\ P_{ij}^{AA} &= c_A^2 + \langle \sigma_i \sigma_j \rangle, \end{aligned} \tag{2.12}$$

since $\langle \sigma_i \rangle = 0$ and any constant may be taken outside the average $\langle \; \rangle$. Thus $\langle \sigma_i \sigma_j \rangle$ is a useful form of the pair correlation parameter, since from equation (2.12) it gives the difference between the local probability of finding an A atom at i and a B atom at j in an alloy and the same probability for a random alloy. Equation (2.12) could be derived directly from this definition since, for example, from equation (2.10),

$$\begin{aligned} \langle \sigma_i \sigma_j \rangle &= \langle (\sigma_i^A - c_A)(c_B - \sigma_j^B) \rangle \\ &= c_A c_B - \langle \sigma_i^A \sigma_j^B \rangle, \end{aligned}$$

and similarly for the other results in (2.12).

Warren (1969) and Cowley (1975) have used a slightly different pairwise correlation parameter α_{ij} in their extensive studies of local atomic configuration which is defined by

$$\langle\sigma_i\sigma_j\rangle = c_A c_B \alpha_{ij}. \tag{2.13}$$

Furthermore, they recognised that in cubic materials the atomic distribution could be represented by shells of atoms about a central atom, for example the jth, and that the central atom could be taken without loss of generality as the origin. This leads to the description of c_i atoms in the ith shell a distance $r_i = |\mathbf{r}_i|$ from the central atom. In this case the appropriate pair correlation parameter is written as

$$\begin{aligned}\langle\sigma_0\sigma_i\rangle &= c_A c_B \alpha_{0i}\\ &= c_A c_B \alpha_i.\end{aligned} \tag{2.14}$$

The pair correlation parameter gives information about the local atomic environment. For example, in a truly random alloy $\langle\sigma_i\sigma_j\rangle$ (or α_i) $=0$ for all $i \neq j$ (although such a state is rarely obtained on the atomic scale due to local statistical fluctuations). In this case the pair probability can be written as a product of the single-site probabilities:

$$\begin{aligned}\langle\sigma_i^A\sigma_j^B\rangle &= \langle\sigma_i^A\rangle\langle\sigma_j^B\rangle\\ &= c_A c_B.\end{aligned} \tag{2.15}$$

If $\langle\sigma_i\sigma_j\rangle$ (or α_i) is positive and i and j are adjacent sites (or i represents the first shell of neighbours), then there is a number of like near neighbours in excess of the random probability. This is described as atomic clustering. If $\langle\sigma_i\sigma_j\rangle$ (or α_i) is negative, then there is an excess of unlike near neighbours. This is short range (SRO) atomic ordering. The variation of $\langle\sigma_i\sigma_j\rangle$ (or α_i) with separation of the sites $\mathbf{r}_i - \mathbf{r}_j$ (or $\mathbf{r}_i$) gives information about the spatial variation of the pairwise atomic correlations and can be used to obtain a model of the structure, as discussed in the next section. If the solid has long range atomic order there will be no decay in the range of atomic correlations with distance (by definition) and so the parameters $\langle\sigma_i\sigma_j\rangle$ will be independent of the separation $\mathbf{r}_i - \mathbf{r}_j$ although they will depend upon the location of sites i, j (whether face centre or cube corner, for example). Cowley (1975) has shown that long range ordered state in an infinite homogeneous crystal is thus described by

$$\lim_{\mathbf{r}_i - \mathbf{r}_j \to \infty} \{\langle\sigma_i\sigma_j\rangle\} = c_A c_B S_{ij}, \tag{2.16}$$

defining now a set of long range order parameters S_{ij}. Determination of the limiting values of $\langle\sigma_i\sigma_j\rangle$ (or S_{ij}) for all ranges of atomic separation then becomes an exercise in number theory (Hall & Christy 1966). As an example, the limiting values for $\langle\sigma_i\sigma_j\rangle$ at large $\mathbf{r}_i - \mathbf{r}_j$ (i.e. the parameters S_{ij}) are $-\frac{1}{3}$ for the face centre sites and $+1.0$ for the cube-corner sites of a

Cu_3Au type structure. For a B_2 structure they are -1 at the body centre and $+1$ at the cube corner. The CuAu structure gives (perhaps slightly unexpectedly) the same results as Cu_3Au because of the averaging over sites (for this reason the idealised superlattice diffraction patterns of the two are indistinguishable). These parameters are directly related to the point approximation LRO parameter S' (equation (2.6)) if the deviations from perfect order are completely random:

fcc (Cu_3Au):

$$\left.\begin{aligned} S_{ij}(\text{face-centre}) &= -\tfrac{16}{9}c_A c_B S'^2 \\ S_{ij}(\text{cube-corner}) &= \tfrac{16}{3}c_A c_B S'^2 \end{aligned}\right\}. \tag{2.17}$$

bcc (B_2):

$$\left.\begin{aligned} S_{ij}(\text{body-centre}) &= -4c_A c_B S'^2 \\ S_{ij}(\text{cube-corner}) &= 4c_A c_B S'^2 \end{aligned}\right\}. \tag{2.18}$$

However, Cowley (1960) has shown that only in the special case $c_A = c_B$ is it sometimes possible to describe the state of the alloy by a single long range order parameter. As the degree order decreases so does the range of atomic correlation (i.e. the 'correlation length') and so the pairwise correlation parameters decay more rapidly with distance, although they still satisfy various inequalities (Christoph & Richter 1980). Some results which illustrate this behaviour in Cu–25%Au are given in Table 2.2. By direct analogy with the pair correlation parameters described above, correlations between groups of 3, 4, ..., N atoms may be described by

$$\langle\sigma_i\sigma_j\sigma_k\rangle, \quad \langle\sigma_i\sigma_j\sigma_k\sigma_l\rangle, \quad \text{etc.}$$

One question that naturally arises is the size of the smallest set of parameters $\{\langle\sigma_i\rangle, \langle\sigma_i\sigma_j\rangle, \langle\sigma_i\sigma_j\sigma_k\rangle\ldots\}$ which adequately describes the atomic configuration. Implicit in this question is the independence of the correlation parameters and whether clusters are irreducible. The answer to these questions will ultimately depend upon the particular property of the solid that is of interest. In problems of alloy theory and phase stability it appears that motifs or clusters containing in excess of four atoms must be considered in order to reproduce some of the subtleties of a real phase diagram and motifs up to and including a double tetrahedron–octahedron are often included (Sanchez & de Fontaine 1978*a*, 1980; Mohri *et al.* 1985). However, as we will show in Chapter 5, the pair correlation (two-particle cluster) completely describes the *diffraction* effects of atomic correlations and so we need only go up to $\langle\sigma_i\sigma_j\rangle$ to determine the effects of atomic correlations on the resistivity. Nevertheless, static atomic displacements due to atomic size differences also cause electron scattering and these will depend upon the higher-order multisite correlation parameters.

Table 2.2. *Pairwise atomic correlation parameters α_i for Cu–25%Au after various heat treatments*

T (°C)	Treatment	α_1	α_2	α_3	α_4	α_5	α_6	α_7	α_8
800	quenched	−0.273	0.450	−0.045	0.240	−0.129	0.149	−0.052	0.092
450	at temperature	−0.195	0.215	0.003	0.077	−0.052	0.028	−0.010	0.036
405	at temperature	−0.218	0.286	−0.012	0.122	−0.073	0.069	−0.023	0.067
Perfect order	theory	$-\frac{1}{3}$	1	$-\frac{1}{3}$	1	$-\frac{1}{3}$	1	$-\frac{1}{3}$	1

Source: After Moss 1964.

2.3 Composition waves

The atomic correlation parameters described in the previous section provide a means of describing the atomic configuration. However, there are alternative approaches. For example, if the deviations from randomness are periodically distributed throughout the alloy it may be more appropriate to employ a composition wave to characterise the structure. This was the approach adopted in early continuum studies of spinodal decomposition by Cahn (1961, 1962, 1968) in which solution of linearised diffusion equations gave the time and spatial dependence of the local composition in the form of a composition wave

$$c(\mathbf{r}, t) = \exp[R(\boldsymbol{\beta})t]\cos(\boldsymbol{\beta}\cdot\mathbf{r}) \tag{2.19}$$

where $\boldsymbol{\beta}$ is the wavevector of a particular composition wave of wavelength $\lambda = |\boldsymbol{\beta}|$ and $R(\boldsymbol{\beta})$ is an amplification factor that determines the rate of growth of a wave of particular wavevector $\boldsymbol{\beta}$ with time. Such linearised equations are useful for predicting the nature of the instability of a solution but are in error in predicting an experimental growth rate due to the importance of the non-linear terms (Langer 1973; Langer *et al.* 1975; Marro *et al.* 1975; Rundman & Hilliard 1967; Ditchek & Schwartz 1980).

From the point of view of a resistivity study it is important to consider not only the amplitude of such composition fluctuations but also the relative sizes of the wave length of decomposition and the conduction electron mean free path length Λ. For example, in the fully developed two-phase structure, if the particle size and spacing (determined largely by the wavelength of the most rapidly growing component) is greater than Λ, interfacial scattering may be the dominant scattering factor. (We return to this problem in Chapter 5.)

In atomic ordering systems, one can define a composition wave having the same periodicity as the ordered structure. The growth in amplitude of this wave then gives a description of the degree and type of homogeneous long range order. For example, a (100) ordering wave of wavelength a_0 as would describe $L1_0$ ordering in a unit cell of dimensions a_0 is shown in Figure 2.4. The degree of order is given by the square of the amplitude of this wave. More complicated structures can be built up from a superposition of different ordering waves. For example, a ternary ABC_2 Heusler alloy can be generated from a (100) wave of wavelength a_0 and a $(\frac{1}{2}\frac{1}{2}\frac{1}{2})$ wave of wavelength $2a_0/\sqrt{3}$ (de Fontaine 1979). In general terms, the atomic distribution of a periodic solid can always be expanded as a Fourier series (i.e. a superposition of static concentration waves) such that the local composition is given by a generalised form

$$c(\mathbf{r}, t) = \bar{c} + \frac{1}{2} \sum_j [A(\boldsymbol{\beta}_j, t) \exp(\mathrm{i}\boldsymbol{\beta}_j \cdot \mathbf{r}) + A^*(\boldsymbol{\beta}_j, t) \exp(-\mathrm{i}\boldsymbol{\beta}_j \cdot \mathbf{r})], \tag{2.20}$$

where $\boldsymbol{\beta}_j$ is a non-zero wavevector defined in the first Brillouin zone of the disordered alloy, $\bar{c}$ the mean alloy composition and $A(\boldsymbol{\beta}_j, t)$ the amplitude of the $\boldsymbol{\beta}_j$ composition wave which may evolve with time. Khachaturyan (1979) gives an excellent discussion of the stability of alloys in terms of concentration waves including descriptions of many different superstructures. He also points out that the static concentrations that describe an ordered structure are completely determined by the diffraction pattern that they generate: the wavevector of each concentration wave is equal to the reciprocal lattice reflection which is within the first Brillouin zone. The amplitude of the concentration wave is proportional to the structure amplitude of this reflection. Note that by using the notation described above, the previous example of spinodal decomposition may be represented by a modulated (000) lattice wave.

2.4 Reciprocal space representation

In the foregoing sections we have taken a description of the atomic structure of the alloy in real space. In many of the computations that follow it will be more convenient to adopt a reciprocal space (or **k**-space) description of the atomic structure. This is done by taking a discrete or continuous Fourier transform depending upon whether the lattice is being considered as discrete or a continuum.

Fig. 2.4. Representation of an $L1_0$ ordered structure by a (100) composition wave of wavelength a_0.

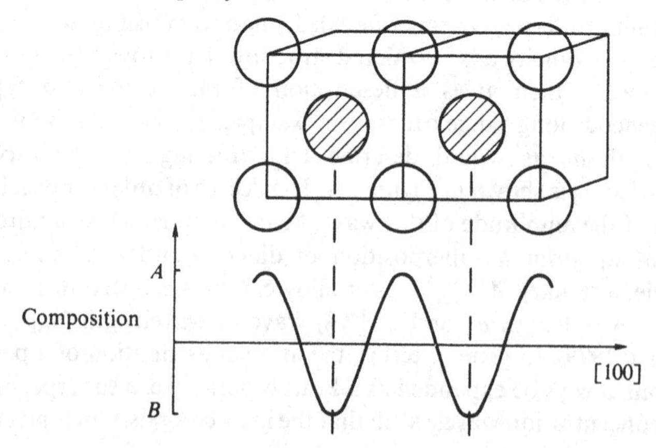

Discrete:

$$F(\mathbf{k}) = \frac{1}{N} \sum_{i=1}^{N} f(\mathbf{r}_i) \exp(-i\mathbf{k} \cdot \mathbf{r}_i), \tag{2.21}$$

continuous:

$$F(\mathbf{k}) = \frac{1}{\Omega} \int f(\mathbf{r}) \exp(-i\mathbf{k} \cdot \mathbf{r}) \, d\mathbf{r}, \tag{2.22}$$

where Ω is the volume over which the integral is taken. If the spectrum of **k** values being considered is discrete (e.g. corresponding to special points in the Brillouin zone) there will be a discrete inverse transform:

$$f(\mathbf{r}_i) = \sum_{\mathbf{k}} F(\mathbf{k}) \exp(i\mathbf{k} \cdot \mathbf{r}_i). \tag{2.23}$$

If **k** is a continuous variable then the inverse transform is given in an integral form:

$$f(\mathbf{r}) = \int F(\mathbf{k}) \exp(i\mathbf{k} \cdot \mathbf{r}) \, d\mathbf{k}. \tag{2.24}$$

Warren (1969) gives a useful discussion of errors associated with Fourier transforms when using a limited set of data. The nature of a Fourier transform shows immediately why any periodically ordered structure can be represented in terms of site occupations in real space or by a composition wave, with the position of the wavevectors in **k**-space (i.e. site occupations in **k**-space) giving the periodicity of the correlations. This is of course why there is a direct link between an ordered structure and its corresponding diffraction pattern. It is also why the Fourier transform of the various correlation parameters and functions will be of direct use in determining their effects on the scattering of conduction electrons. As a general guideline it is found that if the atomic configuration under question is of a short range and possibly inhomogeneous nature than a direct space representation is most useful, but if it has longer range periodicities or is more uniform in distribution then a reciprocal space (i.e. concentration wave) representation is more appropriate.

2.5 Short range atomic configurations

2.5.1 *Mode of decomposition*

The ordering of a random solid solution or decomposition into a two-phase mixture may proceed either by a continuous process (e.g. spinodal ordering or spinodal decomposition) or by a discontinuous process (nucleation and growth). This may be understood by considering the composition dependence of the free energy of a hypothetical binary alloy shown in Figure 2.5. The existence of two minima indicates that alloys having nominal compositions between c_1 and c_2 will exist as a two-phase mixture in the ground state, the composition of each phase being given by the common tangent to c_1 and c_2. Fluctuations of composition of alloys between c_1 and c_1', and c_2 and c_2' will cause an initial increase in free energy, i.e. there is an energy barrier opposing the decomposition. Once a nucleus of suitable size has formed, growth of the second phase can proceed. This is *discontinuous growth*, as it relies on formation of a nucleus. If, however, the nominal composition lies between c_1' and c_2' then composition fluctuations will cause a decrease in free energy. There is no energy barrier and the decomposition is a *continuous* process. If different temperatures are considered, the form of the free energy curve will change due to the entropy contribution. The locii of the points c_1 and c_2 (as the temperature is varied) then trace out the miscibility gap, while the locii of c_1' and c_2' trace out the coherent spinodal.

The morphology of the resultant phase distributions will be influenced by the nature of the decomposition process. For example, in the case of heterogeneous nucleation the distribution of precipitate particles may be quite non-uniform, while a spinodal process will produce a periodic array of precipitates (although one cannot exclude homogeneous nucleation as the origin of such a structure).

Fig. 2.5. Free energy of a hypothetical binary alloy showing the origin of continuous and discontinuous transformations.

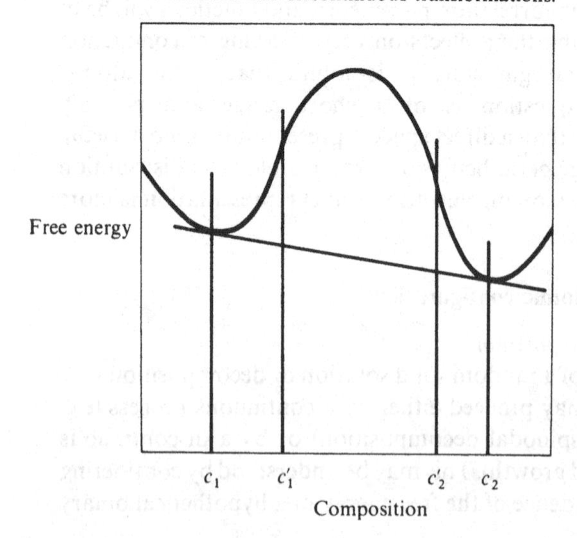

2.5.2 *Phase separation*

(a) *Clustering*

Atomic interactions in some systems will be such that a preference for like nearest neighbours occurs. Such systems will exhibit atomic clustering as a precursor to their decomposition into two separate phases at low temperatures. Such clustered states may be quenched-in or exist as equilibrium fluctuations at temperatures above a two-phase solvus or miscibility gap (Matsubara & Cohen 1983; Wu & Cohen 1983; Wendrock 1983; Chen & Cohen 1979). As clustering specifically implies local deviations from the average concentration, it cannot exist as a homogeneous state and therefore size effects producing interfacial strains may be important.

As noted in Section 2.2, the occurrence of like nearest neighbours implies a positive $\langle\sigma_i\sigma_j\rangle$. Typical values of $\langle\sigma_i\sigma_j\rangle$ for some Cu–Ni alloys are given in Table 2.3. This data was obtained from neutron diffraction experiments using isotopically enriched alloys at compositions chosen so that the intensity of the Bragg peaks was zero, leaving only the diffuse scattering due to atomic correlations. The scattered intensities observed are shown in Figure 2.6 and have been included to give a graphic indication of how the clusters will scatter conduction electrons. It is tempting to interpret the spatial variation of the correlation parameters as a direct indication of the size of clustered regions but, as noted by Cohen (1970), the interpretation of data in this manner is not justified.

Fig. 2.6. Diffuse neutron scattering cross-section $d\sigma/d\Omega$ due to short range atomic correlations in a null-matrix Cu–Ni alloy as a function of scattering wavevector q (or scattering angle θ) (after Wagner *et al.* 1982).

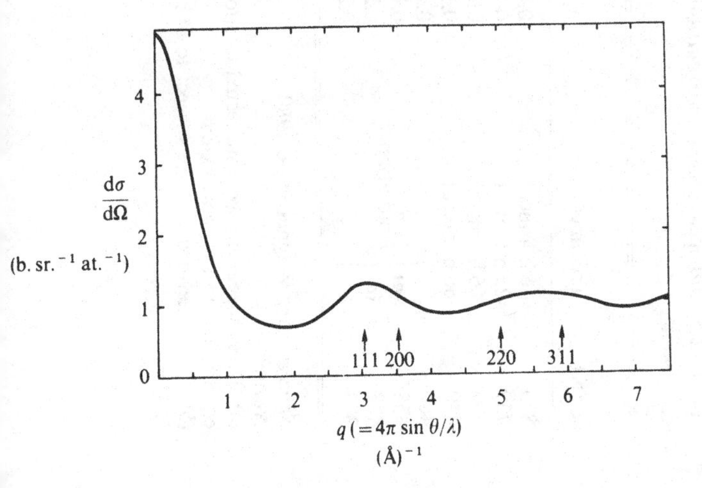

Table 2.3. *Pairwise atomic correlation parameters α_i for Cu–58.6%Ni subjected to various heat and irradiation treatments*

T (K)	Treatment	α_1	α_2	α_3	α_4	α_5 (*b*)	α_6 (*b*)	α_7 (*b*)
870	6 h anneal	0.149	−0.013	0.022	0.025	−0.004	0.001	0.004
739	22 h anneal	0.132	0.014	0.024	0.024	0.005	−0.010	0.008
690	66 h anneal	0.139	0.017	0.027	0.028	0.007	−0.010	0.009
640	66 h anneal	0.147	0.019	0.033	0.037	0.012	−0.012	0.017
480	irradiation	0.206	0.029	0.052	0.044	0.010	0.003	0.016
458	(*a*) irradiation	0.191	0.017	0.043	0.035	0.005	0.005	0.009
423	(*a*) irradiation	0.202	0.001	0.045	0.018	0.005	0.013	0.007
373	irradiation	0.218	0.007	0.053	0.025	0.005	0.023	0.006

Source: After Wagner *et al.* 1980.

Notes:

(*a*) These specimens were subjected to a lower irradiation dose than the 480 K or 373 K specimen and so may not have approached equilibrium to the same extent.

(*b*) The higher-order α_i are subject to uncertainties at least as large as the values given.

However, it is possible to obtain statistical information about cluster sizes and morphologies from the diffuse scattering data by computer modelling using the techniques of Gehlen & Cohen (1965), Williams (1978) and Clapp (1971). Such programs usually adjust the configuration of some suitably large array until the short range order parameters calculated from the array match those produced experimentally. Cohen *et al.* have argued that only parameters for the first few shells $\alpha_1, \alpha_2, \alpha_3, \ldots$, need be considered as the arrays generated tended to also reproduce the higher distance correlation parameters, while Williams argues for consideration of parameters out to the 18th shell. Both of these techniques give real space information about the size and shape of an average cluster. The probability variation method of Clapp differs in that it only gives information about the probability of occurrence of various small cluster configurations. Some examples of these types of analyses are given by Gragg *et al.* (1971) and Cohen & Georgopoulos (1984). Microdiffraction and direct imaging by electron microscopy have recently become more readily available and are being used to obtain information about cluster shapes and sizes, as are atom probe field-ion microscope techniques. Other statistical data may be obtained from the analysis of small angle scattering data using the methods of Levelut and Guinier (see Guinier & Fournet 1955). Small angle X-ray scattering (SAXS) is useful in systems containing components with sufficiently large differences in electron densities, and small angle neutron scattering (SANS) is useful in cases of components having an appreciable difference in neutron scattering length. Some data relating to the size and number of Guinier–Preston (GP) zones in an Al–1.7 at.%Cu alloy obtained by a variety of techniques are given in Table 2.4. In this system the GP zones are coherent platelet clusters one or two atoms thick.

(b) *Precipitation*

After the initial coherent deviation from the random state, discontinuous precipitation or phase separation processes often occur via a sequence of reactions during which there is a progressive loss of coherency with the matrix. For example, if Al–Cu alloys are heat treated below the GP zone solvus decomposition occurs according to the sequence

$$\alpha \rightarrow \text{GP zones} \rightarrow \theta'' \rightarrow \theta' \rightarrow \theta$$

(see e.g. Porter & Easterling 1981). The structures associated with these stages are illustrated in Figure 2.7. The platelet morphology of the GP zones and complicated sequence of structures in this system results from the large size difference of the copper and aluminium atoms and consequentially large interface strain energies. Table 2.5 gives the

Table 2.4. *Size and number of GP zones in Al–1.7%Cu determined by various means*

T (K)	Treatment (s)	Zone diameter (nm)	No. of zones $\times 10^{24}$ (m^{-3})	Technique
RT	4×10^8	3.0 – 4.4	8.9	X-ray diffuse scattering
RT	4×10^8	1.5 – 2.0	—	FIM
RT	4×10^8	4.3 ± 0.3	4.7	TEM
RT	4×10^8	4.5 ± 0.2	5.0	SAXS
353	6×10^5	5.1	4.0	SAXS
313	6×10^5	3.9	6.9	SAXS
373	6×10^5	1.6	17.0	SAXS

Source: From Cohen & Georgopoulos 1984.

Fig. 2.7. Decomposition sequence in Al-rich Al–Cu alloys (after Martin & Doherty 1976).

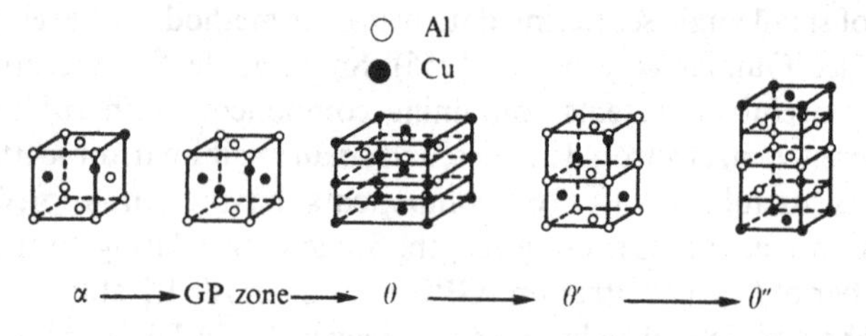

precipitation sequences observed in some other systems. Note that in those systems which have small lattice parameter misfits (e.g. Al–Zn, Al–Ag) the zones tend to have spherical symmetry and the precipitation sequence is less complicated.

Because of the importance of both strain and volume of precipitate particles, their characterisation for the purpose of calculating electrical resistivity is quite difficult, particularly at the later stages of decomposition where the particle size is larger than Λ and the structure of the incoherent interface needs to be taken into account (see Section 2.6).

(c) *Spinodal decomposition*

As noted in Section 2.5.1, decomposition within the spinodal proceeds by continuous growth of the most favoured composition wave leading to microstructures of the form shown in Figure 2.8. Although the

Table 2.5. *Precipitation sequences in some age hardening alloys*

Alloy	Precipitation sequence
Al–Ag	GP zones (spheres) → γ' plates) → γ (Ag_2Al)
Al–Cu	GP zones (discs) → θ'' (discs) → θ' (plates) → θ ($CuAl_2$)
Al–Cu–Mg	GP zones (rods) → S' (laths) → S ($CuMgAl_2$) (laths)
Al–Zn–Mg	GP zones (spheres) → η' (plates) → η ($MgZn_2$) (plates or rods)
Al–Mg–Si	GP zones (rods) → β' (rods) → β (Mg_2Si) (plates)
Cu–Be	GP zones (discs) → γ' → γ (CuBe)
Cu–Co	GP zones (spheres) → β (Co) (plates)
Fe–C	ε-carbide (discs) → Fe_3C (plates)
Fe–N	α'' (discs) → Fe_4N
Ni–Cr–Ti–Al	γ' (cubes or spheres)

Source: From Martin 1968; Porter & Easterling 1981.

Fig. 2.8. Typical microstructure of a spinodally decomposed Fe–Cr–Co alloy (Houghton & Rossiter 1978).

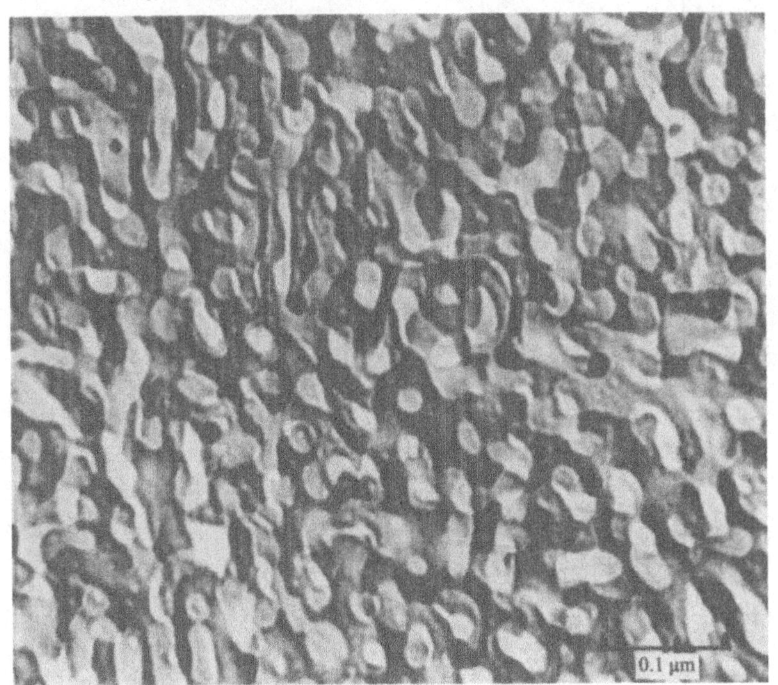

dominant wavelength may change somewhat during decomposition, the resultant microstructure remains periodic in three dimensions with the magnitude of the composition fluctuations steadily increasing. The dominant wavelength of the decomposition depends strongly upon the temperature at which the alloy is aged in relation to the spinodal, and any stress anisotropy or application of a magnetic field during decomposition may result in an anisotropic structure with different dominant composition wavelengths in different directions. However, the initial stages of the decomposition occur very rapidly and the actual decomposition wavelengths encountered are more likely to be determined by a coarsening process. Typical values lie in the range ~1–100 nm, which makes such alloys potentially very interesting for a resistivity study as this range should encompass the conduction electron mean free path length. The periodicity of the decomposing structure produces satellites about the Bragg peaks, as shown in Figure 2.9.

Fig. 2.9. Small angle neutron-scattering cross-section $d\sigma/d\Omega$ for a Cu–Ni–Fe alloy showing the evolution of the spinodal satellite as a function of ageing time. The numbers on the curves refer to the time (hr) that the specimen was aged at 673 K (except for the 2-hr curve which was at 973 K) (after Wagner *et al.* 1984).

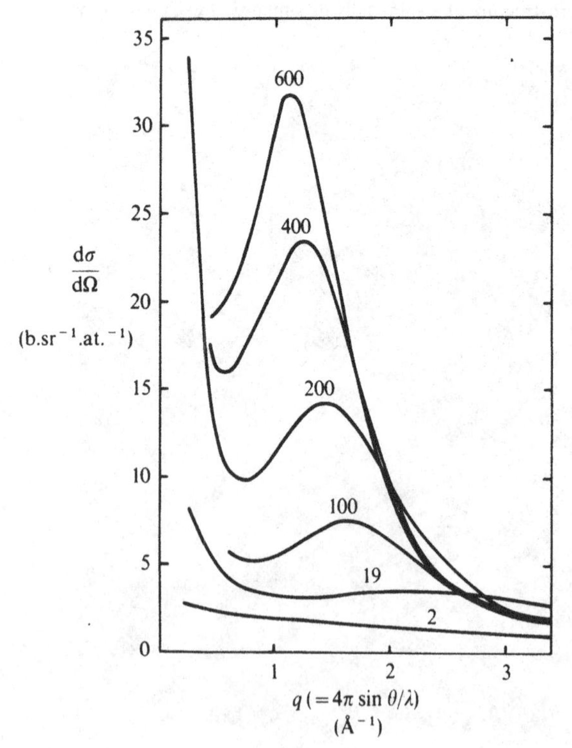

2.5.3 *Atomic ordering*

Atomic ordering effects are observed in systems which exhibit a preference for unlike nearest neighbours. In terms of the parameters introduced in Section 2.2 this means that the pair correlation parameter $\langle\sigma_i\sigma_j\rangle$ will be negative when i and j are nearest neighbours. Some values for these parameters obtained from short range ordered Cu–25%Au by X-ray diffraction were given in Table 2.2. The corresponding diffuse scattered intensity is shown in Figure 2.10 and, as in the clustering example considered earlier, gives an indication of the spatial distribution of the scattering. Note that, unlike the clustering example, the diffuse intensity goes to zero at $q=0$ (i.e. the diffuse intensity peaks at positions away from the Bragg maxima of the point lattice), although in both cases the scattering approaches a constant value at large q.

In contrast to the case of clustering considered above, ordering may take place with no change in the locally averaged composition. However, there are at least three different types of short range order topologies which should be distinguished (see e.g. Kornilov 1974):

(a) *Type I homogeneous (statistical) SRO*

This model of SRO assumes that ordering proceeds uniformly throughout the specimen, i.e. given any site i occupied by an atom of particular type (A) there will be some probability which is independent of the choice of i that adjacent sites will be occupied by atoms of a different

Fig. 2.10. Diffuse scattered intensity I_{SRO} for short range ordered Cu–25%Au calculated from the values given in Table 2.6.

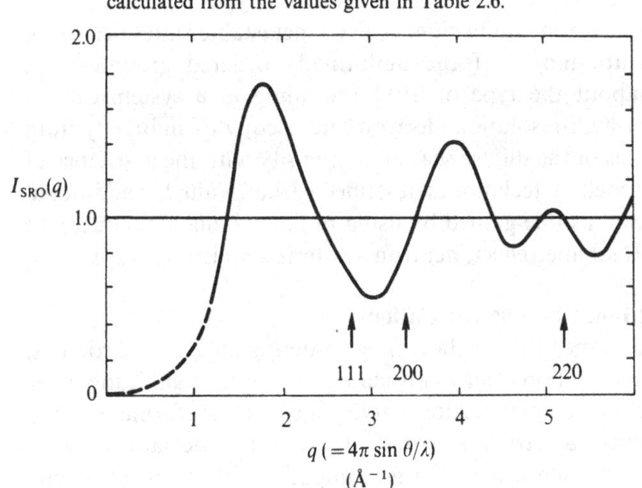

kind (B). However, as there is no correlation between *distant* sites, widely separated regions may or may not contain the same ordering sequence. For example, in a stoichiometric fcc AB system the A atoms may occupy the face-centre positions in one region but they may occupy the cube-corner positions in another region. This structure may develop into a long range ordered structure by ageing beneath the order–disorder solvus and so will lead to the presence of antiphase domain boundaries.

(b) *Type II(a) heterogeneous SRO (microdomain model)*

As an alternative to the scheme described above, SRO may develop by the emergence of small domains (which may or may not differ in composition from the nominal alloy composition) having some equilibrium degree of order. Provided that these domains have dimensions less than the coherence length of the probe (i.e. the electron mean free path), they will only contribute to the diffuse scattering and as such are considered as short range ordering effects. As ordering proceeds, these microdomains grow and finally coalesce into a contiguous network of antiphase domains.

(c) *Type II(b) heterogeneous SRO (antiphase domain model)*

This model assumes that the early stages of ordering are characterised by a very fine random network of antiphase domains, each domain having some equilibrium degree of order. However, if the domain size is less than the coherence length, such ordering will again result in diffuse scattering and be properly classified as short range order.

Type I SRO is likely to represent the equilibrium state of an alloy at temperatures above the order–disorder solvus. Types II(a) and (b) are likely to represent non-equilibrium or even metastable states occurring prior to the attainment of the uniformally ordered ground state. Information about the type of SRO prevailing in a system can be obtained from high-resolution electron microscopy or indirectly from detailed analysis of the diffuse scattered intensity with the assistance of some of the modelling techniques described earlier. Valuable additional information can also be gained by using probes of different coherence length (e.g. Mössbauer effect, neutron scattering and resistivity).

2.6 Long range atomic correlations

As decomposition or short range ordering in an alloy proceeds, the atomic correlations may ultimately become significant over distances much larger than Λ. Alternatively, a phase transformation may directly produce a structure that already has deviations from randomness on a scale which is large compared with Λ (as in eutectic reaction, for example). In accordance with the definitions given in Section 1.4 these structures are described as having long range atomic correlations. In order to understand the electrical properties of such materials it is necessary to characterise the structure of each phase as well as the interphase boundaries. This is by no means a trivial problem and requires application of the concepts of topology and quantitative stereology (see e.g. Underwood 1970; de Hoff & Rhines (1968)). The absolute minimum requirement would appear to be a knowledge of the volume faction and degree of order of each phase in a multiphase solid and the total area of internal (i.e. antiphase and/or interphase) boundaries. Determination of even these basic quantities can often prove difficult. Let us now consider some of the typical microstructures encountered.

2.6.1 *Long range ordering*

Ordered materials will not in general be completely homogeneous because of the way long range order develops from the disordered phase. Furthermore, the antiphase boundaries may be thermodynamically quite stable since there are situations when they do not lead to any increase in enthalpy and may even lead to a lowering of the free energy. For example, it can be shown that if the pairwise interchange energies V_i satisfy the relationship $V_2-4V_3+4V_4=0$ there will be no excess enthalpy associated with an antiphase boundary in $L1_0$ or $L1_2$ superstructures. More generally, Khachaturyan (1979) shows that a homogeneous superlattice can be thermodynamically stable with respect to the formation of antiphase domains only when the Fourier transform of the interchange energy has minima at all superlattice reciprocal lattice points. A typical antiphase domain structure is shown in Figure 2.11. The presence of a long period superlattice (LPS) that may or may not be commensurate with the lattice (i.e. have a period that is an integral number of lattice spacings) can also lead to a decrease in the free energy and so be part of the ground state structure. Formation of such long period structures has been described in terms of either electronic structure effects (Sato & Toth 1965) (which are essentially long range in nature) or competing nearest neighbour and next nearest neighbour interactions (i.e. short range interactions), as in the ANNNI model (Fisher & Selke, 1980, 1981; de Fontaine & Kulik 1985), for example. These LPS structures have been the subject of intense experimental and theoretical investigation in recent years and good reviews are available in the proceedings of a number of conferences including Modulated Structures – 1979 (Cowley *et al.* 1979), Phase Transformations in Solids (Tsakalakos 1984), NATO Advanced Studies Institute on Modulated

Structures (1985), Order–Disorder (Warliamont 1974), Phase Transformation in Solids (Aaronson *et al.* 1982), and a variety of other texts, including Sato & Toth (1965). High-resolution electron microscopy (see e.g. van Tendeloo & Amelinckx 1978, 1984; Schryvers *et al.* 1985; Watanabe & Terasaki 1984; Sinclair *et al.* 1974; Sinclair & Thomas 1975) and atom probe field-ion microscopy (see e.g. Müller & Tsong 1969; Ivchenko & Syutkin 1983) indicates that the antiphase boundaries are only one or at most two atomic layers thick. These structures can be partly characterised in terms of an antiphase domain size or wavelength of a long period modulation of the ordering wave, as shown in Figure 2.12. Some typical results are given in Table 2.6.

If the scale of features is larger than Λ, then from the point of view of scattering of conduction electrons they should be regarded as an

Fig. 2.11. Dark field transmission electron-microscope image of antiphase boundaries in Fe–24 at.%Al × 18 000 (after Allen & Cahn 1979).

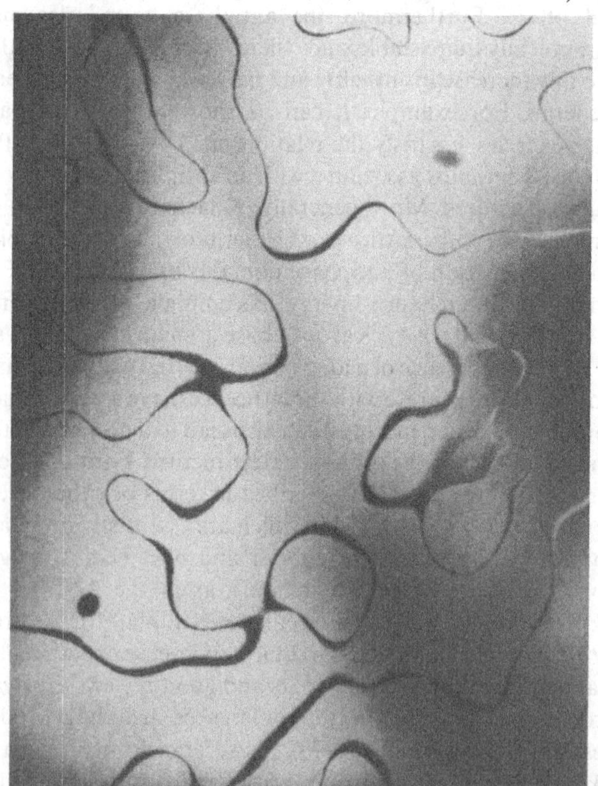

Table 2.6. *Antiphase domain structure of some different long range ordered alloys*

Alloy	Composition	Elongation or contraction of the long period direction	Structure	Periodicity (a_0)
CuAu II	50.0% Au	1.003		5.0
Cu_3Au II	32.2% Au	1.0		8.5
Ag_3Mg	24.5% Mg	1.002		1.8
Au_3Zn H	21.0% Zn	1.02		2.0
Au_3Zn R_1	23.5% Zn	1.057	one dimen-	complex
Au_3Zn R_2	26.0% Zn	1.062	sional	complex
Pd_3Mn	27.4% Mn	0.99		2.0
Cu_3Pt	25.8% Pt	0.99		6.4
CuPd (α)	25.2% Pd	0.987		4.4
CuPd (α)	28.5% Pd	0.992		3.5
		1.0		3.8
Au_3Zn	19.0% Zn	1.009	two dimen-	3.5
		1.003	sional	2.9
Au_3Mn	25.0% Mn	1.012		1.0–1.3
				2.1–2.5

Source: After Owaga 1974.

Fig. 2.12. Long period modulation of order. (*a*) Long period superlattice. (*b*) Ordering wave. (*c*) Modulation wave.

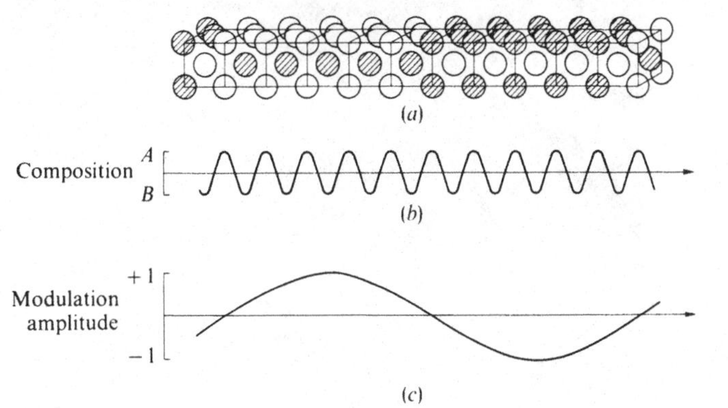

interconnected network of scattering interfaces separated by regions of homogeneous order. Ageing can then produce complicated results with both a change in the degree of order and a change in topology of the antiphase domain structure. If the scale of the features is less than Λ they should be considered as a short range effect, and averaged over the specimen volume.

2.6.2 *Two-phase mixtures*

As in the ordered structures, there is a seemingly endless variety of structures that can result from phase separation. Some of those typically encountered include the Widmanstatten, lamellar and precipitate microstructures shown in Figure 2.13. However, when the scale of these structures is larger than Λ they can all be represented (as in the long range order case) as an array of scattering interfaces separated by relatively homogeneous media of some characteristic composition and structure.

2.6.3 *Some general comments*

In systems that undergo either ordering or phase separation it appears that the interfaces are likely to play an important role. If the electron scattering from them is diffuse then it should be sufficient to simply determine their total area. If, however, the scattering is specular (see Section 5.5.2) then it may be necessary to consider a projected area

Fig. 2.13. Some typical microstructures in phase-separating systems: (*a*) Widmanstatten α in β matrix (Cu–42at.Zn $\times$ 300); (*b*) lamellar Cu–8.4at.% P $\times$ 1300); (*c*) dispersed precipitates (fine and coarse precipitates in Fe–45%Co–6%Ni–6%V $\times$ 132 000) ((*a*), (*b*) courtesy of N. C. Tranter; (*c*) courtesy R. A. Jago).

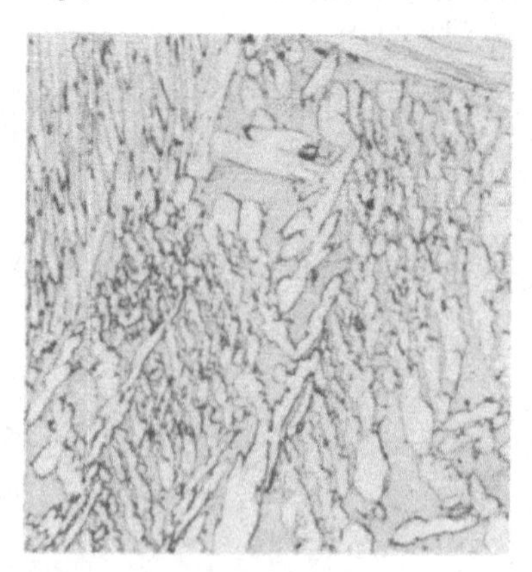

(*a*)

or take into account more details of their topology. In either case the important intrinsic features of the interfaces on a scale $\sim\Lambda$ will be their composition gradients and (as we show in the next section) strain and defect concentrations resulting from lattice parameter miss-match. In this regard the high-resolution (analytical) TEM and atom-probe field-ion microscope (FIM) should prove invaluable, although one should not neglect the possible contributions that may come from more indirect techniques such as position annihilation. As examples, the concentration gradient that occurs across a α–γ boundary in an Fe–Ni alloy determined by microprobe analysis is shown in Figure 2.14 and the results of a FIM study of a Fe–Cr–Co spinodal are shown in Figure 2.15.

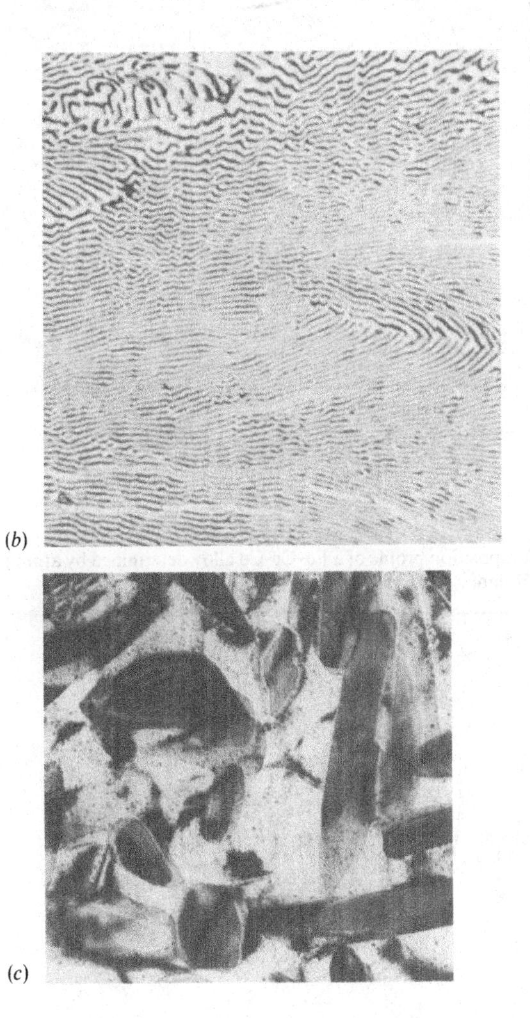

Fig. 2.14. Concentration gradients across a α–γ phase boundary in Fe–Ni determined by microprobe analysis and corrected for instrumental resolution (after Lin *et al.* 1977).

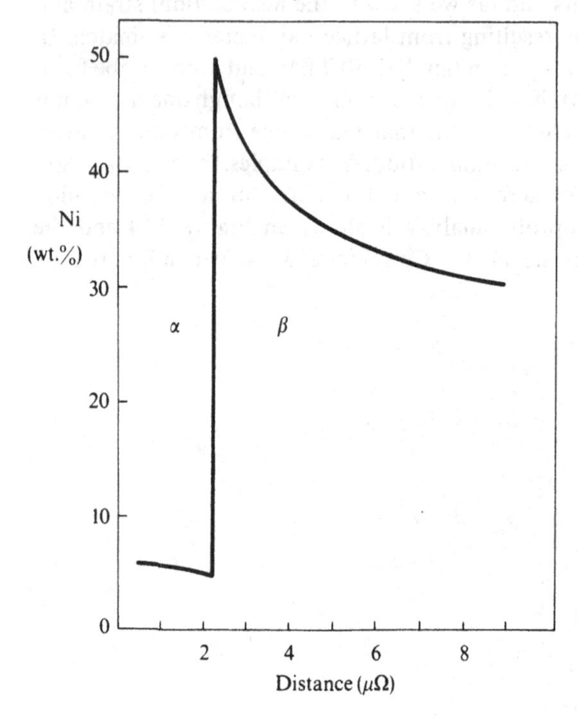

Fig. 2.15. Composition profile of a Fe–Cr–Co alloy determined by atom probe FIM (after Brenner *et al.* 1984).

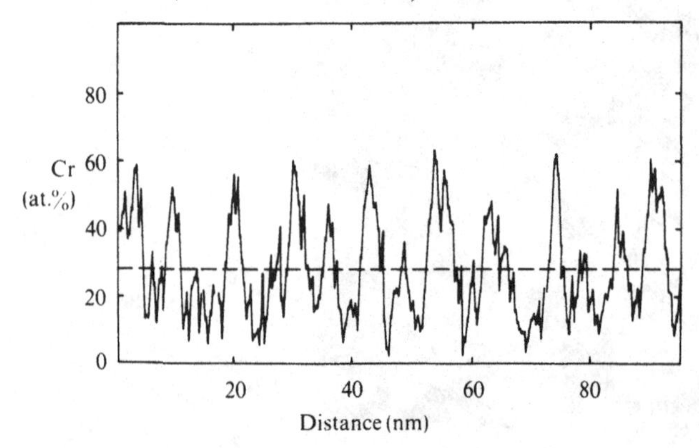

The geometrical configuration of a structure may be modelled as an array of elements, as shown in Figure 2.16. Each element may represent different phases or antiphase or interphase boundaries. Such techniques are discussed in more detail in Section 5.5.

Defects which are likely to be associated with the interfaces (see e.g. Nakagawa & Weatherly 1972; Brown 1977; Knowles & Goodhew 1983*a*,*b*) may also make a significant contribution to the resistivity.

2.7 Atomic displacement effects

In the previous sections we have concentrated mainly upon those aspects of microstructures resulting from the rearrangement of atoms on a rigid periodic lattice. However, the atomic species concerned will in general have different atomic or ionic volumes and so may suffer displacements away from the lattice sites in order to allow better atomic packing, the local increase in strain energy being offset by the decrease in

Fig. 2.16. Various structural models of a complex solid. The shaded and unshaded regions may represent different phases or elements of an interphase or antiphase domain boundary.

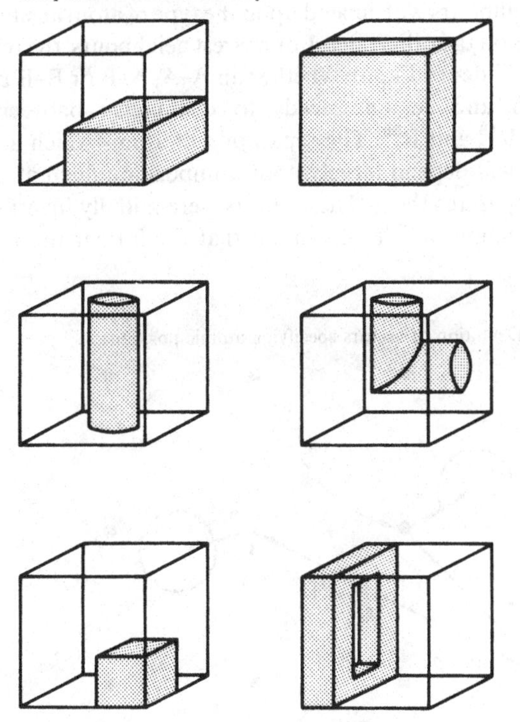

interatomic potential energy. As well as these static atomic displacement effects it is likely that the dynamic properties of the lattice will also be affected by the atomic configuration. Furthermore, we have only been concerned with the atomic distributions associated with replacive phase transitions (i.e. those associated only with the rearrangement of atom type over the lattice). Many metals and alloys also support displacive phase transitions in which a transformation from one structure to another occurs solely by the displacement of atoms away from the lattice sites of the parent phase with no specific change of atom type. All of these displacive effects represent changes in periodicity and so will give rise to electron scattering effects. Let us now consider each of these in more detail.

2.7.1 *Atomic size effects*

The undisplaced positions of atoms in a perfect solid can be defined by a set of vectors $\mathbf{r}_i$ which go over all lattice sites. The atomic positions in the presence of static displacements may then be given by

$$\mathbf{R}_i = \mathbf{r}_i + \mathbf{u}_i, \tag{2.25}$$

as illustrated in Figure 2.17. The actual nature of the displacements $\mathbf{u}_i$ between close neighbours will depend upon the type of atom at site i and the types of atoms on the other sites. For nearest neighbours, the relative atomic spacing will depend upon whether an A–A, A–B or B–B pair is being considered and so one needs to distinguish between the separations $\mathbf{R}_i^{AA}$, $\mathbf{R}_i^{AB}$ and $\mathbf{R}_i^{BB}$. The separation of atoms which are not close neighbours will depend largely upon composition fluctuations in the region that separates them. These effects were initially investigated using X-ray diffraction and it was found that the former produces a

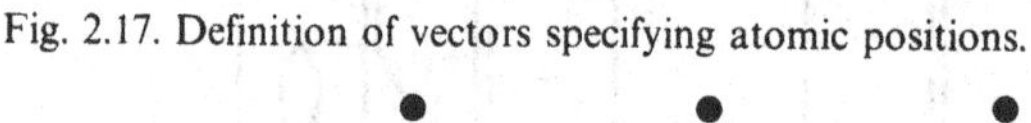

Fig. 2.17. Definition of vectors specifying atomic positions.

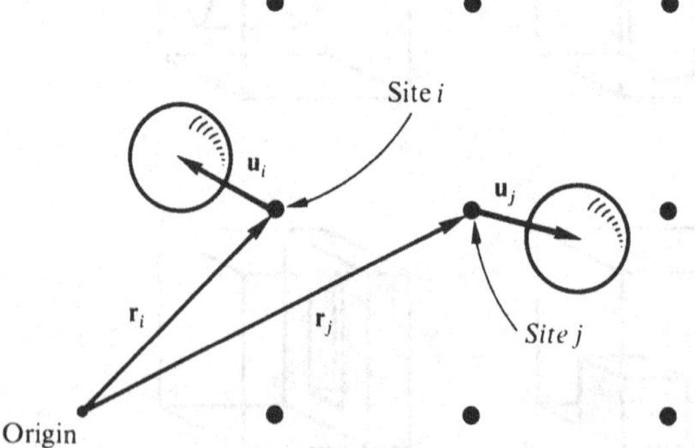

modulation of the diffuse scattering intensity, as shown in Figure 2.18, while the latter decreases the intensity of the Bragg peaks and increases the diffuse background (Borie 1957; Warren 1969, Ch. 12). The analogous conduction electron scattering effects are discussed in Section 5.6.

As an approximation to the real vector displacements, one could assume that the displacements occur along the line joining the near neighbours. Using the shell model described in Section 2.2 then allows these displacements to be written as

$$\begin{aligned} R_i^{AA} &= r_i(1+\varepsilon_i^{AA}) \\ R_i^{AB} &= R_i^{BA} = r_i(1+\varepsilon_i^{AB}) \quad , \\ R_i^{BB} &= r_i(1+\varepsilon_i^{BB}) \end{aligned} \tag{2.26}$$

where r_i is now the average radius of the ith shell. Because of the dependence of the displacements on the atom type there is a useful subsidiary relationship which follows from the definition of the average radius:

$$2\langle\sigma_i^A\sigma_j^B\rangle r_i(1+\varepsilon_i^{AB})+\langle\sigma_i^A\sigma_j^A\rangle r_i(1+\varepsilon_i^{AA})+\langle\sigma_i^B\sigma_j^B\rangle r_i(1+\varepsilon_i^{BB})=r_i. \tag{2.27}$$

This relationship is

$$\langle\sigma_i\sigma_j\rangle(\varepsilon_i^{AA}-2\varepsilon_i^{AB}+\varepsilon_i^{BB})+c_A^2\varepsilon_i^{AA}+2c_Ac_B\varepsilon_i^{AB}+c_B^2\varepsilon_i^{BB}=0. \tag{2.28}$$

Such a model is also useful in describing the lattice dilation or contraction around other defects such as vacancies or interstitials.

Fig. 2.18. Diffuse X-ray intensity produced by size effect modulations of atomic positions (solid line). Also shown is the diffuse intensity due to atomic SRO (dotted line) (after Warren *et al.* 1951).

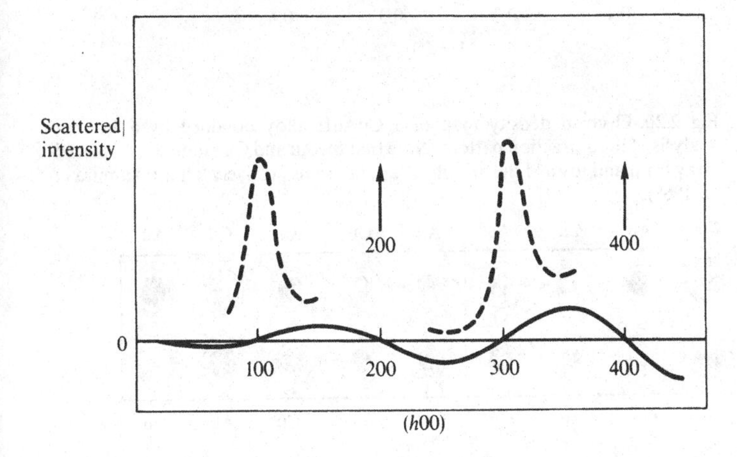

In a long range ordered material the atomic displacements will be periodic, culminating in a tetragonal distortion of $L1_0$ structures which have alternate layers of A and B atoms, for example. In the $L1_2$ structure the symmetry of the atomic environment results in zero distortion in a completely ordered lattice but as the degree of order is reduced this symmetry is broken and atomic displacements will occur, as shown in Figure 2.19. The presence of antiphase domain boundaries will also break this symmetry and give finite atomic displacements in the region of the antiphase boundaries, as shown in Figure 2.20, for example. There is in fact a periodic displacement of the atom planes around an antiphase boundary which gives rise to coherent diffraction effects. There is also a net elongation or contraction in the long period direction, as shown also

Fig. 2.19. Variation of mean square static displacement with $(1\text{–}S^2)$ in a Cu_3Au alloy (after Kozlov *et al.* 1974).

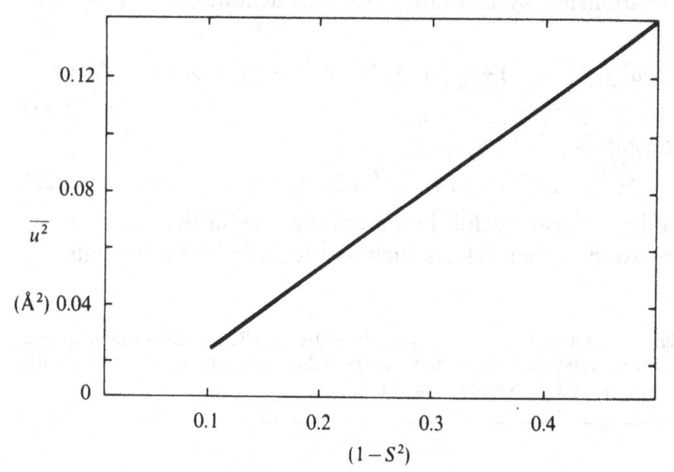

Fig. 2.20. Electron density map of a CuAuII alloy obtained by a Fourier analysis of the diffraction pattern. Note that the Au and Cu atoms are displaced away from and toward the antiphase boundary respectively (after Okamura *et al.* 1968).

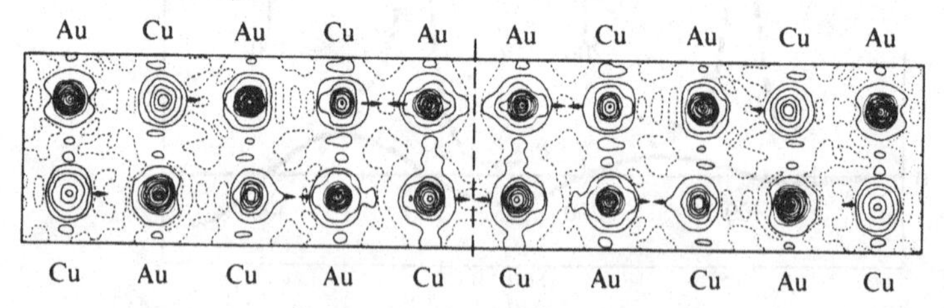

Table 2.7. *Maximum atomic displacements and direction of displacement near an antiphase domain boundary: +, toward boundary; −, away from boundary*

Alloy	Atom	Maximum displacement (Å)	Direction
Au_3Cd	Au	0.02	−
	Cd	0.03	+
CuAu II	Au	0.05	−
	Cu	0.08	+
Cu_3Pd (α)	Pd	0.01	+
	Cu	0.01	−
Pd_3Mn	Pd	0.01	+
	Mn	0.01	−

Source: After Ogawa 1974.

in Table 2.6. A brief summary of some experimentally determined displacements in ordering systems is given in Table 2.7. Similarly, if the constituent atoms in a spinodal system have appreciably different sizes there will be atomic displacements which are periodic and follow the composition wave profile.

2.7.2 *Dynamic atomic displacements*

At temperatures above absolute zero, thermal excitations of the lattice will cause atomic displacements $\boldsymbol{\delta}_i$ so that

$$\mathbf{R}_i = \mathbf{r}_i + \boldsymbol{\delta}_i. \tag{2.29}$$

The 'zero point' displacements that occur at $T = 0$ do not contribute to ρ (see e.g. Dugdale 1977, p. 219). As we show in Section 4.5.4, the adiabatic principle allows calculation of the effects of lattice vibrations simply in terms of the atomic displacements and so it should be sufficient to characterise the effects of atomic configuration on the lattice vibrations in terms of those displacements. However, the wavelike nature of the lattice excitations leads to some complications. Let us now consider two simple descriptions of the lattice vibrations.

(a) *Einstein model*

If the collective excitations of the lattice (i.e. phonons) are neglected, the atoms may be considered to vibrate independently about their mean positions. This is the Einstein model but is only a reasonable approximation at high temperatures ($T \gg \Theta_D$). The equation of motion of an atom of mass M is then

$$M\ddot{x}+bx=0, \tag{2.30}$$

where bx is the harmonic restoring force acting on the atom. We only consider linear displacements since the motion of such a particle is uncorrelated in the three independent coordinate directions x, y, z. The natural frequency of this simple harmonic oscillator is

$$\omega=\left(\frac{b}{M}\right)^{1/2}. \tag{2.31}$$

The mean square displacement at temperature T is then given by the equipartition theorem as

$$\tfrac{1}{2}M\omega^2\overline{x^2}=\tfrac{1}{2}k_B T. \tag{2.32}$$

The natural frequency can be written in terms of a characteristic temperature (the Einstein temperature) Θ_E:

$$\hbar\omega=k_B\Theta_E, \tag{2.33}$$

in which case

$$\overline{x^2}=\frac{\hbar^2 T}{Mk_B\Theta_E^2}. \tag{2.34}$$

At lower temperatures the equipartition theorem is not valid and one has to use the Planck distribution (see equation (2.41)), although the approximation of independent vibrations is then no longer valid and so there is little to be gained by such an extension. As the displacements are uncorrelated they can only give rise to a diffuse electron scattering which will depend upon atomic configuration via the interatomic forces which determine Θ_E, and the masses M. For example, equation (2.34) can be rewritten in terms of b:

$$\overline{x^2}=\frac{k_B T}{b}, \tag{2.35}$$

and the parameter b assumed to be a linear function of the number of A–A, A–B and B–B nearest neighbour bonds:

$$b=\bar{b}+\delta b\langle\sigma_0\sigma_1\rangle, \tag{2.36}$$

giving

$$\overline{x^2}=\frac{k_B T}{\bar{b}+\delta b\langle\sigma_0\sigma_1\rangle}$$
$$\approx\frac{k_B T}{\bar{b}}\left(1-\frac{\delta b}{\bar{b}}\langle\sigma_0\sigma_1\rangle\right), \tag{2.37}$$

if $\delta b \ll \bar{b}$.

(b) *Debye model*

At low and intermediate temperatures the excitation of lattice waves must be taken into account. The displacement of an atom at site $\mathbf{r}_i$ by a lattice wave of wavevector $\mathbf{Q}$ is given by

$$\mathbf{U}_Q=\mathbf{u}_Q\exp[i(\mathbf{Q}\cdot\mathbf{r}_i-\omega_Q t)], \tag{2.38}$$

where ω_Q and $\mathbf{u}_Q$ are the frequency and vector amplitude of that mode respectively. The relation between ω and $\mathbf{Q}$ is called the dispersion relation and is plotted in the form of a dispersion curve, as shown in Figure 2.21.

Given some form of interatomic potential, details of the lattice waves may be found by solving the dynamical matrix (see e.g. Born & Huang 1954; Cochran 1973), which will give one solution or 'branch' for a monatomic one-dimensional crystal and six branches for a diatomic three-dimensional crystal. Each of these solutions are indicated by an index j and the displacements summed over all j. Each branch corresponds to a different mode of vibration: longitudinal (atomic displacement parallel to $\mathbf{Q}$), transverse (atomic displacement perpendicular to $\mathbf{Q}$), acoustic (cooperative motion of all atoms in a unit cell in a general wavelike motion of the lattice) and optical (internal vibrations of atoms relative to each other, e.g. adjacent atoms in a diatomic solid moving in opposition).

The energy of a particular mode $\mathbf{Q},j$ is given by

$$\overline{E_{\mathbf{Q},j}}=\tfrac{1}{2}M\omega_{\mathbf{Q},j}^2|u_{\mathbf{Q},j}|^2. \tag{2.39}$$

Fig. 2.21. Phonon dispersion curves for Al (from Yarnell *et al.* 1965).

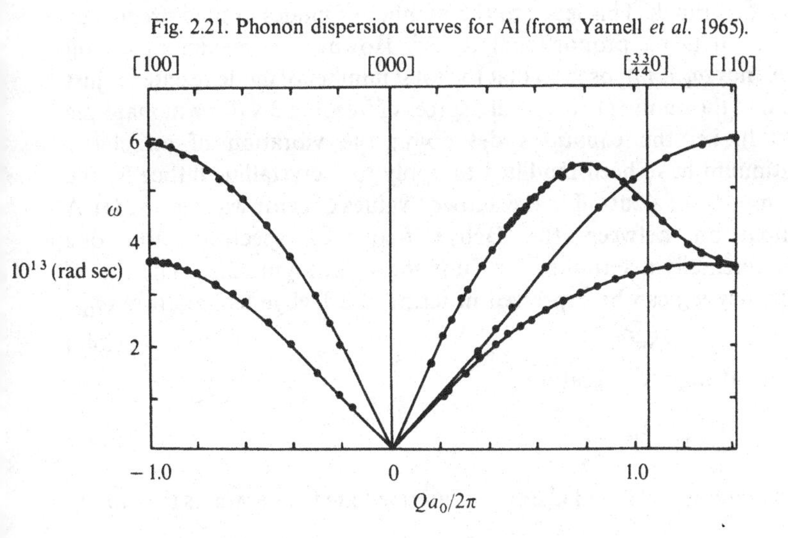

The energy per wave is given as a function of temperature by the quantum theory of a harmonic oscillator

$$E_{\mathbf{Q},j} = (n_{\mathbf{Q},j} + \tfrac{1}{2})\hbar\omega_{\mathbf{Q},j}, \tag{2.40}$$

where $E_{\mathbf{Q},j}$ is the energy of $n_{\mathbf{Q},j}$ phonons of type $\mathbf{Q},j$. The average number of modes excited at any temperature T is given by the Planck distribution

$$\langle n_{\mathbf{Q},j} \rangle = \frac{1}{\exp(\hbar\omega_{\mathbf{Q},j}/k_B T) - 1}, \tag{2.41}$$

leading to

$$\overline{E_{\mathbf{Q},j}} = \left[\frac{1}{\exp(\hbar\omega_{\mathbf{Q},j}/k_B T) - 1} + \frac{1}{2}\right]\hbar\omega_{\mathbf{Q},j}. \tag{2.42}$$

From equations (2.39) and (2.42) one obtains

$$|u_{\mathbf{Q},j}|^2 = \frac{2}{M\omega^2_{\mathbf{Q},j}} \frac{\hbar\omega_{\mathbf{Q},j}}{\exp(\hbar\omega_{\mathbf{Q},j}/k_B T) - 1}. \tag{2.43}$$

At high temperatures one expects

$$\sum_{\mathbf{Q},j} \overline{E_{\mathbf{Q},j}} \cong k_B T, \tag{2.44}$$

and so

$$\sum_{\mathbf{Q},j} |u_{\mathbf{Q},j}|^2 = \frac{2k_B T}{M\omega^2_{\mathbf{Q},j}}. \tag{2.45}$$

In the Debye model the lattice is replaced by an elastic continuum with a linear dispersion relation $\omega \propto Q$ corresponding to a single acoustic branch. This leads to the number of modes excited (frequency spectrum) being proportional to ω^2. However, a maximum cut-off frequency ω_D is imposed so that the total number of modes excited is just equal to the number of classical degrees of freedom $3N$ (for a monatomic crystal), i.e. the equations describing the vibration of an elastic continuum have been modified to apply to a crystalline lattice by the assumption of a cut-off in the allowed values of ω (or wavevector $\mathbf{Q}$). A comparison between the Debye frequency spectrum and one experimentally determined for Cu is shown in Figure 2.22. The cut-off frequency ω_D can be expressed in terms of a Debye temperature Θ_D:

$$\hbar\omega_D = k_B \Theta_D, \tag{2.46}$$

giving, at high temperatures,

$$\sum_{\mathbf{Q},j} |u_{\mathbf{Q},j}|^2 = \frac{2\hbar^2 T}{M k_B \Theta_D^2}, \tag{2.47}$$

(cf. equations (2.33) and (2.34)). At intermediate temperatures the atomic displacements must be found by summing over all modes excited:

$$\boldsymbol{\delta}_i = \sum_{\mathbf{Q},j} [\mathbf{u}_{\mathbf{Q},j} \exp(\mathrm{i}\mathbf{Q}\cdot\mathbf{r}_i) + \mathbf{u}^*_{\mathbf{Q},j} \exp(-\mathrm{i}\mathbf{Q}\cdot\mathbf{r}_i)]. \tag{2.48}$$

While the Debye model represents a rather drastic approximation, the Debye temperature Θ_D remains a useful measure of the energy scale of the vibration of a solid. A full solution of the dynamical matrix (see e.g. Born & Huang 1954; Maradudin *et al.* 1963; Squires 1963) leads to more realistic dispersion curves as shown, for example, in Figure 2.23.

The effects of alloying and atomic configurations on the dispersion relations and phonon spectra follow from the different interatomic potentials (force defects) and masses (mass defects) of the atoms concerned. While a calculation of these effects is not trivial (see e.g. Maradudin 1966, 1969) there have been some attempts to take into account a non-random atomic distribution (Towers 1972; Tibbals 1974; Wu 1984; Hartman 1968; Best & Lloyd 1975; Braeter & Priezzhev 1975; Evseev & Dergachev 1976; Synecek 1962; Punz & Hafner 1985). Neutron spectroscopy is a very useful tool in experimental investigations of phonon spectra since, in order to obtain a neutron wavelength comparable with the lattice spacing for suitable diffraction conditions, neutrons of thermal energies must be used. The energy change due to the absorption or emission of a typical phonon (e.g. equation (2.39)) is then

Fig. 2.22. Comparison of the true phonon frequency spectrum for Cu (after Svensson *et al.* 1967) and the Debye frequency spectrum (scaled so that the areas under the two curves are the same).

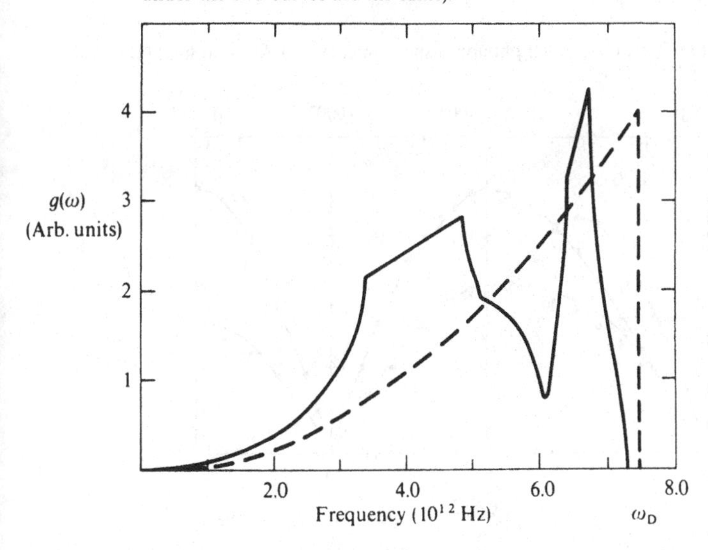

of the same order of magnitude as the energy of the neutron and is thus easily detected. In general it is found (both theoretically and experimentally) that the presence of light foreign atoms leads to new localised vibrational modes which appear at frequencies above the band of perfect crystal frequencies ('out-of-band' modes), while with heavy impurities the band itself may be altered ('in-band' modes). Short range ordering or clustering can lead to a change in frequency and structure of these effects. The overall shape of the dispersion curves are also found to change, particularly as the electron concentration changes (Brockhouse *et al.* 1967) and may show splitting (Kunitomi *et al.* 1973). Studies of long range ordered alloys indicate that in some cases (e.g. β–CuZn) splittings in the dispersion curves at the ordered Brillouin zone boundaries below T_c vanish above T_c (Gilat & Dolling 1965), but in other cases (e.g. Fe_3Al, Ni_3Fe, α–Cu_3Zn) no such effects exist (van Dijk & Bergsma 1968; Hallman & Brockhouse 1969), possibly due to a balance between the combined effects of the force and mass defect reducing the splitting to a value below experimental resolution.

The slopes of the acoustic branches of the phonon dispersion curves as **Q** approaches zero (i.e. long phonon wavelengths) are directly related to the elastic constants. For example, for the longitudinal acoustic branch of a cubic material and **Q** parallel to the cube axis:

$$\left.\frac{\omega_{\mathbf{Q},3}}{Q}\right|_{Q\to 0} = \left(\frac{c_{11}}{D}\right)^{1/2}, \tag{2.49}$$

Fig. 2.23. Calculated phonon dispersion curves of Al (solid line) (after Hafner 1975).

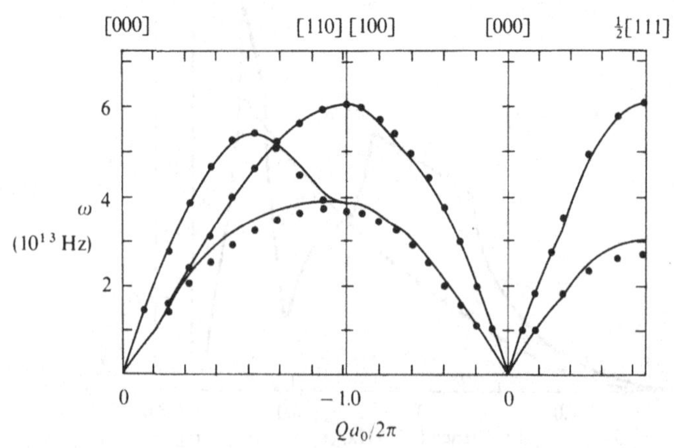

and for the associated transverse acoustic branch:

$$\left.\frac{\omega_{\mathbf{Q},1}}{Q}\right|_{Q\to 0} = \left(\frac{c_{44}}{D}\right)^{1/2}, \tag{2.50}$$

where D is the density of the solid and c_{11} and c_{44} are the usual elastic constants. An example of this behaviour is shown in Figure 2.24.

There have been many studies of the elastic properties of alloys and Young's modulus E is typically found to change by a few percent during precipitation and by larger amounts during long range ordering (see e.g. Muto & Takagi 1955; Silvertsen 1964). It is just in the range $\mathbf{Q} \to 0$ that Umklapp scattering can give rise to large angle electron scattering and so such effects may give a contribution to the electrical resistivity which is not negligible. Changes in the lattice specific heat which are also indicative of changes in the phonon spectrum have also been observed during ordering (see e.g. D. L. Martin 1972, 1976).

2.7.3 *Displacive phase transitions*

Here we consider the transition from one crystal structure to another by atomic displacements specified in relation to the parent lattice. As noted by Cook (1973), the site occupation parameters of the type defined by equation (2.3) can be used to describe transformations

Fig. 2.24. Dispersion curves for Cu–25%Zn (dashed line) and the corresponding elastic constants (solid lines) (from Hallman & Brockhouse 1969).

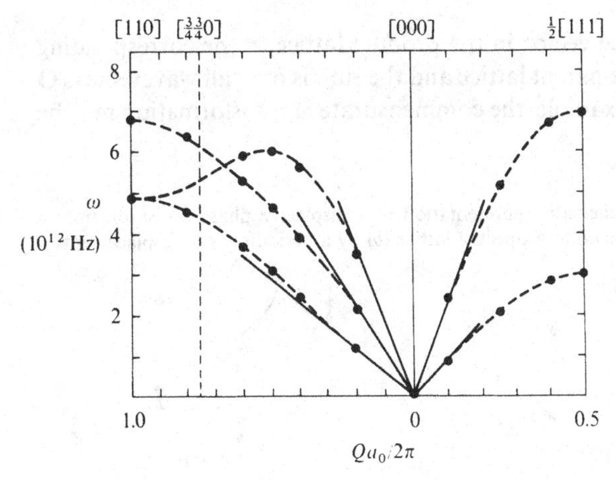

from the following Bravais lattices to the corresponding Bravais or non-Bravais lattices:

$$\text{bcc} \rightarrow \text{fcc}$$

$$\text{fcc} \rightarrow \begin{cases} \text{bcc} \\ \text{fct} \end{cases}$$

$$\left.\begin{matrix} \text{fcc} \\ \text{bcc} \end{matrix}\right\} \rightarrow \text{hcp}$$

$$\text{bcc} \rightarrow \omega$$

$$\left.\begin{matrix} \text{bcc} \\ \text{fcc} \end{matrix}\right\} \rightarrow \text{faulted or dislocated} \begin{cases} \text{bcc} \\ \text{fcc} \end{cases}$$

This type of transformation is shown schematically in Figure 2.25 and emphasises the fact that the displacements must be periodically distributed to produce a new periodic lattice (cf. atomic ordering). The bcc (β) $\rightarrow \omega$ transformation appears to have received the greatest attention. For a given ω variant (i.e. a particular [111] direction), atomic displacements responsible for this transformation are either: forward displacements by one-half of the (111) interplanar spacing; backward by one-half of the (111) spacing; or no displacement. An appropriate choice for the site occupation parameters would then be $+1, -1, 0$ respectively (Sanchez & de Fontaine 1978*b*). The ω transformation can thus be modelled as a ternary displacive alloy.

Such transformations can also be described in terms of static displacement waves (cf. Section 2.3) in which case the displacement $\mathbf{u}_i$ in equation (2.25) is replaced by a periodic term to give

$$\mathbf{R}_i = \mathbf{r}_i + \sum_{\mathbf{Q},j} (\mathbf{u}_{\mathbf{Q},j} \exp(i\mathbf{Q}\cdot\mathbf{r}_i) + \mathbf{u}^*_{\mathbf{Q},j} \exp(-i\mathbf{Q}\cdot\mathbf{r})), \tag{2.51}$$

where $\mathbf{R}_i$ is a lattice vector in the product lattice, $\mathbf{r}_i$ the corresponding lattice vector of the parent lattice and the sum is over all wavevectors $\mathbf{Q}$ and modes j. For example, the commensurate ω transformation may be

Fig. 2.25. Schematic representation of a displacive phase transition from a parent lattice (*a*) to a product lattice (*b*) by a periodic set of displacements δ.

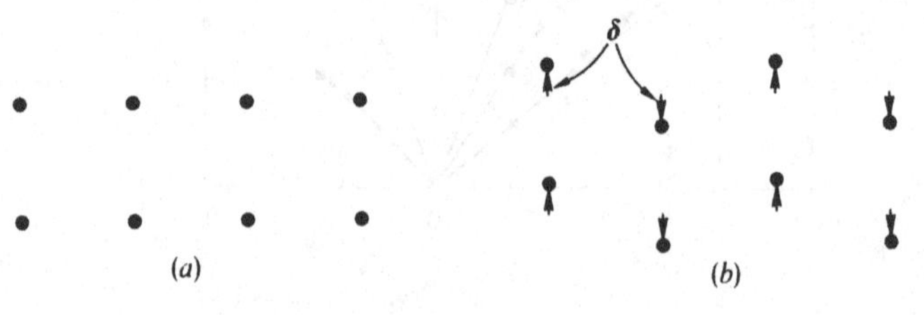

described in terms of a longitudinal static displacement wave of wavevector $\mathbf{Q}_\omega = \frac{2}{3}\langle 111\rangle 2\pi/a_\beta$ and amplitude $u_\omega = a_{\beta/6}$, where a_β is the β lattice spacing (de Fontaine 1970). Incommensurate structures are also encountered in which case $\mathbf{Q}'_\omega = \mathbf{Q}_\omega + \delta\mathbf{q}$ and the magnitude of $\delta\mathbf{q}$ depends upon the solute content, the degree of undercooling and application of pressure. These phases tend to produce diffuse scattering effects rather than the sharp commensurate reflections. An excellent review of ω transformations has been given by Sikka *et al.* (1982). Materials with displacive ferroelectric phase transitions have also been extensively investigated within this framework (see e.g. Cochran 1973). The similarity between equation (2.51) and the phonon equation (2.48) is not accidental, and one proposed mechanism of this type of transformation is a softening of the lattice for the appropriate phonon modes. Such softening can again be investigated by neutron diffraction and spectroscopy techniques, and produces a characteristic 'zero frequency peak' at the wavevector associated with the soft mode. As the displacement wave will by necessity have a short wavelength ($\sim a_0$) it will produce different scattering effects to those resulting from a longer wavelength static lattice modulation in a spinodal system, for example.

Just as in the case of replacive phase transitions, one can apply the concepts of continuous or discontinuous transformations. For example, it appears that both the martensitic and $\alpha \rightarrow \omega$ transformations involve a discontinuous nucleation stage, while the $\beta \rightarrow \omega$ transformation does not require thermal activation and so may be described as a continuous transformation. Similarly one can use the notions of short range and long range correlations depending upon the spatial extent of the 'displaced' regions. This is particularly so at temperatures above the long range transformation temperatures where precursor structures have been observed (Sikka *et al.* 1982; Dubinin *et al.* 1981; Cook 1975; Clapp 1981; Comstock *et al.* 1985; Shapiro 1981), although it appears that in some cases these structures are in fact competing metastable structures rather than truly preminatory effects (Otsuka *et al.* 1981; Hwang *et al.* 1983*a,b*).

2.8 Amorphous alloys

It is possible to produce solid metallic alloys that have no long range structural order by quenching at rates of $\sim 10^5$–10^8 K s^{-1} from the liquid phase, evaporation or sputtering onto a substrate, ion implantation and electro- and chemical deposition (see e.g. Chen 1980 and references therein). Such materials are usually described as 'non-crystalline' or 'amorphous', while terms 'glassy' or 'metallic glass' are often applied to those that have a significant metalloid (Si, B, C, P) content and are prepared by rapid quenching. No such distinction will be made in this text and these descriptions will all be used interchangeably. Systems which are known to form glasses at moderate quench rates ($\sim 10^6$ K s^{-1}) include transition metal or noble metal alloys containing about 10–30% metalloid, alloys of early transition metals (Zr, Nb, Ta, Ti) and late transition metals (Fe, Co, Ni, Cu, Pd) and alloys containing group IIA metals (Mg, Ca, Be) (Takayama, S. 1976). The ability of a system to form a glassy solid can be measured by the critical quench rate R_c which would just allow the cooling liquid to miss the nose of the C-curve that relates to the start of crystallisation (see e.g. Cahn 1982). Presumably, similar kinetic terms could be identified in the other production processes. A low R_c (i.e. a high glass forming ability) is favoured by a large difference in the atomic or ionic radius of the constituent atoms (typically > 10%) and high interatomic bond strength in the liquid (indicated by a large negative heat of mixing), both of these quantities being important in determining the diffusion kinetics and hence the rate of crystallisation. The critical quench rate R_c has also been related to the change in volume on fusion, a small or negative (i.e.

Fig. 2.26. Diffuse X-ray structure factor $S(\mathbf{q})$ for an amorphous Cu–42at.%Ti alloy (after Sakata *et al.* 1980).

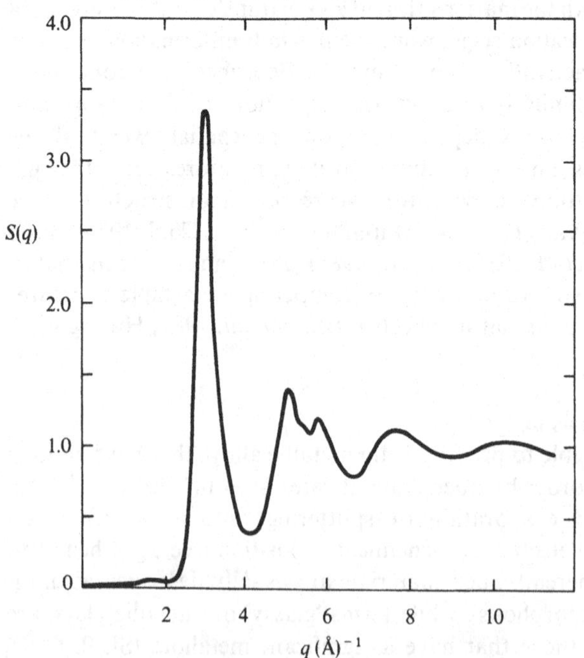

contraction) volume change favouring glass formation. A good review of the formation, structure and properties of amorphous metallic alloys has been prepared by Luborsky (1983).

2.8.1 *Static atomic structure*

While there is no long range structural order in such materials, the well-defined atomic sizes and closest distance of approach (and range of bond angles in a covalent material) result in definite local 'structures'. These local correlations produce diffuse diffraction effects which appear as a limited number of diffuse peaks in X-ray, electron or neutron scattering experiments, as shown in Figure 2.26. A useful statistical characterisation of such a structure is given by the radial distribution function (RDF):

$$R(r)=4\pi r^2 D(r), \tag{2.52}$$

such that $D(r)$ is the average density of atoms at distance r from a reference atom. If there are n different atomic species (as is generally the case for an amorphous solid), a set of n^2 partial RDFs are defined by

$$R_{ij}(r)=4\pi r^2 D_{ij}(r), \tag{2.53}$$

where i and j refer to the different atomic species and $D_{ij}(r)$ the average density of j atoms at distance r from a central i atom. Experimentally determined RDFs in multicomponent systems actually give the weighted average

$$\begin{aligned}\bar{R}(r)&=4\pi r^2 \bar{D}(r)\\ &=4\pi r^2 \sum_i^n \sum_j^n W_{ij} D_{ij}(r)/c_{ij},\end{aligned} \tag{2.54}$$

where the weighting factor W_{ij} depends upon composition and the details of the technique employed. Other useful forms of distribution function are the reduced RDF:

$$G_{ij}(r)=4\pi r^2 (D_{ij}(r)-D_0), \tag{2.55}$$

where D_0 is the average density, and the pair correlation function:

$$P_{ij}(r)=D_{ij}(r)/D_0, \tag{2.56}$$

and their associated averages. The local atomic structure is reflected in these functions, as shown in Figure 2.27, with the positions of the maxima in $R(r)$, $G(r)$ or $P(r)$ indicating frequently occurring interatomic separations. The precise positions of these maxima depend upon the particular function with

$$r_{max}(R)>r_{max}(G)>r_{max}(P),$$

although they differ by, at most, a few percent in typical metallic glasses.

The width of the maxima (corrected for experimental conditions) give information about the range and distribution of neighbour spacings (nearest, next nearest, etc.).

However, because RDFs give only spherically averaged information about correlations in atomic positions, they cannot specify uniquely the positions and type of atoms in the material. Some detailed structural information is available by modelling the structure and comparing the predicted RDFs with experiment. In this regard there are basically two different classes of structural model: one based on continuous random networks and the other on highly ordered clusters (whether crystalline or not) in a less-ordered amorphous matrix (see e.g. Cargill 1975). Recent high-resolution electron microscopy studies would appear to favour the heterogeneous models (Gaskell 1979), although this is by no means universally accepted. Furthermore, there is evidence for considerable atomic (i.e. replacive) short range ordering as well as the structural short range correlations (see e.g. Chieux & Ruppersberg 1980; Wagner & Lee

Fig. 2.27. The functions $\bar{R}(r)$, $\bar{G}(r)$ and $\bar{P}(r)$ for an amorphous solid.

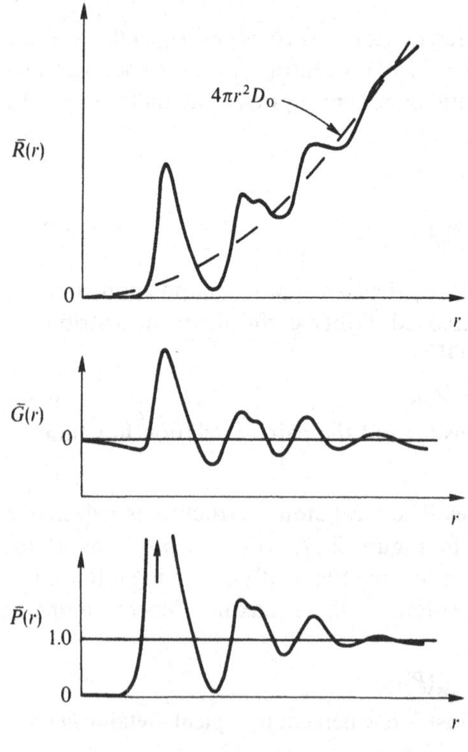

1980; Schaafsma *et al.* 1980; Wong 1981), although again it is not clear whether this is homogeneous or heterogeneous in nature (cf. Section 2.5.3). It appears that short range ordering is also an important factor in determining the R_c of a system and the stability of the resultant glass. This SRO may be described by a spherically averaged pairwise correlation function $\alpha(r)$ (cf. equation (2.13)), where the local composition in a shell of radius r and thickness dr is taken to be a continuous function of r. For example, in a binary alloy containing atom types 1 and 2:

$$\alpha(r)=1-D_{12}(r)/(C_2D_w(r)), \tag{2.57}$$

and the corresponding *chemical* RDF is

$$G_{CC}(r)=4\pi rD_w(r)\alpha(r), \tag{2.58}$$

where

$$D_w(r)=C_2D_1(r)+C_1D_2(r). \tag{2.59}$$

Where both atomic and spatial short range correlations exist, an experimentally derived RDF must be separated into its separate partial contributions G_{NN}, G_{CC} and G_{NC} representing the spatial and chemical correlations (equations (2.55) and (2.58)) and cross-term, respectively

Fig. 2.28. Reduced distribution functions $G(r)$, $G_{NN}(r)$, $G_{NC}(r)$ and $G_{CC}(r)$ for an amorphous Ni–60at.%Ti alloy (after Wagner & Lee 1980).

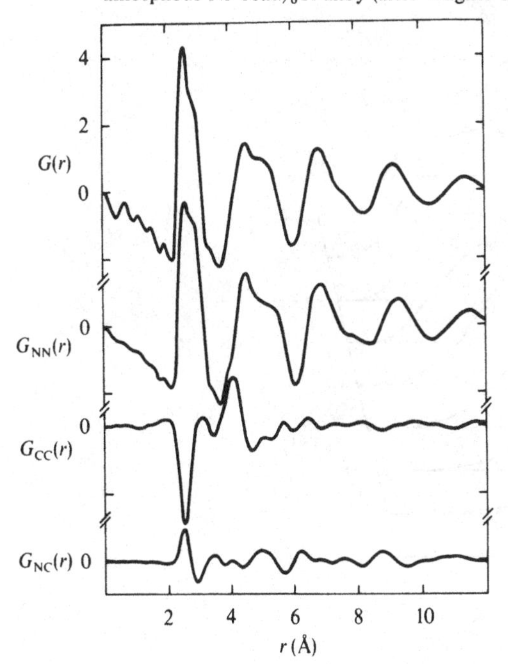

Fig. 2.29. Frequency spectrum $g(\omega)$ and dynamic structure factors $S(q\omega)$ for longitudinal and transverse excitations in a 500 atom model amorphous solid (after von Heimendahl 1979).

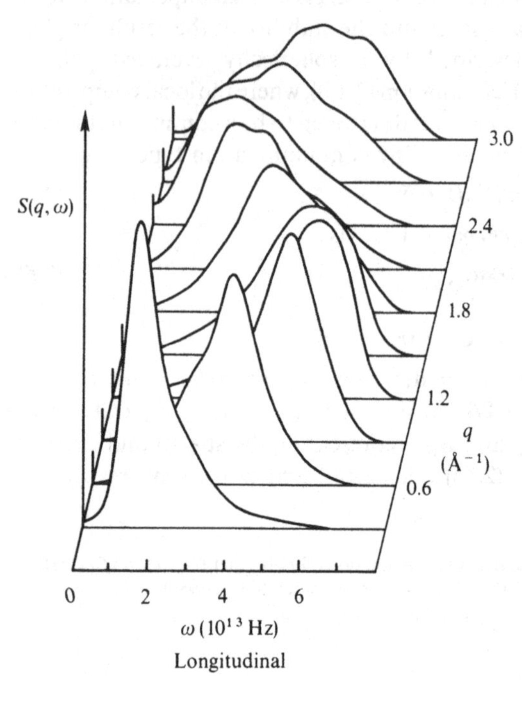

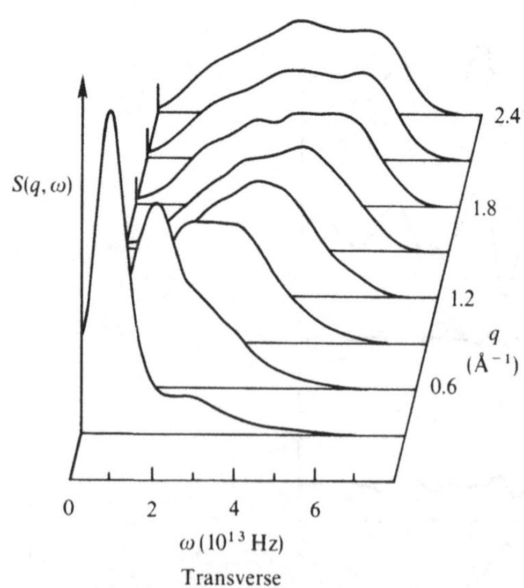

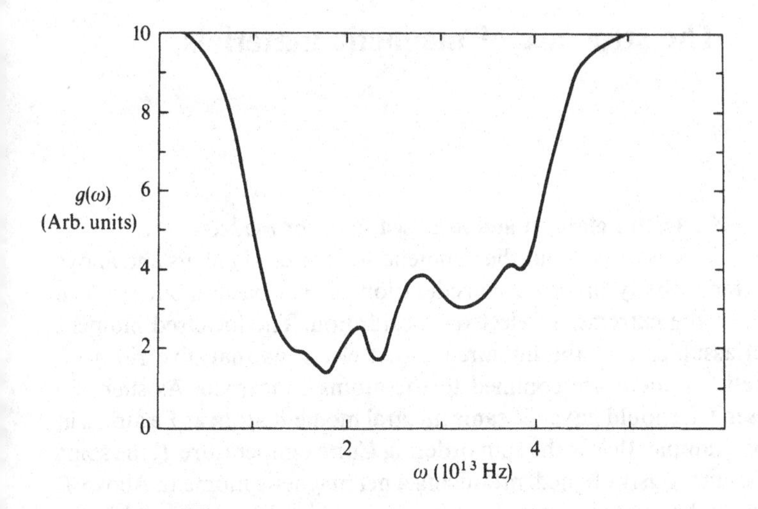

(Wagner 1978). Some data for an amorphous Ni–60 at.%Ti alloy is shown in Figure 2.28. The first broad peak in $G_{NN}(r)$ results from a superposition of the G_{NiTi} and G_{TiTi} partials and the dip in $G_{CC}(r)$ at the same r as the G_{NiTi} peak indicates a strong preference for unlike nearest neighbours. The Fourier transforms of the reduced distribution functions are often quoted, these being the partial structure factors $S_{NN}(\mathbf{q})$, $S_{CC}(\mathbf{q})$ and $S_{NC}(\mathbf{q})$. As these directly enter the resistivity calculations they will be discussed in more detail in Chapter 5.

2.8.2 *Dynamic fluctuations*

Due to the lack of long range structural periodicity the concept of lattice waves (phonons) is of little use in describing the thermal motion of atoms in an amorphous solid, although one could argue that the short range periodicity would allow propagation of well-damped (i.e. diffuse) phonon modes. The normal approach is to replace the static RDF by a dynamical one (or its Fourier transform), taking into account the frequency of atomic vibrations. Neutron spectroscopy can again be useful in determining the frequency distribution and is being complemented with various theoretical calculations (Suck *et al.* 1980; Hafner 1980, 1983). A typical frequency spectrum and dynamical structure factor for a model amorphous solid are shown in Figure 2.29.

3 The structure of magnetic materials

3.1 Collective electron and localised moment models

Discussions about the magnetic structures of metals and alloys have traditionally involved consideration of two basic models which represent the extremes of electron localisation. The localised moment model assumes that the unpaired d or f electrons that give rise to a magnetic moment are confined to the atoms concerned. As such, Fe atoms in Cu should have the same integral moment atom as Fe atoms in Ni, for example. Below the spin ordering Curie temperature T_c the spins are spontaneously aligned, producing a net magnetic moment. Above T_c the thermal energy destroys the cooperative spin alignment resulting in zero net magnetic moment. On the other hand, the collective electron model (also known as the band or itinerant electron model) assumes that these electrons are completely delocalised and exist in narrow bands throughout the material, although the spin density need not be uniformly distributed. Below the Curie temperature the bands are exchange split, the resultant imbalance between the number of up and down spins leading to the observed magnetic moment. Above T_c there should be no splitting and so no moment.

The 3d electrons of metals of the first transition series definitely form narrow bands and their orbital angular momentum is effectively quenched by the crystal field giving a spin-only magnetic moment. These materials do not have integral moments and in fact usually display a moment per atom that varies smoothly with composition. Nevertheless, neutron diffraction studies indicate that the magnetic spin density is highly localised in space at the atomic sites (see e.g. Low 1969; Shull 1968; Cable & Wollan 1973), and that localised spins exist above T_c (see e.g. Lowde *et al.* 1983). An example of the spatial distribution of the magnetic moment in an ordered Ni_3Fe alloy below T_c is shown in Figure 3.1.

In ferromagnetic Fe–Ni alloys, the Fe and Ni atoms carry a different moment, which is relatively composition independent, whereas the addition of Cr or V to ferromagnetic Co or Ni produces a rapid decrease in magnetic moment which is shared between the Co or Ni atoms and the Cr or V impurity atoms (Low & Collins 1963). Some of this behaviour is demonstrated on the Slater–Pauling curve shown in Figure 3.2. In the fcc region, the alloys that fall on the common straight line can be partly understood in terms of collective electron effects with the aid of the rigid-band model, i.e. a rigid 3d-band being progressively filled by the electrons donated by the constituent atoms. (This model and its shortcomings are discussed in more detail in Chapter 6.) The deviations from this line can then indicate a different behaviour in which the constituent atoms do not simply change E_F via electron concentration effects, but which tend to act in a more individual manner. Similarly, in atomically disordered antiferromagnetic Mn–Cu or Mn–Cr alloys each magnetic site acts as though it were occupied by an 'average' Mn–Cu atom with its associated localised magnetic moment. On the other hand, antiferromagnetic behaviour is more often found in ordered alloys (e.g. $MnAu_2$, MnAu, NiMn, $FePt_3$, FeMn, FeRh, MnSe, MnTe, Mn_2As)

Fig. 3.1. Density of magnetic moment in parallel (100) planes of ordered Ni_3Fe. The contour lines give the relative strengths of the moments (after Cable & Wollan 1973).

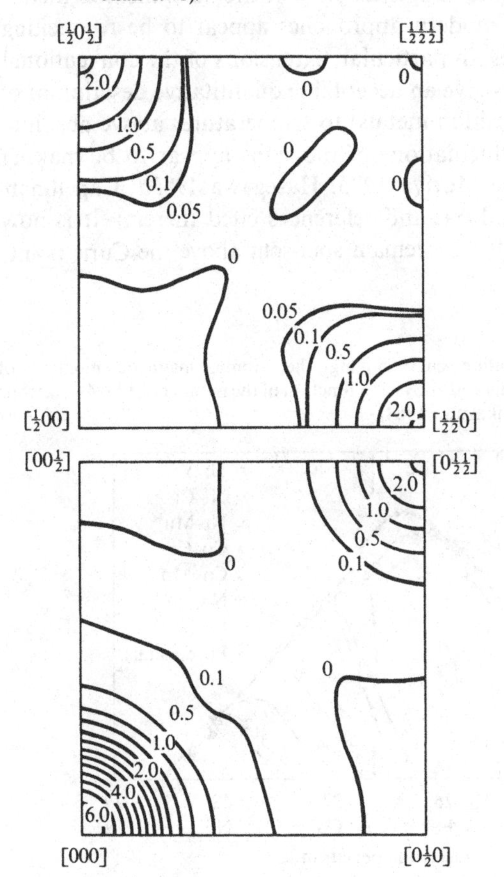

where particular moments are associated with particular atom types at particular lattice sites.

In some other antiferromagnetic metals and alloys (notably Cr and Cr-rich alloys) the periodicity of the magnetic structure is not locked to the crystal lattice and incommensurate spin structures are observed, although the spins still appear to be spatially localised.

The general picture that emerges is one in which the magnitudes of the moments associated with each atom are largely determined by collective electron effects, although in many cases these moments are strongly localised at the atomic sites, the degree of spatial localisation presumably depending upon the amount of overlap between the d-orbitals. However, it is also evident that the moments may exist in a disordered state above T_c. It thus appears that neither the localised moment nor the simple collective electron models are capable of giving a complete description of the magnetic state of the 3d transition metals and their alloys. More modern approaches appear to be reconciling some of these differences. In particular, extensions of the conventional band theory (which does give an acceptable quantitative description of the ground state of transition metals) to temperatures above absolute zero by incorporating fluctuations of the spins appear to be making significant advances (see Moriya 1985; Hasegawa 1983*a*; Capellman 1981; Heine & Samson 1981, and references cited therein). It is now recognised that the bands may remain spin-split above the Curie point,

Fig. 3.2. Slater–Pauling curve giving the atomic magnetic moment of ferromagnetic metals and alloys as a function of the number of 3d + 4s electrons per atom (after Chikazumi 1964).

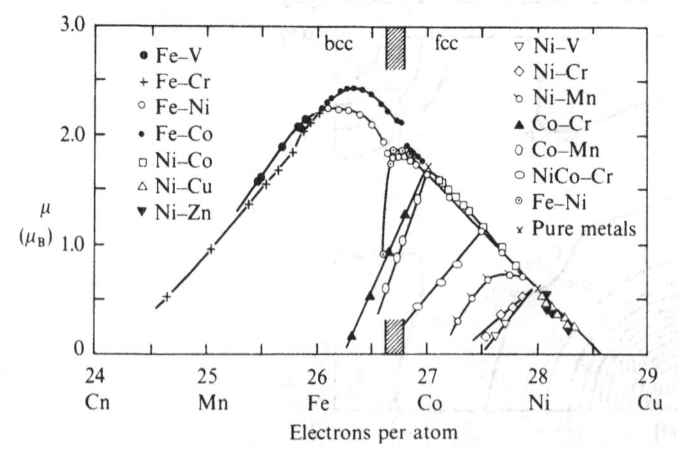

although it is not clear whether there is a large or small degree of magnetic short range order there (see Lowde *et al.* 1983; Hasegawa 1983*b*, 1984). Other theories treat the d electrons as partly itinerant and partly localised (see e.g. Adamowicz 1977, and references therein).

By contrast, the 4f electrons of the rare earths lie deep within the atomic structure and are effectively screened from the crystalline field by a significant outer shell 5s, 5p electron distribution. This means that the moments are well localised and contain contributions from both the spin and orbital angular momenta. Various reviews of the magnetic properties of metals and alloys are available in Bozorth (1951), Chikazumi (1964), Crangle (1977), Cullity (1972), Tebble & Craik (1969), and references contained therein.

In general, the variation of magnetic properties with composition can be quite complex. Current theoretical approaches concerned with electron states in alloys do not make any presumptions about the degree of spatial localisation and treat quite generally the effect of alloying. As an example, Figure 3.3 shows the spin density calculated for ordered FeCo (Schwarz *et al.* 1984). Some of these are discussed in more detail in Chapter 6. However, they are only beginning to grasp the problems associated with inhomogeneities and this forewarns of significant problems in finding a complete theory of electrical resistivity. For the purposes of the largely qualitative consideration of electron transport properties in magnetic materials which follows, we will assume that the generally localised nature of the moments allows the interaction with conduction electrons to be described via a spatially localised short range interaction which depends upon both the electron spin and the magnitude of the (localised) spin **S**. Such an interaction is the exchange interaction described by equation (1.16), for example.

Fig. 3.3. Spin density $D_{\uparrow}(\mathbf{r})$–$D_{\downarrow}(\mathbf{r})$ of ordered FeCo in the (110) plane determined by the linearised augmented plane wave method (after Schwartz *et al.* 1984).

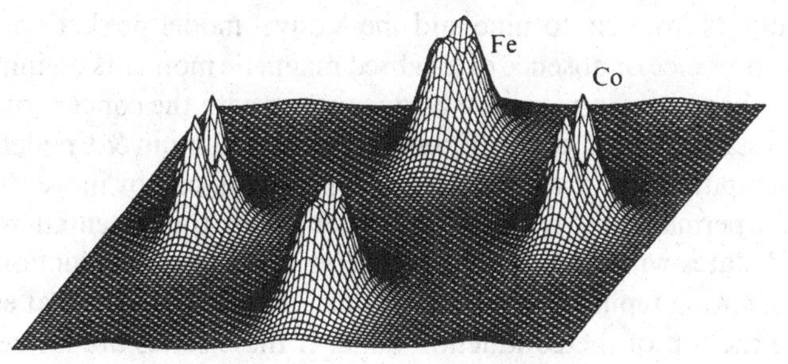

3.2 Magnetic configuration

3.2.1 *Isolated moments*

In the magnetically dilute limit we consider the isolated magnetic ion having a localised magnetic moment when dissolved in an otherwise non-magnetic host metal. Not all transition metals form such localised moments, the archetypes being Cr, Mn or Fe in Cu, Au or Ag. Atoms of Rh, Co and Ni appear to require a sufficiently ferromagnetic environment before they acquire a moment, e.g. at least two Co near neighbours or eight Ni near neighbours when in non-magnetic hosts (Cable 1977; Boucai *et al.* 1971; Haller *et al.* 1981; Hurd & McAlister 1980), although, as in the case of short and long range order, whether or not a moment is observed depends upon the characteristic time of the probe used. We return to the dynamic aspects of spin fluctuations in Section 3.3. Various models of the dependence of the local moment upon the atomic environment have been proposed to reproduce this phenomena. For example, in the models of Jaccarino & Walker (1965) and Perrier *et al.* (1970) it is assumed that the moment on an atom is zero unless there is a certain number p of 'magnetic' nearest neighbours, in which case it adopts some maximum value μ_m. The mean local moment per magnetic atom in a random alloy is then given by

$$\bar{\mu}=\mu_m \sum_{n=p}^{N} \frac{N!}{n!\,(N-n)!} c_A^n c_B^{N-n}, \tag{3.1}$$

where N is the coordination number. In an alternative model due to Kouvel (1969) the mean moment per magnetic atom is expanded as a power series of the number of nearest neighbours of the opposite atom type:

$$\bar{\mu}(A)=a_0+a_1\frac{n_B}{N}+a_2\frac{n_B(n_B-1)}{N(N-1)}+\cdots. \tag{3.2}$$

By way of comparison, the concentration dependence of the mean saturation magnetic moment per Ni atom in an Au matrix is shown in Figure 3.4. In this diagram the experimental values are compared with the prediction of the Perrier *et al.* model for a critical number of nearest neighbours from six to nine and the Kouvel model prediction.

The presence or absence of localised magnetic moments on impurities in an otherwise non-magnetic host is tied up with the concept of virtual bound states, introduced by Friedel (1958) and Blandin & Friedel (1959). If the impurity has a strong attractive potential, one or more electrons will be permanently trapped and localised in the potential well, i.e. bound states will form at energies below that of the conduction band. With a strong repulsive energy, bound states may be formed at energies above the top of the conduction band. If these states are far removed from the top of the conduction band they are of no consequence to the electrical resistivity. They are either completely filled (and so regarded as part of the electronic structure of the impurity), or they are completely empty. However, if the potential of the impurity is such that the localised energy level lies within the conduction band, its interaction with the conduction electrons must be taken into account. Electrons can now enter the impurity state from the conduction band and leak back out into the conduction band. This means that the lifetime of the state is limited. To first order the result is a broadening of the impurity level and a shift in energy from its unperturbed value. The formation of such a virtual bound state is shown schematically in Figure 3.5.

This interaction with the conduction electrons can also be regarded as a scattering problem. As the impurity can nearly support a d bound state there is a resonance in the scattering in the d wave part of the conduction electron wavefunction at energies near those of the impurity d level. We return to this aspect of the problem in Section 6.6.2. For s and p states the broadening of the state is usually so large that the concept of a virtual bound state becomes meaningless.

Whether or not such a virtual bound state can support a magnetic moment is (just like the Heisenberg exchange and RKKY interactions to be discussed in Section 3.2.2) determined by the intra-atomic electrostatic Coulomb interaction U between the electrons occupying that state. That this interaction can have the effect of stabilising a

Fig. 3.4. Concentration dependence of the mean saturation moment per Ni atom in an Au matrix. The isolated points are experimental values; the solid lines the Perrier *et al.* model for a critical number of nearest neighbours from 6 to 9; and the dotted line is the Kouvel model (after Hurd *et al.* 1981).

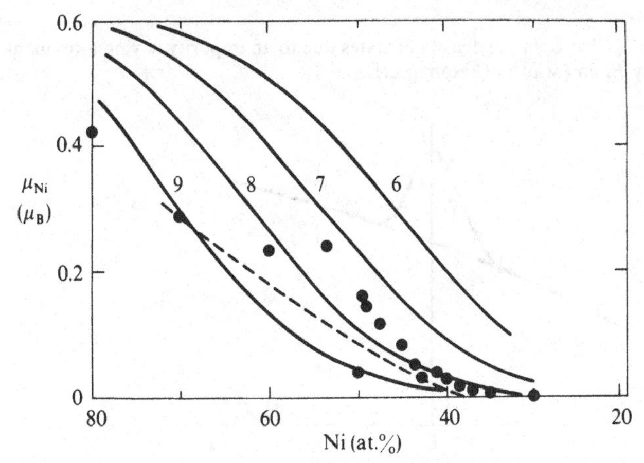

magnetic moment can be seen from the following simple model of Anderson (1961): let us assume that there is a single localised energy level which can be occupied, at most, by two electrons, one having spin-up and the other spin-down. The interaction between these electrons U causes a splitting of the level and, if the higher state then lies above E_F, it will be empty leading to a net spin polarisation. If we neglect the effects of thermal fluctuations, a Hartree–Fock calculation (Stoner 1947, 1951) shows that if

$$UN_d(E_F) > 1$$

where $N_d(E_F)$ is now the (local) density of states at the impurity site, a localised magnetic moment will exist. This model has been extended to include several d orbitals (Dworin & Narath 1970; Yosida *et al.* 1965; Klein & Heeger 1966). The nature of this model makes it appropriate for transition metal impurities in simple metals, e.g. a 3d impurity in Cu or Al. This same problem has been treated within the scattering context by Wolff (1961). In this model the conduction electrons scatter from the impurity potential. The virtual bound state appears as a scattering resonance (as described before) and the formation of a magnetic moment follows from the electron–electron exchange potential which, in a similar way to U, leads to a spin polarisation. This model has also been extended by a number of workers (Appelbaum & Penn 1971; Liu *et al.* 1983) and gives a more reasonable treatment of transition metal impurities in a second transition metal, e.g. a 3d atom in a 4d host. The correspondence between the two models has been discussed in detail by Rivier & Zitkova (1971) and Fischer (1978). The problem with both of these models is the uncertainty in the parameters which characterise the interactions, making a direct comparison with experiment rather difficult. Other

Fig. 3.5. Change in the density of states due to an impurity having a resonant energy E_r and width of resonance Γ.

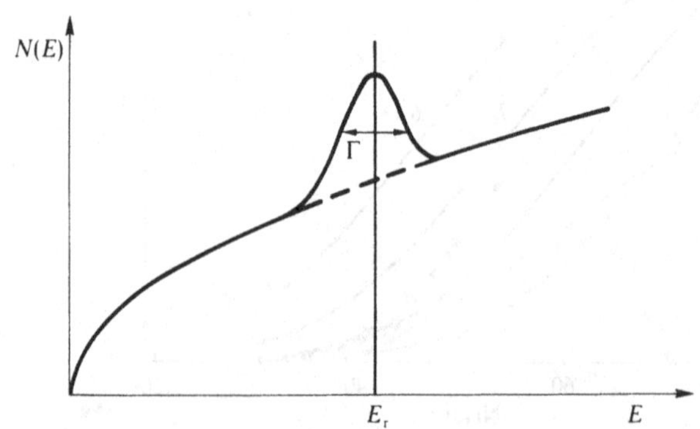

Table 3.1. *Presence or absence of localised moments of transition metals dissolved in non-magnetic hosts on the basis of the Hartree–Fock criterion. Presence is indicated by 'Yes', a question mark indicates uncertainty*

	Host			
Impurity	Au	Cu	Ag	Al
Ti	No	—	—	No
V	?	—	—	No
Cr	Yes	Yes	Yes	No
Mn	Yes	Yes	Yes	?
Fe	Yes	Yes	—	No
Co	?	?	—	No
Ni	No	No	—	No

Source: After Heeger 1969.

approaches based on the KKR–Green's function and CPA schemes have produced results which bear more direct comparison with experiment (see e.g. Hasegawa & Kanamori 1971; Levin *et al.* 1971, 1972*a*,*b*; Gautier *et al.* 1974; Brouers *et al.* 1972, 1976; Ghosh & Bhattacharya 1975; Bhattacharya *et al.* 1976; van der Rest 1977; Jo 1976; Zeller *et al.* 1980; Podloucky *et al.* 1980; Zeller 1981; Deutz *et al.* 1981; Braspenning *et al.* 1984). Furthermore, both of these models attribute the localised magnetic moment to electron spin. There is experimental evidence to suggest that in some systems (e.g. Co or Ni in Pd) there is also a significant contribution from the unquenching of the local orbital moment (see e.g. Le Doug Khoi *et al.* 1976) which should be taken into account. Nevertheless, it is now clear that the Jaccarino–Walker and Perrier *et al.* models discussed above are based on the Hartree–Fock type of condition for the existence of a localised moment. More detailed discussions along these lines are given by Kim (1970) and Benneman & Garland (1971).

The presence (+) or absence (−) of localised moments in the Hartree–Fock sense for a number of dilute alloy systems is indicated in Table 3.1. However, this specification ignores the effects of fluctuations and is also known to overestimate such a tendency toward ferromagnetism. More realistically, as pointed out by Heeger (1969), one should ask whether or not the fluctuations in the spin density can be sufficiently slow that on the timescale of a given experimental probe there appears to be normal magnetic behaviour. We return to this problem in Section 3.3. Until then let us simply assume that (as found experimentally) at finite temperatures some impurities act as though they have a well-defined magnetic moment. In this regime it is assumed that the magnetic impurities are present at concentrations generally below 0.1% and are randomly distributed at substitutional sites in the non-magnetic hosts.

3.2.2 *Spin glasses*

If the concentration of magnetic impurities increases, they may start to interact indirectly via the conduction electrons. The basis of this interaction lies in the exchange interaction (equation (1.16)), proposed by Ruderman & Kittel (1954) and extended by Kasuya (1956) and Yosida (1957), and now known as the RKKY interaction. To understand this interaction consider the case of $J<0$. Equation (1.16) shows that conduction electrons with a spin antiparallel to the ion will have the lowest energy. The result is a polarisation of the conduction electron spins in the vicinity of the impurity and the original uniform distribution of electrons with antiparallel spins changes to one with an oscillatory behaviour which decays with increasing distance from the magnetic moment. Similarly, electrons with a spin parallel to the impurity moment are repulsed by that impurity and also suffer an oscillatory spin distribution. The net effect is to produce an oscillatory spin density which decays with distance, as shown schematically in Figure 3.6. Note that this is a spin effect and that the overall *charge* density remains uniform. If another magnetic ion is situated in the region of this spin density oscillation it will be in a lower energy state with its spin either parallel or antiparallel to the first, depending upon the local state of polarisation. The ions thus interact via the conduction electron spins

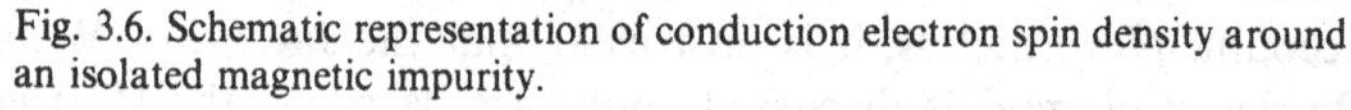
Fig. 3.6. Schematic representation of conduction electron spin density around an isolated magnetic impurity.

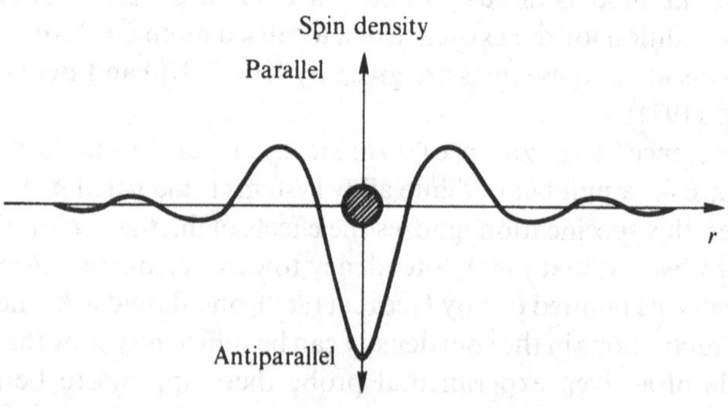

and at large distances the interaction is of the asymptotic form

$$V(r) \sim \frac{\cos(2k_F r)}{r^3}. \tag{3.3}$$

In a random alloy there will thus be a random interaction between spins producing a random or 'spin glass' spin alignment as the magnetic ground state. The absence of long range magnetic order is a direct consequence of the randomness of the positions of the spins and the oscillatory nature of the RKKY interaction. At elevated temperatures this alignment will be 'unfrozen' at temperatures corresponding to $k_B T_0 \approx \Delta_s$, where Δ_s is the average interaction strength and T_0 is the spin glass freezing temperature. It is important to note that the range of this interaction will depend upon the conduction electron mean free path. As shown by de Gennes (1962), a reduction in Λ effectively reduces the strength of the RKKY interaction at distance r by a factor $\exp(-r/\Lambda)$ due to a damping of the spin polarisation oscillation.

Spin glass behaviour is characterised by a sharp peak in the low field magnetic susceptibility at T_0, a broad, almost featureless, maximum in the temperature dependence of the specific heat above T_0 and the onset of irreversibilities and time dependencies with the application of external magnetic fields at temperatures below T_0. Such behaviour is typically observed at impurity concentrations in the range $\sim$0.1 to 10% in systems such as Au–Fe, Au–Cr, Au–Mn, Cu–Mn and Ag–Mn, all of which support well-defined localised moments on the magnetic impurities above the characteristic spin-fluctuation temperature T_{sf} (see Section 3.3). For a general review of spin glasses see Mydosh (1978).

The physical properties of a spin glass are determined by the dynamic spin–spin correlations. At very high temperatures the spins will behave paramagnetically, although there may be some short range spin–spin correlations. As the temperature is reduced, spin clusters emerge from these short range correlations, their size and shape being determined by the distribution of other spins in the immediate neighbourhood (see e.g. Levin *et al.* 1979). Furthermore, the coupling within and between the clusters depends upon the spatial distribution of the spin according to the RKKY interaction. At this stage an experiment would encounter a wide distribution of relaxation times varying from that of the single paramagnetic isolated spin to those of some much larger, more slowly responding clusters associated with favourable concentration fluctuations. As the temperature is further lowered towards T_0, the size and density of the clusters increases until finally at T_0 an 'infinite' glassy cluster of frozen spins forms. However, there will still be many free smaller clusters that are not part of this large cluster and which can respond to external fields maintaining a distribution of relaxation times. As T is lowered well below T_0, more and more of those clusters join the infinite cluster or become blocked due to a lack of thermal activation energy (see next section). The wide spectrum of relaxation times characteristic of excitations over a wide range of energy barriers can make analysis of spin glass behaviour confusing. Again the value of a measured property will depend upon the frequency (or characteristic time) of measurement. This aspect is graphically demonstrated in Figure 3.7 where the reciprocal of the apparent freezing temperature T_0 is plotted as a function of the characteristic frequency of measurement (Maletta 1981). This behaviour results from the fact that the 'infinite' cluster is a body containing many links of varying strength and many dynamical fluctuations and excitations. A high frequency measurement will miss the low frequency fluctuations so that the infinite cluster will appear to have formed at a higher temperature. Cooling through T_0 in a magnetic field induces a frozen preferential spin alignment which is then manifest as an internal uniaxial anisotropy due to intercluster interactions. Fundamental to the spin glass problem is the question of whether or not an equilibrium state is reached in the limit of a very long measurement time. However, details of the dynamical behaviour of spin glasses (particularly at low temperatures $T \ll T_0$) is far from being understood (see e.g. Fischer 1980). The simplest picture is one of strongly damped spin waves, the damping resulting from a lack of translational

Fig. 3.7. Reciprocal of the apparent spin glass freezing temperature T_0 as a function of the characteristic frequency of measurement f_m (after Maletta 1981).

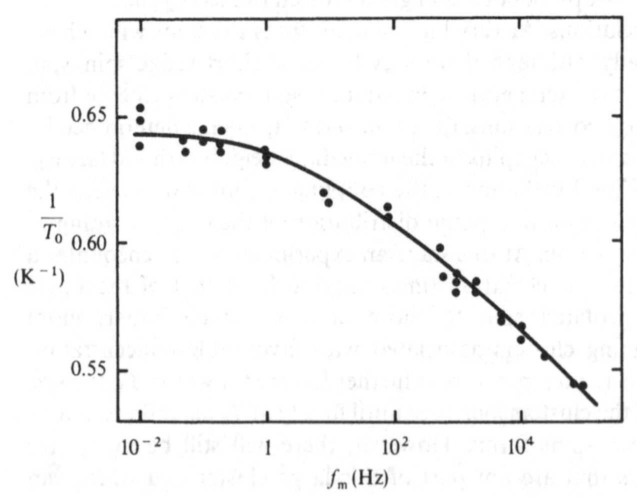

symmetry. However, the excitation spectrum and form of dispersion (see Section 3.2.6) of these diffuse spin waves is not clear.

Implicit in the above discussion is the existence of well-defined local magnetic moments and so one has to consider also the dynamics of the spin fluctuations of the individual spins. From the discussions that follow in Section 3.3 it is clear that spin glass behaviour will only be possible at temperatures above the characteristic spin-fluctuation temperature T_{sf}.

Should T_{sf} not be below or very near to 0 K, the spin structure will be influenced by both the dynamics of the spin fluctuations and the spin glass excitations. Conversely, there is the effect of the local spin correlations on the spin fluctuations. It appears that the RKKY interaction may be sufficiently strong to destroy the Kondo–Anderson resonant state below T_{sf}, although remnants of its existence are observable above T_0. Alternatively, we might argue that the increased RKKY interaction may push T_{sf} to lower and lower temperatures (see e.g. Fischer 1981). We will return to this problem in Section 7.4.

3.2.3 *Magnetic clusters*

At concentrations of magnetic impurity in the range ~10–15% the interactions between the ions may become strong enough to condense small regions into a magnetically aligned state. These regions will generally be associated with composition fluctuations enhancing the local concentration of magnetic species and the spin ordering result from direct (Heisenberg) exchange, local band structure (i.e. collective electron) effects or RKKY interactions, as described in the previous section (the dominant nature of the interaction will depend upon the nature of the atoms concerned). Here there are two separate situations that should be identified. If the magnetic clusters develop from a spin glass state they will interact with the (isolated) spins in the surrounding spin glass matrix, giving the so-called mictomagnetic structure (Beck 1972). Below the spin glass freezing temperature T_0 these clusters will be locked into directions determined by the local frozen spin alignment. Above T_0 they will act as giant paramagnetic (i.e. superparamagnetic) units and at still higher temperatures the magnetic alignment within the clusters will be destroyed (i.e. at the Curie temperature of the clusters). An alternative situation applies to systems in which the impurity atoms only acquire a magnetic moment if they are present in sufficient concentration, e.g. Co or Ni in a Cu matrix. (As in the case of isolated moments (Section 3.2.1) and, as discussed in Chapter 1, it should be remembered that whether or not a cluster is deemed to have a magnetic moment depends upon the rate of spin fluctuation of the cluster in relation to the coherence time of the probe being used.) If the particle concentration is small they will act as independent entities in a non-magnetic matrix. However, if the density of clusters is sufficiently large, the RKKY intercluster interaction may be sufficient to promote freezing of the cluster moments into a cluster glass below some particular temperature. Evidence for this type of behaviour has been found at compositions near the ferromagnetic transition in systems such as Au–Fe (Coles *et al.* 1978; Sarkissian 1981), Cu–Ni (Aitken *et al.* 1981) and Pd–Ni (Cheung & Kouvel 1983). In the case of an alloy system with a nearly magnetic matrix (see Section 3.3) there is also the interesting possibility of polarisation fluctuations in the matrix being enhanced by the particle moments.

Whatever the situation, there will in general always be some local magnetic anisotropy described by an anisotropy constant K (measured in J/m^3). If the volume of each particle is V then the energy barrier that must be overcome before a particle can change its direction of magnetisation is $\Delta E = KV$. The rate of change of magnetisation is given approximately by

$$\frac{\mathrm{d}M}{\mathrm{d}t} = \frac{M}{\tau} = 10^9 M \exp(-\Delta E/k_\mathrm{B}T), \tag{3.4}$$

where the rate of change $\mathrm{d}M/\mathrm{d}t$ has been written as M/τ for a single relaxation time process. This equation allows definition of the superparamagnetic freezing or blocking temperature T_B which gives a relaxation time τ of $\approx 10^2$ s, i.e. a typical period of observation in a bulk magnetic experiment. Application of a field H causes a decrease in the height of the energy barrier to

$$\Delta E = KV\left(1 - \frac{HM_\mathrm{s}}{2K}\right), \tag{3.5}$$

where M_s is the magnetic moment per atom within the cluster. Thus the bulk magnetic behaviour will depend upon time, temperature, field and particle size. The variation of τ with $KV/k_\mathrm{B}T$ (for small H) is given in Table 3.2 and shows that $T_\mathrm{B} \approx KV/25k_\mathrm{B}$. Consideration of the conduction electron relaxation time $\sim 10^{-14}$ s shows that, over all temperatures up to the Curie temperature of the particle, the superparamagnetic particles will appear to be stationary (frozen or blocked) in a resistivity study, a clear example of the caution suggested in Section 1.4. However, the freezing processes involving a glassy phase are fundamentally different to superparamagnetic blocking in a non-magnetic matrix. It has been argued that the former process is the result of a cooperative process involving all of the particles of the system (see

Table 3.2. *Superparamagnetic relaxation time as a function of $\Delta E/k_B T$*

$\Delta E/k_B T$	τ (s)
50.0	5.2×10^{12}
25.0	7.2×10
25.3	10^2
10.0	2.2×10^{-5}
1.0	2.7×10^{-9}

Note: From equation (3.4).

e.g. Edwards & Anderson 1975), whereas the latter is an independent particle effect. Some of the differences have been discussed by Maletta (1981). In most cases the Curie temperature of the cluster or particle will be relatively independent of particle size (Kneller 1958), although it could of course depend upon the chemical composition of the cluster.

3.2.4 *Long range magnetic order, $T < T_c$*

As the concentration of magnetic ions reaches the percolation limit variously quoted in the range 15.6–20.8% for a fcc lattice (Stauffer 1977) an 'infinite' long range ordered magnetic cluster can form by a series of nearest neighbour links, even though there will still be a large number of impurity moments in other large but finite clusters. Thus while the specimen supports a long range magnetic order, the magnetic structure is inhomogeneous as shown schematically in Figure 3.8. With a further increase in the concentration of magnetic ions these clusters link up, finally resulting in uniform long range magnetic order. An increase in temperature of an alloy in this composition range results in a breakdown of long range magnetic order along a similar inhomogeneous path, with some parts retaining ferromagnetic order in superparamagnetic clusters to higher temperatures than others, again as a result of local composition fluctuations.

This inhomogeneous approach to long range magnetic order has been discussed by Coles *et al.* (1978), in relation to the Au–Fe and Cu–Ni systems, and their proposed magnetic phase diagram for Au–Fe is shown in Figure 3.9.

We turn now to alloys that lie well inside the magnetic region. The nature of the long range spin ordering depends upon the nature of the

Fig. 3.8. Schematic representation of the magnetic regions of an alloy with a concentration of magnetic ions just above the percolation limit. The ferromagnetic regions are shown shaded.

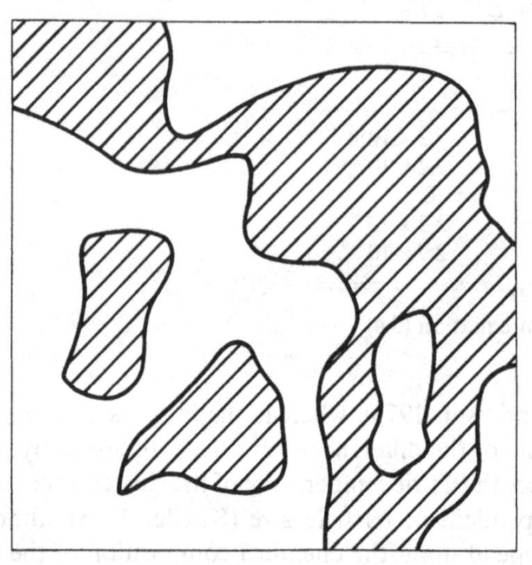

Fig. 3.9. Magnetic phase diagram of Au–Fe solid solutions; *p*: paramagnetic; *sp*: superparamagnetic; *sg*: spin glass; *cg*: cluster glass; *f*: ferromagnetic (after Sarkissian 1981).

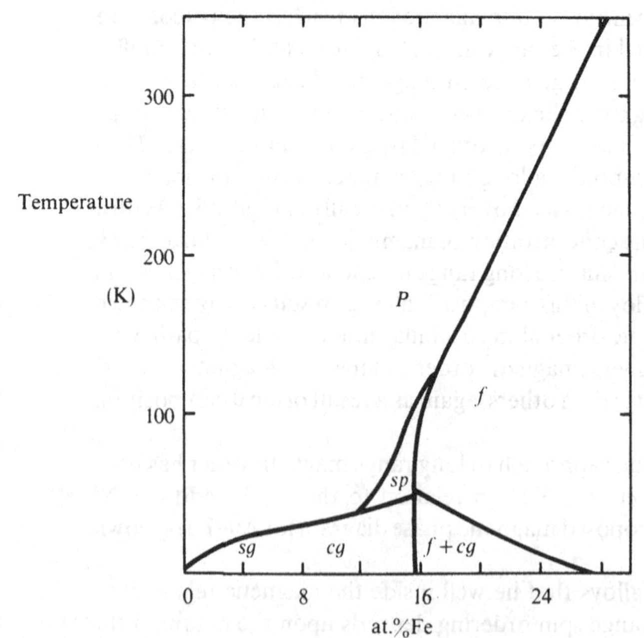

Table 3.3. *Magnetic structures and transition temperatures of some rare-earth metals*

Element	Magnetic structure (transition temperature in K)
Gd	Ferromagnet $\xrightarrow{293}$ Paramagnet
Tb	Ferromagnet $\xrightarrow{220}$ Helix $\xrightarrow{230}$ Paramagnet
Dy	Ferromagnet $\xrightarrow{85}$ Helix $\xrightarrow{179}$ Paramagnet
Ho	Cone $\xrightarrow{20}$ Helix $\xrightarrow{132}$ Paramagnet
Er	Cone $\xrightarrow{20}$ Antiphase cone $\xrightarrow{53}$ Collinear sinusoidal $\xrightarrow{85}$ Paramagnet
Tm	Collinear square wave $\xrightarrow{\sim 35}$ Collinear sinusoidal $\xrightarrow{85}$ Paramagnet
Yb	Paramagnet

Source: After Crangle 1977.

magnetic spin–spin interactions. Direct Heisenberg exchange between nearest neighbours:

$$\mathscr{H} = -2J_{12}\mathbf{S}_1 \cdot \mathbf{S}_2, \tag{3.6}$$

should produce simple parallel (ferromagnetic) or antiparallel (antiferromagnetic) spin alignments depending upon the sign of the exchange integral J_{12}. At the other extreme the longer range RKKY interaction can produce a variety of ground states because of the different strengths and signs of interaction at different interatomic spacings. This interaction is more likely to be operative in the rare-earths due to the lack of overlap between the small (~ 0.3 Å) 4f shells and may explain the wide range of magnetic structures observed as shown in Table 3.3 and illustrated in Figure 3.10. Other possible interactions include collective electron effects and superexchange (Crangle 1977).

Whatever the nature of this ground state our assumption about the localisation of spins makes it possible to define a set of site occupation parameters of the form given in equation (2.4), the state of magnetic order then being given by an average over some suitably defined motif. In the magnetic case the point approximation is known as the molecular (or mean) field approximation where an average over all single sites is taken as the order parameter, this being just the saturation magnetisation M_s, or sublattice magnetisation in the case of an antiferromagnet (Smart 1966). The variation of M_s with temperature for

Ni is shown in Figure 3.11 together with a result calculated on the basis of the molecular field approximation assuming a spin state of $\frac{1}{2}$. Not surprisingly these curves are very similar to those describing the state of chemical order in the Bragg–Williams approximation (Figure 2.3).

A suitable definition of a magnetic long range order parameter normalised to lie between 0 and 1 is given by the long range spin–spin correlation function,

$$M = \frac{\langle \mathbf{S}_o \rangle \cdot \langle \mathbf{S}_i \rangle}{S(S+1)}, \tag{3.7}$$

Fig. 3.10. Schematic representation of the magnetic structures referred to in Table 3.2 and found in the rare earths.

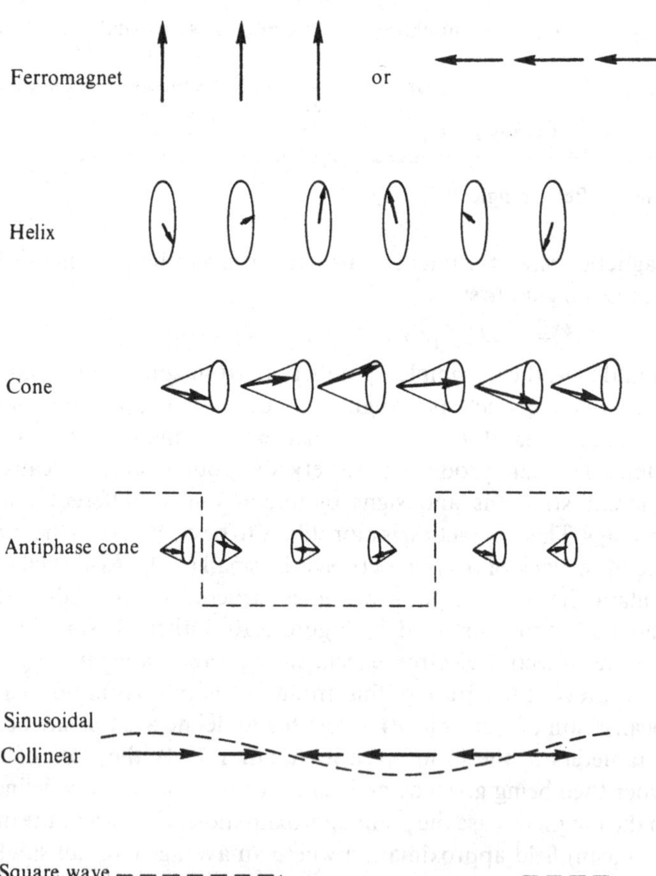

where it is assumed that each site is occupied either by an up spin or a down spin of magnitude S and the angled brackets $\langle\ \rangle$ now represent the van Vleck expectation values. The subscripts o and i refer to the origin site and all other sites surrounding it. By analogy with equations (2.12)–(2.14) the pairwise correlation parameter may be defined by

$$m_i = \frac{\langle \mathbf{S}_o \cdot \mathbf{S}_i \rangle - \langle \mathbf{S}_o \rangle \cdot \langle \mathbf{S}_i \rangle}{S(S+1)}, \tag{3.8}$$

and, as in the case of chemical LRO, will give a limiting set of values which are independent of separation for long range magnetic order. The general variation of M and m_i with temperature are illustrated in Figure 3.12. Those spin structures which have a periodic nature may also be described (cf. Section 2.3) by a static spin polarisation wave of the form

$$\begin{aligned} S^x(\mathbf{r}_i) &= \sum_j u^x(\mathbf{Q}_j^x) \exp(-i\mathbf{Q}_j^x \cdot \mathbf{r}_i) + \mathrm{cc} \\ S^y(\mathbf{r}_i) &= \sum_j u^y(\mathbf{Q}_j^y) \exp(-i\mathbf{Q}_j^y \cdot \mathbf{r}_i) + \mathrm{cc} \end{aligned}, \tag{3.9}$$

where S^x and S^y are the Cartesian components of the spin $\mathbf{S}$, $\mathbf{Q}$ the static spin wave vector and cc represents the complex conjugate.

Because of the large magnetostatic contribution to the free energy in a uniformly magnetised body, most ferromagnetic materials will not be

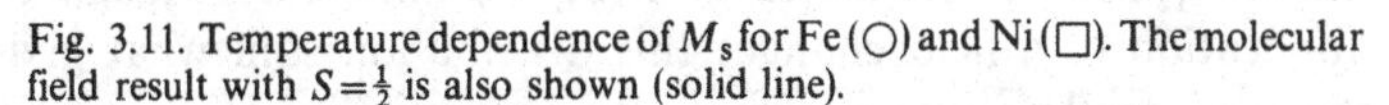

Fig. 3.11. Temperature dependence of M_s for Fe (○) and Ni (□). The molecular field result with $S=\frac{1}{2}$ is also shown (solid line).

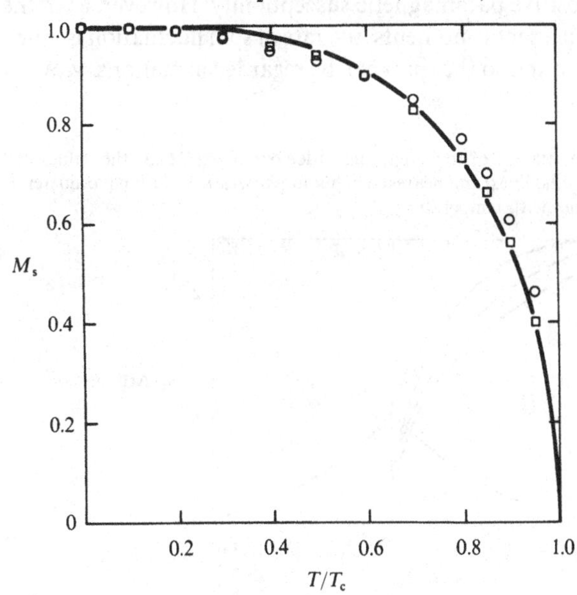

uniformly magnetised but will be divided into magnetic domains, as shown in Figure 3.13. Magnetic domains may also be found in antiferromagnetic and more complicated spin configurations as a result of the development of those structures from the disordered state. Unlike the antiphase boundaries that develop in systems with long range atomic order, the domain walls are quite thick, typically ~300 Å in Fe and ~700 Å in Ni (Cullity 1972) and certainly larger than the electron mean free path Λ in most alloys. As such, they represent a disordered phase of fairly small volume fraction as far as the resistivity is concerned, and their presence may easily be masked by other scattering effects.

3.2.5 *Short range magnetic order*, $T > T_c$

As foreshadowed in Figure 3.12, pairwise spin correlations are expected to exist over a limited range of temperatures above T_c, with the range of correlation decreasing with increasing temperature. This is demonstrated by the results for Fe given in Table 3.4. In a pure element these short range spin correlations will fluctuate in both space and time but in a concentrated alloy they are more likely to be associated with local composition fluctuations. Once the cooperative ferromagnetic order is broken down by thermal excitations of the spin system, the spins adopt random orientations (i.e. they are equally distributed over all allowed quantised spin states) and $\langle \mathbf{S}_i \rangle = 0$ if there is no magnetic field. Application of a field changes this distribution somewhat giving the observed small positive paramagnetic susceptibility. However, as in the case of superparamagnetic moments, the rate of spin fluctuation is much less than τ (10^{-14} s) and so the spins can be regarded as stationary. What

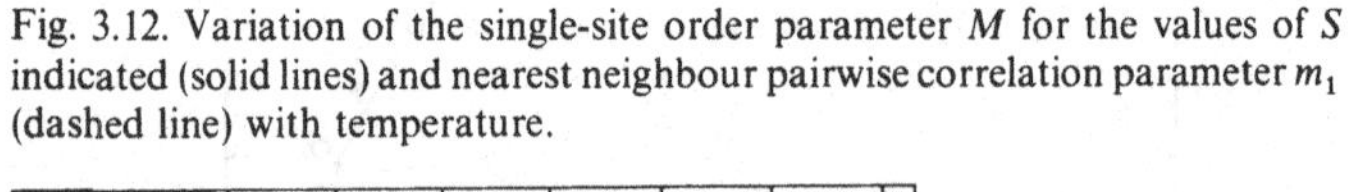

Fig. 3.12. Variation of the single-site order parameter M for the values of S indicated (solid lines) and nearest neighbour pairwise correlation parameter m_1 (dashed line) with temperature.

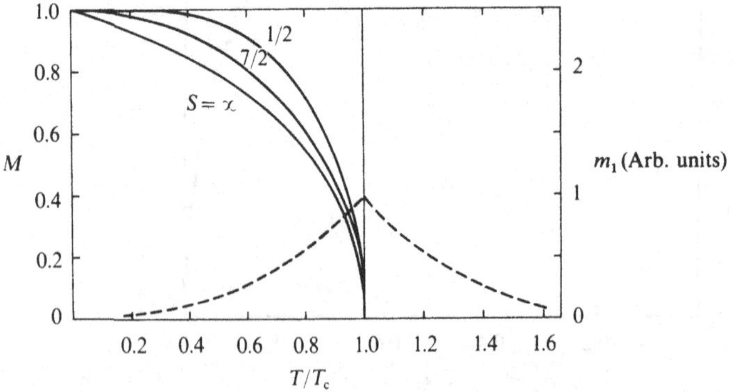

the conduction electrons see is thus a quasistatic disordered array of spins.

3.2.6 Magnons

The description of the spin system in terms of site occupation parameters that can adopt only one of two possible values (i.e. an Ising model) implies that the lowest energy excitation of the system from the ground state is as shown in Figure 3.14(*b*) with one of the spins reversed. However, there is another form of excitation which produces the same reduction in the total spin of the solid but at much lower energy. This is the spin wave or magnon which shares the reduction in spin over all spins by a wavelike procession of the spins about their z direction, as shown in Figure 3.14(*c*). Indeed, it can be shown rigorously (see e.g. Ashcroft & Mermin 1976, p. 704) that the single-spin excitation is not an

Fig. 3.13. (*a*) Magnetic domains in Fe–3%Si imaged in a scanning electron microscope. (*b*) Diagram of the domain structure indicating the directions of magnetisation. (Courtesy J. P. Jacubovics and D. J. Fathers.)

(*a*)

(*b*)

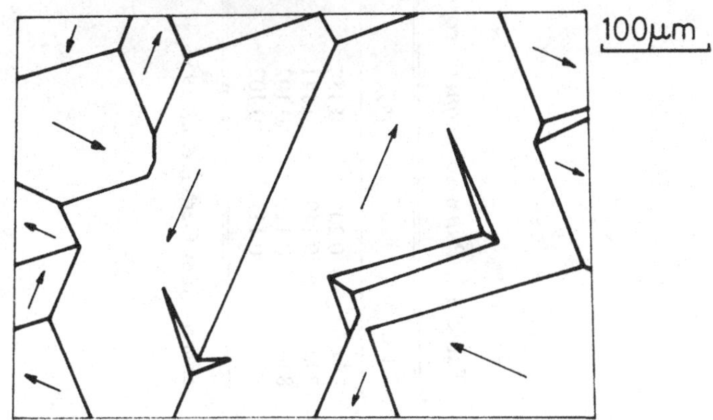

Table 3.4. *Pairwise magnetic correlation parameters of Fe as a function of temperature*

T (°C)	m_1	m_2	m_3	m_4	m_5	m_6	m_7	m_8	m_9	m_{10}
754	0.201	0.184	0.158	0.141	0.137	0.115	0.111	0.107	0.102	0.094
790	0.149	0.141	0.107	0.090	0.090	0.068	0.055	0.055	0.043	0.038
836	0.111	0.102	0.081	0.064	0.060	0.051	0.038	0.038	0.030	0.026
854	0.111	0.102	0.081	0.064	0.060	0.047	0.034	0.034	0.026	0.021

Source: After Gersch *et al.* 1956.

eigenstate of the Heisenberg Hamiltonian whereas the spin wave is, and represents a lowering of the total spin by the same amount. These spin waves are travelling waves with Cartesian components of the form (cf. equation (3.9))

$$\left.\begin{aligned} S^x(\mathbf{r}_i) &= u^x \exp[\mathrm{i}(\mathbf{Q}\cdot\mathbf{r}_i - \omega t)] \\ S^y(\mathbf{r}_i) &= u^y \exp[\mathrm{i}(\mathbf{Q}\cdot\mathbf{r}_i - \omega t)] \end{aligned}\right\}, \tag{3.10}$$

and are characterised by the dispersion relations (Kittel):

$$\left.\begin{aligned} &\text{ferromagnetic:} && \hbar\omega_{\mathbf{Q}} = 2JS_z(1-\gamma_{\mathbf{Q}}) \\ &\text{antiferromagnetic:} && \hbar\omega_{\mathbf{Q}} = 2JS_z(1-\gamma_{\mathbf{Q}}^2)^{1/2} \end{aligned}\right\}, \tag{3.11}$$

where $\gamma_{\mathbf{Q}} = 1/z\sum_i \exp(\mathrm{i}\mathbf{Q}\cdot\mathbf{r}_i)$ and the summation is over the z nearest neighbours on a Bravais lattice. At long wavelengths and in cubic lattices these reduce to the form:

$$\left.\begin{aligned} &\text{ferromagnet:} && \hbar\omega_{\mathbf{Q}} = 2JS(Qa_0)^2 \\ &\text{antiferromagnet:} && \hbar\omega_{\mathbf{Q}} = 4\sqrt{3}\, JSQa_0 \end{aligned}\right\}. \tag{3.12}$$

A simpler deviation of the above in one-dimensional systems is given in Kittel (1976). Some characteristic spin wave spectra measured by inelastic neutron scattering in both a ferromagnet and an antiferromagnet are shown in Figure 3.15 and verify the general form of these equations.

Fig. 3.14. (*a*) Classical picture of the ground state of a simple Ising ferromagnet. (*b*) An elementary excitation of one spin reversed. (*c*) A lower energy elementary excitation: a spin wave that causes the ends of the spin vectors to precess with a fixed phase relationship.

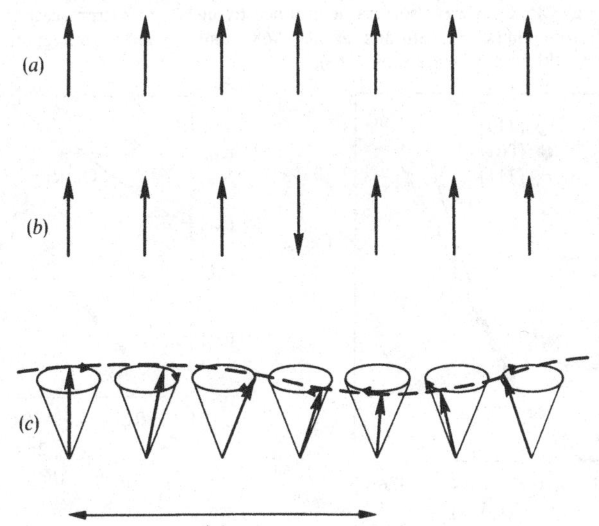

The quantisation of spin waves proceeds as for phonons (Section 2.7.2), leading to the concept of magnons. If the system contains N atoms each of spin S, the allowed values of the total spin quantum number are NS, $NS-1, NS-2, \ldots$, etc. If $n_{\mathbf{Q}}$ magnons of wavenumber $\mathbf{Q}$ are excited, the total spin quantum number is then $NS-\sum_{\mathbf{Q}} n_{\mathbf{Q}}$. The destruction of a magnon increases the total spin quantum number by unity and creation of a magnon decreases this number by unity. The energy of a mode of frequency $\omega_{\mathbf{Q}}$ with $n_{\mathbf{Q}}$ magnons is again given by equation (2.36). Thus to create a magnon of frequency $\omega_{\mathbf{Q}}$ requires an energy of $\hbar\omega_{\mathbf{Q}} \sim kT_{\mathrm{B}}$. As in the case of phonons, this energy is much less than the characteristic conduction electron energy, and so is negligible. Conservation of momentum leads directly to the phonon equation (1.14) with the same implications. However, there is one major difference between the magnon and phonon scattering: the change in the total spin quantum number is balanced by a change in spin of the conduction electron from $+\frac{1}{2}$ to $-\frac{1}{2}$ or $-\frac{1}{2}$ to $+\frac{1}{2}$ to maintain spin conservation. This is why the scattering from magnons is always inelastic and has a significant effect on the resistivity. Magnons have been found over a wide range of temperatures in both ferromagnetic and antiferromagnetic systems (Kittel 1976).

3.3 Nearly magnetic metals – spin fluctuations

The short range magnetic correlations that occur in a magnetic material above the spin-ordering temperature are in some ways

Fig. 3.15. Spin wave spectra determined by inelastic neutron scattering in (*a*) ferromagnetic Fe (Shirane *et al.* 1968); and (*b*) antiferromagnetic Fe_2O_3 (Brockhouse & Watanabe 1963).

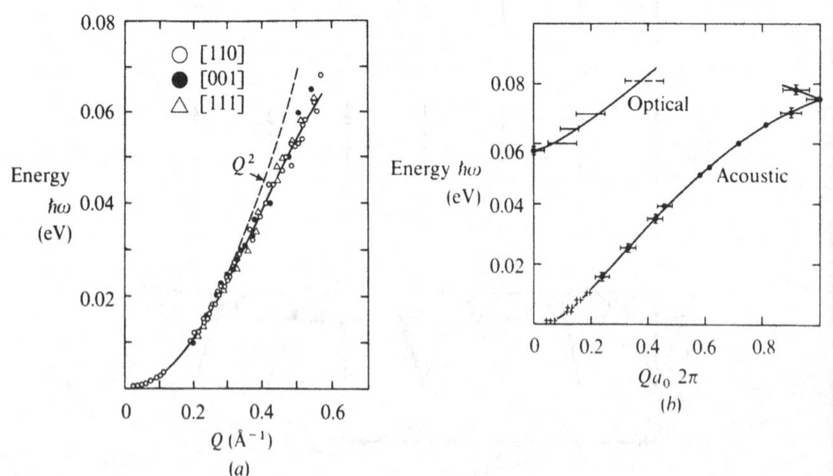

analogous to those found in strongly paramagnetic (i.e. nearly magnetic) materials due to exchange enhancement effects. These effects may occur in pure metals such as Pd or Pt, or in nearly magnetic alloys containing a concentration of magnetic species just less than that required to support a local magnetic moment. In the absence of impurities these correlations may be regarded as extensive regions of spin polarisation which fluctuate rapidly in time and space ('paramagnons'). In concentrated nearly magnetic alloys there will be the complication of local composition fluctuations which will give rise to a local susceptibility enhancement and localised spin fluctuations. As we might expect, understanding the origin of these fluctuations is tied up with the origin of localised moments, discussed in Section 3.2.1. If U is just too weak to produce the situation $UN_d(E_F) > 1$, the system will still have large fluctuations into the magnetic state, these being the spin fluctuations. This behaviour may be presented in the form of a locally enhanced static Pauli magnetic susceptibility (Rivier & Zuckermann 1968):

$$\chi = \frac{\chi_0}{1 - N_d(E_F)U}, \tag{3.13}$$

where χ_0 is the Pauli susceptibility of the material in the absence of the extra Coulomb interaction. The quantity $(1 - N_d(E_F)U)^{-1}$ is known as the Stoner enhancement factor (Stoner 1946). The enhancement is at a maximum when the virtual bound state is half-full, e.g. between Cr and Mn impurities in Al. One can see from (3.13) that as the magnetic instability is approached $\chi \to \infty$. However, these fluctuations only exist over a mean lifetime (Rivier & Zuckermann 1968)

$$\tau_{sf} = \frac{\pi N_d(E_F)}{1 - UN_d(E_F)}. \tag{3.14}$$

Note, however, that this differs from the lifetime of the virtual bound state $\tau_\Gamma = \hbar/\Gamma$, where Γ is the width of the virtual bound state (Heeger 1969; Rivier *et al.* 1969). The time of the memory of its spin (e.g. the lifetime of the spin fluctuation τ_{sf}) is longer because of the Coulomb repulsion U between electrons of opposite spin at the impurity. In fact, $\tau_{sf} \to \infty$ as $UN_d(E_F) \to 1$. The spin fluctuation lifetime may be written in terms of a characteristic spin fluctuation temperature

$$T_{sf} = \hbar/k_B\tau_{sf}. \tag{3.15}$$

At temperatures below T_{sf}, rapid temporal spin fluctuations occur, i.e. the spin *polarisation* fluctuates rapidly with time. At temperatures above T_{sf} the spin polarisation exists for a time that is long compared with the thermally induced *spatial* fluctuations of the moment. A Curie–Weiss law is then expected and the behaviour is indistinguishable from that of a

Table 3.5. *Approximate characteristic temperatures (in K) for magnetic impurities in some non-magnetic hosts in the dilute limit*

Impurity	Host							
	Au	Cu	Ag	Al	Zn	Pt	Ir	Rh
V	290[a]							
Cr	3[c]	3[e]		1200[b]		200[a]		
Mn	1[e]	1[e]	1[e]	530[a]	1[a]			
Fe	5[e]	12[a]	10[e]		80[a]		225[c]	12[d]
Co	100[e]							

Notes:
[a] Rizzuto *et al.* 1973.
[b] Caplin & Rizzuto 1968.
[c] Sarachik 1968.
[d] Rusby 1974.
[e] See Daybell & Steyert 1968.

localised magnetic moment. This is most important, so let us state it another way. Above the characteristic temperature T_{sf} the impurity behaves exactly as though it had a well-defined magnetic moment. Below that temperature the spins fluctuate rapidly compared with their thermal motion and so the impurity appears to be non-magnetic. This applies to all impurities in non-magnetic hosts that produce a virtual bound state in the vicinity of the Fermi surface and so in this sense there should be a '+' at all entries in Table 3.1, although in some cases the extrapolated characteristic temperature may lie above the melting point. As we will show in Section 7.3, the rapid relaxation time of the conduction electrons makes the resistivity a nearly ideal probe. We should also note that those systems predicted to have a magnetic moment in the Hartree–Fock model (i.e. the '+' entries in Table 3.1) are subject to exactly the same arguments. A rigorous analysis shows that sufficiently below some characteristic temperature they will all become non-magnetic! In that sense none of the impurities shown will have a magnetic moment at 0 K. This is the widely investigated Kondo–Nagaoka behaviour (Rivier & Zuckermann 1968). In the dilute limit the characteristic temperature in independent of the concentration of the impurity and some characteristic values are given in Table 3.5. At higher concentrations, groups of 3, 4, ... impurity atoms will have their own characteristic spin fluctuation lifetimes, each becoming larger as more and more atoms contribute to the local spin enhancement. For example, if each nearest neighbour impurity atom changes the Coulomb interaction by δU, the fluctuation temperature of an impurity atom with p nearest neighbours will be

$$T_{sf}(p)=\frac{\hbar}{k_B}\frac{1-(U+p\,\delta U)N_d(E_F)}{\pi N_d(E_F)}. \tag{3.16}$$

Clearly, if p exceeds a critical value the impurity cluster will acquire a permanent moment.

Chouteau *et al.* (1974) have used an analysis of this form and found that isolated Ni atoms in a Pd matrix have a Curie–Weiss temperature θ_1 of -25 K, whereas that for Ni-pairs was $\theta_2 \approx -13$ K ($\theta \approx -4T_{sf}$). By extrapolation they found $\theta_3 \approx 0$, i.e. groups of three or more Ni atoms should be magnetic. Similarly, Boucai *et al.* (1971) proposed $T_{sf}=190$ K for isolated Co atoms in Au and $T_{sf}=23$ K for pairs, groups of three or more neighbouring Co atoms being magnetic. However, near neighbours do not always lead to a lowering of T_{sf}. In Av–V, for example, Claus *et al.* (1967) have proposed that isolated V atoms would be 'magnetic', nearest neighbours 'non-magnetic' and second or third nearest neighbours 'more magnetic' than the isolated V atoms (see also Narath & Gossard 1969). Star (1972) has suggested that if isolated V atoms had $T_{sf} \sim 300$ K, first and second nearest neighbours $T_{sf} > 1000$ K and third nearest neighbours $T_{sf} \sim 100$ K, the observed concentration dependence of the effective overall T_{sf} could be explained. Similar effects have also been discussed in Cu–Ni (Cornut *et al.* 1971), Cu–Co (Tournier & Blandin 1970), Cu–Fe (Tholence & Tournier 1970; Star *et al.* 1972*a*), Pd–Cr and Pt–Cr (Star *et al.* 1972*b*) and Fe impurities in Pd–Ni (Chouteau *et al.* 1974) and Cu–Ni (Tholence & Tournier 1970; Bennett *et al.* 1969). See also Coles (1974), Fischer (1978), and references cited therein.

A similar Coulomb interaction term can be applied to electrons within a narrow d band to explain the presence of magnetic moments in pure metals and concentrated alloys (see e.g. Hasegawa 1983*a*,*b*, 1984). However, should the metal be such that $UN_d(E_F)$ is just less than unity, it will not be magnetic in the Hartree–Fock sense, although there will again be large spin fluctuations into the magnetic state. This situation arises in Pd, for example, where $UN_d(E_F) \sim 0.9$ (see e.g. Lederer & Mills 1968). Such a metal has a large exchange enhancement, even in the absence of impurities. Nevertheless if an impurity causes a local increase in U, then there will be a further enhancement of the amplitude of the spin fluctuations and the spin polarisability of the matrix near the impurity (Lederer & Mills 1968). Thus in such cases, even though there may be no impurity scattering resonance in the band structure near E_F, there can still be large localised spin fluctuations due to the electronic structure of the host. The 'giant' polarisation clouds associated with the impurities lead to a very large apparent magnetic moment per impurity atom. For example, the diffuse neutron scattering data for Pd–Fe (Low & Holden 1966; Hicks *et al.* 1968) indicate that the polarisation cloud extends over about 10 Å and accounts for $\frac{2}{3}$ of moment associated with each Fe atom (see also Diamond 1972; Medina & Parra 1982). A similar effect has been found in more concentrated Pt and Pd alloys but in such cases the host is an enhanced *alloy* with nearly magnetic single impurity atoms or atom pairs etc. (Aldred *et al.* 1970). Similar effects are also found in other concentrated alloys just below ferromagnetic concentrations (see e.g. Amamou *et al.* 1974, 1975). However, we should note that the interpretation of experimental data is by no means unambiguous and at the time of writing there appears to be a trend away from these giant cloud, local environment models in favour of critical scattering models (see e.g. Ododo 1985; Loram & Mirza 1985).

The concept of localised spin fluctuations thus replaces the artificially sharp boundary between magnetic and non-magnetic impurities of the Hartree–Fock model with a smooth transition between slow and fast localised spin fluctuation regimes. We have preferred this description of the low temperature behaviour of 'magnetic' impurities to the Kondo–Nagoaka theories which assume the presence of a well-defined spin, interacting weakly with the conduction electrons via the s–d exchange interaction (see e.g. Kondo 1969; Fischer 1970), as it allows for a much more transparent picture of the physical processes involved. We can also see why detection of a magnetic moment depends upon the coherence time of the probe being used τ_P in relation to the characteristic spin fluctuation times τ_{sf}. This is why isolated Co atoms in Au appear to be magnetic in a resistivity study ($\tau_P < \tau_{sf}$) and non-magnetic in a susceptibility study ($\tau_P > \tau_{sf}$) (see e.g. Henger & Korn 1981). A more detailed description of spin fluctuation necessarily involves the dynamical (local) susceptibility $\chi(\mathbf{Q}, \omega)$ which includes both the frequency and wavenumber response of the magnetic spins (see e.g. Friedel 1969). A collection of data relating to Kondo and spin fluctuation systems has been given by Fischer (1981). Further discussions about these fluctuations will be reserved for Chapter 7.

3.4 Effects of atomic rearrangements

We have considered the magnetic structure of metals and their alloys in some detail because of the sensitivity of the resistivity to the magnetic state. The previous sections of this chapter have shown how the magnetic properties can depend critically upon composition. (For a more detailed discussion of experimental data see Bozorth 1951; Tebble & Craik 1969.) We now want to show that the state of order of a phase containing magnetic or nearly magnetic ions can also affect the magnetic state. This will be important because any change in composition or atomic arrangement will in general then produce a two-fold contribution to ρ: a direct contribution from the atomic distribution and an indirect contribution via changes in the magnetic state. A full discussion of the effects of atomic distribution on magnetic properties is beyond the scope of this book but we give below some examples to illustrate the point.

3.4.1 *Long range effects*

Because the nature of the magnetic interaction between nearest neighbours in a particular alloy depends upon the actual type of atoms concerned, long range ordering usually influences the magnetic properties, sometimes quite dramatically. For example, ordered Au_4Mn is ferromagnetic and ordered Pd_3Mn and Au_3Mn are both antiferromagnetic but all three are mictomagnetic in the disordered state (Beck 1972). In the case of $MnNi_3$, MnBi, MnAu and the Heusler alloys Cu_2MnSn and Cu_2MnAu the alloys are ferromagnetic when ordered and paramagnetic when disordered. By contrast, FeCo and $FeNi_3$ are both ferromagnetic in either the ordered or disordered state, the effect of a long range ordering being only to produce a slight increase in the saturation magnetisation (Cullity 1972). Directional ordering effects can also produce marked changes in magnetic anisotropy (Chikazumi 1964).

Structural transformations can also result in a dramatic change in magnetic properties, the classic example being the non-magnetic fcc Fe-rich Austenites transforming to the ferromagnetic bcc phases at lower temperatures. Similarly, the composition changes that result from a phase separation can give rise to dramatic effects, as evident in the Alnico or Fe–Cr–Co permanent magnets, for example. In both of these cases the high temperature bcc α phase is non-magnetic and separates by spinodal decomposition into a fine mixture of a strongly ferromagnetic α_1 phase and a weakly ferromagnetic or paramagnetic α_2 phase. The spontaneous magnetisation of a two-phase alloy is given by the sum of the independent contributions from the different phases weighted according to their volume fractions. An example of the type of behaviour observed is given in Figure 3.16 and shows that the different contributions may be easily identified if they have markedly different Curie temperatures.

3.4.2 *Short range effects*

Short range ordering or clustering can also have a pronounced effect on the magnetic properties of an alloy, particularly if its nominal composition lies near a non-magnetic magnetic phase boundary. Alloys containing Ni or Co atoms in a non-magnetic matrix provide a clear example of this type of behaviour. As noted in Section 3.2.2, in both cases the ions will only acquire a magnetic moment over any given period of observation if they are in a sufficiently Ni- or Co-rich environment. As an example we consider Cu-rich Cu–Ni alloys. One proposed form of the dependence of the static moment at a Ni site upon

Fig. 3.16. Saturation magnetisation of a two-phase Fe–C alloy and the components due to the Fe and Fe_3C phases (after Crangle 1977).

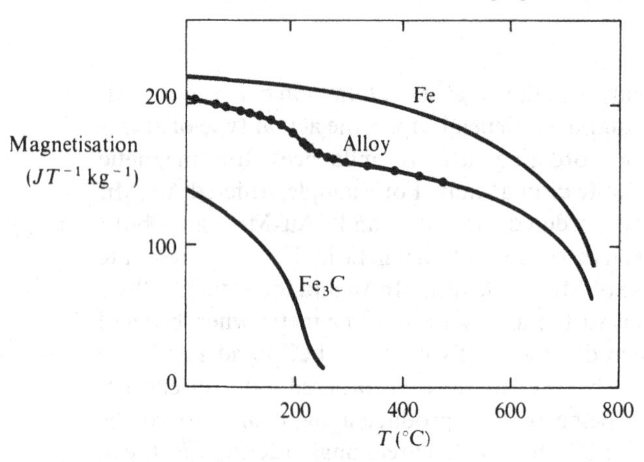

Fig. 3.17. Dependence of the local static atomic magnetic moment upon the number of Ni, nearest neighbours n and number of Ni second nearest neighbours m (after Robbins *et al.* 1969).

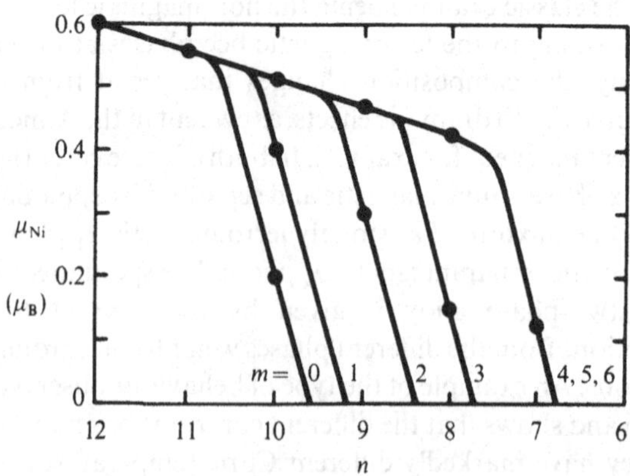

Table 3.6. *Fractional change in mean magnetic moment of a Ni atom in Cu–Ni alloys as the first two SRO parameters change from $\alpha_1 = \alpha_2 = 0$ to $\alpha_1 = 0.05$, $\alpha_2 = -0.05$*

Composition (% Ni)	$\delta\bar{\mu}/\bar{\mu}$
10	16
20	3
30	1

Note: From equation 3.17.

the local Ni concentration has been given by Robbins *et al.* (1969) and is shown in Figure 3.17. This moment may be evaluated by assuming that the atomic distribution is given by a modified binomial (cf. 3.1):

$$\bar{\mu} = \sum_{n=0}^{12} \sum_{m=0}^{6} P_{nm}\mu_{nm}, \tag{3.17}$$

where μ_{nm} is the mean magnetic moment of a Ni atom having n first and m second near neighbours that are also Ni atoms, and P_{nm} is given by

$$P_{nm} = c_{\mathrm{Ni}} \frac{12!\,6!}{(12-n)!\,(6-m)!\,n!\,m!} y^n(1-y)^{12-n} z^m (1-z)^{6-m} \tag{3.18}$$

and

$$\left.\begin{aligned} y &= (1-c_{\mathrm{Ni}})\alpha_1 + c_{\mathrm{Ni}} \\ z &= (1-c_{\mathrm{Ni}})\alpha_2 + c_{\mathrm{Ni}} \end{aligned}\right\} \tag{3.19}$$

where c_{Ni} is the atomic fraction of Ni and α_1, α_2 are the first two Warren–Cowley SRO parameters. The fractional change in this mean magnetic moment as the alloy changes from an ideally random state $\alpha_1 = \alpha_2 = 0$ to a clustered state $\alpha_1 = 0.05$, $\alpha_2 = -0.05$ is given in Table 3.6 for different nominal Ni concentrations. These results show that the statistical probability of finding a sufficiently Ni-rich cluster in a Cu–10%Ni alloy is small if the alloy is random, but much larger if clustering occurs. An alloy with 30% Ni already has a large probability of finding such a cluster and so a change in the degree of clustering then produces a relatively smaller effect. More realistic computer modelling of these atomic distributions lead to similar results (Rossiter 1981*b*; Vrijen 1977).

Such an increase in the 'giant' cluster moments with clustering or short range ordering has also been reported in Pd–Cr, Cu–Mn, Au–Fe and Au–Mn alloys (Beck 1972). However, following on from the discussion of spin fluctuations in Section 3.3 we can now see that the effects of the local environment should really be considered in terms of the characteristic spin fluctuation times (or temperatures). A change in the short range atomic correlations will change the number of near neighbours p and from (3.16) produce a change in $T_{sf}(p)$. Thus a localised region can change from being non-magnetic to magnetic at any given measuring temperature T_m by a change in p that reduces T_{sf} to below T_m. The simple models discussed above are one way of representing this change but it must again be remembered that the number of near neighbours (or degree of short range correlation) required to produce a 'magnetic' state will depend upon the characteristic time of the probe being used and will be much less for a resistivity study than in a static susceptibility measurement, for example. More realistic calculations based on the cluster extensions to the CPA will be briefly discussed in Chapter 6.

4 Electrons in simple metals and alloys

4.1 Scattering potentials and electron wavefunctions

The scattering equation introduced in Chapter 1 (equation (1.2)) requires knowledge of the incident and scattered wavefunctions $\Phi_{\mathbf{k}}$ and $\Psi_{\mathbf{k}'}$ and of the scattering potential $V(\mathbf{r})$. In this chapter we will be concerned mainly with crystalline materials in which case the symmetry of the lattice must be reflected in these quantities. This means that the lattice potential must be periodic and satisfy the relationship

$$V(\mathbf{r}) = V(\mathbf{r} + \mathbf{T}), \tag{4.1}$$

where $V(\mathbf{r})$ is the potential at position $\mathbf{r}$, and $\mathbf{T}$ is a Bravais lattice translation vector. As a consequence the electronic charge density must also be periodic, so that

$$|\Psi(\mathbf{r})|^2 = \Psi(\mathbf{r} + \mathbf{T})|^2. \tag{4.2}$$

Wavefunctions of the form proposed by Bloch:

$$\psi_{\mathbf{k}}(\mathbf{r}) = U_{\mathbf{k}}(\mathbf{r}) \exp(\mathrm{i}\mathbf{k} \cdot \mathbf{r}) \tag{4.3}$$

satisfy requirement (4.2) provided that the Bloch functions $U_{\mathbf{k}}(\mathbf{r})$ also satisfy the relation

$$U_{\mathbf{k}}(\mathbf{r}) = U_{\mathbf{k}}(\mathbf{r} + \mathbf{T}), \tag{4.4}$$

(see e.g. Ashcroft & Mermin 1976, p. 134). It is convenient at this stage to be able to introduce a pictorial representation of these functions. We do this by first considering the wavefunctions of an isolated atom which is obtained by solving the time-independent Schrödinger equation

$$\frac{\hbar^2}{2m} \frac{\mathrm{d}^2\psi_{\mathbf{k}}(\mathbf{r})}{\mathrm{d}r^2} + (E(\mathbf{k}) - V(\mathbf{r}))\psi_{\mathbf{k}}(\mathbf{r}) = 0. \tag{4.5a}$$

Note that this equation can also be written in the form

$$\left(-\frac{\hbar^2}{2m}\nabla^2 + V(\mathbf{r})\right)\psi_{\mathbf{k}}(\mathbf{r}) = E(\mathbf{k})\psi_{\mathbf{k}}(\mathbf{r}), \tag{4.5b}$$

which may be abbreviated to

$$\mathscr{H}\psi_{\mathbf{k}}(\mathbf{r}) = E(\mathbf{k})\psi_{\mathbf{k}}(\mathbf{r}), \tag{4.5c}$$

where $\mathscr{H}$ is the Hamiltonian which appears in the round brackets in equation (4.5b). The electrostatic interaction between an electron and the nucleus is adequately represented by a Coulomb attraction

$$V(r) = -\frac{Ze^2}{4\pi\varepsilon_0 r}, \tag{4.6}$$

where Ze is the charge on the nucleus. However, as the nucleus is surrounded by more than one electron the effective interaction must include also the repulsion due to the other $Z-1$ electrons (i.e. these electrons will 'screen' part of the charge of the nucleus). To solve this problem it is necessary to calculate the charge distribution of the $Z-1$ electrons and so obtain the actual electrostatic potential of the Zth electron and hence its wavefunction from the Schrödinger equation. However, it is also necessary to take into account the effect of the Zth electron on the $Z-1$ other electrons. It is thus necessary to find some form of self-consistent solution. One approach is to adopt the Hartree self-consistent field approximation (Harrison 1970, p. 73): assume an approximate set of wavefunctions for the $Z-1$ electrons, compute the potential due to the nucleus and these electrons, solve the Schrödinger equation using this potential and obtain enough solutions to accommodate all Z electrons in accordance with the Pauli exclusion principle. The new electron states are then used to recalculate the potential and the process is repeated until it converges onto a self-consistent potential that reproduces itself. However, even though this solution is self-consistent it is still only an approximation since it assumes that the total wavefunction can be written as the product of independent one-electron wavefunctions, each depending only upon the coordinates of one electron. In reality the electrons all interact with each other and, because of this interaction, the Schrödinger equation is not separable (i.e. the Hamiltonian cannot be written as the sum of one-electron Hamiltonians). Nevertheless, the Hartree self-consistent field approximation yields a set of plausible one-electron wavefunctions and energies so that the total energy of the system is given by

$$E = \sum_i \varepsilon_i - \frac{1}{2}\frac{1}{4\pi\varepsilon_0}\sum_{i\neq j}\sum \iint \frac{e^2}{\mathbf{r}_i - \mathbf{r}_j} |\psi_i(\mathbf{r}_i)|^2 |\psi_j(\mathbf{r}_j)|^2 \, d\mathbf{r}_i \, d\mathbf{r}_j, \tag{4.7}$$

where ε_i are the one-electron energies obtained from the separated equations $H_i\psi_i(\mathbf{r}_i) = \varepsilon_i\psi_i(\mathbf{r}_i)$, and the second term arises because $\sum_i \varepsilon_i$ counts the interaction between each pair of electrons twice. The real part of the wavefunction $\psi(\mathbf{r}_i)$ is shown schematically in Figure 4.1, and by summing $4\pi r^2|\psi_i|^2$ over all occupied states gives the charge distribution shown schematically in Figure 4.2.

When the atoms are combined into a periodic array we need some means of transforming the atomic electron states into the energy levels of the solid. However, we know that the wavefunction must satisfy equations (4.3) and (4.4) and so is represented by an atomic-like function for $U_{\mathbf{k}}(\mathbf{r})$ repeated with the periodicity of the crystal and modulated by a plane wave $\exp(i\mathbf{k}\cdot\mathbf{r})$. This is shown in Figure 4.3, together with a representation of the periodic potential $V(\mathbf{r})$. Note that only the real parts of the functions are plotted and that the charge density $|\psi(\mathbf{r})|^2$ will actually be the same at each site.

Fig. 4.1. Schematic representation of the real part of the electron wavefunction for a many-electron atom.

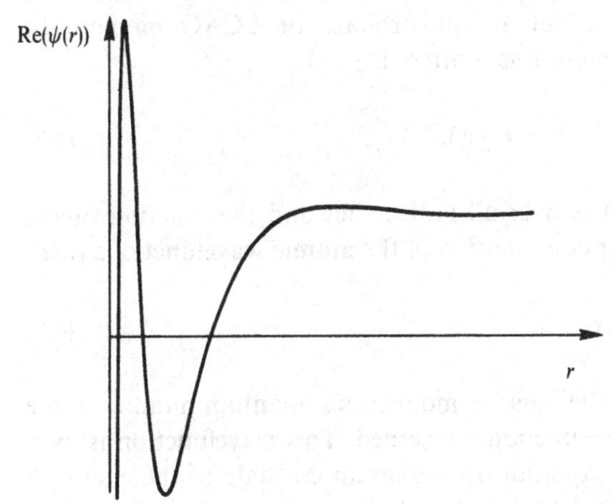

Fig. 4.2. Schematic representation of the radial charge distribution about a many-electron atom.

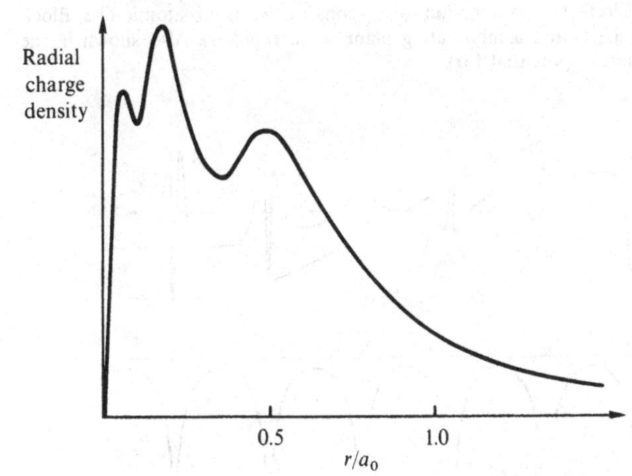

In order to proceed we need some means of recalculating the wavefunction and potential for the periodic solid. As in the free atom case there are problems associated with screening and interactions between electrons and most methods are again based on an independent electron approximation using self-consistent field techniques. The tight binding approximation is conceptually one of the simplest methods (Ziman 1972, p. 91; Ashcroft & Mermin 1976, p. 176; Harrison 1970, p. 151; Bullett 1980). It assumes that when the atoms are brought together into a solid, the electron states can be obtained directly from the atomic states, the wavefunction being derived from a linear combination of atomic wavefunctions. For this reason the technique is also known as the 'linear combination of atomic orbitals' or LCAO method. In particular, the wavefunction is written as

$$\psi_{\mathbf{k}}(\mathbf{r}) = \sum_i \exp(i\mathbf{k}\cdot\mathbf{r}_i)\phi(\mathbf{r}-\mathbf{r}_i), \tag{4.8}$$

where the summation is over all lattice sites and the function $\phi(\mathbf{r})$ is obtained from a linear combination of the atomic wavefunctions $\psi_n(\mathbf{r})$:

$$\phi(\mathbf{r}) = \sum_n a_n \psi_n(\mathbf{r}). \tag{4.9}$$

The index n refers to the angular momentum quantum number of the particular atomic wavefunction concerned. This wavefunction is used with the Schrödinger equation to obtain an estimate of the energy E which is then minimised by suitable choice of the coefficients a_n. The

Fig. 4.3. Electron wavefunction $\psi_{\mathbf{k}}(\mathbf{r})$ constructed from atomic-like Bloch functions $U_{\mathbf{k}}(\mathbf{r})$ and a modulating plane wave $\exp(i\mathbf{k}\cdot\mathbf{r})$. Also shown in the periodic lattice potential $V(\mathbf{r})$.

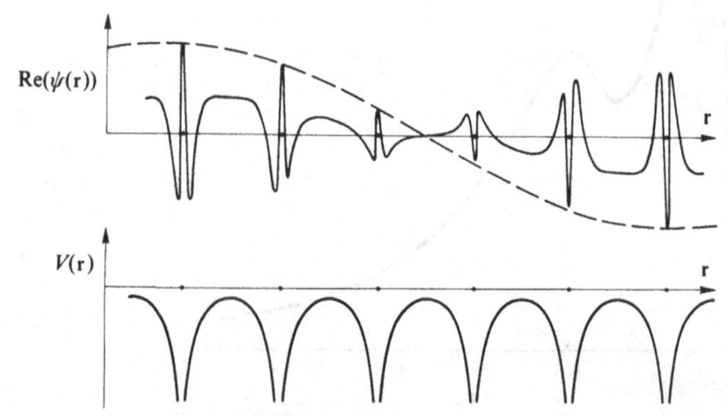

potential in this case is given by a superposition of potentials centred on the individual atoms:

$$V(\mathbf{r})=\sum_i v(\mathbf{r}-\mathbf{r}_i). \tag{4.10}$$

However, while the wavefunction (4.8) can be matched reasonably well to the atomic-like states of the inner electron shells it cannot adequately represent the Bloch states in the interstitial regions which, as we shall show below, are more like combinations of free electron plane waves.

The tight binding method and the free electron approximation thus represent extreme limits: on the one hand the electron states are constructed from localised atomic states and on the other they are assumed to be completely delocalised plane waves. The electron states of most metals lie somewhere between these two limits and it is convenient at this stage to distinguish between the core electrons, which are treated as almost completely localised, and the valence electrons, which are assumed to go into Bloch states.

Advances beyond the tight binding approach came from the realisation that many metals do not deviate a great deal from free electron behaviour and so the wavefunction must be essentially more like free electron plane waves in the interstitial regions. In the augmented plane wave method (APW) this feature is taken explicitly into account by placing small spheres around each ion and within each sphere using an atomic core wavefunction $\psi_{\mathbf{k}}^{c}(r)$ which joins smoothly onto a plane wavefunction between the spheres (Loucks 1967; Dimmock 1971). Because of the nature of this model it is usually used with a 'muffin-tin' (mt) potential that has spherical symmetry within the sphere and is constant elsewhere. The associated wavefunctions and potential are shown in Figure 4.4 and defined by

$$\left.\begin{aligned}\psi_{\mathbf{k}}(\mathbf{r})&=\psi_{\mathbf{k}}^{c}(\mathbf{r})\\ V(\mathbf{r})&=V_{\mathrm{mt}}(\mathbf{r})\end{aligned}\right\}\text{within sphere}$$
$$\left.\begin{aligned}\psi_{\mathbf{k}}(\mathbf{r})&=\exp(\mathrm{i}\mathbf{k}\cdot\mathbf{r})\\ V(\mathbf{r})&=\text{const}\end{aligned}\right\}\text{outside sphere} \tag{4.11}$$

Other approaches have been based on expanding the wavefunction entirely in terms of normalised plane waves $\Omega^{-1/2}\exp(\mathrm{i}\mathbf{k}\cdot\mathbf{r})$, where Ω is the volume of the crystal, and taking as the potential a superposition of atomic potentials (as in equation (4.10)). This is known as the plane wave method but suffers from the problem that a large number of high-frequency components are required to reproduce the rapid oscillations in the core. An improvement on this method is the orthogonalised plane wave method (OPW). Here the basis set of wavefunctions used in the expansion already contain the rapid oscillations in the core region by requiring them to be orthogonal to the core states. However, they still retain the free electron character in the interstitial regions. This is achieved by taking as the wavefunction

$$\psi_{\mathbf{k}}(\mathbf{r}) = \exp(i\mathbf{k}\cdot\mathbf{r}) + \sum_{c} b_c \psi_{\mathbf{k}}^{c}(\mathbf{r}), \tag{4.12}$$

where the sum is over all of the core levels with wavevector **k**. The core wavefunctions are usually obtained by tight binding combinations of the atomic levels and the constants b_c obtained from the orthogonality condition:

$$\int \psi_{\mathbf{k}}^{c*}(\mathbf{r})\phi_{\mathbf{k}}(\mathbf{r})\,d\mathbf{r} = 0, \tag{4.13}$$

i.e.

$$b_c = -\int \psi_{\mathbf{k}}^{c*}(\mathbf{r})\exp(i\mathbf{k}\cdot\mathbf{r})\,d\mathbf{r}. \tag{4.14}$$

The core oscillations contained in $\psi_{\mathbf{k}}^{c}(\mathbf{r})$ are thus imposed by equation (4.14) onto $\phi_{\mathbf{k}}(\mathbf{r})$, although the summation in (4.12) is small outside the core region leaving the required plane wave form in the interstitial regions. These wavefunctions are shown schematically in Figure 4.5 and can be used with a variety of potentials. This method has been reviewed by Reitz (1955) and Woodruff (1957).

Fig. 4.4. Schematic representation of augmented plane wave (*a*); and muffin-tin potential (*b*).

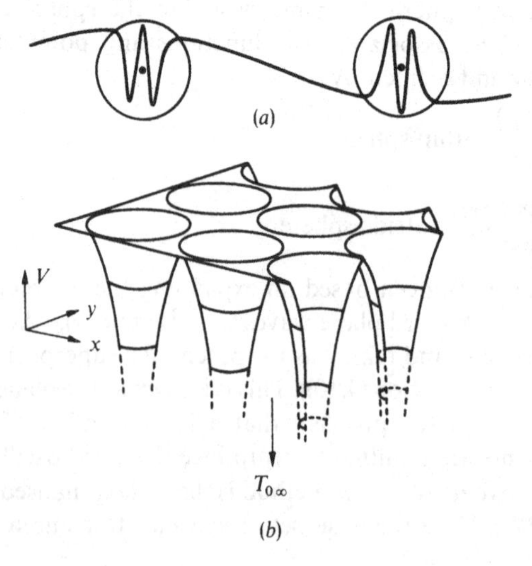

In all of the above cases the trial wavefunctions are constructed from the basis states (APW, plane wave, OPW):

$$\psi_{\mathbf{k}}(\mathbf{r}) = \sum_{n} c_n \phi_{\mathbf{k}+\mathbf{g}_n}(\mathbf{r}), \tag{4.15}$$

where the summation is over the reciprocal lattice vectors $\mathbf{g}_n$ to satisfy the Bloch condition. The function $\psi_{\mathbf{k}}(\mathbf{r})$ is used as a trial solution to the Schrödinger equation, and the variational principle is used to minimise the expectation value of the energy E and hence determine the coefficients c_n (Ziman 1972, p. 104; Ashcroft & Mermin 1976, p. 201).

One final method of approximation will be discussed as it is being extensively used to determine the band structures of transition metal alloys (see Chapter 5). This is the Green's function method of Korringa, Kohn & Rostoker (KKR) which starts with the integral form of the Schrödinger equation:

$$\psi_{\mathbf{k}}(\mathbf{r}) = -\int \frac{\exp(\mathrm{i}K|\mathbf{r}-\mathbf{r}'|)}{4\pi|\mathbf{r}-\mathbf{r}'|}, \tag{4.16}$$

where $K^2 = 2mE/\hbar^2$. Using a muffin-tin potential for $V(\mathbf{r}')$ and the Bloch condition leads to (Ziman 1972, p. 106; Ashcroft & Mermin 1976, p. 202)

$$\psi_{\mathbf{k}}(\mathbf{r}) = \int G_{\mathbf{k},E}(\mathbf{r}-\mathbf{r}')V(\mathbf{r}')\psi_{\mathbf{k}}(\mathbf{r}')\,\mathrm{d}\mathbf{r}', \tag{4.17}$$

where the integration is now over a single muffin-tin cell. All details of the crystal structure are contained in the Green's function:

$$G_{\mathbf{k},E}(\mathbf{r}-\mathbf{r}') = \sum_{i} \frac{\exp(\mathrm{i}K|\mathbf{r}-\mathbf{r}'+\mathbf{r}_i|)}{|\mathbf{r}-\mathbf{r}'+\mathbf{r}_i|}\exp(\mathrm{i}\mathbf{k}\cdot\mathbf{r}_i), \tag{4.18}$$

where the summation is over all lattice sites $\mathbf{r}_i$. Solution proceeds by constructing a functional which gives equation (4.18) on variation, and selecting a trial core wavefunction of the form:

$$\psi_{\mathbf{k}}(\mathbf{r}) = \sum_{l,m} A_{lm} Y_{lm}(\theta,\phi) R_l(\mathbf{r},E), \tag{4.19}$$

Fig. 4.5. Orthogonal plane wave constructed from the core wavefunction $\psi_{\mathbf{k}}^{\mathrm{c}}(\mathbf{r})$ and plane wave $\exp(\mathrm{i}\mathbf{k}\cdot\mathbf{r})$.

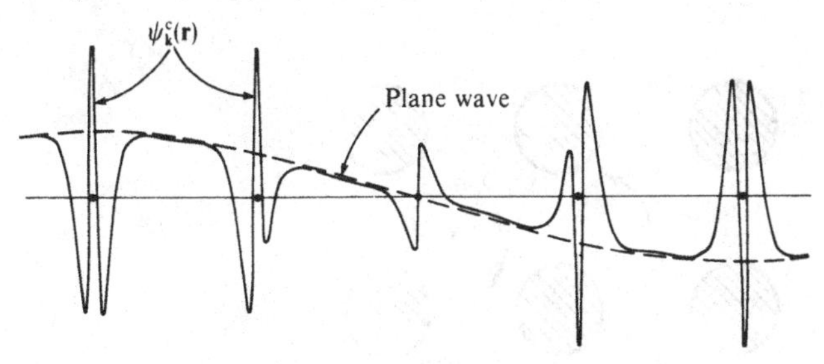

where $Y_{lm}(\theta, \phi)$ is a spherical harmonic for the direction (θ, ϕ) of the vector $\mathbf{r}$ and R_l satisfies the radial equation

$$-\frac{1}{r^2}\frac{\partial}{\partial r}\left(r^2\frac{\partial R_l}{\partial r}\right)+\left(\frac{l(l+1)}{r^2}+V(r)\right)R_l=ER_l. \tag{4.20}$$

In effect the atoms are replaced by 'black boxes' characterised entirely by the scattering properties at the surface of the muffin-tin cell, as shown in Figure 4.6. The variational condition ultimately gives the coefficients A_{lm} and the relationship between E and $\mathbf{k}$. As the number of spherical harmonics contained in the summation (4.20) must be limited, the approximation in this method lies in the wavefunction in the core region. (For reviews of this method and Green's functions see Segall & Ham 1968; Csanak *et al.* 1971; and Doniach & Sondheimer 1974).

Lloyd (1965) has shown that the OPW, APW and KKR methods can all be derived from the general principle that the band structure depends only upon the scattering properties of the core potential. Heine (1969) and Ziman (1971) also give a general comparison of these methods.

Finally, we mention the cluster methods of electronic structure determination in which it is assumed that the solid can be constructed by replication of some particular smaller cluster. The problem then reduces to determining the electronic structure of the cluster (Johnson 1973; Keller & Smith 1972; Faulkner *et al.* 1974; Faulkner 1977; Slater & Johnson 1972; Johnson *et al.* 1979; Braspenning *et al.* 1984). While this type of approach has proved very useful for molecular calculations, its applicability to bulk metallic solids remains somewhat uncertain (Friedel 1973) because of the large number of surface atoms. Nevertheless, it does appear that quite small clusters can reproduce the electronic structure of the bulk reasonably well (see the references cited

Fig. 4.6. Plane wave electrons propagating in the weak potential between the atomic 'black boxes' which are characterised by their scattering properties.

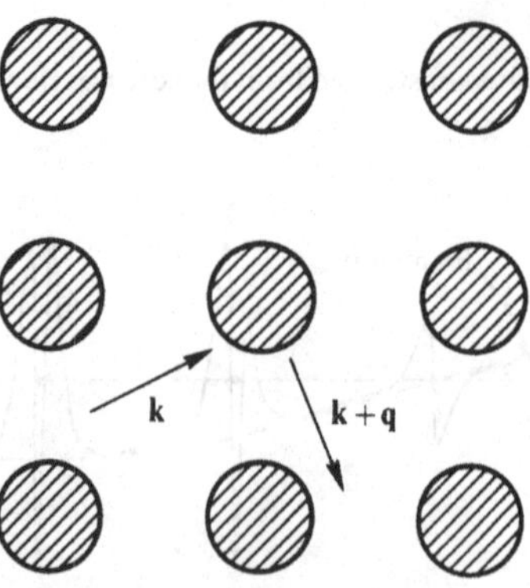

in Johnson *et al.* 1979). One possible way of minimising the effect of the surface atoms is to embed the cluster in some suitable host environment. We will describe some calculations along these lines in Chapter 6.

4.2 Pseudo- and model potentials

While the techniques described above are capable of generating realistic band structures, they all require expertise in dealing with electron states and considerable computing power. As such they really lie in the domain of the band structure specialist and are not as yet suitable general-purpose tools for understanding resistivity behaviour. However, these methods to suggest an approach which is quite amenable to simple calculation. In most cases the equations can be rearranged so that the difficult core terms are combined with the real potential to give a much weaker pseudopotential $W(\mathbf{r})$. The corresponding pseudo-Schrödinger or pseudopotential equation is

$$\left(-\frac{\hbar^2}{2m}\nabla^2+W(\mathbf{r})\right)\phi_{\mathbf{k}}^{\mathrm{v}}=E_{\mathbf{k}}^{\mathrm{v}}\phi_{\mathbf{k}}^{\mathrm{v}}. \tag{4.21}$$

where the pseudo-wavefunctions $\phi_{\mathbf{k}}^{\mathrm{v}}$ are just the plane wave (valence) part of the trial wavefunction expansion. In effect what has happened is that in going to a smoother wavefunction (e.g. the plane wave part of the OPW) the loss in kinetic energy in the atomic region has been compensated by a change to a weaker potential $W(\mathbf{r})$. This behaviour is shown schematically in Figure 4.7. The earlier first-principles calculation of pseudopotentials were largely supplanted by model potential methods (to be discussed below) as these tended to produce more reliable electronic properties (Harrison 1966; Cohen & Heine 1970). More recently there has been a trend back to ab-initio calculations based directly on information about the atoms concerned (Cohen 1984). We return to discussion of these when considering pseudopotentials in alloys (Section 4.5.3).

It should be noted that, unlike the local potential $V(\mathbf{r})$, which depends only upon position, the pseudopotential is non-local: it depends upon $\mathbf{k}, \mathbf{k}'$ and the angle between them, as well as the actual energy of the state concerned. However, this is not a problem in determining the resistivity since we are only interested in conduction electrons at the Fermi energy. By defining the scattering vector

$$\mathbf{q}=\mathbf{k}-\mathbf{k}', \tag{4.22}$$

this means that the pseudopotential appropriate to E_{F} depends only upon q/k_{F} (although it is only defined in the range $q/k_{\mathrm{F}}=0$ to 2) and can be treated as a local potential (Harrison 1966, p. 60). It should also be noted that the pseudopotential is not unique. In fact its very origin lies in the fact that one potential (e.g. the true lattice potential) can be replaced by another (e.g. the pseudopotential) without changing the scattering of conduction electrons. As we show in Section 4.4, some pseudopotentials seem to give better results than others, depending upon the actual property concerned.

The other significant fact which emerges from the pseudopotential concept is that the scattering potential is much weaker than the actual ion potential, thereby explaining why the band structures and Fermi surfaces of the simple metals are so nearly free electron in character. The weakness of this potential allows the use of perturbation theory and opens the way for the application of pseudopotentials to an enormous variety of solid state problems. The unperturbed pseudowavefunction is then just the free electron plane wave (or one-OPW). We can now see that the free electron Fermi surface shown in Figure 1.8 is in fact the unperturbed one-OPW Fermi surface, and that if $W(\mathbf{r})=0$ the free electron model (sometimes called the 'empty lattice') is reproduced exactly. Such a perturbation approach is not justified for the transition metals, rare-earths or actinides with incomplete inner d- or f-shells because of the greater coupling between the core and valence states. This aspect of the problem is considered in more detail in Chapter 6.

Heine & Abarenkov (1964) employed an alternative formulation of essentially the same problem. This is the model-potential method in

Fig. 4.7. Schematic representation of the change in going from a real potential $V(\mathbf{r})$ and wavefunction $\psi(\mathbf{r})$ to a pseudopotential $W(\mathbf{r})$ and pseudowavefunction $\phi(\mathbf{r})$.

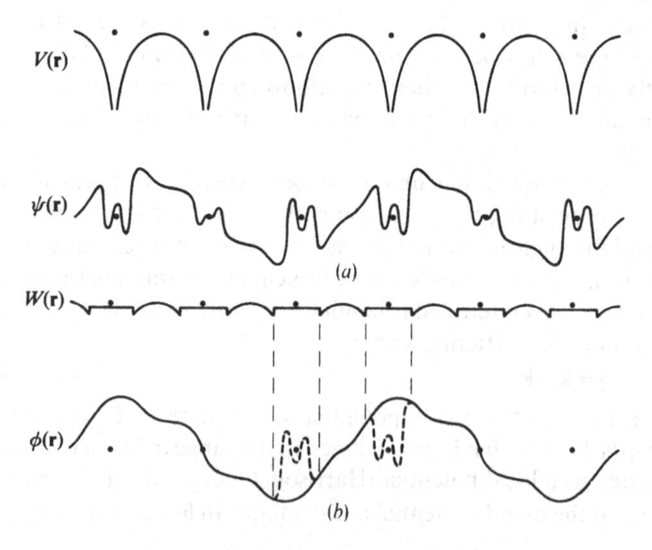

which the actual potential within a small sphere about each ion is replaced by a model potential. The value and slope of the wavefunction calculated for the model potential is matched at the surface of the sphere to the valence wavefunction outside the sphere. The potential is chosen so that there are no nodes in the wavefunction outside the sphere and, like the pseudopotential, is non-local and energy dependent. Thus the same wavefunction and energy of the valence electrons is reproduced even though the true ion potential has been replaced by a weak model potential. The corresponding model- or pseudowavefunction is then smooth and free-electron-like throughout the crystal, as shown in Figure 4.8. The advantage of this method lies in the fact that the parameters of the model potential may be obtained by fitting to experimental data such as the spectroscopic levels of the free atom or liquid metal resistivities, thereby avoiding the complexity and uncertainty of computing core wavefunctions and self-consistent potentials. There have been an enormous number of model potentials proposed, some of which are:

Point-ion (δ-function):

$$w(r) = -\frac{Ze^2}{4\pi\varepsilon_0 r} + \beta\,\delta(r), \tag{4.23}$$

Fig. 4.8. The model potential $W(r)$ is adjusted so that the pseudowavefunction ϕ within the atomic sphere matches the true wavefunction ψ at the surface of the sphere.

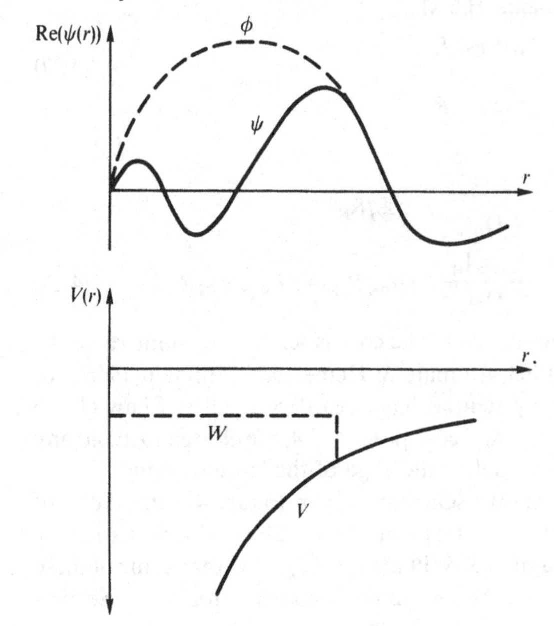

The corresponding 'form factor' to be discussed in Section 4.5.2 is

$$\langle \mathbf{k}+\mathbf{q}|w(\mathbf{r})|\mathbf{k}\rangle = \frac{1}{\Omega_0}\left(-\frac{Ze^2}{\varepsilon_0 q^2}+\beta\right), \tag{4.24a}$$

where Ω_0 is the effective volume per ion Ω/N. In this model the ion is reduced to a point charge and only the Coulomb potential due to the net ion charge Ze remains. The repulsion of the core states is represented by the delta-function of strength β. This potential is often modified by a damping term $\exp(-r/R_\mathrm{d})$ so that it tends to zero at large q as required for a true pseudopotential (Harrison 1966):

$$\langle \mathbf{k}+\mathbf{q}|w(\mathbf{r})|\mathbf{k}\rangle = \frac{1}{\Omega_0}\left(-\frac{Ze^2}{\varepsilon_0 q^2}+\frac{\beta}{(1+q^2R_\mathrm{d}^2)^2}\right). \tag{4.24b}$$

Empty-core (Ashcroft):

$$\begin{aligned} w(r) &= 0 && \text{for } r<R_\mathrm{c} \\ &= -\frac{Ze^2}{4\pi\varepsilon_0 r} && \text{for } r>R_\mathrm{c}, \end{aligned} \tag{4.25}$$

and

$$\langle \mathbf{k}+\mathbf{q}|w(\mathbf{r})|\mathbf{k}\rangle = -\frac{1}{\Omega_0}\frac{Ze^2}{\varepsilon_0 q^2}\cos(qR_\mathrm{c}). \tag{4.26}$$

Here the pseudopotential of the core is set equal to zero (Ashcroft, 1966).

Heine–Abarenkov–Animalu (HAA):

$$\begin{aligned} w(r) &= -A_\mathrm{H} && \text{for } r<R_\mathrm{H} \\ &= -\frac{Ze^2}{4\pi\varepsilon_0 r} && \text{for } r>R_\mathrm{H}, \end{aligned} \tag{4.27}$$

and

$$\begin{aligned} \langle \mathbf{k}+\mathbf{q}|w(\mathbf{r})|\mathbf{k}\rangle = &-\frac{Ze^2}{\Omega_0\varepsilon_0 q^2}\cos(qR_\mathrm{H}) \\ &-\frac{A_\mathrm{H}}{\Omega_0\varepsilon_0 q^2}(\sin(qR_\mathrm{H})-qR_\mathrm{H}\cos(qR_\mathrm{H})). \end{aligned} \tag{4.28}$$

In this case the pseudopotential of the core is set at a constant value A_H (Heine & Abarenkov 1964; Animalu & Heine 1965; Animalu 1966). An 'optimised' form of this potential has been described by Shaw (1968) whereby a cut-off radius R_s and core potential A_s are chosen to avoid any discontinuity in the potential at the edge of the 'core' region.

These potentials are shown schematically in Figure 4.9 and tables of parameters are available in Harrison (1966), Shaw (1968), Cohen & Heine (1970), and Appapillai & Williams (1973) (note that some of these relate to screened potentials rather than bare-ion potentials – see Section

4.3), and some of the different models have been discussed by Gohel *et al.* (1984). The effect of different form factors $\langle \mathbf{k}+\mathbf{q}|w(\mathbf{r})|\mathbf{k}\rangle$ on the calculated resistivity will be discussed in Chapter 5. Note that, in their full non-local forms, the parameters characterising the above potentials will in general depend upon the angular quantum number and energy of the particular electron states concerned.

4.3 Electron–electron interactions

In Section 4.1 we introduced the concept of Hartree screening whereby the effective charge of the nucleus as seen by an electron in an outer shell was reduced (screened) by the inner shell negative charge distribution. This was taken into account in the Hartree self-consistent field calculation. However, this correction describes the effects of other electrons only through their average positions weighted by their wavefunctions, and does not take into account the exchange and correlation effects that result directly from electron–electron interactions. Furthermore, in a metallic conductor the screening effect of the conduction electrons will need to be considered. Fortunately these complicated many-body effects can be incorporated in an approximate but fairly straightforward manner.

Fig. 4.9. The Heine–Abarenkov–Animalu (HAA) (*a*), Shaw (*b*), point-ion (*c*) and empty-core (Ashcroft) (*d*) model potentials.

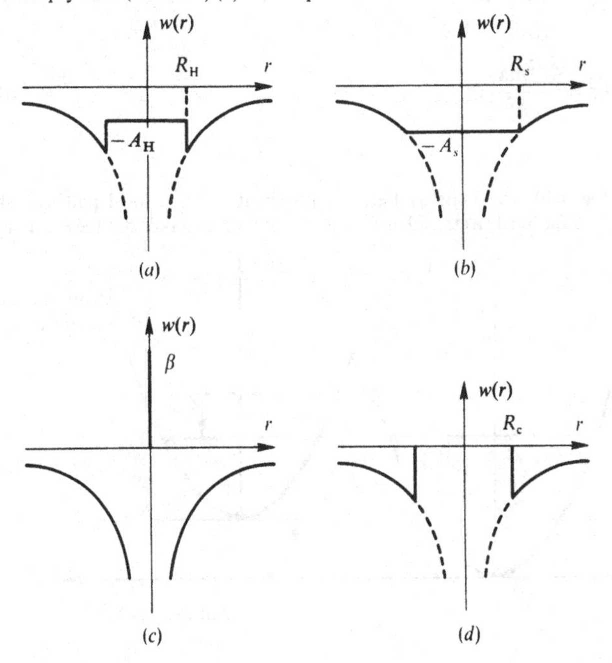

4.3.1 *Screening in metals*

The effects of screening by conduction electrons can be readily visualised with the aid of the Thomas–Fermi approximation for a free electron gas. If a local repulsive potential is present in the electron gas, the local electron density will be decreased to maintain a uniform Fermi energy since it is a chemical potential and must be the same at all places. This behaviour is shown in Figure 4.10.

In general, the effects of screening may be included within the self-consistent field framework by dividing the Fourier components of the bare unscreened potential $V_b(\mathbf{q})$ by the dielectric constant $\varepsilon(\mathbf{q})$ (Ashcroft & Mermin 1976, p. 337; Harrison 1970, p. 280; Ziman 1972, p. 146):

$$V_s(\mathbf{q}) = \frac{V_b(\mathbf{q})}{\varepsilon(\mathbf{q})}, \tag{4.29}$$

(we only consider here 'static', i.e. zero frequency perturbations). If the potential is a slowly varying function of position (on a scale of the Fermi wavelength), the Thomas–Fermi approach gives

$$\varepsilon(\mathbf{q}) = 1 + \frac{k_0^2}{q^2}, \tag{4.30}$$

where k_0 is the Thomas–Fermi screening parameter (or wavevector) given by

$$k_0^2 = 4\pi e^2 N(E_F). \tag{4.31a}$$

For free electrons:

$$k_0^2 = \frac{4e^2 m k_F}{\pi \hbar^2}. \tag{4.31b}$$

Fig. 4.10. The Thomas–Fermi approximation for a local potential showing how the local electron density will adjust itself to screen out the extra potential.

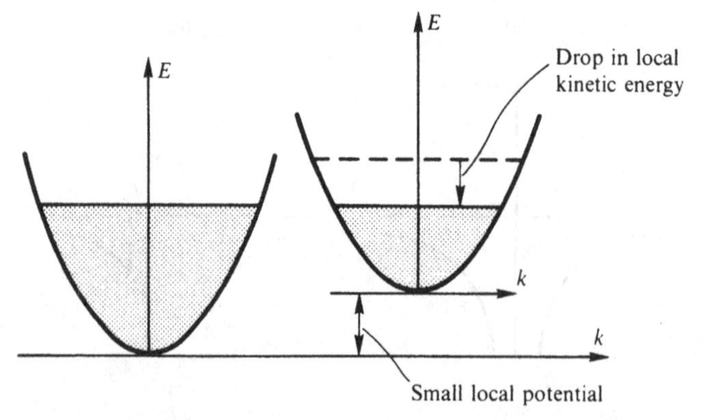

If q and k_F are given in Å, k_0^2 has the value $2.679k_F$ (Å^2). As an illustration we consider the Coulomb potential of an electron at distance r from a point charge of magnitude Ze:

$$V_b(\mathbf{r})=\frac{Ze^2}{4\pi\varepsilon_0|\mathbf{r}|}. \tag{4.32}$$

Fourier transforming (with the aid of equation (2.22) and a convergence factor $\exp(-\mu r)$ which may later be taken equal to unity) gives

$$V_b(\mathbf{q})=\frac{Ze^2}{\Omega_0\varepsilon_0 q^2}. \tag{4.33}$$

Dividing by the dielectric constant gives the screened potential

$$V_s(\mathbf{q})=\frac{Ze^2}{\Omega_0\varepsilon_0(q^2+k_0^2)}. \tag{4.34}$$

Fourier transforming back into direct space then gives

$$V_s(\mathbf{r})=\frac{Ze^2}{4\pi\varepsilon_0|\mathbf{r}|}\exp(-k_0 r), \tag{4.35}$$

i.e. the effective potential is the original Coulomb potential damped by the term $\exp(-k_0 r)$. This is called a screened Coulomb (or Yukawa) potential. The limit of the slow rate of change in potential in the Thomas–Fermi method renders the results reliable only at small values of q ($<k_F$). If the potential is weak the Lindhard method (or random phase approximation, RPA) can be used to obtain an exact solution for Hartree screening by an electron gas up to terms of linear order in the potential. First-order perturbation theory then leads to a more general expression for $\varepsilon(q)$ at $T=0$ known variously as the static Hartree, random phase approximation (RPA) or Lindhard dielectric function

$$\varepsilon(q)=1-4\pi e^2\frac{\chi(q)}{q^2}, \tag{4.36}$$

where $\chi(q)$ is the perturbation characteristic given by

$$\chi(q)=-\frac{N(E_F)}{2}\left[\frac{1-x^2}{2x}\ln\left|\frac{1+x}{1-x}\right|+1\right], \tag{4.37}$$

and $x=q/2k_F$. This dielectric function is shown plotted in Figure 4.11 for a monovalent fcc metal and has a logarithmic singularity at $q=2k_F$. As a result of this singularity the screened potential of a point charge at large distances (and $T=0$) goes as (Harrison 1970, p. 300):

$$V(r)\sim\frac{\cos(2k_F r)}{r^3}. \tag{4.38}$$

These oscillations (known as Friedel or Ruderman–Kittel oscillations) will be damped by electron scattering and should decay as $\exp(-r/\Lambda)$.

Furthermore, the singularity in $\varepsilon(q)$ arises from the abrupt energy cut-off at $T=0$ but at finite temperatures the Fermi surface is not so sharp and this will also damp the oscillations (Khannanov 1977). Note that this fine structure is absent in the semiclassical Thomas–Fermi screening parameter. While this discussion has been concerned with weak local potentials, Harrison shows that it may be applied equally well to local and non-local pseudopotentials (Harrison 1970, p. 301).

4.3.2 *Exchange and correlation*

As noted above the Hartree approximation neglects any direct electron–electron interactions. In particular, the neglect of the requirement of the Pauli exclusion principle for antisymmetric wave-functions is corrected in the Hartree–Fock approximation, which gives an extra 'exchange' term in the energy (cf. equation (4.7)). Correlation effects arise from the Coulomb repulsion between electrons, leading to the concept of a 'correlation hole' around each electron which excludes other electrons. As in the case of screening, there have been a number of calculation schemes proposed. A number of simple forms incorporate such corrections into the potential via a modified dielectric function (see e.g. Kleinman 1967, 1968):

$$\varepsilon(q) = 1 - \frac{4\pi e^2}{q^2}(1 - f(q))\chi(q), \tag{4.39}$$

where $f(q)$ is the exchange and correlation correction. Some of the proposed forms for $f(q)$ are given in Table 4.1. However, as will be shown in Chapter 5, the resistivity is not especially sensitive to the form

Fig. 4.11. Static Hartree (RPA, Lindhard) dielectric function $\varepsilon(q)$. There is a logarithmic singularity at $q = 2k_F$.

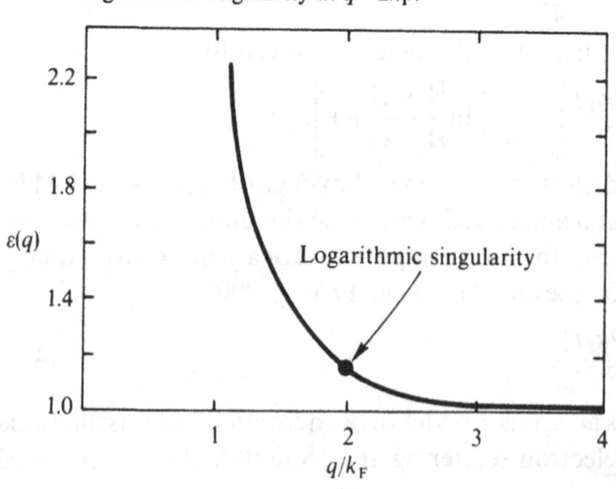

Table 4.1. *Various forms of the exchange and correlation correction $f(q)$*

Lindhard: $f(q)=0$
(RPA, Hartree)
Hubbard–Sham: $f(q)=\frac{1}{2}\left(\frac{q^2}{q^2+k_F^2+k_0^2}\right)$
(Hubbard 1957*a,b*; Sham 1965)
Kleinman: $f(q)=\frac{1}{4}\left[\frac{q^2}{q^2+k_F^2+k_S^2}+\frac{q^2}{q^2+k_S^2}\right]$
(Kleinman 1967, 1969; Langreth 1969)
Singwi: $f(q)=A\left[1-\exp\left[-B\left(\frac{q}{k_F}\right)^2\right]\right]$
$A=0.8994$
$B=0.3401$
(Singwi *et al.* 1968, 1970)
If q is given in terms of $Å^{-1}$: $k_0^2=2.679k_F$ $(Å^{-2})$
$k_S^2=1.95k_F$ $(Å^{-2})$

of correction used, particularly when the uncertainty in some of the other parameters involved in the calculation (such as the pseudopotential itself) are considered.

Calculation of atomic properties which depend more directly upon the core states, such as phonon dispersion curves or the effective pair potentials, will be more sensitive to exchange and correlation effects. In this respect we mention the $X\alpha$ local statistical exchange approximation of Slater (1972) in which the complicated Hartree–Fock exchange contribution is replaced by a local potential:

$$V_{X\alpha}(\mathbf{r})=-6\alpha(\tfrac{3}{8}\pi\rho(\mathbf{r}))^{1/3}, \tag{4.40}$$

where $\rho(\mathbf{r})$ is the local electronic charge density and α a scaling parameter (see also Hafner 1975*a*). As such this could also be called a 'local charge density' approximation. In systems that support the splitting of bands into spin-up and spin-down sub-bands, the effects of exchange and correlation are often taken into account by a generalisation of this approach, called the local spin density approximation (Gunnarson & Lundqvist 1976; Hohenberg & Kohn 1964; Poulson *et al.* 1976; Moruzzi *et al.* 1977). These approximations have been mentioned in passing as the reader may encounter such terms in association with the description of potentials in some studies.

Other aspects of the problem of exchange and correlation are discussed by Harrison (1970, ch. 4), Singwi & Tosi (1981) and Callaway & March (1984). The influence of different forms of the exchange and correlation correction have been investigated by Hafner (1973, 1975*b*), King & Cutler (1970) and Behari (1973).

4.4 Nearly free electron theory

We are now in a position to establish the formal framework for the scattering of the valence electrons by the lattice potential. We adopt the pseudopotential concept of planewave electron states and a weak scattering potential. The weakness of the potential allows for a perturbation expansion of the energy and wavefunctions:

$$\psi = \psi_0 + \psi_1 + \psi_2 + \cdots, \tag{4.41}$$

$$E = E_0 + E_1 + E_2 + \cdots, \tag{4.42}$$

where the zeroth-order terms ψ_0 and E_0 are the free electron values, the first-order terms ψ_1 and E_1 are comparable with the potential $W(\mathbf{r})$, the second-order terms ψ_2 and E_2 are comparable with $W(\mathbf{r})^2$, and so on. The wavefunctions and energies have been written here without the argument $\mathbf{k}$ in order to simplify notation. These may be substituted into the Schrödinger equation (equation (4.5b)) and the result separated by orders, assuming that $W(\mathbf{r})$ is of first-order smallness in comparison with the kinetic energy $(-\hbar^2/2m)\,\nabla^2\psi_0$:

$$-\frac{\hbar^2}{2m}\nabla^2\psi_0 = E_0\psi_0 \tag{4.43a}$$

$$-\frac{\hbar^2}{2m}\nabla^2\psi_1 + W(\mathbf{r})\psi_0 = E_0\psi_1 + E_1\psi_0 \tag{4.43b}$$

$$-\frac{\hbar^2}{2m}\nabla^2\psi_2 + W(\mathbf{r})\psi_1 = E_0\psi_2 + E_1\psi_1 + E_2\psi_0. \tag{4.43c}$$

Solution of the zeroth-order equation (4.43a) gives immediately the free electron result

$$\psi_0 = \Omega^{-1/2}\exp(i\mathbf{k}\cdot\mathbf{r}), \tag{4.44}$$

$$E_0 = \frac{\hbar^2 k^2}{2m}. \tag{4.45}$$

To obtain the first-order terms, the wavefunction is expanded in planewaves

$$\psi = \sum A_n \exp(i\mathbf{k}_n\cdot\mathbf{r}), \tag{4.46}$$

where the wavevector $\mathbf{k}_n = \mathbf{k} + \mathbf{g}_n$ and $\mathbf{g}_n$ is a reciprocal lattice vector (cf. equation (4.15)). The $n = 0$ term is just ψ_0 if A_0 is taken as unity. The remaining coefficients are determined by substitution into the first-order equation (4.43b), multiplying on the left by the complex conjugate $\exp(-i\mathbf{k}_n\cdot\mathbf{r})$ and integrating to give

$$A_n = \frac{\int_\Omega \psi_n^* W(\mathbf{r})\psi_0\, d\mathbf{r}}{E_0 - E_n}, \tag{4.47}$$

and hence

$$E_1 = \int_\Omega \psi_0^* W(\mathbf{r})\psi_0\, d\mathbf{r}. \tag{4.48}$$

The zero- and first-order wavefunctions are substituted into the second-order equation (4.43c) to obtain E_2 leading to the result

$$E(\mathbf{k}) = \frac{\hbar^2 k^2}{2m} + \int_\Omega \psi_0^* W(\mathbf{r})\psi_0\, d\mathbf{r} + \sum_n{}' \frac{\int \psi_0^* W(\mathbf{r})\psi_n\, d\mathbf{r} \int \psi_n^* W(\mathbf{r})\psi_0\, d\mathbf{r}}{E_0 - E_n}. \tag{4.49}$$

This is more conveniently written in the Dirac notation

$$E(\mathbf{k}) = \frac{\hbar^2 k^2}{2m} + \langle\mathbf{k}|W(\mathbf{r})|\mathbf{k}\rangle + \sum_n{}' \frac{\langle\mathbf{k}+\mathbf{g}_n|W(\mathbf{r})|\mathbf{k}\rangle\langle\mathbf{k}|W(\mathbf{r})|\mathbf{k}+\mathbf{g}_n\rangle}{E_0 - E_n}, \tag{4.50}$$

where

$$\langle\mathbf{k}+\mathbf{g}_n|W(r)|\mathbf{k}\rangle = \frac{1}{\Omega}\int_\Omega \exp[-i(\mathbf{k}+\mathbf{g}_n)\cdot\mathbf{r}]W(\mathbf{r})\exp(-i\mathbf{k}\cdot\mathbf{r})\, d\mathbf{r}. \tag{4.51}$$

In both equations (4.49) and (4.50) the prime on the summation indicates that $n \neq 0$.

Note also that $W(\mathbf{r})$ is in general a non-local operator and so the factors in the integrands to not commute. However, since this pseudopotential must have the same periodicity as the lattice it can be expressed as a Fourier series

$$W(\mathbf{r}) = \sum_n W_n \exp(-i\mathbf{g}_n\cdot\mathbf{r}), \tag{4.52}$$

where the summation is over all reciprocal lattice sites and the Fourier components W_n are given by the inverse transform

$$W_n = \frac{1}{\Omega}\int_\Omega W(\mathbf{r})\exp(i\mathbf{g}_n\cdot\mathbf{r})\, d\mathbf{r}. \tag{4.53}$$

Substitution of (4.52) into (4.50) gives, after some manipulation (see e.g. Wilkes 1973, p. 331),

$$E(\mathbf{k}) = E_0 + W_0 + \sum_n{}' \frac{W_n^* W_n}{E_0 - E_n}. \tag{4.54}$$

The second term in (4.49) or (4.50) has been written as W_0 since it simply represents a shift in the mean energy of the system . There are two conditions that must be satisfied for this expansion to be valid: (i) the summation should rapidly converge and (ii) there should be no degeneracy of the form $E_0 = E_n$. The first condition requires the potential to be weak and reasonably smooth, as is likely to be the case with a screened pseudopotential. The second condition may be understood by considering a one-dimensional model, as shown in Figure 4.12. Free electron E–k curves are drawn around each reciprocal lattice site and, provided that $k \neq \mathbf{g}_n/2$, the difference $E_0 - E_n$ is large. This argument is easily extended to three dimensions and shows that $E_0 = E_n$ when $|\mathbf{k}| = |\mathbf{k} \pm \mathbf{g}_n|$, i.e. $\mathbf{k}$ lies on the perpendicular bisector of $\mathbf{g}_n$ which, as shown in Figure 4.13, is just the Brillouin zone boundary.

It is interesting to note that this is exactly the Ewald construction for coherent Bragg diffraction for an electron of wavevector $\mathbf{k}$ to be diffracted into the direction $\mathbf{k}' = \mathbf{k} - \mathbf{g}_n$. This shows that the valence electrons are indeed diffracted by the crystal as suggested in Chapter 1. It is common to extend these equations to show that energy gaps $2|E_n|$ are introduced into the free electron E–k curves at the Brillouin zone boundaries, as shown in Figure 4.14, and that the incident and diffracted beams give a stationary (i.e. not propagating) state (see e.g. Ziman 1972, p. 80). These results show that the one-OPW Fermi surfaces shown in

Fig. 4.12. E–k curves drawn around reciprocal lattice sites to illustrate the connection between k and $E_0 - E_n$.

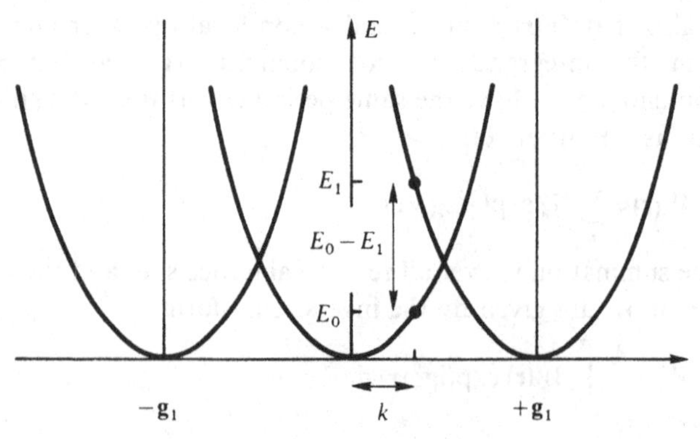

Figure 1.8 would suffer distortions at the Brillouin zone boundaries if considered to second order in the perturbation expansion. We might also note that exactly the same formulas and constructions arise in the dynamical theory of electron diffraction.

4.5 The scattering matrix

We come now to a more detailed consideration of the form of the scattering matrix. In Chapter 1 this was introduced as $\langle\Psi_{\mathbf{k}'}|V(\mathbf{r})|\Phi_{\mathbf{k}}\rangle$ where $\Phi_{\mathbf{k}}$ is the wavefunction of an electron that was in some initial state $\Psi_{\mathbf{k}}$ but which is now in the presence of the scattering potential $V(\mathbf{r})$ and so is given by the solution of the time-independent Schrödinger equation, while $\Psi_{\mathbf{k}'}$ is the wavefunction of the freely propagating scattered wave. In a periodic solid $\Psi_{\mathbf{k}'}$ and $\Psi_{\mathbf{k}}$ must be Bloch states. However, because of the effects of the perturbation $V(\mathbf{r})$, $\Psi_{\mathbf{k}'}$ and $\Phi_{\mathbf{k}}$ cannot both be basis vectors in the same representation and so

Fig. 4.13. Construction for the divergence in perturbation series.

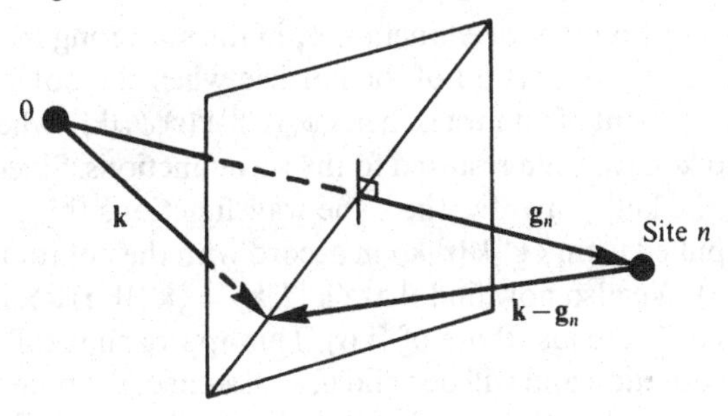

Fig. 4.14. Nearly free electron energies in one dimension. The solid curve gives the energies in the extended zone scheme and the dotted curve shows the upper branches redrawn in the reduced zone scheme.

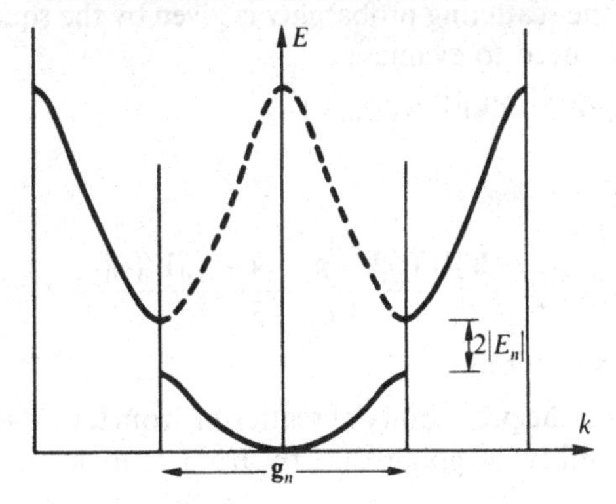

$\langle\Psi_{\mathbf{k}'}|V(\mathbf{r})|\Phi_{\mathbf{k}}\rangle$ is strictly not a matrix element of the potential $V(\mathbf{r})$ (see e.g. Messiah 1961, p. 806). While this may be overcome by the assumption that the potential is weak, it is opportune to introduce at this stage the so-called Transition or *T*-matrix which is defined by

$$\begin{aligned} T_{\mathbf{kk}'} &= \langle\Psi_{\mathbf{k}'}|T|\Psi_{\mathbf{k}}\rangle \\ &= \langle\Psi_{\mathbf{k}'}|V(\mathbf{r})|\Phi_{\mathbf{k}}\rangle. \end{aligned} \tag{4.55}$$

It is assumed that the scattering is elastic so that $\Psi_{\mathbf{k}'}$ and $\Psi_{\mathbf{k}}$ are both Bloch states of the same energy. We now investigate the approximations that follow from the assumption of a weak scattering potential $V(\mathbf{r})$.

4.5.1 *The first and second Born approximations*

From equations (4.46) and (4.47), the wavefunction $\psi_{\mathbf{k}}$ perturbed by the weak potential $W(\mathbf{r})$ and written as $\phi_{\mathbf{k}}$ is given to first order in the perturbation expansion by

$$\phi_{\mathbf{k}} = \psi_0 + \sum_n{}' \psi_n \frac{\int \psi_n W(\mathbf{r}) \psi_0}{E_0 - E_n} \, \mathrm{d}\mathbf{r}. \tag{4.56}$$

As a first approximation the wavefunction $\psi_{\mathbf{k}}$ in the scattering matrix is replaced by ψ_0, the wavefunction of the particle when the potential is zero. The matrix element of interest is then $\langle\psi_0(\mathbf{k}')|W(\mathbf{r})|\psi_0(\mathbf{k})\rangle$ where the arguments $\mathbf{k}$ and $\mathbf{k}'$ have been restored to the wavefunctions. Since both the bra $\langle\ |$ and the ket$|\ \rangle$ involve the same wavefunctions this may be written in a simplified form $\langle\mathbf{k}'|W(\mathbf{r})|\mathbf{k}\rangle$ in accord with the notation used in equation (4.51). We also now find that $\langle\mathbf{k}'|T|\mathbf{k}\rangle = \langle\mathbf{k}'|W(\mathbf{r})|\mathbf{k}\rangle$, i.e. the matrix elements of T are just those of $W(\mathbf{r})$. This approach is called the first Born approximation and will be valid if, as assumed, the potential is much weaker than the 'kinetic energy' of the electrons $(-\hbar^2/2m)\nabla^2\psi_0(\mathbf{k})$. However, there are some special applications where it is necessary to go to the second Born approximation which includes the second term in equation (4.50). Since the scattering probability is given by the square of the matrix element, we need to evaluate

$$\langle\psi_0(\mathbf{k}')|W(\mathbf{r})|\psi(\mathbf{k})\rangle\langle\psi(\mathbf{k})|W(\mathbf{r})|\psi_0(\mathbf{k}')\rangle, \tag{4.57}$$

which gives

$$\begin{aligned} &|\langle\mathbf{k}'|W(\mathbf{r})|\mathbf{k}\rangle|^2 \\ &\quad + \left[\langle\mathbf{k}'|W(\mathbf{r})|\mathbf{k}\rangle \sum_n{}' \frac{\langle\mathbf{k}'|W(\mathbf{r})|\mathbf{k}+\mathbf{g}_n\rangle\langle\mathbf{k}+\mathbf{g}_n|W(\mathbf{r})|\mathbf{k}\rangle}{E_0 - E_n} + \mathrm{cc}\right] \\ &\quad + \text{higher-order terms} \end{aligned} \tag{4.58}$$

This result indicates that the probability of scattering from $\mathbf{k}$ to $\mathbf{k}'$ is now the sum of the probability of going directly from $\mathbf{k}$ to $\mathbf{k}'$ plus the probability of going indirectly via the intermediate states $\mathbf{k}_n$. If any of these intermediate states involve a vanishing energy denominator then a different approach is required, as discussed in Section 4.3 of Harrison (1966). When we consider the pseudopotential of an alloy in the next section we will find that the second approximation becomes even more complicated although there is valuable information to be gained.

4.5.2 *Factorisation of the matrix elements*

An electron at any point $\mathbf{r}$ will interact with all ions situated at the lattice sites $\mathbf{r}_i$ and so the pseudopotential $W(\mathbf{r})$ can be written as the sum or superposition of the spherically symmetric individual ion pseudopotentials (cf. equation (4.10)):

$$W(\mathbf{r}) = \sum_i w(\mathbf{r} - \mathbf{r}_i). \tag{4.59}$$

In the first Born approximation the required matrix elements are then given by (see e.g. equation (4.51))

$$\langle \mathbf{k}' | W(\mathbf{r}) | \mathbf{k} \rangle = \frac{1}{\Omega} \int \exp(-i\mathbf{k}' \cdot \mathbf{r}) \sum w(\mathbf{r} - \mathbf{r}_i) \exp(-i\mathbf{k} \cdot \mathbf{r})\, d\mathbf{r}. \tag{4.60}$$

After interchanging the summation and the integration this gives

$$\begin{aligned} \langle \mathbf{k}' | W(\mathbf{r}) | \mathbf{k} \rangle = & \frac{1}{N} \sum_i \exp[-i(\mathbf{k}' - \mathbf{k}) \cdot \mathbf{r}_i] \\ & \times \frac{1}{\Omega_0} \int \exp[-i\mathbf{k}' \cdot (\mathbf{r} - \mathbf{r}_i)] w(\mathbf{r} - \mathbf{r}_i) \\ & \times \exp[i\mathbf{k} \cdot (\mathbf{r} - \mathbf{r}_i)]\, d\mathbf{r} \end{aligned} \tag{4.61}$$

where we have used the relation $\Omega = N\Omega_0$. The final integral now contains a single potential and plane waves with all positions measured from the position of that ion. This expression may now be written as

$$\langle \mathbf{k} + \mathbf{q} | W(\mathbf{r}) | \mathbf{k} \rangle = S(\mathbf{q}) \langle \mathbf{k} + \mathbf{q} | w(\mathbf{r}) | \mathbf{k} \rangle, \tag{4.62}$$

where $S(\mathbf{q})$ is the structure factor

$$S(\mathbf{q}) = \frac{1}{N} \sum_{\mathbf{r}} \exp(-i\mathbf{q} \cdot \mathbf{r}). \tag{4.63}$$

and $\langle \mathbf{k} + \mathbf{q} | w(\mathbf{r}) | \mathbf{k} \rangle$ is the form factor.

This represents an enormous simplification. We have found that the scattering matrix for simple metals can be written as the product of a structure factor, that depends only upon the ion positions and will be given in terms of the parameters introduced in Chapter 2, and a form factor that is dependent only upon the ion type. This form factor is often referred to simply as the pseudopotential, written $w(\mathbf{q})$, and is available from a variety of first principle pseudopotential or model potential studies. As noted in Section 4.2, we are only concerned with electron states at the Fermi surface, in which case the form factor is given by a single continuous curve (as it would be for a local potential) and depends only upon q/k_F.

We will refer to some calculations of these in relation to specific metals in Chapter 5 (see, for example, Figure 5.30). The effects of screening, exchange and correlation are now simply included by dividing the form factor by the dielectric function, in accordance with equation (4.29).

4.5.3 *The pseudopotential in alloys*

The extension of the pseudopotential technique to alloys follows naturally from the definition of the pseudopotential in equation (4.59). If more than one atomic species is present the sum of ionic potentials becomes

$$W(\mathbf{r}) = \sum_{\mathbf{r}_i(\mathrm{A})} w_\mathrm{A}(\mathbf{r}-\mathbf{r}_i) + \sum_{\mathbf{r}_i(\mathrm{B})} w_\mathrm{B}(\mathbf{r}-\mathbf{r}_i) + \cdots, \tag{4.64}$$

where the summations are over the sites containing A, B and other atoms respectively. In what is essentially the virtual crystal approximation an average lattice potential is defined by

$$\bar{w}(\mathbf{r}) = c_\mathrm{A} w_\mathrm{A}(\mathbf{r}) + c_\mathrm{B} w_\mathrm{B}(\mathbf{r}). \tag{4.65}$$

With the aid of a difference or deviation potential defined by

$$w^\mathrm{d}(\mathbf{r}) = w_\mathrm{A}(\mathbf{r}) - w_\mathrm{B}(\mathbf{r}), \tag{4.66}$$

equation (4.59) then becomes

$$W(\mathbf{r}) = \sum_{\mathbf{r}_i} \bar{w}(\mathbf{r}-\mathbf{r}_i) + c_\mathrm{A} \sum_{\mathbf{r}_i(\mathrm{A})} w^\mathrm{d}(\mathbf{r}-\mathbf{r}_i) - c_\mathrm{B} \sum_{\mathbf{r}_i(\mathrm{B})} w^\mathrm{d}(\mathbf{r}-\mathbf{r}_i), \tag{4.67}$$

where the first summation is over the whole lattice. In effect this means that the actual potential at any site is constructed by adding an appropriate difference potential $c_\mathrm{B} w^\mathrm{d}(\mathbf{r})$ for an A atom or $-c_\mathrm{A} w^\mathrm{d}(\mathbf{r})$ for a B atom to the virtual crystal average potential at that site. It is these deviations in potential which give rise to a non-zero electrical resistivity. The required matrix elements can then be written as

$$\langle \mathbf{k}+\mathbf{q}|W(\mathbf{r})|\mathbf{k}\rangle = S(\mathbf{q})\langle \mathbf{k}+\mathbf{q}|\bar{w}(\mathbf{r})|\mathbf{k}\rangle + S^\mathrm{d}(\mathbf{q})\langle \mathbf{k}+\mathbf{q}|w^\mathrm{d}(\mathbf{r})|\mathbf{k}\rangle, \tag{4.68}$$

where the deviation lattice structure factor $S^\mathrm{d}(\mathbf{q})$ is given in terms of the partial structure factors $S^\mathrm{A}(\mathbf{q})$ and $S^\mathrm{B}(\mathbf{q})$:

$$S^\mathrm{d}(\mathbf{q}) = c_\mathrm{B} S^\mathrm{A}(\mathbf{q}) - c_\mathrm{A} S^\mathrm{B}(\mathbf{q}), \tag{4.69}$$

which describe the distribution of the A and B atoms and are defined by:

$$S^\mathrm{A}(\mathbf{q}) = \sum_{\mathbf{r}_i(\mathrm{A})} \exp(-i\mathbf{q}\cdot\mathbf{r}_i(\mathrm{A})), \tag{4.70}$$

and similarly for $S^{\mathrm{B}}(\mathbf{q})$. The first term in (4.68) is the pseudopotential of the perfect 'average' lattice with structure factor given by equation (4.63) which involves a sum over all ions, just as for a pure metal, and which is non-zero only for $\mathbf{q}=\mathbf{g}_n$. At such values of $\mathbf{q}$ the second term becomes

$$\left(\frac{c_{\mathrm{B}}}{N}\right)Nc_{\mathrm{A}}w^{\mathrm{d}}(\mathbf{g}_n)-\left(\frac{c_{\mathrm{A}}}{N}\right)Nc_{\mathrm{B}}w^{\mathrm{d}}(\mathbf{g}_n)=0, \tag{4.71}$$

and so gives no extra contribution to the scattering matrix. The matrix elements of the difference or deviation pseudopotential are consequently non-zero only when $\mathbf{q}\neq\mathbf{g}_n$. Thus when the square of the matrix element $|\langle\mathbf{k}+\mathbf{q}|W(\mathbf{r})|\mathbf{k}\rangle|^2$ is evaluated, two independent terms evolve: one due to the average lattice when $\mathbf{q}=\mathbf{g}_n$ and another due to the difference lattice when $\mathbf{q}\neq\mathbf{g}_n$. However, as shown in Section 5.6, the cross terms do not vanish if the atoms are allowed to be displaced from the lattice sites.

The above result may also be derived in terms of the site occupation parameters (equation (2.2)) by writing the pseudopotential as

$$W(\mathbf{r})=\sum_i\sum_{\kappa=1}^{n}\sigma_i^{\kappa}w_{\kappa}(\mathbf{r}-\mathbf{r}_i), \tag{4.72}$$

where κ refers to the atom type. For example, for a binary A–B alloy

$$W(\mathbf{r})=\sum_i[\sigma_i^{\mathrm{A}}w_{\mathrm{A}}(\mathbf{r}-\mathbf{r}_i)+\sigma_i^{\mathrm{B}}w_{\mathrm{B}}(\mathbf{r}-\mathbf{r}_i)]. \tag{4.73}$$

Following (4.60)–(4.62) the matrix elements of the pseudopotential may be written as

$$\langle\mathbf{k}+\mathbf{q}|w(\mathbf{r})|\mathbf{k}\rangle=\frac{1}{N}\sum_i\exp(-\mathrm{i}\mathbf{q}\cdot\mathbf{r}_i)[\sigma_i^{\mathrm{A}}w_{\mathrm{A}}(\mathbf{q})+\sigma_i^{\mathrm{B}}w_{\mathrm{B}}(\mathbf{q})], \tag{4.74}$$

where the form factors $w_{\mathrm{A}}(\mathbf{q})$ and $w_{\mathrm{B}}(\mathbf{q})$ are defined by analogy with (4.61) by

$$\begin{aligned}w_{\mathrm{A}}(\mathbf{q})&=\langle\mathbf{k}+\mathbf{q}|w_{\mathrm{A}}(\mathbf{r})|\mathbf{k}\rangle\\&=\frac{1}{\Omega_0}\int\exp[-\mathrm{i}(\mathbf{q}+\mathbf{k})\cdot\mathbf{r}]w_{\mathrm{A}}(\mathbf{r})\exp(\mathrm{i}\mathbf{k}\cdot\mathbf{r})\,\mathrm{d}\mathbf{r},\end{aligned} \tag{4.75}$$

and similarly for $w_{\mathrm{B}}(\mathbf{q})$. As before, this may be written in terms of a perfect average lattice and a difference lattice. If the average lattice potential is defined by

$$\begin{aligned}\langle\mathbf{k}+\mathbf{q}|\bar{w}(\mathbf{r})|\mathbf{k}\rangle&=\sum_i\langle\sigma_i^{\mathrm{A}}w_{\mathrm{A}}(\mathbf{q})+\sigma_i^{\mathrm{B}}w_{\mathrm{B}}(\mathbf{q})\rangle\exp(-\mathrm{i}\mathbf{q}\cdot\mathbf{r}_i)\\&=\sum_i[\langle\sigma_i^{\mathrm{A}}\rangle w_{\mathrm{A}}(\mathbf{q})+\langle\sigma_i^{\mathrm{B}}\rangle w_{\mathrm{B}}(\mathbf{q})]\exp(-\mathrm{i}\mathbf{q}\cdot\mathbf{r}_i)\\&=(c_{\mathrm{A}}w_{\mathrm{A}}(\mathbf{q})+c_{\mathrm{B}}w_{\mathrm{B}}(\mathbf{q}))\sum_i\exp(-\mathrm{i}\mathbf{q}\cdot\mathbf{r}_i),\end{aligned} \tag{4.76}$$

the difference potential is then

$$\langle \mathbf{k}+\mathbf{q}|w^{\mathrm{d}}(\mathbf{r})|\mathbf{k}\rangle = \langle \mathbf{k}+\mathbf{q}|w_{\mathrm{A}}|\mathbf{k}\rangle - \langle \mathbf{k}+\mathbf{q}|w_{\mathrm{B}}|\mathbf{k}\rangle, \tag{4.77}$$

i.e.

$$\left.\begin{aligned} w_{\mathrm{A}}(\mathbf{q}) &= \bar{w}(\mathbf{q}) + c_{\mathrm{B}} w^{\mathrm{d}}(\mathbf{q}) \\ w_{\mathrm{B}}(\mathbf{q}) &= \bar{w}(\mathbf{q}) - c_{\mathrm{A}} w^{\mathrm{d}}(\mathbf{q}) \end{aligned}\right\}, \tag{4.78}$$

where $\bar{w}(\mathbf{q})$ and $w^{\mathrm{d}}(\mathbf{q})$ are defined in the same way as $w_{\mathrm{A}}(\mathbf{q})$ and $w_{\mathrm{B}}(\mathbf{q})$. From the above we find

$$\begin{aligned} \langle \mathbf{k}+\mathbf{q}|W(\mathbf{r})|\mathbf{k}\rangle &= \frac{1}{N}\sum_i \exp(-\mathrm{i}\mathbf{q}\cdot\mathbf{r}_i)\langle \mathbf{k}+\mathbf{q}|\bar{w}(\mathbf{r})|\mathbf{k}\rangle \\ &\quad + \sum_i \exp(-\mathrm{i}\mathbf{q}\cdot\mathbf{r}_i)(\langle \sigma_i^{\mathrm{B}}\rangle - \sigma_i^{\mathrm{B}})\langle \mathbf{k}+\mathbf{q}|w^{\mathrm{d}}(\mathbf{r})|\mathbf{k}\rangle \\ &= \frac{1}{N}\sum_i \exp(-\mathrm{i}\mathbf{q}\cdot\mathbf{r}_i)\langle \mathbf{k}+\mathbf{q}|\bar{w}(\mathbf{r})|\mathbf{k}\rangle \\ &\quad + \sum_i \exp(-\mathrm{i}\mathbf{q}\cdot\mathbf{r}_i)\sigma_i\langle \mathbf{k}+\mathbf{q}|w^{\mathrm{d}}(\mathbf{r})|\mathbf{k}\rangle. \end{aligned} \tag{4.79}$$

Again the first term is only non-zero for $\mathbf{q}=\mathbf{g}_n$, while the second term is only non-zero for $\mathbf{q}\neq\mathbf{g}_n$ and so the square of the matrix element becomes

$$\begin{aligned} &|\langle \mathbf{k}+\mathbf{q}|W(\mathbf{r})|\mathbf{k}\rangle|^2 \\ &= \frac{1}{N^2}\sum_i\sum_j \exp(-\mathrm{i}\mathbf{q}\cdot(\mathbf{r}_i-\mathbf{r}_j))|\langle \mathbf{k}+\mathbf{q}|w(\mathbf{r})|\mathbf{k}\rangle|^2 \\ &\quad + \frac{1}{N^2}\sum_i\sum_j \exp(-\mathrm{i}\mathbf{q}\cdot(\mathbf{r}_i-\mathbf{r}_j))(\sigma_i\sigma_j)|\langle \mathbf{k}+\mathbf{q}|w^{\mathrm{d}}(\mathbf{r})|\mathbf{k}\rangle|^2. \end{aligned} \tag{4.80}$$

This may be expressed as a sum over all terms involving pairs of atoms having the same spacing and then summing over all spacings to give

$$\begin{aligned} &|\langle \mathbf{k}+\mathbf{q}|W(\mathbf{r})|\mathbf{k}\rangle|^2 \\ &= \frac{1}{N}\sum_{ij} \exp(\mathrm{i}\mathbf{q}\cdot(\mathbf{r}_i-\mathbf{r}_j))|\langle \mathbf{k}+\mathbf{q}|w(\mathbf{r})|\mathbf{k}\rangle|^2 \\ &\quad + \frac{1}{N}\sum_{ij} \exp(-\mathrm{i}\mathbf{q}\cdot(\mathbf{r}_i-\mathbf{r}_j))\langle \sigma_i\sigma_j\rangle|\langle \mathbf{k}+\mathbf{q}|w^{\mathrm{d}}(\mathbf{r})|\mathbf{k}\rangle|^2, \end{aligned} \tag{4.81}$$

where $\langle\sigma_i\sigma_j\rangle$ is now the average of the product $(\sigma_i\sigma_j)$, as discussed in Section 2.2.

Screening is again taken into account by dividing the appropriate form factors by the dielectric function, remembering that it must now be evaluated using parameters (in particular, k_{F}) appropriate to the *alloy* of interest. Note that some pure metal form factors are only available already screened. If these were directly used to derive the difference pseudopotential for an alloy the screening parameters would be incorrect as they relate to the atomic volume and Fermi wavevector of the pure metal. They can be approximately 'rescreened' for the alloy by multiplying by $\varepsilon(q)^m/\varepsilon(q)^a$ where $\varepsilon(q)^m$ and $\varepsilon(q)^a$ are the Lindhard dielectric constants appropriate to the pure metal and alloy respectively. Similarly, corrections to the atomic volume may be taken into account by multiplying by Ω^m/Ω^a. However, there is still a problem in choosing suitable form factors (even if unscreened), since these should also be determined for atomic environments appropriate to the alloy rather than the pure metal constituents. This is because the core states will in general be influenced by the electron density, particularly if they are not well separated in energy from the valence states (Ball & Islam 1980), and because of the complicated effects of charging whereby the electron wavefunction deposits different amounts of electron charge in the vicinity of different atoms due to their different valencies and atomic volumes. This leads to further complications in determination of the many-body electron–electron effects (Pines 1955; Falicov & Heine 1961; Stern 1966, 1968, 1970). As a consequence, a complete recalculation of the form factor should be performed for each alloy composition (Gupta 1968; Hafner 1976). This represents a severe limitation to the application of such *ab-initio* form factors, although, provided that the differences in valence and atomic volume between the constituents are small, simply taking the pure metal form factors may be a reasonable first approximation. If a study is aimed at determining the effects of a *small* change in atomic distribution, a better approach may be to model the difference potential $w^d(\mathbf{r})$ directly using one of the forms described in Section 4.2, for example, and adjust the parameters of the model so that it gives results consistent with the homogeneous disordered solid solution (Hillel 1970, Tanigawa & Doyama 1973). Note that a difference potential so obtained may depend upon the actual physical property being investigated, so that a resistivity study may produce a different $w^d(\mathbf{r})$ to an elastic modulus study, for example (Cohen & Heine 1970; Gohel *et al.* 1984). However, there is still the complication that a large change in atomic distribution is likely to affect the scattering form factor. In addition to the reasons described above this will also occur because the mean virtual crystal potential $\bar{w}(\mathbf{r})$ (and hence the scattering potential $w^d(\mathbf{3})$) must be evaluated over a region comparable in size with Λ. If a precipitate of this size and containing mostly Zn atoms forms within an Al-rich Al–Zn matrix, for example, then the choice of the same almost pure Al matrix average potential within the almost pure Zn particle would be quite incorrect. In such cases the actual potential should be determined for each particular environment, these being the matrix, the particle and the matrix-particle interface regions in the example cited above. We will consider this problem in more detail in the next chapter.

Finally, we return briefly to the second Born approximation as it applies to alloys. There are now two different types of higher-order terms that should be considered: terms which involve one matrix element from the perfect (average) crystal and one from the difference lattice and terms in which both matrix elements arise from the difference lattice. These latter terms simply represent the correction to the scattering by the 'defect' in going to the next order in perturbation. In general, such a correction will be negligible but we might note that inclusion of such terms to all orders would give the exact scattering rate. Terms of the first type are particularly interesting as they would include corrections to the scattering from band structure effects. However, there does not seem to have been much development along these lines as yet.

4.5.4 *The pseudopotential in a deformed lattice*

As suggested in Chapter 2, we will be interested in determining the effects of static and dynamic atomic displacements on the electrical resistivity, and so we need to determine the effects of such displacements upon the lattice potential. Historically there were two approaches to this problem: the deformable-ion model of Bloch (1928) and the rigid-ion model of Houston (1929) and Nordheim (1931). In the former model it is supposed that the change in potential due to the displacement of an ion is restricted entirely to the Wigner–Seitz cell of that ion. In the rigid-ion model, the potential surrounding each ion is assumed to be rigidly attached to it and move bodily with it. A schematic comparison between these two models is given in Figure 4.15 and reveals a fundamental flaw in the deformable-ion model. It shows that there is no change in the potential midway between two displaced ions in the deformable-ion model whereas the rigid-ion model predicts a substantial change. In fact, this change is real and leads to a change in electron energies described in

Fig. 4.15. The deformable-ion (*a*) and rigid-ion (*b*) atomic potentials in a distorted lattice.

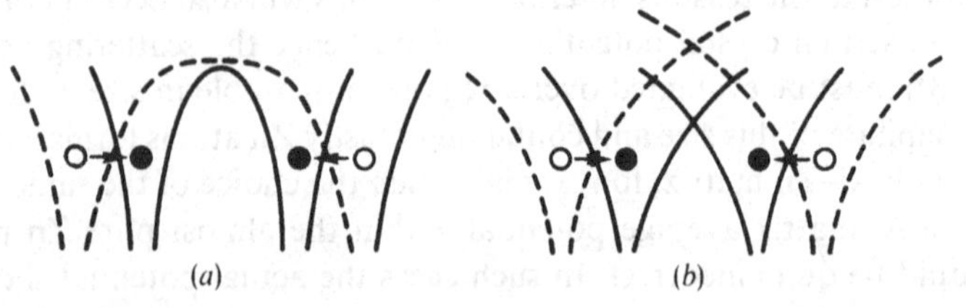

terms of a deformation potential which is proportional to the local lattice dilation (Ziman 1960, p. 186). For this reason the deformable-ion model is regarded as unsatisfactory, although it has historical importance as it represented the first serious attempt to elucidate the electron–phonon interaction (see Ziman 1960, p. 188). We will thus adopt the rigid-ion model as the basis for discussion of the effects of static and dynamic atomic displacements. However, we need to consider also the effects of screening since the conduction electron cloud is not static but capable of collective motion to shield out the local electrostatic perturbations caused by the displacements of the ions. Here we appeal to the adiabatic (or Born–Oppenheimer) approximation, whereby it is assumed that the motion of the ions is always slow enough to enable all of the conduction electrons to adjust adiabatically to the ionic positions. This means that the ground state of the electrons is a function only of the instantaneous position of the ions and does not contain any intrinsic time dependence. It is as if the electron regarded the ions as being frozen at any instant in some particular configuration. In metals the electron velocity at the Fermi surface is typically $\sim 10^6$ m s^{-1} whereas the ion velocity (i.e. the velocity of sound) $\sim 10^3$ m s^{-1} and so, from a casual examination, it would appear that the approximation is reasonable. Ziman (1960, p. 212) gives a more comprehensive discussion and it appears that the condition $\Lambda k_F > 1$ must be satisfied, this being the same condition stipulated in relation to the concept of wavepackets, as discussed in Section 1.8.1. He noted also an additional requirement $\Lambda > \lambda_p$, where λ_p is the phonon wavelength, if the approximation is to be applied to the electron–phonon interaction. However, in simple metals at normal temperatures both of these conditions are essentially satisfied and the adiabatic approximation regarded as valid. With this approximation the effects of screening can be easily incorporated into the rigid-ion model. Provided that the ion cores do not overlap, the bare pseudopotential $w_b(\mathbf{r})$ can again be written as a superposition of the individual ion pseudopotentials, as in equation (4.59). As before, this potential is screened by dividing each Fourier component by the dielectric function $\varepsilon(q)$. Because of the assumed linear nature of the screening, the operations of summing over ion positions and screening can be interchanged so that the local ion pseudopotentials $w(\mathbf{r})$ are first screened and then summed to give the total screened pseudopotential $W_s(\mathbf{r})$. This means that the solid can be regarded as an assembly of neutral 'pseudoatoms' each having its ionic potential screened by a localised cloud of electrons (Ziman 1964, 1972, p. 153). Provided that the assembly remains close packed with a reasonably uniform density, the superposition of the screening clouds then gives the total screening charge density at any point. In this model, the effects of lattice distortions will appear mainly via the structure factors. Other approaches which incorporate the effects of strain around an impurity by introducing a correction to the valency of the impurity that is proportional to the local strain will be briefly discussed in Section 6.3.

5 Electrical resistivity of simple metals and alloys

We now want to consider the electrical resistivity of simple metallic alloys in terms of the pseudopotential formalism developed in Chapter 4 for the scattering potential and using the methods developed in Chapter 2 to describe the structure of the alloys of interest. In most cases we will assume that this scattering process can be adequately described in terms of the relaxation time approximation discussed in Chapter 1. Some of the alternative methods which are applicable to non-simple metals are discussed in Chapter 6, although it is noted that these are equally suited to the study of simple metals as evidenced by some of the examples discussed in that chapter.

The alloy features that can be considered within this framework include atomic correlations and their associated static atomic displacements, vacancies and interstitials and their associated static atomic displacements and thermally induced displacements. Of these, the effects of atomic correlations, vacancies, interstitials and their associated displacement fields are dealt with directly in terms of the deviation lattice described in Chapter 4. This is because the defects are adequately described in terms of the site occupation parameters (or their averages) and the scattering potentials can be expressed in terms of a deviation from the average potential. In the case of atomic correlations this is just the difference in the atomic form factors $w^{A}(\mathbf{r})-w^{B}(\mathbf{r})$. In the case of vacancies it is the absence of the atomic potential at one site (or the addition of the negative of the atomic potential) giving a deviation in potential to the average host potential, and in the case of interstitials it is the additional extra atomic potential at the site (for a self interstitial in a pure metal) or the addition of the appropriate deviation in potential $\bar{w}-w^{I}$ for a foreign interstitial of type I. Associated with these defects will be the appropriate structure factor $S^{d}(\mathbf{q})$ which takes into account the nature of the correlations or position of the defect and the associated atomic displacement fields. Actually, a similar situation arises with respect to thermally induced displacements: we will be able to identify an average lattice which gives the usual Bragg scattering and again consider the corrections to this introduced by the deviation lattice terms (the defects now being displacements) and which give rise to the observed temperature dependence of resistivity. (We might note that in an earlier study by the author (Rossiter 1977), the defect and thermally induced strains were not explicitly considered and so their separation from the average lattice was not carried out.)

5.1 A general resistivity expression

The first step is to derive a general expression for the resistivity in terms of the pseudopotential formalism and scattering theory described in the previous chapters. Equation (1.32) gives $\partial f_{\mathbf{k}}/\partial t|_{\text{scatt}}$ in terms of the deviation $g_{\mathbf{k}}$ from the equilibrium distribution function. This can be written in terms of the function $\Phi_{\mathbf{k}}$ (equation (1.48)) as

$$\left.\frac{\partial f_{\mathbf{k}}}{\partial t}\right|_{\text{scatt}} = -\frac{\Omega}{(2\pi)^3}\Phi_{\mathbf{k}}\frac{\mathrm{d}f_0}{\mathrm{d}E_{\mathbf{k}}}\int Q_{\mathbf{kk}'}\left(\frac{\Phi_{\mathbf{k}'}}{\Phi_{\mathbf{k}}}-1\right)\mathrm{d}\mathbf{k}'. \tag{5.1}$$

Comparing this with the definition of $\tau_{\mathbf{k}}$ in the relaxation time approximation (equation (1.36)) leads immediately to the result

$$\frac{1}{\tau_{\mathbf{k}}} = \frac{\Omega}{(2\pi)^3}\int Q_{\mathbf{kk}'}\left(1-\frac{\Phi_{\mathbf{k}'}}{\Phi_{\mathbf{k}}}\right)\mathrm{d}\mathbf{k}'. \tag{5.2}$$

Since $\tau_{\mathbf{k}}$ cannot then depend upon the direction of $\mathbf{k}$, a general expression for $\Phi_{\mathbf{k}}$ is

$$\Phi_{\mathbf{k}} = \alpha\mathbf{k}\cdot\hat{\mathbf{x}}, \tag{5.3}$$

where $\hat{\mathbf{x}}$ is a unit vector in the direction of the electric field (Ashcroft & Mermin 1976, p. 325; Blatt 1968, p. 186; Jones 1956, p. 239; Ziman 1960, p. 284) and α is a coefficient which will cancel in (5.2). Substitution of (5.3) into (5.2) then gives

$$\frac{1}{\tau_{\mathbf{k}}} = \frac{\Omega}{(2\pi)^3}\int Q_{\mathbf{kk}'}(1-\cos\theta)\,\mathrm{d}\mathbf{k}', \tag{5.4}$$

where θ is the angle between $\mathbf{k}$ and $\mathbf{k}'$, as discussed in Section 1.2. The scattering probability $Q_{\mathbf{kk}'}$ is given to first order in perturbation theory (Fermi's Golden Rule) (Schiff 1968, pp. 285, 314; Landau & Lifschitz 1958) as

$$Q_{\mathbf{kk}'} = \frac{2\pi}{\hbar}|\langle\mathbf{k}+\mathbf{q}|W(\mathbf{r})|\mathbf{k}\rangle|^2. \tag{5.5}$$

The integral over the volume element $\mathrm{d}\mathbf{k}'$ is then transformed to an integral over constant energy shells (Appendix B) so that (5.4) and (5.5) give

$$\frac{1}{\tau_{\mathbf{k}}} = \frac{\Omega}{4\pi^2\hbar|\mathrm{d}E_{\mathbf{k}'}/\mathrm{d}\mathbf{k}'|}\int|\langle\mathbf{k}+\mathbf{q}|W(\mathbf{r})|\mathbf{k}\rangle|^2(1-\cos\theta)\,\mathrm{d}S'. \tag{5.6}$$

The element of area $\mathrm{d}S'$ can be taken to be a ring on the Fermi surface (now assumed to be spherical) with an area $2\pi k'^2\sin\theta\,\mathrm{d}\theta$, and $\mathrm{d}E_{k'}/\mathrm{d}k'$

given from equation (1.10) as $\hbar^2 k_F/m$ so that (5.6) becomes

$$\frac{1}{\tau_{\mathbf{k}}}=\frac{mk_F\Omega}{2\pi\hbar^3}\int|\langle\mathbf{k}+\mathbf{q}|W(\mathbf{r})|\mathbf{k}\rangle|^2(1-\cos\theta)\sin\theta\,d\theta. \tag{5.7}$$

It is more convenient to write the integral in terms of the scattering vector $\mathbf{q}$ (see Figure 1.2 and equation (4.22)):

$$\frac{1}{\tau_{\mathbf{k}}}=\frac{m\Omega}{4\pi\hbar^3k_F^3}\int_0^{2k_F}|\langle\mathbf{k}+\mathbf{q}|W(\mathbf{r})|\mathbf{k}\rangle|^2q^3\,dq. \tag{5.8}$$

With equations (1.5) and (4.62), this gives the required result (often known as the Ziman equation):

$$\rho=CN\int_0^{2k_F}|\langle\mathbf{k}+\mathbf{q}|W(\mathbf{r})|\mathbf{k}\rangle|^2q^3\,dq \tag{5.9a}$$

$$=CN\int_0^{2k_F}|S(\mathbf{q})|^2|\langle\mathbf{k}+\mathbf{q}|w(\mathbf{r})|\mathbf{k}\rangle|^2q^3\,dq, \tag{5.9b}$$

where

$$C=\frac{3\pi m^2\Omega_0}{4\hbar^3e^2k_F^6}. \tag{5.10}$$

As this expression will form the basis for most of the calculations which follow in this chapter it is worth recapping the approximations that it implies. These are: the scattering is elastic and given by the relaxation time approximation so that $\tau_{\mathbf{k}}$ does not depend upon the direction of $\mathbf{k}$; both $\mathbf{k}$ and $\mathbf{k}'$ lie on the Fermi surface which is assumed to be spherical; the scattering matrix is given by the first Born approximation; the screening is linear; the electron mean free path Λ is larger than the de Boglie wavelength $\lambda_{\mathbf{k}}$. As such, its applicability is limited to nearly free electron simple metals and alloys. However, as we will now demonstrate, this still includes many materials of interest.

5.2 The resistivity of alloys with short range atomic correlations

As shown in Chapter 4, the pseudopotential for an alloy can be written as the sum of a (virtual crystal) average lattice potential and a deviation in potential that depends specifically upon the atomic distribution (equation (4.67)). This allows the square of the matrix elements required in (5.9) to be written as the sum of two contributions, one from the virtual crystal average lattice and the other from the deviation lattice. The matrix elements of the average lattice are

$$S(\mathbf{q})\langle\mathbf{k}+\mathbf{q}|\bar{w}(\mathbf{r})|\mathbf{k}\rangle, \tag{5.11}$$

where

$$S(\mathbf{q})=\frac{1}{N}\sum_i\exp(-i\mathbf{q}\cdot\mathbf{r}_i), \qquad \text{(equation (4.63))}$$

and

$$\bar{w}(\mathbf{r}) = c_A w_A(\mathbf{r}) + c_A w_A(\mathbf{r}). \qquad \text{(equation (4.65)}$$

The deviation lattice has the matrix elements

$$S^d(\mathbf{q})\langle \mathbf{k}+\mathbf{q}|w^d(\mathbf{r})|\mathbf{k}\rangle, \tag{5.12}$$

where

$$S^d(\mathbf{q}) = c_B S^A(\mathbf{q}) - c_A S^B(\mathbf{q}), \qquad \text{(equation (4.69))}$$

and

$$S_A(\mathbf{q}) = \frac{1}{N} \sum_{\mathbf{r}_i(A)} \exp[-i\mathbf{q}\cdot\mathbf{r}_i(A)], \qquad \text{(equation (4.70))}$$

and similarly for $S_B(\mathbf{q})$, and

$$w^d(\mathbf{r}) = w_A(\mathbf{r}) - w_B(\mathbf{r}). \qquad \text{(equation (4.66))}$$

As also noted in Chapter 4, provided that the atoms are at rest on the lattice sites, we can evaluate the square of the structure factors $|S(\mathbf{q})|^2 = S^*(\mathbf{q})S(\mathbf{q})$ and $|S^d(\mathbf{q})|^2 = S^{d*}(\mathbf{q})S^d(\mathbf{q})$ independently because the first is non-zero only for $\mathbf{q} = \mathbf{g}_n$ whereas the second is non-zero only for $\mathbf{q} \neq \mathbf{g}_n$ (atomic displacement effects are discussed in Section 5.6). The square of the average lattice structure factor is simply

$$|S(\mathbf{q})|^2 = \frac{1}{N^2} \sum_i \sum_j \exp[-i\mathbf{q}\cdot(\mathbf{r}_i - \mathbf{r}_j)], \tag{5.13}$$

and is zero except when $\mathbf{q} = \mathbf{g}_n$, in which case it has the value of unity. This term gives the Bragg scattering of conduction electrons but does not by itself contribute to the resistivity because of the integration over $\mathbf{q}$. The square of the deviation lattice structure factor is

$$\begin{aligned}
|c_B S^A(\mathbf{q}) - c_A S^B(\mathbf{q})|^2 \\
&= \frac{1}{N^2}[c_B^2 S^{A*}(\mathbf{q})S^A(\mathbf{q}) \\
&\quad - c_A c_B (S^{A*}(\mathbf{q})S^B(\mathbf{q}) + S^A(\mathbf{q})S^{B*}(\mathbf{q})) + c_A^2 S^{A*}(\mathbf{q})S^B(\mathbf{q})] \\
&= \frac{c_B^2}{N^2} \sum_{\mathbf{r}_i^A} \sum_{\mathbf{r}_j^A} \exp[i\mathbf{q}\cdot(\mathbf{r}_i^A - \mathbf{r}_j^A)] + \frac{c_A^2}{N^2} \sum_{\mathbf{r}_i^B} \sum_{\mathbf{r}_j^B} \exp[i\mathbf{q}\cdot(\mathbf{r}_i^B - \mathbf{r}_j^B)] \\
&\quad - \frac{c_A c_B}{N^2} \left[\sum_{\mathbf{r}_i^A} \sum_{\mathbf{r}_j^B} \exp[i\mathbf{q}\cdot(\mathbf{r}_i^A - r_j^B)] + \sum_{\mathbf{r}_i^B} \sum_{\mathbf{r}_j^A} \exp[i\mathbf{q}\cdot(\mathbf{r}_i^B - \mathbf{r}_j^A)] \right].
\end{aligned} \tag{5.14}$$

In order to simplify this expression we group together all terms which have the same relative spacing $\mathbf{R}_{ij}^{nm} = \mathbf{r}_i^n - \mathbf{r}_j^m$ and then sum over $\mathbf{R}_{ij}$. The required averaging over all possible configurations is then carried out by weighting each term by the number of such pairs given by N times the configurationally averaged probability of occurrence of each pair as defined in Chapter 2. This gives

$$|S^{\mathrm{d}}(\mathbf{q})|^2 = \frac{1}{N}\sum_{ij} [c_{\mathrm{B}}^2 P_{ij}^{\mathrm{AA}} \exp(-\mathrm{i}\mathbf{q}\cdot\mathbf{R}_{ij}^{\mathrm{AA}}) + c_{\mathrm{A}}^2 P_{ij}^{\mathrm{BB}} \exp(-\mathrm{i}\mathbf{q}\cdot\mathbf{R}_{ij}^{\mathrm{BB}}) - 2c_{\mathrm{A}}c_{\mathrm{B}} P_{ij}^{\mathrm{AB}} \exp(-\mathrm{i}\mathbf{q}\cdot\mathbf{R}_{ij}^{\mathrm{AB}})], \tag{5.15}$$

where the summation is over all sites having separations $\mathbf{R}_{ij} = \mathbf{r}_i - \mathbf{r}_j$. These probabilities are given directly in terms of the pairwise correlation parameters (equation (2.12)) or the Warren–Cowley SRO parameters (equation (2.13)) and the resistivity becomes

$$\rho = CN\int_0^{2k_{\mathrm{F}}} |S^{\mathrm{d}}(\mathbf{q})|^2 |\langle\mathbf{k}+\mathbf{q}|w^{\mathrm{d}}(\mathbf{r})|\mathbf{k}\rangle|^2 q^3\,\mathrm{d}q. \tag{5.16}$$

We now consider application of this equation to various situations.

5.2.1 *Homogeneous atomic correlations*

In the absence of any atomic displacements
$\mathbf{R}_{ij}^{\mathrm{AB}} = \mathbf{R}_{ij}^{\mathrm{BB}} = \mathbf{R}_{ij}^{\mathrm{AA}} = \mathbf{R}_{ij}$.

Equations (5.15), (5.16) and (2.12) then give

$$|S^{\mathrm{d}}(\mathbf{q})|^2 = \frac{1}{N}\sum_{ij} \langle\sigma_i\sigma_j\rangle \exp(-\mathrm{i}\mathbf{q}\cdot\mathbf{R}_{ij}). \tag{5.17}$$

This result could have been obtained more directly by working from the pseudopotential as defined in (4.81). The resistivity due to atomic correlations in an alloy can then be found by substituting (5.17) into (5.12) and (5.9b) to give

$$\rho = C\int_0^{2k_{\mathrm{F}}} \sum_{ij} \langle\sigma_i\sigma_j\rangle \exp(-\mathrm{i}\mathbf{q}\cdot\mathbf{R}_{ij}) |\langle\mathbf{k}+\mathbf{q}|w^{\mathrm{d}}(\mathbf{r})|\mathbf{k}\rangle|^2 q^3\,\mathrm{d}q, \tag{5.18}$$

since the average lattice term does not contribute to ρ. This expression can be further simplified by noting that the integral does not depend upon the direction of $\mathbf{q}$ and so only that part of the integral invariant to rotations of $\mathbf{q}$ can contribute. As $\langle\mathbf{k}+\mathbf{q}|w(\mathbf{r})|\mathbf{k}\rangle$ depends only upon $|\mathbf{q}|$, only that part of the exponential invariant to such rotations can contribute. This invariant part can be found by expansion of the exponential in spherical Bessel functions and Legendre polynomials (Gibson 1956; Watson 1944, p.128) or by simply averaging over all directions of $\mathbf{q}$ (Rossiter & Wells 1971). Both approaches give

$$\langle\exp(-\mathrm{i}\mathbf{q}\cdot\mathbf{r}_i)\rangle_{\mathbf{q}\cdot\mathbf{r}_i} = j_0(qr_i) = \frac{\sin(qr_i)}{qr_i}. \tag{5.19}$$

Furthermore the pairwise correlation parameter can be expressed in terms of an average over the c_i atoms in a shell i to give

$$\rho = Cc_A c_B \int_0^{2k_F} \sum_i c_i \alpha_i \frac{\sin(qr_i)}{qr_i} |\langle \mathbf{k}+\mathbf{q}|w^d(\mathbf{r})|\mathbf{k}\rangle|^2 q^3 \, dq, \tag{5.20}$$

where the summation is now over the shells of atoms and we have introduced the Warren–Cowley SRO parameter defined by equation (2.13). This formula (or variations of it) has been derived by a number of people using a variety of different forms of scattering potential. (Murakami 1953; Gibson 1956; Asch & Hall 1963; Wang & Amar 1970; Rossiter & Wells 1971; Aubauer 1978). It is interesting to note that the square of the structure factor in (5.20) $|S^d(\mathbf{q})|^2 = 1/N \sum_i c_i \alpha_i \sin(qr_i)/(qr_i)$ is exactly the same as that used to describe the diffuse scattering of X-rays or neutrons from a polycrystalline specimen. It is thus possible to substitute for $|S^d(\mathbf{q})|^2$ an experimentally determined diffuse intensity, suitably corrected for background, Bragg peaks and other unwanted diffuse scattering and scaled so that $|S^d(\mathbf{q})|^2 = 1$ as $q \to 0$ such as shown in Figures 2.6 or 2.10 (see e.g. Wells & Rossiter 1971). Note, however, that this approach will only be valid if the correlations are homogeneous and short range. Otherwise the different averaging procedures involved in the different techniques will lead to errors (see also the discussion in Section 5.7.5).

For a random alloy, $\langle \sigma_0 \sigma_0 \rangle = c_A c_B$ and $\langle \sigma_i \sigma_j \rangle = 0$ for all $i, j \neq 0$ (i.e. $\alpha_0 = 1$ and $\alpha_i = 0$ for all $i > 0$) and so equation (5.20) becomes

$$\rho = Cc_A c_B \int_0^{2k_F} |\langle \mathbf{k}+\mathbf{q}|w^d(\mathbf{r})|\mathbf{k}\rangle|^2 q^3 \, dq, \tag{5.21}$$

giving the familiar parabolic composition dependence (provided that the form factor has no dependence upon composition). For small concentrations c_A of an impurity A, $c_A c_B \approx c_A$ and (5.21) predicts a linear composition dependence in the dilute alloy limit.

There may be situations where it would be more appropriate to use a composition wave description of the atomic distribution (see Section 2.3), in which case the appropriate expression follows directly from (5.16). The appropriate structure factor is then the Fourier transform of the composition fluctuation wave (cf. (5.17) which is just the discrete Fourier transform of the pairwise correlation). For example, if we take a simple composition wave of the form

$$c(\mathbf{r}_i) = \frac{c_A + c_B}{2} + A(\boldsymbol{\beta}) \exp(-i\boldsymbol{\beta} \cdot \mathbf{r}_i), \tag{5.22}$$

the deviation lattice structure factor is given by (see equation (2.21))

$$S^{\mathrm{d}}(\mathbf{q})=\frac{1}{N}\sum_{i}^{N}A(\boldsymbol{\beta})\exp(-\mathrm{i}\boldsymbol{\beta}\cdot\mathbf{r}_i)\exp(-\mathrm{i}\mathbf{q}\cdot\mathbf{r}_i)$$

$$=\frac{1}{N}\sum_{i}^{N}A(\boldsymbol{\beta})\exp[-\mathrm{i}(\mathbf{q}+\boldsymbol{\beta})\cdot\mathbf{r}_i]. \tag{5.23}$$

This summation will only be non-zero at $\mathbf{q}+\boldsymbol{\beta}=\mathbf{g}_n$ and so will give rise to satellites about the Bragg positions. If the wavelength $2\pi/|\boldsymbol{\beta}|$ is much less than Λ, this scattering will not contribute to the resistivity. However, any deviations from the perfect periodicity implied by (5.22) will give a diffuse scattering contribution. For example, spatial fluctuations $\delta\boldsymbol{\beta}(\mathbf{r}_i)$ in the wavenumber lead to

$$S^{\mathrm{d}}(\mathbf{q})=\frac{1}{N}\sum_{i}^{N}A(\boldsymbol{\beta})\exp[-\mathrm{i}(\mathbf{q}+\boldsymbol{\beta})\cdot\mathbf{r}_i]\exp[-\mathrm{i}\,\delta\boldsymbol{\beta}(\mathbf{r}_i)\cdot\mathbf{r}_i]. \tag{5.24}$$

The square of the structure factor then becomes

$$|S^{\mathrm{d}}(\mathbf{q})|^2=\frac{1}{N^2}\sum_{i}^{N}\sum_{j}^{N}A(\boldsymbol{\beta})^2\exp[-\mathrm{i}(\mathbf{q}+\boldsymbol{\beta})\cdot(\mathbf{r}-\mathbf{r}_i)]$$

$$\times\exp[-\mathrm{i}(\delta\boldsymbol{\beta}(\mathbf{r}_i)-\delta\boldsymbol{\beta}(\mathbf{r}_j))\cdot(\mathbf{r}_i-\mathbf{r}_j)]. \tag{5.25}$$

We require a spatial average over the fluctuations $\langle\exp[-\mathrm{i}(\delta\beta'_i-\delta\beta'_j)R_{ij}\rangle$, where we have used $\delta\beta'_i$ to indicate the component $\delta\boldsymbol{\beta}(\mathbf{r}_i)$ along the direction of $\mathbf{r}_i$ and $|\mathbf{r}_i-\mathbf{r}_j|=R_{ij}$. This average may be written as (see Appendix C) $\exp[-\frac{1}{2}\langle(\delta\beta'_i-\delta\beta'_j)^2\rangle R_{ij}^2]$. It will produce terms of the form $\exp[-\frac{1}{2}(\langle\delta\beta_i'^2\rangle+\langle\delta\beta_j'^2\rangle)R_{ij}^2]$ which simply reduce the intensity of the Bragg peaks, in direct analogy with the Debye–Waller factor. It will also give terms of the form $\exp(\frac{1}{2}\langle\delta\beta'_j\rangle^2R_{ij}^2)$ which will give a diffuse scattering (and hence a contribution to the resistivity) and which will depend upon the spatial correlation and size of the fluctuations. Local fluctuations in amplitude or phase of the composition wave could be mapped into equivalent wavelength fluctuations, leading to similar results.

In both of the approaches described above we have assumed that the scale of atomic correlations or composition inhomogeneity is small in relation to Λ. The problems that arise when this is not so (such as long range ordering, precipitation and spinodal decomposition) will be discussed with reference to the particular situations later in this chapter.

5.2.2 *Inhomogeneous atomic correlations*

We consider now the case of a small region with a high degree of atomic correlation in a matrix having a different degree of correlation. This is the situation of type II(a) heterogeneous SRO (see Section 2.5.3) or the pre-precipitation stages of GP zone formation. We treat both cases together and we use the word 'zone' to mean either a small highly ordered region or a clustered region. In many alloys that undergo GP zone formation or inhomogeneous short range ordering, the resistivity initially increases in the small zone limit, reaches a peak at a zone size of ~10–20 Å and then decreases. Our theory must explain these essential features. Here we must explicitly consider the effects of a finite mean free path Λ and whether the inhomogeneities occur on a scale small or large compared with Λ. We should also note that Λ is a locally determined quantity which we now write as Λ_l, and will vary from place to place, i.e. Λ_l in a disordered region will be different to Λ_l in an ordered or clustered region. We proceed as before by determining the scattering matrix $\langle \mathbf{k}+\mathbf{q}|W(\mathbf{r})|\mathbf{k}\rangle$ but acknowledge the fact that the scattering is being caused by deviations from the average perfectly periodic potential on a size scale $\sim\Lambda_l$ (Rossiter & Wells 1971). This assertion is in complete accord with the damping of coherent interference oscillations of the RKKY interaction by a term $\exp(-r/\Lambda)$ (de Gennes 1962; Heeger *et al.* 1966), with the idea that the electronic structure at some point $\mathbf{r}_i$ depends mainly upon the conditions inside a volume $\sim\frac{4}{3}\pi\Lambda^3$ about $\mathbf{r}_i$ (Heine 1980; Kohn & Olson 1972), and with the many discussions about a size effect whereby metal specimens show substantially enhanced resistivities when any of their dimensions become of the order of, or less than, the conduction electron mean free path (Chambers 1969). This means that the Boltzmann equation should be solved within each characteristic region to obtain a local mean free path Λ_l. However, the scattering term in the Boltzmann equation depends upon the scattering matrix, which in turn depends upon the size of region over which the potential is averaged. Guided by the other investigations, the proper course to pursue would involve introduction of a damping term $D(r)$ such as

$$D(r)=\exp(-r_i/\Lambda) \tag{5.26}$$

when determining the average potential in the particular region of interest. This would then require a self-consistent solution of Boltzmann's equation and give rise to an extra term $\exp(-R_{ij}/\Lambda)$ in the square of the structure factor, as proposed by Fisher & Langer (1968) with respect to multiple scattering effects in correlated magnetic spin systems. The modification of the resistivity by such a term was also deduced by Chambers (1969) in relation to size effects. He introduced the effects of scattering from a surface directly into the Boltzmann equation via the distribution function $g_{\mathbf{k}}(\mathbf{r})$, which then becomes a function of position. The same result follows if one takes into account the definition of the scattering potential which is then strongly localised at the surface of an otherwise pure specimen. In order to illustrate the general principles involved, we consider the two limiting situations which are shown in Figure 5.1. Either the zone is smaller than Λ_l (Figure 5.1(*a*)) or larger than Λ_l (Figure 5.1(*b*)). In the latter case, the locally averaged scattering potential will be different in the matrix (i), at the zone boundary (ii), or inside the zone (iii). It would be quite wrong to use the same scattering potential in each case.

We will suppose for the moment that the damping function is a simple cut-off:

$$D(r) = 1, \quad r_i < \Lambda_l$$
$$= 0, \quad r_i > \Lambda_l. \tag{5.27}$$

This is compared with the exponential form (5.26) in Figure 5.2.

Let us now consider the change in resistivity that will occur as the zones grow.

(a) *Small zone limit*

The behaviour during the initial formation of zones may be obtained from the results derived previously, but now the pseudopotential $w^d(\mathbf{r})$ (or $w^d(\mathbf{q})$) which appears in the equations must be constructed in accordance with (4.66) with proper regard to the volume $\sim\frac{4}{3}\pi\Lambda_l^3$ being sampled. From (4.68) the scattering matrix may then be written as

$$\langle \mathbf{k}+\mathbf{q}|W(\mathbf{r})|\mathbf{k}\rangle = \frac{\bar{w}(\mathbf{q})}{N'}\sum_i^{N'}\exp(-i\mathbf{q}\cdot\mathbf{r}_i) + \frac{w^d(\mathbf{q})}{N'}$$
$$\times\left[c_B'\sum_{\mathbf{r}_i(A)}\exp(-i\mathbf{q}\cdot\mathbf{r}_i) - c_A'\sum_{\mathbf{r}_i(B)}\exp(-i\mathbf{q}\cdot\mathbf{r}_i)\right], \tag{5.28}$$

Fig. 5.1. The two limiting cases of zone size in relation to Λ_l:(*a*) zone much smaller than Λ_l; (*b*) zone much larger than Λ_l.

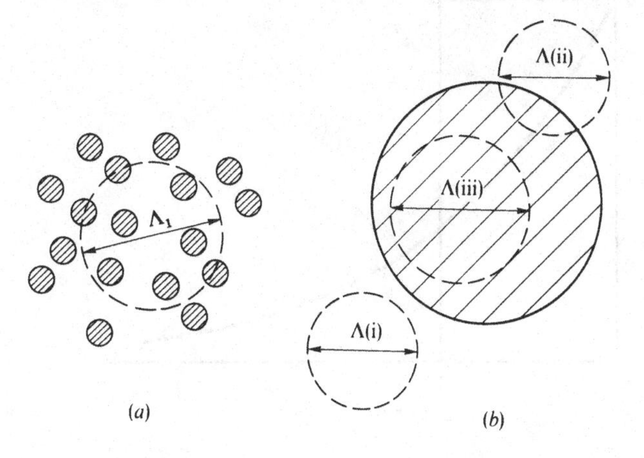

where N' is the number of atoms in the region of volume $\frac{4}{3}\pi\Lambda_l^3$ and c'_A and c'_B are now the average *local* compositions given by

$$c'_A = \left(\frac{N' - N_z\Gamma}{N'}\right)c_A^m + \frac{N_z\Gamma}{N'}c_A^z$$
$$c'_B = \left(\frac{N' - N_z\Gamma}{N'}\right)c_B^m + \frac{N_z\Gamma}{N'}c_B^z, \quad (5.29)$$

where the superscripts m and z on the compositions refer to the matrix and zone respectively and N_z is the number of zones, each zone containing Γ atoms. If one wanted to include a depletion region around the zone then it would be necessary to consider at least three separate contributions. In the completely general case one could also make the size and concentration of the depletion region a function of the size and concentration of the particle and take into account correlations between the three regions. As this extension is quite straightforward but results in more complicated equations it will not be considered any further here. Furthermore, in order to keep the equations simple at this stage we will neglect scattering from the boundary. This will be corrected when we consider the growth of the zones to larger sizes. Note that in the limit of very small zones, $\Gamma \to 0$, and so $c'_A \approx c_A^m$, $c'_B \approx c_B^m$. Equation (5.28) then leads to the expression for homogeneous short range correlations (5.20).

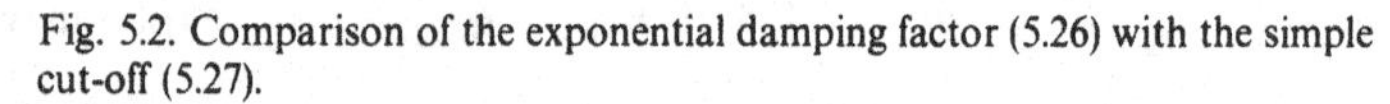
Fig. 5.2. Comparison of the exponential damping factor (5.26) with the simple cut-off (5.27).

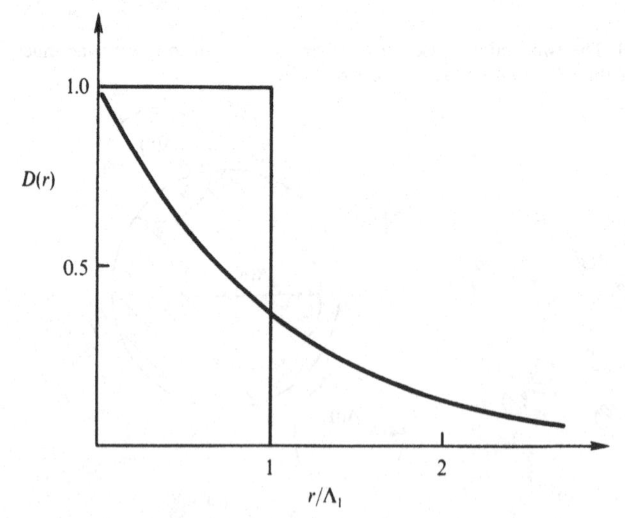

(b) *Intermediate zone size*

The basic physical reason for the resistivity behaviour in systems having massive (i.e. three-dimensional) clusters is now clear. As the zone radius approaches Λ_l (i.e. $N_z\Gamma$ approaches N'), $c'_A \rightarrow c^z_A$ and $c'_B \rightarrow c^z_B$. If the zones are rich in B atoms, c'_B is then large (≈ 1) but the summation over A atoms is small (≈ 0) since there are few A atoms in the zone. While the summation over B atoms is then large ($\approx N'$), the multiplying factor c'_A is small (≈ 0). Thus if the zone fully occupies the region of interest there is only small scattering since there is virtually no disorder *in that region*. So, while the presence of a small zone causes an initial increase in the resistivity, by the time $R_z \approx \Lambda_l$ the contribution from the zone has become small again and the resistivity passes through a maximum at some smaller zone size. We can demonstrate this behaviour with a simple calculation by writing the square of the scattering matrix as

$$|\langle \mathbf{k}+\mathbf{q}|w^d(\mathbf{r})|\mathbf{k}\rangle|^2 = |w^d(\mathbf{q})|^2 \frac{c'_A c'_B}{N'} \sum c_i \alpha_i \exp(-i\mathbf{q}\cdot\mathbf{r}_i). \tag{5.30}$$

The product $c'_A c'_B$ is given by

$$c'_A c'_B = c^m_A c^m_B + \frac{\Gamma}{N'} c^m_A (c^m_A - c^m_B) - c^{m2}_A \frac{\Gamma^2}{N'^2}, \tag{5.31}$$

where we have assumed for the sake of simplicity that $c^z_B = 1$, $c^z_A = 0$, $N_z = 1$. This is shown plotted as a function of zone size in Figure 5.3(*a*). If the parameters $w^d(\mathbf{q})$ and α_i are such that the resistivity initially increases, the factor $c'_A c'_B$ will ensure that it goes through a maximum approximately when $\Gamma \approx N'/2$ if there is only a single zone in the region of interest (i.e. when the zone radius $R_z \approx 0.8\Lambda$). For very small clusters $c'_A \approx c_A$, $c'_B \approx c_B$ and (5.30) reduces to the normal expression for homogeneous atomic correlations leading to a resistivity given by equation (5.20). Note that $c'_A c'_B$ depends only upon the normalised cluster size Γ/N'. This implies that one should be able to superimpose plots of residual resistivity scaled to the peak value as a function of reduced cluster size (or ageing time, again scaled to the value at the peak in ρ). Some results for Al–10.2 wt.%Zn are shown in Figure 5.3(*b*) and support this conclusion. However, if the precipitate is in the form of a very thin platelet, even if the linear dimensions of the plate are comparable with Λ_l, the majority of atoms in the local region of interest will still be matrix atoms (i.e. $\Gamma \ll N'$), and so by this mechanism alone such plates produce only a very weak mean free path effect. This follows from the fact that (5.31) depends upon the ratio Γ/N'. Consideration of the strain field around such a platelet can alter this situation, as we will show later on.

There is an additional complication that arises as the zones evolve, since the zone scattering will change from being completely diffuse in the small zone limit to more Bragg-like as the zone size approaches Λ_l. As pointed out by Hillel *et al.* (1975), this results in a marked anisotropy in the conduction electron scattering and the assumption of an isotropic relaxation time breaks down. Spherically averaged terms of the form (5.20) are then no longer applicable since they give only $\langle 1/\tau_k \rangle$ and one has to resort to a calculation of the full partial structure factor for the

Fig. 5.3. (*a*) Variation of $c'_A c'_B$ with Γ/N' ($c_A^m = 0.95$) from (5.31). (*b*) Variation of resistivity of Al–10.2wt.%Zn normalised to the maximum value $\Delta\rho_m$ with ageing time scaled to the time to reach the maximum t_m. The ageing temperatures are as indicated (after Panseri & Federighi 1960).

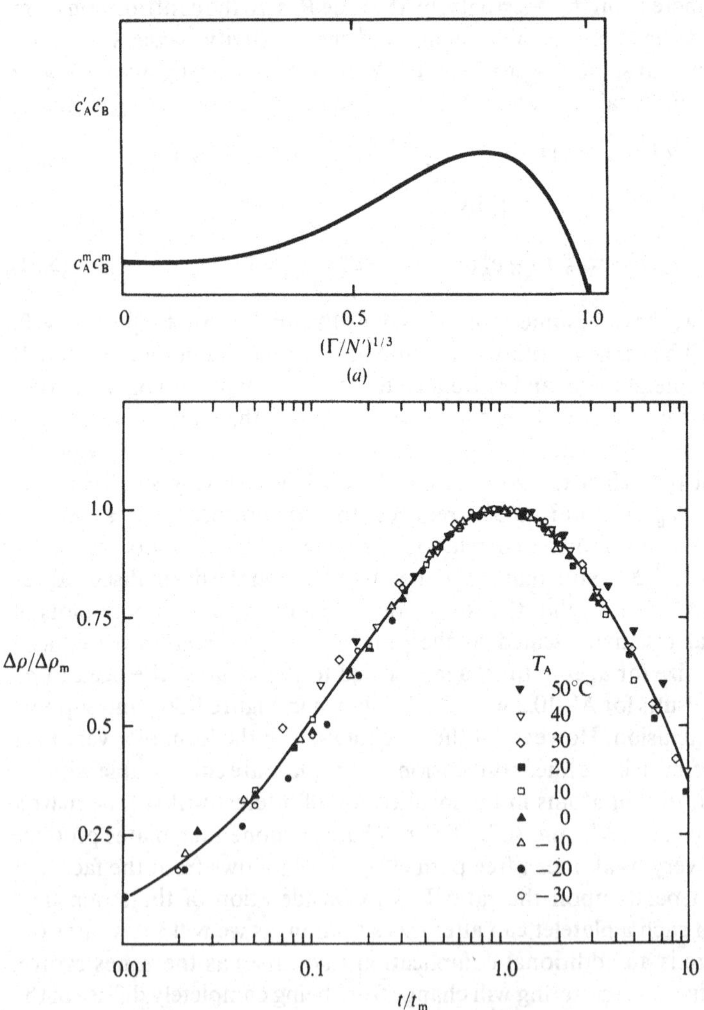

zone (see e.g. Yonemitsu & Matsuda 1976; Edwards & Hillel 1977),

$$S^z(\mathbf{q}) = \frac{1}{N'} \sum_i^{\text{zone}} \exp(-i\mathbf{q}\cdot\mathbf{r}_i), \tag{5.32}$$

in order to determine $1/\langle\tau_{\mathbf{k}}\rangle$. The basic effect of this anisotropy is as follows (Hillel & Rossiter 1981). In the small zone limit the pairwise correlation parameter $\langle\sigma_i\sigma_j\rangle$ decays rapidly with distance $\mathbf{R}_{ij}$ and so the electron scattering is diffusely spread over the Fermi surface. Equations of the form (5.18) are then quite appropriate since then $1/\langle\tau(\mathbf{k})\rangle \cong \langle 1/\tau(\mathbf{k})\rangle$. However, as the clusters grow $\langle\sigma_i\sigma_j\rangle$ increases for larger $\mathbf{R}_{ij}$ and the diffuse scattering becomes increasingly concentrated in Bragg-like maxima on those parts of the Fermi surface in contact with the Brillouin zone boundaries. The relaxation time $\tau(\mathbf{k})$ is now strongly $\mathbf{k}$-dependent (i.e. anisotropic) since not all electrons on the Fermi surface are affected equally by the scattering. The width of a Bragg-like peak in $\mathbf{k}$-space is of the order R_z^{-1}, where R_z is the zone radius. If this is much less than the mean separation Δk of the Fermi surface from the Brillouin zone boundaries, then only a small fraction of the electrons will be strongly scattered. Thus one expects this anisotropy mechanism to become important when the typical zone radius reaches a size

$$R_z \sim (\Delta k)^{-1}. \tag{5.33}$$

In terms of the separation of the scattering matrix elements into average lattice and deviation lattice terms, this means that there is a continuous redistribution in equation (5.28) of scattering weight from the zone terms of the deviation lattice into the average lattice term as R_z increases. This change of weight was incorporated into a phenomenological model by Hillel *et al.* (1975) by allowing for the fact that only a fraction of the Fermi surface f_N was involved in Bragg-like scattering. The anisotropy is then incorporated by assuming a Bragg-like scattering rate over a fraction f_N of the Fermi surface and a different diffuse scattering rate over the remaining fraction $1-f_N$ of the Fermi surface. One may regard the two regions of the Fermi surface as conductors connected in parallel (Osamura *et al.* 1982) so that their conductivities are added:

$$\frac{1}{\rho} = \frac{1-f_N}{\rho_{is}} + \frac{f_N}{\rho_{Br}+\rho_{is}}, \tag{5.34}$$

where ρ_{is} is the isotropic contribution to the resistivity which includes the effects of scattering for single solute atoms in the matrix and from phonons, while ρ_{Br} is the anisotropic particle-size-broadened Bragg-like scattering associated with zones.

(c) *Large zone limit*

We turn now to the large zone limit. There are three characteristic cases to consider, as shown in Figure 5.1(*b*), depending upon whether the region of interest lies in the matrix, in the zone, or includes part of the zone boundary. We now have to evaluate the scattering matrix separately for each region. In the matrix the scattering elements are just those given by a homogeneous alloy, as in equation (5.20) if there are short range atomic correlations, or (5.21) if the matrix is random, with the appropriate matrix parameters $\alpha_i^{\rm m}, c_{\rm A}^{\rm m}, c_{\rm B}^{\rm m}$. (The case of long range order will be considered in the next section.) If the region of interest lies entirely within the zone then we again have a homogeneous situation treated with equations (5.20) and (5.21), but now with the appropriate zone parameters $\alpha_i^{\rm z}, c_{\rm A}^{\rm z}, c_{\rm B}^{\rm z}$. If either the matrix or zone are pure metals (a rather unlikely event) then of course $S^{\rm d}(\mathbf{q})=0$ and they will give no resistivity contribution. When the region of interest includes a zone boundary we have essentially the case of a decomposition where the decomposition wavelength is much larger than $\Lambda_{\rm l}$. If the boundary is diffuse (on a scale of $\Lambda_{\rm l}$) it may be regarded simply as a region having large composition fluctuations producing a large diffuse scattering. The average concentrations of atoms within the region $\bar{c}_{\rm A}$ and $\bar{c}_{\rm B}$ can again be found by determining the number of A and B atoms within the zone and averaging over all locations of the boundary within the region of interest. If the boundary is modelled by symmetric function, as shown in Figure 5.4, the average concentration of A and B atoms in the boundary $c_{\rm A}^{\rm b}$ and $c_{\rm B}^{\rm b}$, respectively, are given by the simple relationships

$$c_{\rm A}^{\rm b}=\frac{c_{\rm A}^{\rm m}+c_{\rm A}^{\rm z}}{2}$$
$$c_{\rm B}^{\rm b}=\frac{c_{\rm B}^{\rm m}+c_{\rm B}^{\rm z}}{2} \qquad (5.35)$$

The scattering matrix elements are then simply

$$|\langle\mathbf{k}+\mathbf{q}|W(\mathbf{r})|\mathbf{k}\rangle|^2_{\rm boundary}=|w^{\rm d}(\mathbf{q})|^2\frac{1}{N'}c_{\rm A}^{\rm b}c_{\rm B}^{\rm b}. \qquad (5.36)$$

However, if the width of the boundary δ is much less than $\Lambda_{\rm l}$, then the scattering may not be completely diffuse and one should resort to a full calculation of the matrix elements from an expression of the form

$$|S^{\rm b}(\mathbf{q})|^2=\frac{1}{N'^2}\sum_{ij}^{\rm boundary}\exp(-i\mathbf{q}\cdot\mathbf{R}_{ij}). \qquad (5.37)$$

In general, this will give rise to a scattering rate that depends upon the angle of incidence (i.e. the angle between the electron velocity $\mathbf{v}(\mathbf{k})$ and the plane of the boundary). The total boundary contribution of the resistivity will depend upon a fraction of the surface area of the boundary that lies somewhere between the total surface area of the particles if the scattering is entirely diffuse and the projected area if the scattering has a marked angular dependence.

A complete expression for the resistivity for all zone sizes may be found by collecting together the zone, boundary and matrix terms. If the zones are not much larger than Λ_l and if we neglect any cross-correlations between the different regions (which are thought to be relatively unimportant; Luiggi *et al.* 1980), this leads to an expression of the form

$$\begin{aligned}
&|\langle \mathbf{k}+\mathbf{q}|W(\mathbf{r})|\mathbf{k}\rangle|^2 \\
&= \frac{|\bar{w}(\mathbf{q})|^2}{N'^2}\sum_{ij}^{N'} \exp(-i\mathbf{q}\cdot\mathbf{R}_{ij}) + \frac{|w(\mathbf{q})|^2}{N'} \\
&\quad \times \Bigg[c'_A c'_B C^z \sum_{ij}^{\text{zone}} \alpha^z_{ij} \exp(-i\mathbf{q}\cdot\mathbf{R}_{ij}) \\
&\qquad + \frac{(c^m_A + c^z_A)}{2}\frac{(c^m_B + c^z_B)}{2} D^b \sum_{ij}^{\text{bdy}} \alpha^b_{ij} \exp(-i\mathbf{q}\cdot\mathbf{R}_{ij}) \\
&\qquad + V^m c^m_A c^m_B \sum_{ij}^{\text{matrix}} \exp(-i\mathbf{q}\cdot\mathbf{R}_{ij}) \Bigg],
\end{aligned} \tag{5.38}$$

Fig. 5.4. Model profile of a symmetrical concentration gradient across a zone boundary. The thickness of the boundary region is δ.

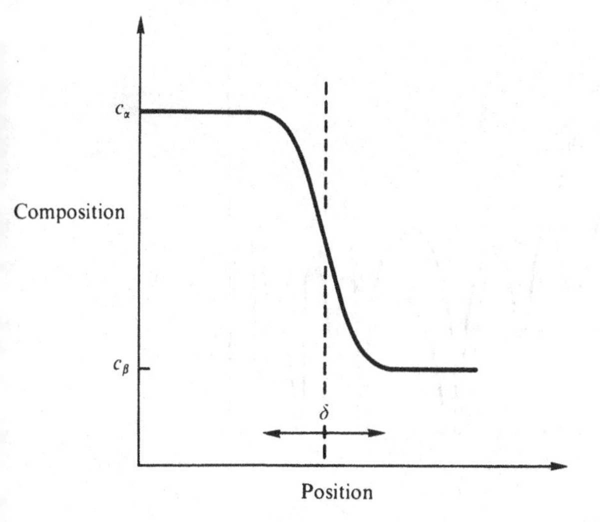

where the summations are over regions $\sim\Lambda_l$ which contain either a zone, a boundary or neither, and C^z is the zone concentration $= N_z/N$, D^b the boundary density $(\sim C^z R_z^2/a_0^2)$ and V^m the volume fraction of the matrix. This separation into the various terms is only approximate since the zone term assumes that the boundary is sharp (i.e. $\delta \ll \Lambda_l$), whereas the boundary term could allow for a thicker boundary. Furthermore, these terms overlap in the case of small zones $(R_z < \Lambda_l)$. However, this should not introduce much error since C^z is then likely to be large while D^b would be small. In the large zone limit there is no problem since the zone term is then zero. If the zone size increases at constant volume fraction D^B will steadily decrease with increased ageing time. If the volume fraction increases during ageing, then D^b may then increase or decrease, the actual behaviour depending upon the particular nucleation and growth processes involved. A more accurate expression than the above would involve a more rigorous evaluation of the structure factors. Equation (5.38) can be written as a sum of partial structure factors:

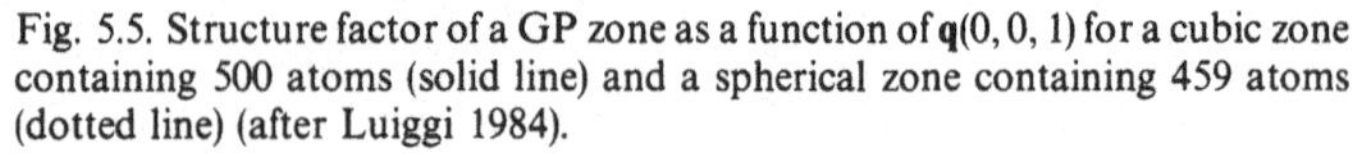

$$|\langle \mathbf{k}+\mathbf{q}|W(\mathbf{r})|\mathbf{k}\rangle|^2 = |\bar{w}(\mathbf{q})|^2|S(\mathbf{q})|^2 + |w^d(\mathbf{q})|^2[C^z|S^z(\mathbf{q})|^2 + D^b|S^b(\mathbf{q})|^2 + V^m|S^m(\mathbf{q})|^2]. \quad (5.39)$$

Although the average lattice term in equation (5.28) involves a sum over a relatively small number of lattice sites N', when the contributions from

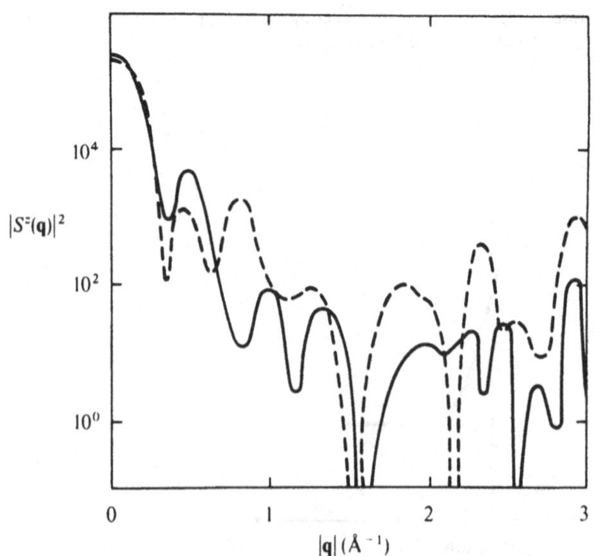

Fig. 5.5. Structure factor of a GP zone as a function of $\mathbf{q}(0, 0, 1)$ for a cubic zone containing 500 atoms (solid line) and a spherical zone containing 459 atoms (dotted line) (after Luiggi 1984).

the various regions are summed we get back exactly the complete perfect lattice term in equation (5.38).

In an isotropic model, all of the exponentials are replaced by their spherical averages to give

$$\left\langle\frac{1}{\tau(\mathbf{k})}\right\rangle=\left\langle\frac{1}{\tau^{z}(\mathbf{k})}\right\rangle+\left\langle\frac{1}{\tau^{b}(\mathbf{k})}\right\rangle+\left\langle\frac{1}{\tau^{m}(\mathbf{k})}\right\rangle, \tag{5.40}$$

whereas in an anisotropic calculation the averaged anisotropic relaxation time $\langle\tau(\mathbf{k})\rangle$ should be obtained from an appropriate solution of the Boltzmann equation for each of the partial structure factors and

$$\frac{1}{\langle\tau(\mathbf{k})\rangle}+\frac{1}{\langle\tau^{z}(\mathbf{k})\rangle}+\frac{1}{\langle\tau^{b}(\mathbf{k})\rangle}+\frac{1}{\langle\tau^{m}(\mathbf{k})\rangle}. \tag{5.41}$$

In general, the matrix scattering may be regarded as isotropic since the matrix is usually disordered. As described above, the boundary term has not yet been investigated in much detail but as an approximation it may also be regarded as diffuse and described by a slowly varying function of ageing time. The zone term can be expressed as a sum over atoms within the zone (1) and another sum (2) over those outside the zone but still in the region of interest giving

$$\begin{aligned}&\frac{c'_{A}c'_{B}}{N'}\sum_{ij}^{\text{zone}}\alpha_{ij}\exp(-i\mathbf{q}\cdot\mathbf{R}_{ij})\\&\quad=\frac{N_1}{N'}c'_{A}c'_{B}\sum_{ij}^{1}\alpha_{ij}\exp(-i\mathbf{q}\cdot\mathbf{R}_{ij})+\frac{N_2}{N'}c'_{A}c'_{B},\end{aligned} \tag{5.42}$$

where we have again assumed the matrix is disordered. For a three-dimensional zone in the form of a regular parallelepiped, the exponential term may be written as (Warren 1969, p. 28)

$$\prod_{j=1}^{3}\left[\frac{\sin^{2}[\frac{1}{2}(\frac{1}{4}\Gamma)^{1/3}\mathbf{q}\cdot\mathbf{a}_{j}]}{\sin^{2}(\frac{1}{2}\mathbf{q}\cdot\mathbf{a}_{j})}\right], \tag{5.43}$$

where $\mathbf{a}_j$ are the primitive cubic lattice vectors and $(\frac{1}{4}\Gamma)^{1/3}$ is the number of simple cubic cells in a fcc zone of Γ sites. This can be evaluated for a particular regular zone geometry (or (5.42) could be evaluated for an arbitrary zone geometry and averaged over all directions of $\mathbf{q}$) leading to the variation of $\langle|S^{z}(\mathbf{q})|^{2}\rangle$ with $\mathbf{q}(0, 0, 1)$ shown in Figure 5.5. For a regular platelet zone the exponential factor becomes (Hillel 1970)

$$\prod_{j=1}^{2}\left[\frac{\sin^{2}[(\Gamma)^{1/2}\mathbf{q}\cdot\mathbf{a}]}{\sin^{2}(\frac{1}{2}\mathbf{q}\cdot\mathbf{a}_{j})}\right]. \tag{5.44}$$

If the effects of anisotropy are neglected, the exponential term may be written in the form (5.19) or the calculated structure factors averaged. A comparison of the structure factors averaged over all directions of $\mathbf{q}$ for two-dimensional and three-dimensional precipitates containing ~200 atoms is given in Figure 5.6.

However, as shown by Hillel & Rossiter (1981), the replacement of $1/\langle\tau(\mathbf{k})\rangle$ by $\langle 1/\tau(\mathbf{k})\rangle$ leads to an error of a factor ~2 in the peak zone resistivity of an Al–10%Zn alloy. Furthermore, in the case of ideal platelet clusters, the isotropic scattering approximation would not produce any peak in the resistivity at all. However, platelets form because of large coherency strains. While we have not as yet considered the effects of static atomic displacements (see Section 5.6.3) they can produce a large conduction electron scattering and so the actual thickness of such a zone seen by a conduction electron may be quite large.

The extension of these equations to include the exponential damping term rather than the sharp cutoff at Λ_1 is quite straightforward. All summations are rewritten to include the term $\exp(-R_{ij}/\Lambda_1)$. The pseudopotential is then determined essentially by a highly localised group of atoms and one may again define a number of atoms N' that have the greatest effect. This may be determined by adopting some criterion such as excluding all atoms whose contribution to the pseudopotential is less than 0.2 times that of an atom at the centre of the region of interest, corresponding to a radius of $\sim 1.6\Lambda_1$. However, as

Fig. 5.6. Average structure factors $|S^z(\mathbf{q})|^2$ for a flat zone (*a*) and a spherical zone (*b*), both containing ~200 atoms (after Hillel 1970).

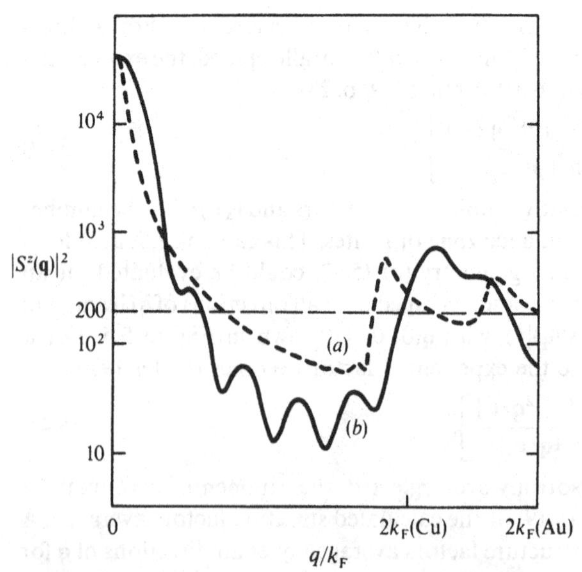

stated earlier, a complete calculation of the effects of inhomogeneous atomic correlations requires a self-consistent solution of the Boltzmann equation to obtain the local mean free path, and the method of solution should properly allow for an anisotropic relaxation time or mean free path, as discussed in Section 1.8.5. The self-consistency may be incorporated by using an initial estimate of $\Lambda^0 = \Lambda V^z$ (Hillel & Rossiter 1981) where Λ is an estimate of the bulk mean free path (~ 100 Å in Al–10%Zn) and V^z the volume fraction of zones, use this to determine the scattering potential and hence the first calculation of Λ_z which we write as Λ_z^1, replace Λ_z^0 with Λ_z^1 and redetermine the scattering potential to obtain Λ_z^2. The process is then repeated until $\Lambda_z^{n+1} = \Lambda_z^n$.

As the zone size becomes much larger than Λ it will be necessary to add the zone, boundary and matrix contributions, taking proper account of the distribution and volume fraction of each of these 'phases', although, if the volume fraction of zone is small, a simple law of mixtures may suffice. Some more general models are described in Section 5.5.1.

For some years there was a controversy over the relative importance of the local mean free path effect (the so-called RW mechanism) and the anisotropy effect (the HEW mechanism) in clustering materials. We can now see that both effects are important, although to varying degrees depending upon the morphology of the precipitate (or ordered region). If the zone is essentially three-dimensional (i.e. spherical or parallelepiped) the local mean free path effect gives the resistivity peak, but the magnitude of the peak is in error unless the relaxation time anisotropy in the peak region is included. It may be noted that the anisotropy mechanism alone can give rise to a peak in the resistivity but the resistivity beyond the peak is then incorrectly predicted to vary as $\Gamma^{4/3}$ for each cluster (Hillel *et al.* 1975) rather than the surface scattering dependence of $\Gamma^{2/3}$. Furthermore, the physical basis of such a peak is dubious since the wrong scattering potential is then implied. This behaviour is illustrated in Figure 5.7, the results being based on the semi-phenomenological model of Hillel & Edwards (1977) and Hillel (1983), using the zone data of Osamura *et al.* (1973, 1982). However, if the zone is essentially two-dimensional (i.e. platelet) the anisotropy mechanism is likely to be more dominant, depending upon the extent of the strain fields around the particle. In both cases, the initial increase is adequately described by a homogeneous model such as (5.20) when the appropriate values of α_i and $w^d(\mathbf{q})$ are employed.

It is interesting to compare the effects of measuring temperature T_m on these two mechanisms. If the HEW mechanism were acting alone, the maximum in ρ should occur when the effects of anisotropy became significant, i.e. when $\rho_{Br} \approx \rho_{is}$ in equation (5.34). Since ρ_{is} increases with T_m

(due to the phonon contribution) an increase in T_m would delay the onset of significant anisotropy so that the maximum in ρ would be displaced to larger particle sizes (i.e. longer ageing times) and would be larger in magnitude. On the basis of the RW mechanism alone, an increase in T_m would decrease Λ_l (again because of the effects of phonon scattering), so that the maximum in ρ would be displaced to smaller zone sizes (shorter ageing times) and would be smaller in magnitude. Thus in the case of three-dimensional clusters the two mechanisms produce an opposite effect as T_m is varied. However, for two-dimensional clusters the RW mechanism is probably less significant (again depending upon the magnitude and extent of the strain fields) and the observed behaviour may become dominated by the HEW mechanism. Hillel (1983) has performed a semiphenomenological calculation of these effects for an Al–10 at.%Zn alloy, again using the experimental data of Osamura *et al.*

Fig. 5.7. The variation in residual resistivity for an Al–10at.%Zn alloy aged at 25 °C. Curve *A*: no mean free path or anisotropy effects; curve *B*: with anisotropy but no mean free path; curve *C*: mean free path but no anisotropy; curve *D*: both mean free path and anisotropy effects (after Hillel 1983). Also shown (●) are the experimental results of Osamura *et al.* (1973) measured at 77 K.

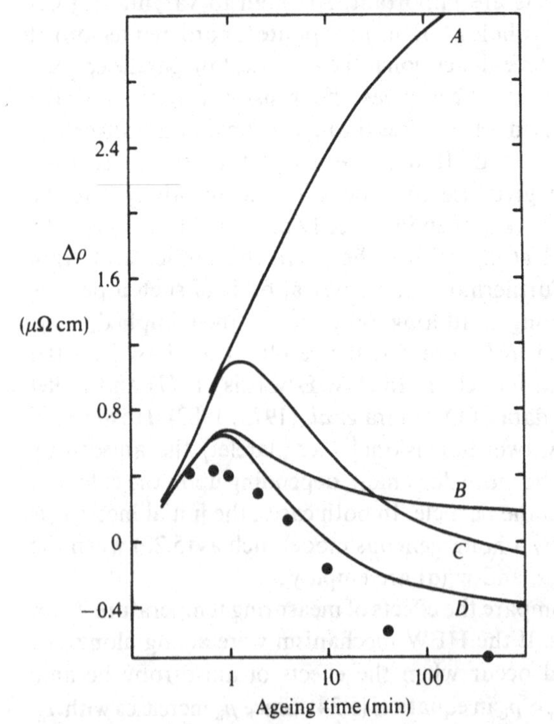

(1973, 1982) for the number density, size of zones and distribution of zone sizes as a function of ageing time. His results are shown in Figure 5.8 and confirm that the trend is as indicated above, although the effect is quite small for the RW mechanism. This might be expected since the local mean free path is given approximately by:

$$\frac{1}{\Lambda_l}=\frac{1}{\Lambda_m}+\frac{1}{\Lambda_p}+\frac{1}{\Lambda_z}. \qquad (5.45)$$

From experiment we know that $\Lambda_z \approx 10$ Å at the peak in resistivity (Osamura *et al.* 1973). This quantity is given very roughly by $\Lambda_z \approx V^z\Lambda$ (Hillel & Rossiter 1981), where Λ is the bulk mean free path ≈ 100 Å and $V^z \sim 0.1$. The corresponding local mean free paths (and hence the particle sizes) at the peak in resistivity are thus given as a function of temperature by the values shown in Table 5.1. The effect of incorporating both mechanisms is to virtually eliminate any shift in the peak with measuring temperature. One might note that the HEW mechanism alone gives a quite unrealistic result at $T_m = 318$ K. However, as noted above, any two-dimensional platelet clusters will usually be associated with large local strains making the particle effectively less two dimensional. These strain effects may be incorporated into the calculation by properly evaluating the structure factor terms, as discussed in Section 5.6.3. There is also the possibility of multilayer cluster formation.

Fig. 5.8. The temperature dependence of the change in resistivity of an Al–10at.%Zn alloy calculated on the basis of only anisotropy (*a*) and only finite local mean free path (*b*) effects. The full line corresponds to $T_m = 77$ K and the dashed line to $T_m = 318$ K (after Hillel 1983).

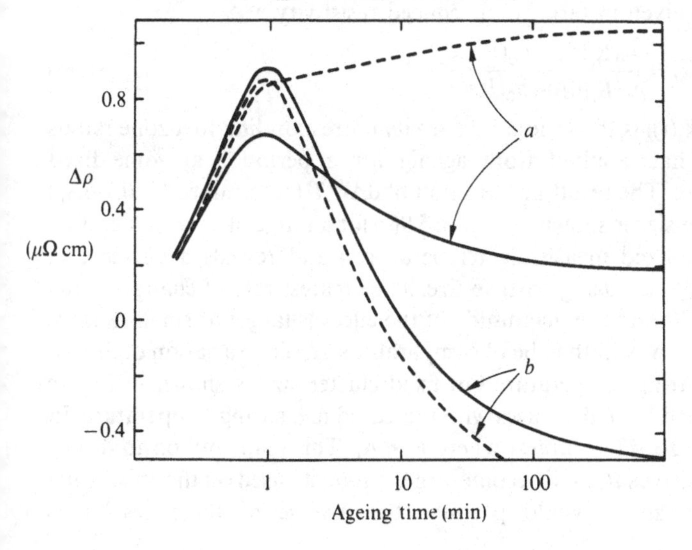

Table 5.1. *The approximate local mean free path for an Al–Zn alloy at various measuring temperatures* T_m

T_m (K)	Λ_m (Å)	Λ_z (Å)	Λ_p (Å)	Λ_l (Å)
0	240	10	∞	9.6
100	240	10	560	9.4
200	240	10	175	9.1
300	240	10	90	8.7

Equation (5.45) suggests another form of unusual behaviour. If the zone size is kept fixed so that Λ_z is constant, varying the temperature at which the resistivity is determined (and hence Λ_l) may lead to a change in the temperature coefficient of resistivity. This is because Λ_l may initially be larger than R_z at low temperatures, but at high temperatures we may find $\Lambda_l < R_z$. The resistivity behaviour is then changing from the small zone limit to the large zone limit as the temperature increases. Similarly, the effects of scattering anisotropy will be delayed to larger zone sizes at higher temperatures. Thus, while there may be no change in the phonon scattering, there will be a contribution to ρ which depends upon measuring temperature and hence a change in the temperature coefficient of resistivity. This effect has been investigated by Aubauer & Rossiter (1981) using a model which included a simple cut-off of the form (5.27), but which neglected the effects of anisotropy (Aubauer 1978). The results are given in terms of a reduced resistivity

$$\Delta(t) = \frac{\rho_0(R_z(t)) - \rho_0(R_z(\infty))}{\rho_0(R_z(0)) - \rho_0(R_z(\infty))} \tag{5.46}$$

where $\rho_0(R_z(t))$ is the residual resistivity corresponding to a zone radius R_z which has resulted from ageing for a period t at some fixed temperature. The resulting variation of $\mathrm{d}\Delta/\mathrm{d}T$ (determined by $\mathrm{d}\Delta/\mathrm{d}\Lambda_p$) with cluster size is shown in Figure 5.9(*a*) for a range of fixed values of Λ_l (and hence fixed measuring temperatures) and reveals a decrease in $\mathrm{d}\Delta/\mathrm{d}T$ with increasing particle size. The greatest rate of change occurs when $\Lambda_l \approx R_z$ and the magnitude of the effect is larger at small Λ_l since both Λ_p and Λ_z will then be of comparable size. The variation of $\mathrm{d}\Delta/\mathrm{d}T$ with measuring temperature but fixed cluster size is shown in Figure 5.9(*b*). Here $\mathrm{d}\Delta/\mathrm{d}T$ decreases with increased measuring temperature. In both cases $\mathrm{d}\Delta/\mathrm{d}T$ is largest where $R_z < \Lambda_l$. This contribution to $\mathrm{d}\rho/\mathrm{d}T$ should vanish as R_z or T become large. A model based on the anisotropy mechanism alone would produce the inverse of these results as schematically indicated by the dotted lines in Figures 5.9(*a*) and (*b*). It should be noted that this effect is just another way of looking at the variation in the resistivity peak with measuring temperature and both of them may be regarded as a 'size-induced' deviation from Matthiessen's rule. The two representations are brought together in Figure 5.10.

All of the above discussion has assumed that there is no intrinsic difference in the temperature coefficient of resistivity of the zones or the matrix. This is generally true if only simple metals are involved, but care

Fig. 5.9. (*a*) $d\Delta/dT$ as a function of reduced precipitate size $r_0 = R_z/\Lambda_1$ for various values of reduced mean free path $l_0 = k_F\Lambda_1$; (*b*) as a function of reduced temperature T_0 (given by R_z/Λ_1) for various zone sizes R_z. Solid lines: mean free path model calculations (after Aubauer & Rossiter 1981). Dotted line: behaviour expected of the anisotropy mechanisms (schematic).

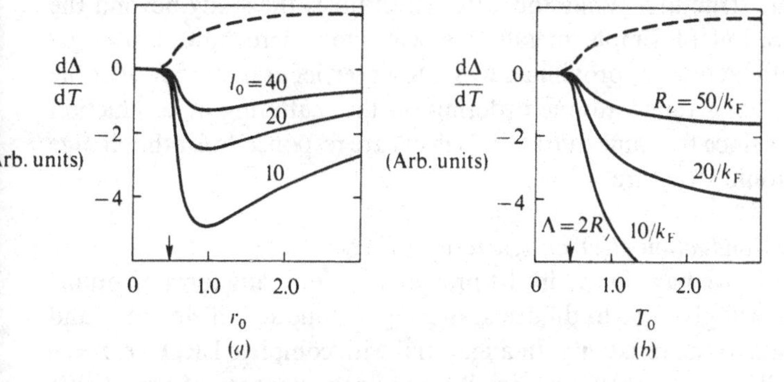

Fig. 5.10. Schematic representation of the variation of resistivity with temperature for specimens having different zone sizes (*a*), and the corresponding resistivity as a function of zone size for different measuring temperatures (*b*).

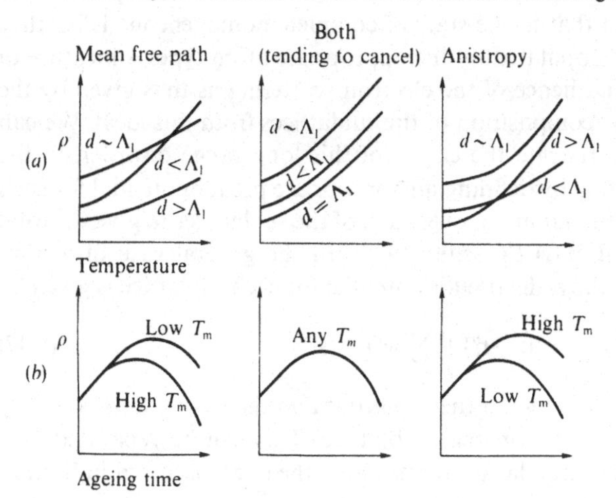

must be exercised in analysing systems that contain transition metals, as it is then possible to find a large difference in the intrinsic temperature coefficients which can also markedly affect the behaviour observed at different measuring temperatures. For example, in Fe–Ni–Cr stainless steels containing $\gamma'Ni_3(Ti, Al)$ precipitates, these effects result in a larger peak in the resistivity at higher measuring temperatures (see e.g. Wilson & Pickering 1968; Raynor & Silcock 1968, 1970; Silcock 1970*a*,*b*), while the opposite effect occurs in some maraging steels (Jones *et al.* 1971).

5.3 Homogeneous long range atomic ordering

Homogeneous long range atomic order will affect the electrical resistivity of an alloy through both the effects of conduction electron scattering and the effects of diffraction by the superlattice on the electronic structure. While the latter effect takes the study beyond the consideration of simple metals, we will show here how it may be incorporated in an approximate way. However, let us firstly consider the effects of long range atomic ordering on the scattering of conduction electrons since the same diffraction effects are responsible for the change in electronic structure.

5.3.1 *Conduction electron scattering effects*

As we have found in the previous sections, any form of atomic disorder will give rise to diffuse scattering of conduction electrons and contribute to the resistivity. In a material with complete LRO (i.e. $S=1$) there will be no such scattering. We require a measure of the atomic disorder due to incomplete LRO. This may be obtained by remembering that (see Section 2.2) the superlattice may be decomposed into sublattices such that in the state of complete homogeneous LRO they will be entirely populated by the appropriate atom type. A measure of the disorder, and hence of the electron scattering, is thus given by the deviation of the composition of the sublattices from this ideal. We can then easily incorporate the effects of this long range disorder on the electron scattering by defining an appropriate pseudopotential for each sublattice and the summing over each of the sublattices to give the total pseudopotential $W(\mathbf{r})$ (Rossiter 1979). In the general case of a non-stoichiometric alloy, the pseudopotential for each sublattice is given by

$$W^n(\mathbf{r}) = \sum_{\mathbf{r}_i^n(\mathrm{A})} w^A(\mathbf{r}-\mathbf{r}_i^n) + \sum_{\mathbf{r}_i^n(\mathrm{B})} w^B(\mathbf{r}-\mathbf{r}_i^n), \tag{5.47}$$

where the index n refers to the sublattice and the summations are only over A and B atoms on that sublattice. This can be separated into average and deviation lattice terms along the lines discussed in Section

4.5.3. The total pseudopotential is then:

$$W(\mathbf{r}) = \sum_n W^n(\mathbf{r}) \tag{5.48}$$

$$= \sum_n \bar{w}(\mathbf{q}) S^n(\mathbf{q}) + \sum_n w^{\mathrm{d}}(\mathbf{q}) S^{\mathrm{d},n}(\mathbf{q})$$

$$= \bar{w}(\mathbf{q}) \sum_n S^n(\mathbf{q}) + w^{\mathrm{d}}(\mathbf{q}) \sum_n S^{\mathrm{d},n}(\mathbf{q}), \tag{5.49}$$

where

$$S^n(\mathbf{q}) = \frac{1}{N^n} \sum_{\mathbf{r}_i^n} \exp(-i\mathbf{q}\cdot\mathbf{r}_i), \tag{5.50}$$

$$S^{\mathrm{d},n}(\mathbf{q}) = \frac{P_{\mathrm{B}}^n}{N^n} \sum_{\mathbf{r}_i^n(\mathrm{A})} \exp(-i\mathbf{q}\cdot\mathbf{r}_i) - \frac{P_{\mathrm{A}}^n}{N^n} \sum_{\mathbf{r}_i^n(\mathrm{B})} \exp(-i\mathbf{q}\cdot\mathbf{r}_i), \tag{5.51}$$

and N^n is the number of atoms on the nth sublattice and P_{A}^n the probability of finding an A atom on the nth sublattice. The scattering probability will thus be determined by $|\sum_n S^{\mathrm{d},n}(\mathbf{q})|^2 |\langle \mathbf{k}+\mathbf{q}|w^{\mathrm{d}}|\mathbf{k}\rangle|^2$. Expanding the square of the deviation structure factor gives

$$\left|\sum S^{\mathrm{d},n}(\mathbf{q})\right|^2 = \sum_n \sum_m \frac{1}{N^n N^m}$$
$$\times \left[P_{\mathrm{B}}^n P_{\mathrm{B}}^m S_{\mathrm{A}}^{n*}(\mathbf{q}) S_{\mathrm{A}}^m(\mathbf{q}) + P_{\mathrm{A}}^n P_{\mathrm{A}}^m S_{\mathrm{B}}^{n*}(\mathbf{q}) S_{\mathrm{B}}^m(\mathbf{q})\right.$$
$$\left. - P_{\mathrm{A}}^n P_{\mathrm{B}}^m [S_{\mathrm{A}}^{n*}(\mathbf{q}) S_{\mathrm{B}}^m(\mathbf{q}) + S_{\mathrm{A}}^n(\mathbf{q}) S_{\mathrm{B}}^{m*}(\mathbf{q})]\right], \tag{5.52}$$

which involves terms of the form

$$\frac{1}{N^n N^m} \left[\sum_{\mathbf{r}_i^n(\mathrm{A})} \sum_{\mathbf{r}_i^m(\mathrm{B})} \exp[-i\mathbf{q}\cdot(\mathbf{r}_i^n(\mathrm{A}) - \mathbf{r}_i^m(\mathrm{B}))] \right]. \tag{5.53}$$

As in the case of homogeneous short range ordering, these may be rearranged to collect together all terms having the same separation $\mathbf{R}_{ij\mathrm{AB}}^{nm} = \mathbf{r}_i^n(\mathrm{A}) - \mathbf{r}_j^m(\mathrm{B})$ and the summation carried out over all i, j. We then replace the double summation over specific sites by a summation over all pairs multiplied by the configurationally averaged number of such pairs (as for (5.15)) expressed in terms of the site occupation parameters (Chapter 2)

$$P_{ij\mathrm{AB}}^{nm} = \langle \sigma_i^{\mathrm{A}} \sigma_j^{\mathrm{B}} \rangle^{nm} \tag{5.54}$$

$$= P_{\mathrm{A}}^n P_{\mathrm{B}}^m - \langle \sigma_i \sigma_j \rangle^{nm}. \tag{5.55}$$

If we assume for the moment

$$\mathbf{R}_{ij\mathrm{AB}}^{nm} = \mathbf{R}_{ij\mathrm{AA}}^{nm} = \mathbf{R}_{ij\mathrm{BB}}^{nm} = \mathbf{R}_{ij}^{nm}, \tag{5.56}$$

and $N^n = N^m$, equations (5.52) and (5.53) give

$$\left|\sum_n S^{\mathrm{d},n}(\mathbf{q})\right|^2 = \frac{1}{N} \sum_n \sum_m \sum_{ij} \langle \sigma_i \sigma_j \rangle^{nm} \exp(-i\mathbf{q}\cdot\mathbf{R}_{ij}^{nm}). \tag{5.57}$$

This is just the sublattice analog of 5.17.

(a) *Bragg–Williams model*

If we now adopt the Bragg–Williams model for LRO, the parameter S' (or S) determines the partition of atoms between 'right' and 'wrong' sites with respect to the perfectly ordered structure. However, as noted in Chapter 2, it also assumes that the detailed distribution of atoms over the sublattice is random and so $\langle\sigma_i\sigma_j\rangle^{nm}=0$ for $i\neq j$, $n\neq m$. Equation (5.57) then becomes

$$\left|\sum_n S^{\mathrm{d,n}}(\mathbf{q})\right|^2=\frac{1}{N}\sum_n\langle\sigma_i\sigma_j\rangle^{nn}y^n$$
$$=\frac{1}{N}\sum_n\langle\sigma^{\mathrm{A}}\rangle^n\langle\sigma^{\mathrm{B}}\rangle^n y^n, \tag{5.58}$$

where y^n is the fraction of sites in the nth sublattice. From equations (2.7), (2.8) and (5.58) we find for an fcc $L1_0$ structure:

$$\left|\sum_n S^{\mathrm{d,n}}(\mathbf{q})\right|^2=\frac{1}{N}c_{\mathrm{A}}c_{\mathrm{B}}(1-\tfrac{16}{3}c_{\mathrm{A}}c_{\mathrm{B}}S'^2), \tag{5.59}$$

and for a bcc B_2 structure:

$$\left|\sum_n S^{\mathrm{d,n}}(\mathbf{q})\right|^2=\frac{1}{N}c_{\mathrm{A}}c_{\mathrm{B}}(1-4c_{\mathrm{A}}c_{\mathrm{B}}S'^2). \tag{5.60}$$

The variation of the maximum value of S' with composition may be deduced from equation (2.6) since, if $y^n>c_{\mathrm{A}}$, the maximum value of $\langle\sigma^{\mathrm{A}}\rangle^n$ is c_{A}/y^n and similarly, if $y^m>c_{\mathrm{B}}$ the maximum value of $\langle\sigma^{\mathrm{B}}\rangle^m$ is c_{B}/y^m. This variation is shown in Figure 5.11 for both the $L1_0$ and B_2 structures.

Given a suitable form of the pseudopotential $w^{\mathrm{d}}(\mathbf{q})$, the resistivity due to homogeneous long range ordering may be determined from (5.59) or (5.60) and

$$\rho=CN\int_0^{2k_{\mathrm{F}}}\left|\sum_n S^{\mathrm{d,n}}(\mathbf{q})\right|^2|w^{\mathrm{d}}(\mathbf{q})|^2q^3\,\mathrm{d}q, \tag{5.61}$$

which follows directly from equation (5.9). The general variation of resistivity as $1-S^2$ in stoichiometric alloys is in agreement with the earlier study of Muto (1936). (In the pseudopotential representation, $E=1$ and terms linear in S in Muto's study vanish.) For non-stoichiometric alloys, the Muto result predicts an order dependence of the LRO term on c_{A}^2 (or c_{B}^2) rather than the product $c_{\mathrm{A}}^2c_{\mathrm{B}}^2$ found here.

(b) *Coexisting long and short range ordering*

The Bragg-Williams concept of LRO specifically excludes consideration of the detailed distribution of atoms over the sublattices. As such it overestimates the degree of atomic disorder, particularly on a scale of Λ. Any correlation between the atom positions either within or between sublattices will affect the electrical resistivity and so must be taken into account in determining the total order-dependent resistivity. To this end we start with the general expression for homogeneous correlations given by equation (5.18). However, the correlations $\langle\sigma_i\sigma_j\rangle$ now contain a periodic component due to the long range ordering. This component is responsible for the superlattice Bragg scattering and will not contribute to the resistivity. However, with the aid of the LRO parameters defined by equation (2.16), we may write (5.18) as

$$|S^{\mathrm{d}}(\mathbf{q})|^2 = \frac{1}{N}\sum_{ij} c_{\mathrm{A}}c_{\mathrm{B}}S_{ij}\exp(-\mathrm{i}\mathbf{q}\cdot\mathbf{R}_{ij}) + \frac{1}{N}\sum_{ij}(\langle\sigma_i\sigma_j\rangle - c_{\mathrm{A}}c_{\mathrm{B}}S_{ij})\exp(-\mathrm{i}\mathbf{q}\cdot\mathbf{R}_{ij}). \tag{5.62}$$

The first summation in this expression is just the periodic component mentioned above and, as it is the sum over all sites in a perfectly periodic superlattice, we can see why it gives no contribution to ρ. The terms included in the second summation decay with distance and the resulting

Fig. 5.11. The maximum value of the long range order parameter S' as a function of composition for binary alloys having the $L1_0$ (full curve) and B_2 (dashed curve) structure (after Rossiter 1979).

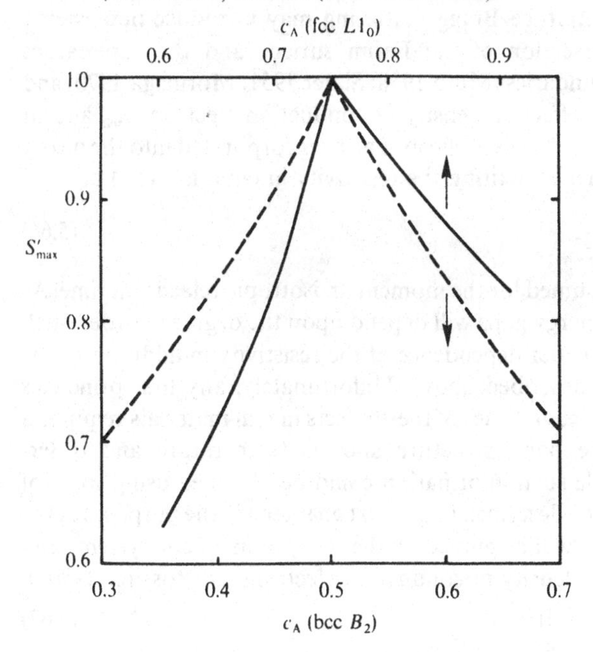

diffuse scattering will contribute to ρ. If short range correlations are omitted, (5.62) gives

$$|S^{d}(\mathbf{q})|^{2}=\frac{1}{N}c_{A}c_{B}(1-S_{ij}). \tag{5.63}$$

Substitution of the appropriate values of S_{ij} (assuming an origin at the cube corner, equations (2.17) and (2.18)) then immediately gives equations (5.59) and (5.60). This derivation is in some ways more appealing than that initially used to derive (5.59) and (5.60) as it avoids the rather complicated sublattice prescription. More generally, if we define an effective SRO parameter for temperatures below T_c by (Rossiter 1979)

$$c_{A}c_{B}\alpha'_{ij}=\langle\sigma_{i}\sigma_{j}\rangle-c_{A}c_{B}S_{ij}, \tag{5.64}$$

then the calculation of ρ due to LRO and SRO is identical to that used for homogeneous short range atomic correlations leading to

$$\rho=c_{A}c_{B}C\int_{0}^{2k_{F}}|w^{d}(\mathbf{q})|^{2}\sum_{i}c_{i}\alpha'_{i}\frac{\sin(qR_{i})}{qR_{i}}q^{3}\,dq. \tag{5.65}$$

The $i=0$ term gives the long range disorder contribution while the remaining terms give the correction due to short range atomic correlations.

5.3.2 *Electron band structure effects*

The superlattice Bragg scattering may introduce new energy gaps at the intersection of the Fermi surface and the superlattice Brillouin zone boundaries (Muto 1938; Slater 1951; Morinaga 1970) and hence change the effective density of conduction electrons n_{eff} and/or their effective mass m^*. These effects may be incorporated into the nearly free electron model by writing the resistivity in equation (1.5) as

$$\rho=\frac{m^*}{n_{eff}e^{2}\tau}, \tag{5.66}$$

where we have assumed for the moment an isotropic relaxation time. As the size of these energy gaps will depend upon the degree of order, such effects lead to an order dependence of the resistivity in addition to the scattering effects described above. Unfortunately, any first principles calculation of the magnitudes of these effects in real materials requires a knowledge of the band structure and its temperature and order-dependence. While such information could be obtained using some of the techniques to be described in the next chapter, for the purposes of the discussion here we will be guided by the study of magnetic systems and write the effective density of conduction electrons as (Rossiter 1980*a*)

$$n_{eff}=n_{0}(1-A|S(q)|^{2}), \tag{5.67}$$

where n_0 is the number of conduction electrons per unit volume in the disordered state and the coefficient A depends upon the relative positions of the Fermi surface and the superlattice Brillouin zone boundaries. The appropriate structure factor is given by equation (5.59) or (5.60) and so, for a stoichiometric alloy,

$$n_{\text{eff}} = n_0(1 - AS^2). \tag{5.68}$$

From (5.59) or (5.60) we also get the order dependence of the relaxation time

$$\frac{1}{\tau(S)} = \frac{1}{\tau(0)}(1 - S^2), \tag{5.69}$$

where we have indicated the explicit dependence of τ on the degree of LRO S. From (5.66), (5.68) and (5.69)

$$\frac{\rho_0(S)}{\rho_0(0)} = \frac{1 - S^2}{1 - AS^2}. \tag{5.70}$$

The temperature coefficient of resistivity will also depend upon n_{eff} via equation (5.66) and so

$$\frac{\partial \rho_{\text{p}}(T)}{\partial T} = \frac{B}{n_{\text{eff}}(S)}. \tag{5.71}$$

If the measuring temperature is above Θ_{D}, the phonon contribution to the resistivity may be approximately written as

$$\rho_{\text{p}}(T) = T\frac{\partial \rho_{\text{p}}(T)}{\partial T}. \tag{5.72}$$

The total resistivity of a specimen of order S at some temperature T is

$$\rho(S, T) = \rho_0(S) + \rho_{\text{p}}(T), \tag{5.73}$$

and so, from (5.70)–(5.73),

$$\rho(S, T) = \rho_0(0)\left(\frac{1 - S^2}{1 - AS^2}\right) + \frac{TB}{n_0(1 - AS^2)}. \tag{5.74}$$

In order to show the general form of this behaviour we can assume that $\rho_0(0) = 1$ and choose a value of B/n_0 which gives a phonon contribution to the resistivity at T_c that is equal to the residual resistivity of a specimen quenched from T_c. This latter behaviour is typical of ordering alloys and gives $B/n_0 = 1/T_c$ (in the more detailed comparison with experiment given in Section 5.7.4 these parameters will actually be derived from experimental data appropriate to the specific alloys under consideration). With these assumptions the values deduced from equation (5.74) are shown in Figure 5.12 for a specimen whose resistivity was determined at the temperature of interest and assuming an equilibrium degree of order at all temperatures (curve A) or at 0 K after quenching from the temperature of interest (curve B). In all cases the temperature dependence of S was obtained from a simple Bragg–Williams calculation for a fcc A_3B alloy (see e.g. Bragg & Williams 1935). In Figure 5.12(*a*) the constant A has been given a range of positive values corresponding to a decrease in n_{eff} with increasing LRO. These results show that, if A is large enough, the electrical resistivity determined at-temperature may actually increase with decreasing temperature over a range of temperatures below T_c (curve A), although the residual resistivity (curve B) always decreases with decreasing temperature below T_c. Negative values of A (i.e. increased ordering giving an increase in n_{eff}) lead to the results shown in Figure 5.12(*b*) and simply accentuate the $1-S^2$ behaviour expected from the electron scattering effects embodied in $\tau(S)$. Short range order may affect both the electron scattering terms and size of the energy gaps (see e.g. Woolley & Mattuck 1973). The former could be ascertained from equations (5.64) and (5.65) and the correction terms added to (5.69) to give

$$\rho(S,T)=\frac{c_A c_B C}{1-AS^2}\int_0^{2k_F}|w^d(\mathbf{q})|^2\sum_i c_i\alpha_i'\frac{\sin(qR_i)}{qR_i}q^3\,dq + \frac{\rho_p(T)}{1-AS^2}, \tag{5.75}$$

where $\rho_p(T)$ is the phonon contribution to be discussed in Section 5.6.2. The latter would add correction terms to the factor $(1-AS^2)$, although, in view of the rather empirical nature of this equation, such a modification is probably not justified at this stage. A first-principle's

Fig. 5.12. (*a*) Reduced electrical resistivity as a function of reduced temperature for various values of the parameter A as indicated on the curves. Curve A refers to at-temperature results, i.e. $T=T_m$ and equilibrium LRO is assumed at each temperature, while curve B relates to a specimen quenched from a temperature T (thus retaining that degree of LRO) and the resistivity measured at $T=0$ K. (*b*) As for (*a*) but with the negative values of A as indicated (after Rossiter 1980).

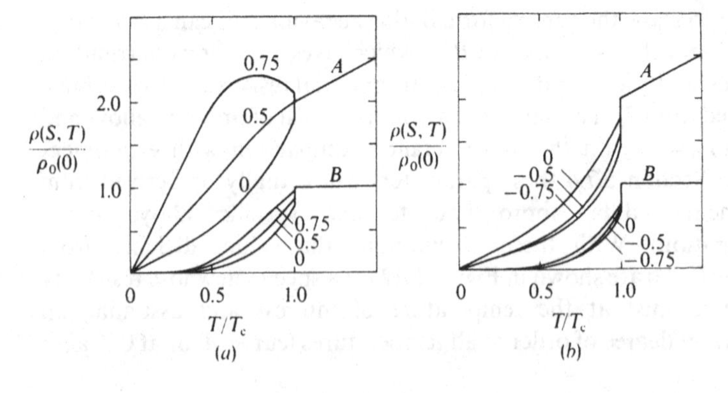

calculation of the band structure term would require a full determination of the band structure including the effects of long and short range correlations. Some of the current approaches to this problem will be discussed in the next chapter.

5.4 Inhomogeneous long range ordering

When the long range atomic order is no longer homogeneous we again have to consider whether the scale of the inhomogeneities is larger or less than Λ_l. Indeed, the whole concept of LRO in finite crystals and those containing grain or antiphase domain boundaries requires careful consideration. Here we will adopt an approach originally given by Cowley (1965) as part of a general study of order parameters to the case of a finite Λ_l (Rossiter 1979). In order to simplify the argument we will consider a one-dimensional stoichiometric AB alloy. The variation of $\langle\sigma_i\sigma_j\rangle$ with distance R_{ij} for an infinite crystal with perfect LRO is as shown in Figure 5.13(*a*). In the case of a finite crystal with N unit cells of dimension a_0 this is modified, as shown in Figure 5.13(*b*). The intercept of $\langle\sigma_i\sigma_j\rangle$ for $R_{ij} \to 0$ may then be taken as a measure of the degree of LRO and, in accordance with the one-dimensional equivalent of (2.17) or (2.18), is proportional to the square of the Bragg–Williams order parameter. If $Na_0 \gg \Lambda$, there will be no deviation in the periodicity of the potential on a scale of Λ, the electrons will undergo Bragg scattering and the order-dependent contribution to the resistivity is zero, the mean free path then being determined by other defects and thermal motion of the atoms. This is consistent with the theory presented in the previous section, since the LRO contribution to ρ is given by the $\mathbf{R}_{ij}=0$ term in the series expansion of the structure factor (equation (5.63)). This term is the same as the $R_{ij}=0$ intercept in the infinite crystal case and the LRO is still 'perfect'. The other terms in the expansion must be truncated for $R_{ij} > \Lambda$ and over the range $R_{ij} < \Lambda$ they differ from the infinite crystal case by only a very small amount and so the short range terms in (5.65) are negligible. If $Na_0 < \Lambda$ the LRO may still be perfect but the potential is no longer periodic on a scale $\sim\Lambda$ and the short range terms in (5.65) are now appreciable. This is just the contribution usually called the 'size effect'. If the alloy has a periodic antiphase domain structure with domain dimensions εa_0 (i.e. a long period superlattice) and if $Na_0 > \Lambda$, the pairwise correlations are as shown in Figure 5.13(*c*). If the antiphase domain structure is not truly periodic the features become smeared, as in Figure 5.13(*d*). Whether or not the antiphase domain effects are important in determining the resistivity depends upon whether $\varepsilon a_0 >$ or $< \Lambda$, and upon the effects of the antiphase boundaries on the electron scattering. If $\varepsilon a_0 \gg \Lambda$, the material appears to be perfectly ordered, except in the region within a distance Λ_l of the boundary. This effect is due to the deviation in the periodicity of the potential at the boundary and gives a resistivity proportional to the area of the boundaries, as discussed in Section 5.2. If $\varepsilon a_0 \ll \Lambda$ the scattering will be determined by the amount that the pairwise correlations (and hence the lattice potential) deviate from a periodic function. If the long range order is not perfect (i.e. $S<1$) the values of $\langle\sigma_i\sigma_j\rangle$ may differ appreciably from the limiting LRO values, as shown in Figure 5.13(*e*). The value of the LRO parameter must then be redefined as the intercept on the R_{ij} axis by extrapolation from

Fig. 5.13. Variation of the pairwise correlation parameter $\langle\sigma_i\sigma_j\rangle$ with distance R_{ij} for a one-dimensional *AB* alloy; (*a*) infinite crystal with perfect LRO ($S=1$); (*b*) finite crystal with *N* unit cells of dimensions a_0; (*c*) finite crystal with periodic antiphase domains of dimensions εa_0; (*d*) as for (*c*) but with irregular domain structure; (*e*) finite crystal with imperfect LRO ($S<1$).

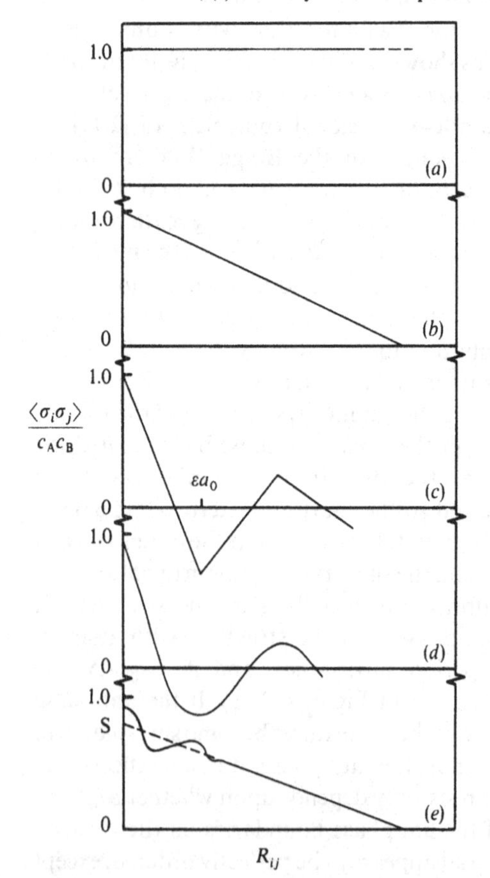

large R_{ij} back through the oscillations in $\langle\sigma_i\sigma_j\rangle$. This is of course only possible if the domain or crystallite size is large enough so that the spatial oscillations in $\langle\sigma_i\sigma_j\rangle$ due to SRO may be separated from those due to the superlattice or finite crystal size. The LRO and SRO contributions to ρ are then calculated as before. The extension of this argument to non-stoichiometry and three dimensions is relatively straightforward.

Let us now consider the effects of an antiphase boundary in somewhat greater detail. Such a boundary may be regarded as a step change in the phase of the ordering wave (see Section 2.3). This discontinuity in the periodicity of the lattice potential will cause electron scattering and may be treated in the same way as the boundary term in (5.38), although there is now the same mean composition on either side of the boundary so that the contribution to the resistivity from the boundaries becomes

$$\rho_{apb} = D^{b}c_{A}c_{B}C\int_{0}^{2k_{F}} |w^{d}(\mathbf{q})|^{2}\sum\alpha_{ij}\exp(-i\mathbf{q}\cdot\mathbf{R}_{ij})q^{3}\,dq, \tag{5.76}$$

where the coefficient D^{b} will depend upon the total antiphase domain boundary area. If the domains are large compared with Λ, the resistivity will be given by the sum of (5.75) and (5.76). If the antiphase domain boundary dimensions are comparable with Λ then the resistivity will be given simply by (5.75), their effect being to drastically reduce the degree of LRO and hence the significance of the band gap at the superlattice Brillouin zone. If the antiphase domain structure is completely periodic (i.e. a long period superlattice structure, LPS), and if the domain size determined now by the period of the LPS is less than Λ, the only effect will be to introduce an extra Bragg scattering component corresponding to the period of the LPS. This scattering will not by itself contribute to the resistivity. If the period of the LPS is greater than Λ, the main effect will simply be the introduction of the extra boundary scattering term (5.76). Finally, if a specimen is aged so that antiphase domains or regions of order which are initially smaller than Λ_l increase in size and become larger than Λ_l, the resistivity will pass through a peak value if the short range correlations are such that they lead to an initial increase in ρ. This is directly analogous to the behaviour during GP zone growth as discussed in Section 5.2.2. It may also give rise to an anomalous change in the temperature coefficient of resistivity as also discussed in that section.

5.5 Long range phase separation

Long range phase separation occurs at the advanced stages of precipitation (i.e. the large zone limit considered briefly in the previous section), at the advanced stages of spinodal decomposition where the composition profile becomes saturated in each phase and in structures that have formed from invariant reactions of the form $\alpha \rightarrow \beta + \delta$, such as eutectics or eutectiods. It is also possible to fabricate multiphase materials by a variety of powder metallurgy and composite techniques. The physical properties of such materials have been investigated for many years and, regardless of whether the actual property being considered is the magnetic permeability, dielectric constant, elastic modulus, or electrical or thermal conductivity, the general formalism is the same. Results from many different studies may thus be applied to the problem of the electrical conductivity or resistivity. As discussed in Section 2.6, the important structural parameters that characterise the system are the volume fractions of each phase, the geometrical distribution of the phases (i.e. whether the phases are regularly or randomly dispersed), the size and size distribution of the phases and the topology of the phases (whether they are regular forms such as spheres, spheroids, ellipsoids, cylinders or irregular shapes). By considering series expansions in the local fluctuations of a property (such as the conductivity) it appears that all of the above information is required to completely determine that property (Brown Jr 1955; Herring 1960). However, such information is usually not available and so we need to know how accurately the conductivity can be calculated given a more limited set of data. Conversely we want to find out if the measurement of a bulk property can give information about the structure of a multiphase material. There will be additional complications if any of the characteristic dimensions of the phase distribution become comparable with Λ. We start by considering the case where all dimensions are larger than Λ.

5.5.1 *Scale of phase separation* $\gg \Lambda$

The absolute minimum information required to characterise the electrical resistivity of a multiphase solid is the volume fraction and conductivity (or resistivity) of each phase. However, given only this information it is not possible to determine the resistivity of the composite exactly, although it is possible to determine upper and lower bounds. Let us consider a solid containing m different phases, each phase having a volume fraction v_t and conductivity σ_t. Using the variational approach of Hashin & Shtrickman (1962) it may be shown that the conductivity of the composite σ^* is bounded by

$$\left.\begin{aligned} &\sigma^* > \sigma_0 + \frac{A}{1-\alpha A} \quad \text{for } \sigma_t < \sigma_0 \\ \text{and} \quad & \\ &\sigma^* < \sigma_0 + \frac{A}{1-\alpha A} \quad \text{for } \sigma_t > \sigma_0, \end{aligned}\right\} \tag{5.77a}$$

where

$$\alpha=(3\sigma_0)^{-1}, \tag{5.77b}$$

and

$$A=\sum_{t=1}^{m}\frac{v_t}{(\sigma_t-\sigma_0)^{-1}+(3\sigma_0)^{-1}}. \tag{5.77c}$$

The (as yet unspecified) conductivity σ_0 is chosen to obtain the highest lower bound and the lowest upper bound. The worst possible choices for σ_0 are 0 and ∞, leading to the *lowest* lower bound and the *highest* upper bound (i.e. the most extreme bounds). In this case (5.77) becomes

$$\sum_{t=1}^{m} v_t\sigma_t>\sigma^*>\left|\sum_{t=1}^{m}\frac{v_t}{\sigma_t}\right|^{-1}. \tag{5.78}$$

Such bounds correspond to the series and parallel alignment of the phases with respect to the electric field, as shown in Figure 5.14. However, for a macroscopically homogeneous and isotropic distribution of phases these are not the best possible bounds. Since the right-hand side of (5.77a) is a monotonically increasing function of σ_0, the highest lower bound is given by $\sigma_0=\sigma_1$ where σ_1 is the smallest of the σ_t and the lowest upper bound by $\sigma_0=\sigma_m$ where σ_m is the largest of the σ_t. Accordingly

$$\sigma_1^*=\sigma_1+\frac{A_1}{1-\alpha_1 A_1}, \tag{5.79a}$$

where

$$\alpha_1=(3\sigma_1)^{-1}, \tag{5.79b}$$

$$A_1=\sum_{t=2}^{m}\frac{v_t}{(\sigma_t-\sigma_1)^{-1}+\alpha_1}, \tag{5.79c}$$

and

$$\sigma_m^*=\sigma_m+A_m(1-\alpha_m A_m), \tag{5.80a}$$

Fig. 5.14. 'Worst case' parallel and series alignment of the phases with respect to the electric field yielding the maximum and minimum conductivities respectively.

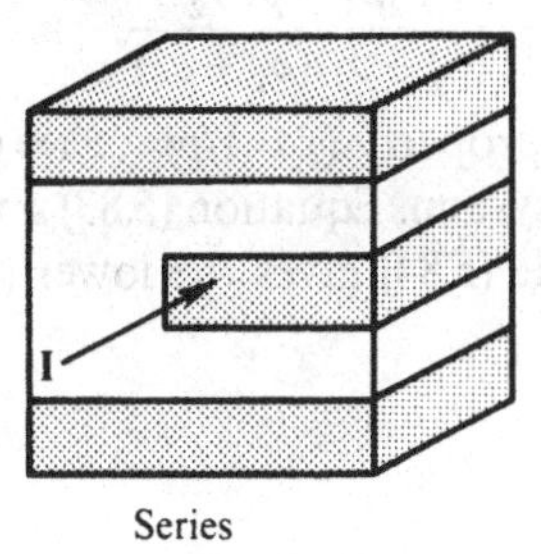

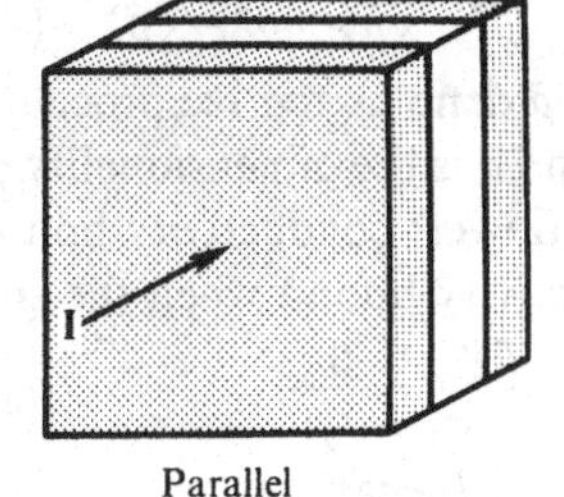

where

$$\alpha_m = (3\sigma_m)^{-1}, \tag{5.80b}$$

$$A_m = \sum_{t=1}^{m-1} \frac{v_t}{(\sigma_t - \sigma_m)^{-1} + \alpha_m}. \tag{5.80c}$$

These results are quite general and apply to as many phases as required. They can also be used to take account of the diffuse scattering from interfaces simply by treating the interface as a separate phase of thickness $\sim\Lambda$, provided that they might be regarded as isotropically distributed. For a two-phase solid without significant interface scattering, (5.79) and (5.80) become

$$\sigma_1^* = \sigma_1 + \frac{v_2}{\dfrac{1}{\sigma_2 - \sigma_1} + \dfrac{v_1}{3\sigma_1}} \tag{5.81}$$

$$\sigma_m^* = \sigma_2 + \frac{v_1}{\dfrac{1}{\sigma_1 - \sigma_2} + \dfrac{v_2}{3\sigma_2}}, \tag{5.82}$$

where we have assumed that $\sigma_1 < \sigma_2$. The worst bounds (5.78) and best bounds (5.81) and (5.82) are shown in Figure 5.15 as a function of volume fraction for a hypothetical two-phase solid with $\rho_1(=1/\sigma_1) = 10\mu\Omega$ cm and $\rho_2(=1/\sigma_2) = 1\mu\Omega$ cm. Other limiting expressions for the conductivity of random two-phase solids have been derived by Brown Jr. (1955) and Herring (1960).

To improve on these bounds requires additional information about the statistics of the spatial distribution of the phases. Bergman (1978) has shown that such information may be obtained through a knowledge of some other effective material constant p of the same composite (e.g. magnetic permeability or thermal conductivity), provided that the values are known for the composite as well as for the pure components. This allows the bounds to be written as

$$\frac{p_2 - p_1}{p^* - \langle p \rangle} - \frac{\sigma_2 - \sigma_1}{\sigma^* - \langle \sigma \rangle} = \frac{d}{v_1 v_2} \frac{p_2\sigma_1 - p_1\sigma_2}{(p_2 - p_1)(\sigma_2 - \sigma_1)}, \tag{5.83}$$

$$\frac{\tilde{p}_2 - \tilde{p}_1}{\tilde{p}^* - \langle \tilde{p} \rangle} - \frac{\rho_2 - \rho_1}{\tilde{\rho}^* - \langle \tilde{\rho} \rangle} = \frac{d}{(d-1)v_1 v_2} \frac{\tilde{p}_2\rho_1 - \tilde{p}_1\rho_2}{(\tilde{p}_2 - \tilde{p}_1)(\rho_2 - \rho_1)}, \tag{5.84}$$

where $\tilde{p}$ denotes the reciprocal of the property ($\tilde{p} = 1/p$), $\langle p \rangle = v_1 p_1 + v_2 p_2$ and d is the dimensionality of the system. Equation (5.83) gives the upper (lower) conductivity bound while (5.84) gives the lower (upper) conductivity bound when the sign of

$$\frac{p_2\sigma_1 - p_1\sigma_2}{\sigma_1 - \sigma_2} \tag{5.85}$$

is positive (negative).

An alternative approach is to make some assumption about the shape and distribution of the phases. Reynolds & Hough (1957) have shown that for a random two-phase distribution, most of the formulae can then be written in one of the general forms

$$\sigma^* = \sigma_2 + (\sigma_1 - \sigma_2) v_1 f_1 \tag{5.86}$$

or

$$(\sigma^* - \sigma_1) v_1 f_1 + (\sigma^* - \sigma_2) v_2 f_2 = 0, \tag{5.87}$$

where the parameter f is given by the ratio of the electric field averaged over all volume to that averaged within phase 1 (which is assumed to be isotropic)

$$f_1 = \frac{\bar{E}_1}{\bar{E}}, \tag{5.88}$$

where $\bar{E} = v_1 \bar{E}_1 + v_2 \bar{E}_2$. Equations (5.86) and (5.87) are only equivalent if the field ratios can be obtained without approximation. Most of the published formulae differ in the approximations made to enable determination of this field ratio, since the average value of the field within a phase will depend upon its shape, the conductivity of that phase and the conductivity of the surrounding medium. The usual approximation for the external field condition is to assume that the 'particle' is immersed

Fig. 5.15. Upper and lower bounds as a function of volume fraction v_2 for a two-phase solid with $\rho_1 = 10 \mu\Omega$ cm, $\rho_2 = 1 \mu\Omega$ cm. Dashed curves are the worst bounds given by equation (5.78) and the solid curves are the best bounds given by equations (5.81) and (5.82).

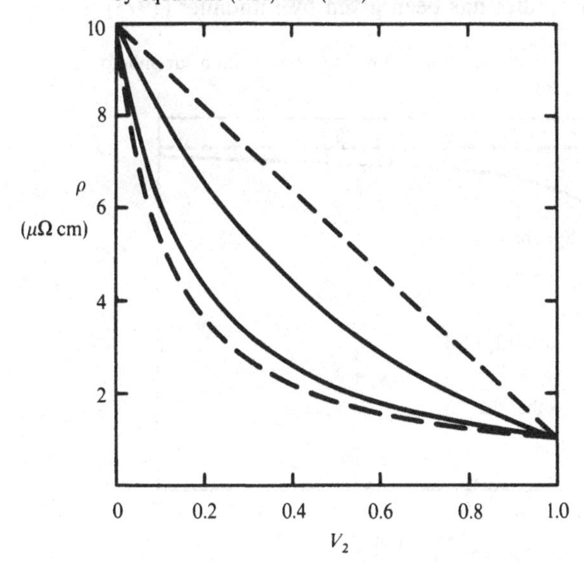

in a homogeneous medium of conductivity σ'. For a very dilute dispersion ($v_1 \to 0$) the interaction between particles may be ignored and $\sigma' = \sigma_2$. For higher concentrations, this interaction must be taken into account. The simplest method is then to take $\sigma' = \sigma^*$, i.e. by embedding the particle in an effective medium that is constructed self-consistently. The problem is then reduced to determining the field inside in the particle immersed in such a homogeneous medium. However, this is generally only calculable for the simple case of ellipsoidal particles including the limiting cases of infinitely long rods and lamellae. The field ratio for an ellipsoid is given by (see e.g. Stratton 1941)

$$f_1 = \sum_{i=1}^{3} \frac{\cos^2 \alpha_i}{1 + A_i\left(\dfrac{\sigma_1}{\sigma'} - 1\right)}, \tag{5.89}$$

where α_i are the angles between the ellipsoid axes and the applied field and the A_i depend upon the axial ratios of the ellipsoid subject to the condition $\sum_{i=1}^{3} A_i = 1$. For a spheroid, $A_2 = A_3 = A$ and $A_1 = 1 - 2A$. The variation of A with the axial ratio of the spheroid is given in Figure 5.16. For a random orientation of spheroids $\cos^2 \alpha_1 = \cos^2 \alpha_2 = \cos^2 \alpha_3 = \frac{1}{3}$, and for the case of long particles with aligned axes

$$\cos^2 \alpha_1 = \cos^2 \alpha_2 = \tfrac{1}{2}, \quad \cos^2 \alpha_3 = 0.$$

A compilation of many of the different formulae proposed is given in Tables 5.2 and 5.3, together with an indication of the nature of the approximation applied to σ' and equation of origin. A good review of some of the earlier studies has been given by Landauer (1978).

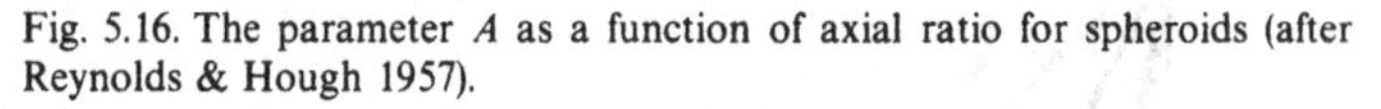
Fig. 5.16. The parameter A as a function of axial ratio for spheroids (after Reynolds & Hough 1957).

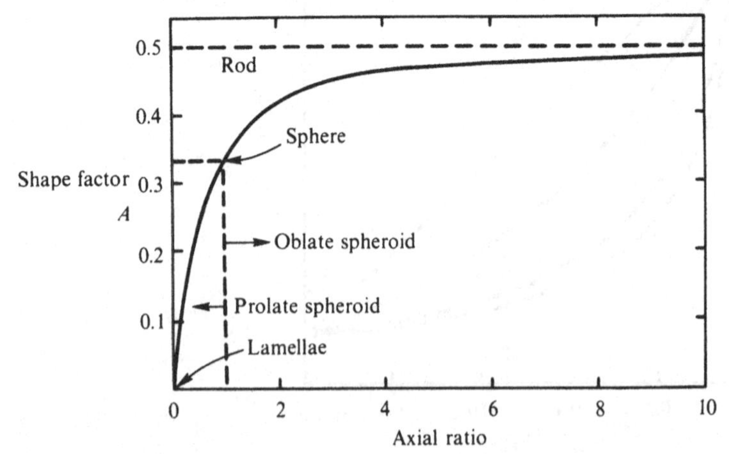

Table 5.2. Formulas for the conductivity of randomly oriented particles and nature of the approximation involved

Particle Shape	Formula		From equation	Factors in equation (5.89) A	σ'
Spheres	$\dfrac{\sigma^*-\sigma_2}{\sigma^*+2\sigma_2}=v_1\dfrac{\sigma_1-\sigma_2}{\sigma_1+2\sigma_2}$	$(T1)$	(5.87)	$\frac{1}{3}$	σ_2
Spheres	$\dfrac{\sigma^*-\sigma_2}{3\sigma_2}=v_1\dfrac{\sigma_1-\sigma_2}{\sigma_1+2\sigma_2}$	$(T2)$	(5.86)	$\frac{1}{3}$	σ_2
Spheres	$v_1\left(\dfrac{\sigma_1-\sigma^*}{\sigma_1+2\sigma^*}\right)+v_2\left(\dfrac{\sigma_2-\sigma^*}{\sigma_2+2\sigma^*}\right)=0$	$(T3)$	(5.87)	$\frac{1}{3}$	σ^*
Spheres	$\dfrac{\sigma^*-\sigma_2}{3\sigma^*}=v_1\dfrac{\sigma_1-\sigma_2}{\sigma_1+2\sigma^*}$	$(T4)$	(5.86)	$\frac{1}{3}$	σ^*
Spheroids	$\sigma^*=\sigma_2+\dfrac{v_1}{3(1-v_1)}\sum_{i=1}^{3}\dfrac{\sigma_1-\sigma^*}{1+A_i\left(\dfrac{\sigma_1}{\sigma_2}-1\right)}$	$(T5)$	(5.87)	A	σ_2
Spheroids	$\sigma^*=\sigma_2+\dfrac{v_1}{3}\sum_{i=1}^{3}\dfrac{\sigma_1-\sigma_2}{1+A_i\left(\dfrac{\sigma_1}{\sigma^*}-1\right)}$	$(T6)$	(5.86)	A	σ^*
Lamellae	$\sigma^{*2}=\dfrac{2(\sigma_1 v_1+\sigma_2 v_2)-\sigma^*}{\sigma_1/v_1+\sigma_2/v_2}$	$(T7)$	(5.87)	0	σ^*
Rods	$5\sigma^{*3}=(5\sigma'_p-4\sigma_p)\sigma^{*2}-(v_1\sigma_1^2+4\sigma_1\sigma_2+v_2\sigma_2^2)-\sigma_1\sigma_2\sigma_p=0$ where $1/\sigma_p=v_1/\sigma_1+v_2/\sigma_2$, $1/\sigma'_p=v_1/\sigma_2+v_2/\sigma_1$	$(T8)$	(5.86)	$\frac{1}{2}$	σ^*

Source: After Reynolds & Hough 1957.

Table 5.3. *Formulae for the conductivity of particles with some degree of orientation and nature of the approximation involved*

Particle shape	Formula		From equation	Factors in equation (5.89): A	σ'	$\cos\alpha 1 = \cos\alpha 2$	$\cos\alpha 3$
Parallel cylinders	$\frac{\sigma^* - \sigma_2}{\sigma^* + \sigma_2} = v_1 \frac{\sigma_1 - \sigma_2}{\sigma_1 + \sigma_2}$	($T9$)	(5.87)	$\frac{1}{2}$	σ_2	$\frac{1}{2}$	0
Parallel cylinders	$v_1 \frac{\sigma_1 - \sigma^*}{\sigma_1 + \sigma^*} + v_2 \frac{\sigma_2 - \sigma^*}{\sigma_2 + \sigma^*} = 0$	($T10$)	(5.87)	$\frac{1}{2}$	σ^*	$\frac{1}{2}$	0
Parallel lamellae	$\sigma^{*2} = \frac{\sigma_1 v_1 + \sigma_2 v_2}{\sigma_1 / v_1 + \sigma_2 / v_2}$	($T11$)	(5.87)	0	σ^*	$\frac{1}{2}$	0
Parallel lamellae with current perpendicular to the lamellae	$\frac{1}{\sigma^*} = \frac{v_1}{\sigma_1} + \frac{v_2}{\sigma_2}$	($T12$)	(5.87)	0	σ^*	0	1
Parallel lamellae with current parallel to the lamellae	$\sigma^* = \sigma_1 v_1 + \sigma_2 v_2$	($T13$)	(5.87)	0	σ^*	1	0
Spheroids with axes aligned and current parallel to one axis	$\sigma^* = \sigma_2 + \frac{v_1(\sigma_1 - \sigma_2)}{1 + A[(\sigma_1/\sigma_2) - 1]}$	($T14$)	(5.86)	A	σ_2	0	1
Spheroids with axes aligned and current parallel to one axis	$\frac{\sigma^*}{\sigma_2} = 1 + \frac{v_1}{[(\sigma_1/\sigma_2) - 1]^{-1} + A(1 - v_1)}$	($T15$)	(5.87)	A	σ_2	0	1

Source: After Reynolds & Hough 1957.

As an example of the significance of the particle shape we consider the results of Fricke (1924) which may be written in the general form

$$\frac{\sigma^* - \sigma_2}{\sigma^* + x\sigma_2} = \frac{v_1(\sigma_1 - \sigma_2)}{\sigma_1 + x\sigma_2}, \tag{5.90}$$

where the subscript 2 again refers to the matrix. Since it uses the approximation $\sigma' = \sigma_2$ it is only applicable to dilute dispersions. The parameter x is a function of the ratio σ_1/σ_2 and of the ratio of the length of the axis of symmetry of the spheroids to the other axis c/b and is shown in Figures 5.17 and 5.18 for oblate and prolate spheroids respectively.

Fig. 5.17. The parameter x as a function of the ratio σ_1/σ_2 and c/b for oblate spheroids (after Fricke 1924): (i) $a = b$; (ii) $a/b = 2$; (iii) $a/b = 3$; (iv $a/b = 4$.

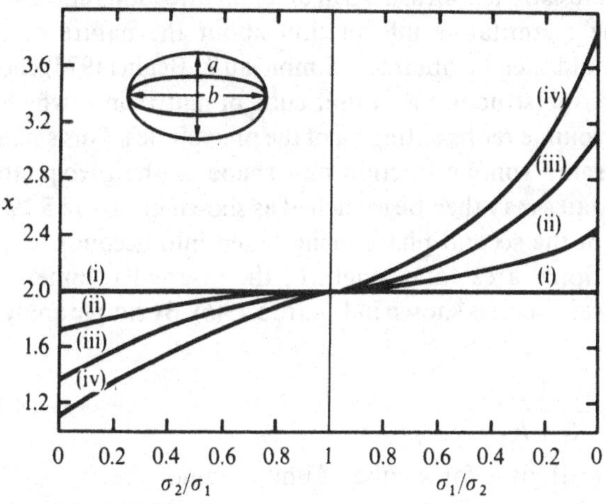

Fig. 5.18. The parameter x as a function of the ratio σ_1/σ_2 and c/b for prolate spheroids (after Fricke 1924): (i) $a = b$; (ii) $a/b = \frac{1}{2}$; (iii) $a/b = \frac{1}{3}$; (iv) $a/b = \frac{1}{4}$.

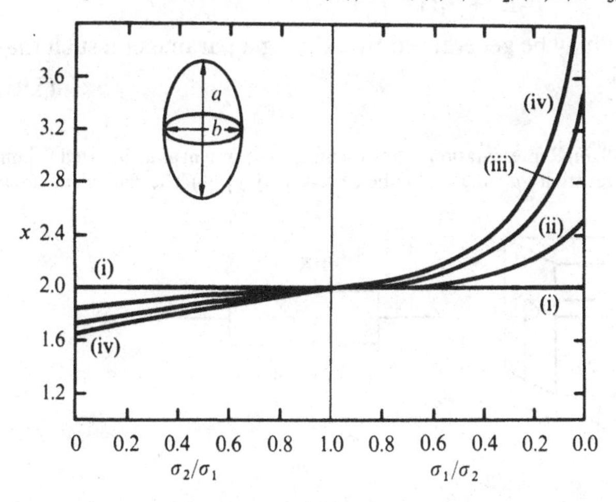

For constant volume fraction of particles (and no significant interfacial scattering) the conductivity is thus independent of the size of the particles, but depends upon their shape if the difference between the particle and matrix conductivities is large. For spherical particles $x=2$ for all values of σ_2/σ_1 and (5.90) reduces to the classical Lorenz–Lorentz result for the refractive index and the Claussius–Mossotti theory for the dielectric constant. It is interesting to note that this result then corresponds exactly to the upper bound (5.82).

Another approach to the problem is to derive an electrical analogue of the structure and use the usual circuit analysis techniques to derive an expression for the resistivity. By adjusting the parameters of the model so that the predicted resistivities measured in different directions agree with experiment, some quantitative information about the nature of the microstructure could then be obtained. Simoneau & Bégin (1973) chose to model a composite structure as a unit cube of matrix into which is inserted another volume representing all of the precipitates. This smaller volume has the same volume fraction and shape as the precipitates. Dispersed precipitates may then be modelled as shown in Figure 5.19(*a*), the morphology of the second phase being taken into account by the relative cross-sectional areas and length of the inserted volume. The equivalent electrical circuit is shown in Figure 5.19(*b*). By simple analysis we find

$$\frac{1}{R}=\frac{1}{R_1}+\frac{1}{R_1'+R_2}, \tag{5.91}$$

and so equation (1.1) gives for a cube of unit volume

$$\frac{1}{\rho}=\frac{1-A_2}{\rho_1}+\frac{A_2}{\rho_2 L_2+\rho_1(1-L_2)}. \tag{5.92}$$

This expression may be generalised by defining a parameter n such that

$$A_2=v_2^n. \tag{5.93}$$

Fig. 5.19. (*a*) Representation of a two-component system as an inserted volume v_2 and resistivity ρ_2 in a unit cube of resistivity ρ_1. (*b*) The equivalent circuit for (*a*).

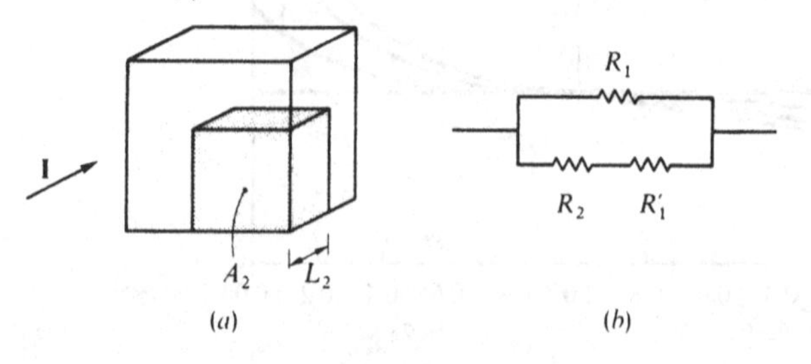

This parameter effectively quantifies the amount of the second phase that is in parallel with the current. The expression for the resistivity then becomes

$$\frac{1}{\rho}=\frac{1-v_2^n}{\rho_1}+\frac{v_2^n}{\rho_2 v_2^{1-n}+\rho_1(1-v_2^{1-n})}. \tag{5.94}$$

If $n=0$ this expression reduces to the series case and if $n=1$ it reduces to the parallel case (see equation (5.78) and Figure 5.14). If the particles are isotropically dispersed, $A_2=L_2^2$ and so $A_2=v_2^{2/3}$, i.e. $n=\frac{2}{3}$. For the case of parallel fibres of square cross-section, $n=1$ when the current lies parallel to the fibre axis and $n=0.5$ when it is perpendicular to the fibre axis, and the corresponding resistivities derived from equation (5.94) are identical to those derived by Liebmann & Miller (1963) for just such a case. The perpendicular resistivity $\rho_\perp$ predicted then lies slightly nearer to the ideal series limit than the predictions of equations (5.79)–(5.80). Departures of the value of n from $\frac{2}{3}$ are thus caused by some sort of anisotropy in the distribution of the second phase with respect to the direction of the current flow. Note, however, that the simple geometry implied by Figure 5.19 restricts the approach to situations in which the matrix is continuous throughout the specimen and the precipitates can be modelled by a square included volume. As such, cylindrical fibres cannot be properly modelled when the cylinder axis is perpendicular to the current flow. Lord Rayleigh (1892) derived an analytical solution for such a situation. His solution may be written as (see e.g. Watson *et al.* 1975)

$$\frac{1}{\rho_\perp}=\frac{1}{\rho_{\mathrm{m}}}\left[1-\frac{2v_{\mathrm{f}}}{v'+\dfrac{b}{a}2\pi v_{\mathrm{f}}\left[\dfrac{-1}{A_n^2}+\dfrac{1}{6}\right]}\right], \tag{5.95}$$

where

$$v'=\frac{\rho_{\mathrm{f}}+\rho_{\mathrm{m}}}{\rho_{\mathrm{f}}-\rho_{\mathrm{m}}},$$

$$A_n=\sum_n \sinh\left(n\pi\frac{b}{a}\right),$$

the subscripts f and m refer to the fibre and matrix and the dimensions b and a are as defined in Figure 5.20. This expression appears to give a better fit to experimental data than any of the analog models proposed for such a situation (see e.g. Springer & Tsai 1967; Watson *et al.* 1975) and so the latter will not be discussed any further.

Composites containing parallel discontinuous fibres with transverse interconnections may also be modelled. If the fibres have a rectangular cross-section then the Simon & Bégin model may be applied. If we consider the general structure shown in Figure 5.21, for example, the resistivity along the direction indicated is given by

$$\frac{1}{\rho}=\frac{A_2}{\rho_2}+\frac{A'_2}{\rho_2 L_2+\rho_1(1-L_2)}+\frac{1-(A_2+A'_2)}{\rho_1}, \tag{5.96}$$

where again the cube is assumed to have unit volume. Comparison with the unbranched case (5.91) reveals the extra term due to the branching. If the branches form a continuous network, the anisotropy of the resistivity will be reduced. The extension to other branched structures with continuous or discontinuous fibres or branches is straightforward. Similar considerations apply to platelet precipitates with holes (Digges & Tauber 1971).

Watson *et al.* (1975) considered extensions to these simple analog models to allow incorporation of discontinuous cylindrical fibres by determining the approximate resistivity contributions from the Rayleigh equation (5.95). However, rather than assuming a particular model and equivalent circuit, they argue for the adoption of multiple models and equivalent circuits, the range of solutions representing an uncertainty characteristic of the mathematical analysis itself. This approach may be demonstrated by considering the three-dimensional fibre model of unit

Fig. 5.20. An ordered array of parallel cylindrical fibres as considered by Lord Raleigh, defining the dimensions a and b.

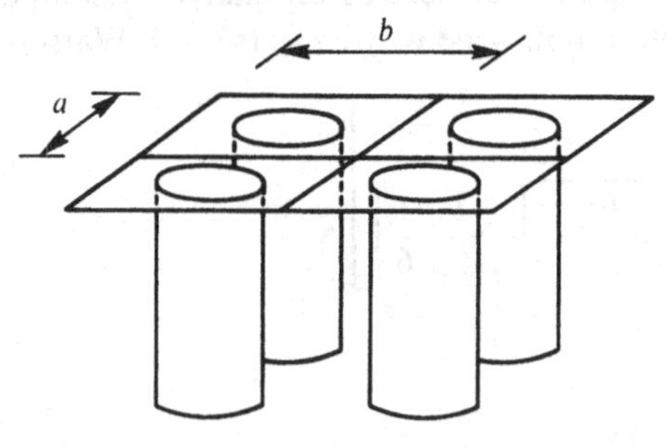

Fig. 5.21. A generalised branched fibre structure and equivalent circuit.

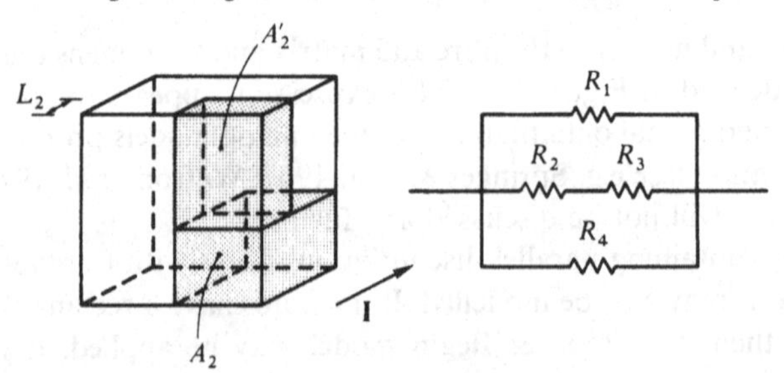

volume shown in Figure 5.22. The parameters X and Y denote the length of the transverse interconnection and the extent of the discontinuity in the longitudinal direction respectively. Such a region may be analysed with the aid of any of the four fibre models shown in Figure 5.23, together with their equivalent circuits for the longitudinal direction. The solution to be applied to each element containing more than one phase is indicated by the appropriate equation number in parentheses. Each of these models will give a slightly different variation of resistivity with X and Y. However, the analytical uncertainty may be reduced by determining the resistivity in two orthogonal directions. Only the values of X and Y that give a simultaneous fit to experiment in the two directions are then acceptable. Other modifications to this model to allow for different fibre and branch diameters, etc., are relatively straightforward although there is a danger of introducing too many unknowns. The reliability of such models can only be ascertained by detailed sectioning of specimens and careful measurement.

In the extreme case of one phase having a much higher conductivity than the other (a highly porous conducting film, superconducting fibres in a normal matrix or insulators filled with conducting particles, for example) the resistivity will be largely determined by the connectivity of the highly conducting phase (see e.g. Jernot *et al.* 1982; Stroud 1980; Benjamin *et al.* 1984). Leaving aside quantum effects such as normal tunnelling, Josephson tunnelling or proximity effects, the system can be regarded as a resistor network which continually branches from each site on a lattice. For any particular coordination number of the lattice z (i.e. number of possible links to adjacent sites), when the number or 'concentration' of resistors just exceeds a critical value v_c there is likely to

Fig. 5.22. A three-dimensional model of a discontinuous, branched fibre. The longitudinal and transverse directions are as indicated (after Watson *et al.* 1975).

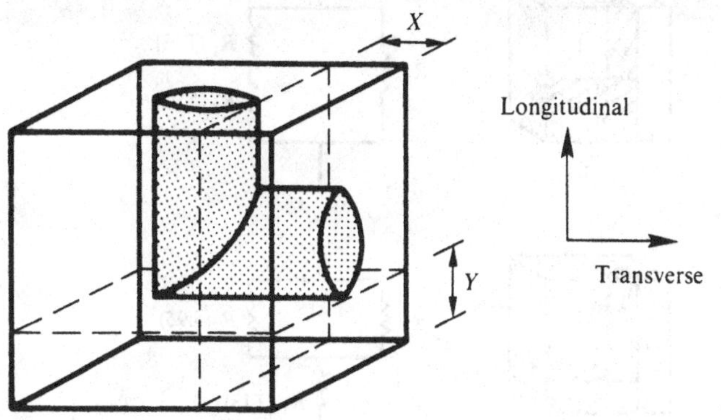

be just one infinite link joining one site to all other sites. For $z=6$, $v_c=0.2$ whereas for $z=4$, $v_c=0.33$. This critical value is known as the percolation threshold. The most likely resistor arrangement is then a long chain of resistors in series, branching infrequently into smaller parallel segments. For a lower concentration of resistors, this infinite link, and hence the conductivity, vanishes. At larger concentrations, the more frequent branching into parallel links reduces the resistivity. It is interesting to note that the Bruggeman equation (Table 5.2, *T*3) predicts such a threshold if the conductivity of one of the phases is zero. This occurs at a volume fraction of the conducting phase of $\frac{1}{3}$. A modified equation for the

Fig. 5.23. The generalised fibre models (sectioned in half for clarity) and assumed equivalent circuits for the longitudinal directions. The method of solution for each element is indicated by the equation numbers in parentheses (after Watson *et al.* 1975).

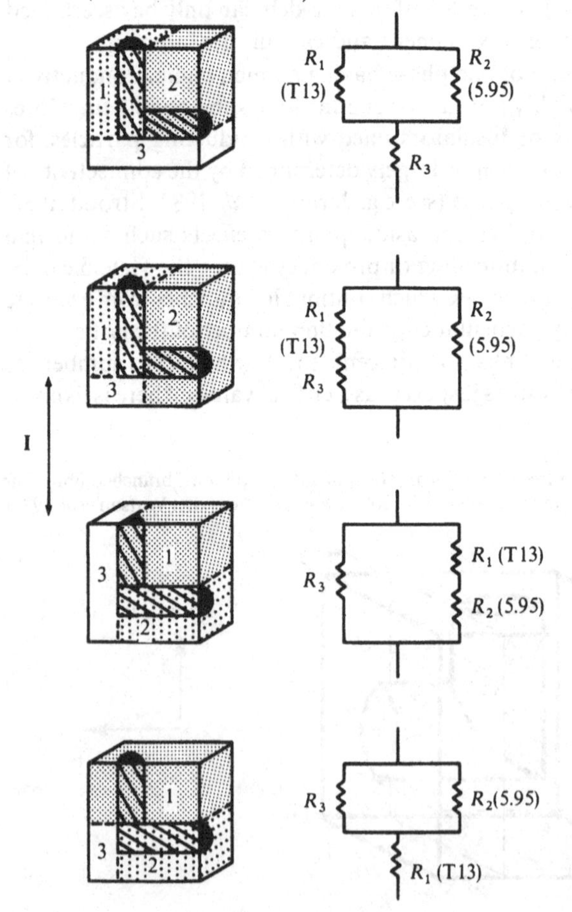

conductivity σ^* of such a percolating network two-phase system expressed in terms of the effective medium theory (as discussed above) has been given by Kirkpatrick (1973):

$$v_1 \frac{\sigma^* - \sigma_1}{\sigma_1 + \left(\dfrac{1}{v_c} - 1\right)\sigma^*} + v_2 \frac{\sigma^* - \sigma_2}{\sigma_2 + \left(\dfrac{1}{v_c} - 1\right)\sigma^*} = 0. \tag{5.97}$$

where v_c is the percolation threshold when $\sigma_2 = 0$. The variations of conductivity with concentration in the effective medium theory and the resistor network analogy are then of the form shown in Figure 5.24. There have been many other experimental and theoretical studies of this percolation phenomenon in both isotropic (see e.g. Shante & Kirkpatrick 1971) and anisotropic systems (Balberg & Binenbaum 1983); in the later case the percolation threshold being found to depend markedly upon the angular distribution of the conducting phase. The problem may also be considered by removing the nodes or sites, rather than the resistors, at random (Watson & Leath 1974; Bernasconi & Wiesmann 1976; Davidson & Tinkham 1976; Granqvist & Hunderi 1978). Nakamura (1984) has shown that the effective medium theories can be brought into good agreement with these site percolation studies if the conductivities σ_1 and σ_2 are replaced by the upper and lower bounds

Fig. 5.24. Conductivity as a function of volume fraction of conducting phase in a percolating system. v_c is the percolation threshold. Solid line: effective medium theory (equation (5.97), $v_c = 0.33$). Solid circles: computer simulation (after Kirkpatrick 1973).

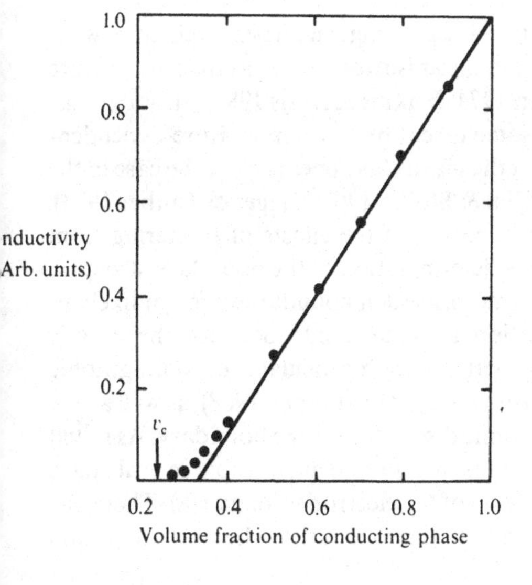

(5.78), (5.81) or (5.82). It may be seen that the basic effective medium theory works well except near the percolation threshold. This might be expected since it assumes that every resistor is in an identical environment, an assumption which will be poor near the threshold since some resistors may be isolated or at the end of a chain whereas others will be linked at both ends. Some general reviews of percolation theory are given by Kirkpatrick (1973), Essam (1972, 1980), and Stauffer (1979).

5.5.2 *Scale of phase separation* $\sim\Lambda$

We turn now to the case of a microstructure in which one or more of the characteristic dimensions is $\sim\Lambda$. As in the case of inhomogeneous short range ordering or clustering considered above, such a situation will affect the resistivity because of the way the scattering potential is determined (both physically and mathematically). Furthermore, as such a structure is likely to result from a fine dispersion of phases, there is likely to be an additional significant contribution to the resistivity from the size-dependent scattering from internal interfaces. Such a situation is often encountered in eutectic alloys in which one of the phases has a relatively low impurity concentration, and hence a large Λ_l. Furthermore, since the mean free path depends upon temperature, the magnitude of this contribution will also depend upon temperature. Such scattering is directly comparable with the size effect scattering widely investigated in thin wires and films, although it is possible that internal surfaces will be smoother than external surfaces and so the angle dependence of the scattering (i.e. the specularity of the scattering) will differ.

The size-dependent interfacial scattering manifests itself in a well-aligned eutectic as an increase in the anisotropy $\rho_\perp/\rho_\parallel$ as the temperature decreases (Simoneau & Bégin 1974; Jenkins & Arajs 1983), although this increase may be masked to some extent by any temperature-dependent anisotropy of the resistivity of the matrix, as appears to be the case in the Bi–Ag and Bi–Mn systems (Yim & Stofko 1967; Digges & Tauber 1971), for example. A proper determination of the effects of scattering from internal boundaries requires a determination of the boundary structure factor given by (5.37). Furthermore, as such boundaries are not likely to be coherent, this determination must take into account the atomic displacements associated with the particular crystallographic relationship between the phases (see e.g. Hogan *et al.* 1968), as well as the composition gradients and atomic disorder over the boundary. As noted in Section 5.2.2, such scattering is likely to be anisotropic in real space and so depend upon the direction of the electron velocity $\mathbf{v}(\mathbf{k})$. There has not as yet been any such general calculations, although structure factors and scattering effects have been determined for stacking faults (Harrison 1966, p. 156; Howie 1960; and Freeman 1965), grain boundaries (Lormond 1979, 1982) and twin boundaries (Shatzkes *et al.* 1973). It turns out that in the case of twin boundaries and stacking faults there is no contribution to the transition rate in first-order perturbation theory for monovalent metals since the set of wavenumbers with non-vanishing structure factors will be outside the Fermi surface. The scattering is then given by higher-order perturbation theory. For higher valences, first-order scattering is possible although the scattering rate is likely to depend upon the angle between $\mathbf{v}(\mathbf{k})$ and the boundary (Harrison 1966, p. 162). In the case of grain boundaries there will be a local displacive disorder and possibly local dilation as well as the misorientation effects leading to a more complicated situation. The scattering effects of such boundaries have been considered with reference to a range of models varying in sophistication. They have been regarded as regions of low ionic density (Seeger and Shottky 1959) and as a wall of regularly spaced cylindrical voids (van der Voort and Guyot 1971) although such studies tend to give a resistivity much lower than observed values. Brown (1977, 1982) has modelled them as an array of dislocations with somewhat better success. The importance of the change in orientation at the boundary has been stressed by Ziman (1960, p. 244) and Guyot (1970), producing larger resistivities than the effects of reduced density. Calculations based on structure factor determinations take into account the detailed positions of the atoms at the boundary and give a good agreement with experimental data (see e.g. Lormand 1982).

Mayadas & Shatzkes (1970) have modelled grain boundaries simply as an array of δ-function scattering potentials and introduced the notion of a reflection coefficient R which in some way describes the fraction of electrons which are scattered by the boundary. This led to an expression for the resistivity

$$\frac{\rho_\infty}{\rho} = 1 - \frac{3}{2}\alpha + 3\alpha^2 - 3\alpha^3 \ln\left(1 + \frac{1}{\alpha}\right), \tag{5.98}$$

where

$$\alpha = \frac{\Lambda_\infty}{d}\left(\frac{R}{1-R}\right), \tag{5.99}$$

and d is the mean grain diameter. The subscript ∞ indicates the value at infinite boundary spacing (i.e. the bulk value).

An empirical specular transmission coefficient t which gives the fraction of electrons whose velocity in the direction of $\mathbf{E}$ is unaffected by the grain boundaries (all others being diffusely scattered) has been introduced by Tellier *et al.* (1979). This parameter is isotropic (i.e. it does not depend upon the angle of incidence of an electron) and leads to an expression for the resistivity for a three-dimensional array of boundaries having a mean spacing d (Pichard *et al.* 1980):

$$\rho_g(v) = \rho_\infty + \rho_\perp(v) + 2\rho_\parallel(v), \tag{5.100}$$

where the resistivities $\rho_\perp(v)$ perpendicular and $\rho_\parallel(v)$ parallel to the current flow are given by

$$\frac{\rho_\perp(v)}{\rho_\infty} = \left[\frac{3}{2}v - 3v^2 + 3v^3 \ln\left(1 + \frac{1}{v}\right)\right]^{-1} - 1, \tag{5.101}$$

$$\frac{\rho_\parallel(v)}{\rho_\infty} = \left[\frac{3}{2}v^2 - \frac{3}{4}v + \frac{3}{2}v(1 - v^2)\ln\left(1 + \frac{1}{v}\right)\right]^{-1} - 1, \tag{5.102}$$

and

$$v = \frac{d}{\Lambda_\infty \ln(1/t)}. \tag{5.103}$$

The resistivities $\rho_g(v)/\rho_\infty$, $\rho_\perp(v)/\rho_\infty$ and $\rho_\parallel(v)/\rho_\infty$ are plotted as a function of v in Figure 5.25. We have given this form of results which relies on the

Fig. 5.25. The normalised grain boundary resistivities as a function of the grain size dependent parameter v. Solid line $\rho_g(v)/\rho$; dotted line $\rho_\perp(v)/\rho_\infty$; dashed line $\rho_\parallel(v)/\rho_\infty$.

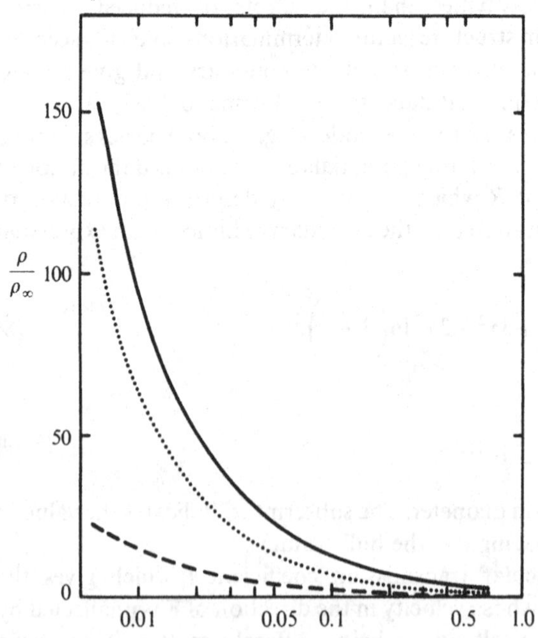

assumption (5.100) rather than a more exact form (Pichard *et al.* 1979) as it allows us to introduce the parallel and transverse contributions independently. We might note in passing that since Λ_∞ depends upon temperature, if $d \sim \Lambda_\infty$ there will be a size effect contribution to the temperature coefficient of resistivity analogous to that discussed in Section 5.2.2.

A similar boundary scattering problem occurs at the external surfaces of conductors and is particularly important in thin wires and films. Calculations of the effects of such scattering by Fuchs (1938) and Sondheimer (1950) for films and Dingle (1950) for wires were based on the assumption that the surface scattering was isotropic but that only a fraction p of electrons were scattered in a specular (i.e. non-diffuse) manner. This fraction is called the specularity parameter and for most thin foils and wires is close to zero (i.e. the surface scattering is mainly diffuse). The results found are

Fuchs–Sondheimer (foils):

$$\frac{\rho_\infty}{\rho} = 1 - \kappa\left[\frac{3}{2}(1-p)\int_1^\infty \left(\frac{1}{t^3} - \frac{1}{t^5}\right)\left(\frac{1-\exp(-\kappa t)}{1-p\exp(-\kappa t)}\right)\right], \tag{5.104}$$

Dingle (wires):

$$\frac{\rho_\infty}{\rho} = 1 - \frac{12}{\pi\kappa}\int_0^{\pi/2} \cos^2\theta \sin^2\theta \, d\theta \int_0^{\pi/2} \sin\psi \, d\psi \times \frac{(1-p)[1-\exp(-\kappa \sin\psi/\sin\theta)]}{1-p\exp(-\kappa\sin\psi/\sin\theta)}, \tag{5.105}$$

where

$$\kappa = \frac{d}{\Lambda_\infty}, \tag{5.106}$$

and d refers either to the foil thickness or wire diameter. As shown by Chambers (1969), such diffuse scattering results in a modification of the electron distribution function of the form

$$g_{\mathbf{k}}(\mathbf{r}) = -e\frac{\partial f_0}{\partial E_{\mathbf{k}}}\mathbf{E}\cdot\mathbf{v}(\mathbf{k})\tau_{\mathbf{k}}\left(1-\exp\left(-\frac{d_s}{\Lambda_l}\right)\right), \tag{5.107}$$

where d_s is the distance from the surface to point $\mathbf{r}$. This shows that the effects of the surface decrease exponentially with distance from the surface measured on a scale of Λ_l. This is exactly the same behaviour discussed in Section 5.5.2: the scattering potential (in this case due to the discontinuity at the surface) is determined by an average over a region $\sim\Lambda_l$. However, if such an isotropic form is assumed it becomes necessary to postulate a temperature dependence of p to allow agreement with experiment over a wide range of temperatures (see e.g. Gaidukov &

Kadletsova 1970*a*,*b*). This apparent temperature dependence can be removed by assuming an angle-dependent p such that $p \approx 0$ if $\mathbf{v}(\mathbf{k})$ is normal to the surface whereas $p \approx 1$ if $\mathbf{v}(\mathbf{k})$ is parallel to the surface (Sambles & Elsom 1980, Sambles *et al.* 1982).

Various mechanisms have been proposed as the origin of surface scattering, including microscopic surface roughness and localised surface charges (Greene 1964; Ziman 1960, p. 459). Brändli & Cotti (1965) and Parrott (1965) used as an analogue the optical reflection from rough surfaces and so proposed a specularity parameter that was a step function of angle

$$\begin{aligned} P(\theta) &= 1 \quad \theta_c < \theta < 90^\circ, \\ &= 0 \quad 0^\circ < \theta < \theta_c, \end{aligned} \tag{5.108}$$

where θ_c is a critical cut-off angle and θ the angle to the surface normal. The surface charge model was devised mainly with reference to charges residing in the surface states of semiconductors. Nevertheless, it has some relevance to the internal incoherent phase boundaries which may be modelled as an array of effective charges, in much the same way as the dilation around a vacancy (see e.g. Section 6.3). Greene & O'Donnell (1966) have determined p for such a situation and found that it depended upon the angle of incidence as shown in Figure 5.26 for two different surface charge densities.

Ziman (1960, p. 459) has proposed a model of surface scattering in which the phase of a reflected wave is advanced or retarded by an

Fig. 5.26. The specularity parameters given by equations (5.108) (dashed line), (5.109) (dotted line), and the surface charge model (solid line), as a function of angle of incidence. Parameters for the latter are (*a*) ($n = 10^{16}/\text{m}^3$, $N = 6.5 \times 10^9/\text{cm}^2$; (*b*) $n = 10^{18}/\text{cm}^3$, $N = 3.8 \times 10^{-12}\,\text{cm}^2$, where n is the density of charge carriers and N the density of localised surface charge (after Greene & O'Donnell 1966).

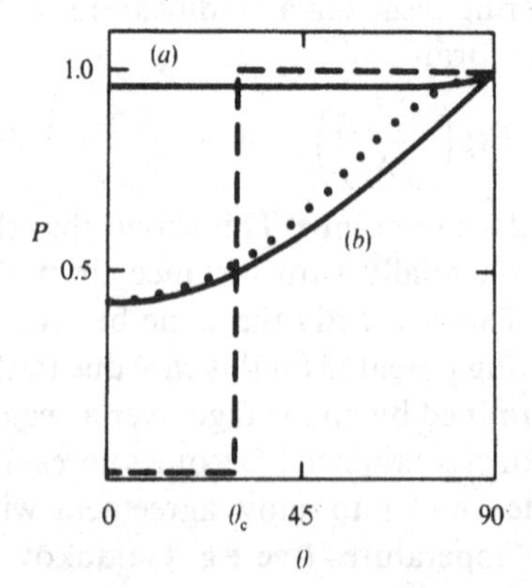

amount depending upon the local height of the surface h measured with respect to some reference surface. For large h/λ_k (λ_k is the de Broglie wavelength of state k) the waves gain or lose phase at random (assuming random fluctuations in h) giving $p \approx 1$. This approach also introduces a dependence of p upon the angle of incidence since the changes in phase will be smaller at large θ (i.e. at glancing incidence). Soffer (1967) has refined this model to satisfy the conservation of flux and obtained the result

$$P(\theta) = \exp(-\xi^2 \cos^2 \theta), \tag{5.109}$$

where

$$\xi = \frac{4\pi h'}{\Lambda_k}, \tag{5.110}$$

and h' is the root mean square height deviation. This is also shown in Figure 5.26 for values of h'/λ_k that bring it into agreement with the surface charge model for $\theta = 0°$. The appropriate expressions for the resistivity are then (Sambles & Preist 1982):

foils:

$$\frac{\rho_\infty}{\rho} = 1 - \frac{3}{2\kappa}\int_0^1 du(u-u^3)\frac{(1-p(u))[1-\exp(-\kappa/u)]}{1-p(u)\exp(-\kappa/u)}, \tag{5.111}$$

wires:

$$\frac{\rho_\infty}{\rho} = 1 - \frac{12}{\pi\kappa}\int_0^{\pi/2} \cos^2\theta \sin^2\theta \, d\theta \times \int_0^{\pi/2} \sin\psi \frac{[1-p(\cos\Theta)][1-\exp(-\kappa\sin\psi/\sin\theta)]}{1-p(\cos\Theta)\exp(-\kappa\sin\psi/\sin\theta)} d\psi, \tag{5.112}$$

where $\cos\Theta = \sin\theta \sin\psi$. In both cases $p(u) = \exp(-\xi^2 u^2)$ and ξ is given by (5.110). Specularity parameters which depend upon angle of incidence have also been found from first principles calculation by Leung (1984) and noted by Lormand (1982).

Another mechanism which could lead to an anomalous temperature dependence has been proposed by Olsen (1958). He suggested that a low angle scattering event near a surface may be sufficient to cause a collision with that surface and hence an anomalously large contribution to the resistivity. Subsequent analysis (Blatt & Satz 1960; Azbel & Gurzhi 1962) have shown that for a wire of radius D this mechanism leads to an additional phonon scattering contribution of the form:

$$\rho_s(D, T) = [8\pi(\rho_\infty \Lambda_\infty)^2]^{1/3} \left[\frac{T}{\Theta_D}\right]^{2/3} [\rho_\infty(T)]^{1/3} D^{2/3}, \tag{5.113}$$

if $D < \Lambda_\infty$. However, in most conducting alloys $\rho_\infty \Lambda_\infty \sim 10^{-5} \mu\Omega\,cm^2$,

$T/\theta_D \sim 1$ and $\rho_\infty(T) < 10\mu\Omega$ cm at room temperature, and so $\rho_s(D, T) \sim 3 \times 10^{-3}(D^{-2/3})\mu\Omega$ cm. This value is typically some orders of magnitude less than the normal boundary scattering and so may usually be neglected. It is also likely that surfaces will affect the electron–electron scattering in thin specimens (see e.g. Yu *et al*. 1984; De Gennaro & Rettori 1985). However, in the alloys of interest here it is likely that the impurity scattering would mask any such effects.

Given some form of the specularity parameter, the resistivity of a conducting foil or filament can thus be written in one of two forms:

$$\frac{\rho(d, T)}{\rho_\infty} = 1 + A(p, \Lambda_\infty, d)\frac{\Lambda_\infty}{d}, \tag{5.114a}$$

or

$$\frac{\sigma(d, T)}{\sigma_\infty} = \frac{\rho_\infty}{\rho(d, T)} = 1 - A'(p, \Lambda_\infty, d)\frac{\Lambda_\infty}{d}, \tag{5.114b}$$

where the parameters $A(p, \Lambda_\infty, d)$ or $A'(p, \Lambda_\infty, d)$ depend upon the particular model used and d is the filament diameter or foil thickness. Some of the limiting values of $A(p, \Lambda_\infty, d)$ and $A'(p, \Lambda_\infty, d)$ for $d \ll \Lambda_\infty$ and $d \gg \Lambda_\infty$ are given in Table 5.4. The dependence of $A'(p, \Lambda_\infty, d)$ upon d/Λ_∞ at intermediate values of d/Λ_∞ is shown in Figure 5.27 for the Fuchs–Sondheimer and Dingle models and the Soffer model for foils and wires in Figure 5.28 (Sambles *et al.* 1982, Sambles & Elsom 1980) for various values of p or h/λ_k. For $d/\Lambda_\infty > 1$ and diffuse scattering ($p \approx 0$) the differences between these two sets of data is not large but for smaller d/Λ_∞ the results will depend upon the particular model chosen. Nevertheless, the results do show that if the range of d/Λ_∞ being

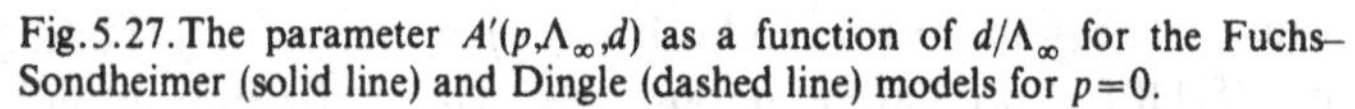
Fig.5.27.The parameter $A'(p,\Lambda_\infty,d)$ as a function of d/Λ_∞ for the Fuchs–Sondheimer (solid line) and Dingle (dashed line) models for $p=0$.

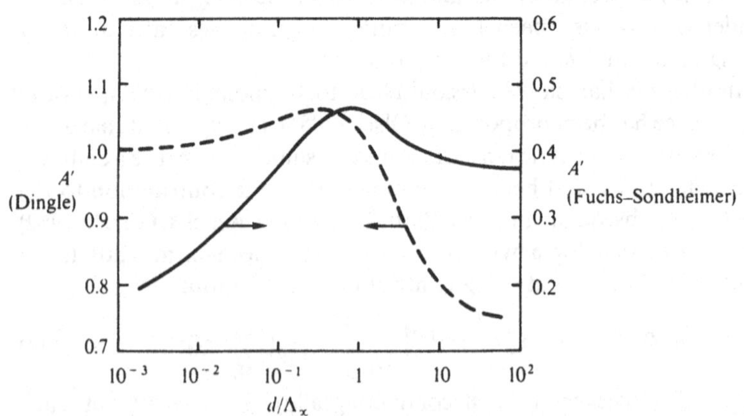

investigated is no more than about one decade, the parameter $A(p, \Lambda_\infty, d)$ may be regarded as a constant. However, for a wider range of d/Λ_∞ (as might be encountered over a wide range of measuring temperatures) the variation of A must be taken into account to avoid the necessity for assuming a drastic temperature dependence of p or $\rho_\infty \Lambda_\infty$. In this regard, the Soffer model allows fitting to data over a wider range of temperature than the Fuchs–Sondheimer or Dingle models (Sambles & Elsom 1980).

Fig. 5.28. The parameter $A'(p,\Lambda_\infty,d)$ as a function of d/Λ_∞ for foils (*a*), and wires (*b*) derived from the Soffer model for the values of h/λ_k indicated. Also shown (dotted lines) are the Fuchs–Sondheimer and Dingle results for the values of p indicated (after Sambles & Elsom 1980; Sambles *et al.* 1982).

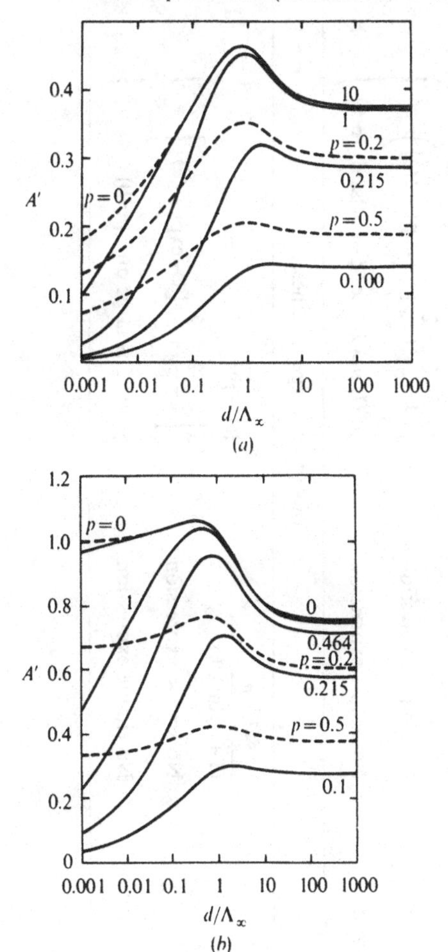

Table 5.4. *Limiting expressions for $A(p, \Lambda_\infty, d)$ or $A'(p, \Lambda_\infty, d)$ for wires, foils and grain boundaries*

Constant	$d<\Lambda$	$d>\Lambda$	Reference
	Wires (d = average diameter)		
A	1	1	Nordheim (1931)
A	$\frac{1-p}{1+p}\frac{d}{\Lambda_\infty}$	$\frac{3}{4}(1-p)$	Dingle (1950)
A'	No simple expression	$\frac{3}{8}\left(1-\frac{2}{\xi^2}+\frac{2}{\xi^4}[1-\exp(-\xi^2)]\right)$	Soffer (1967); Sambles & Preist (1982)
A	No simple expression	$\frac{3}{8}$ (If $h>\lambda_k$ (i.e. $\xi\to\infty$) and $p\simeq 0$)	Soffer (1967); Sambles & Preist (1982)
	Foils (d = average thickness)		
A	$\frac{\frac{4}{3}(1-p)}{(1+p)\ln(1/k)}\frac{d}{\Lambda_\infty}$	$\frac{3}{8}(1-p)$	Fuchs (1938); Sondheimer (1952)
A'	No simple expression	$\frac{3}{8}\left[1-\frac{2}{\xi^2}+\frac{2}{\xi^4}[1-\exp(-\xi^2)]\right]$	Soffer (1967); Sambles & Preist (1982)
A	No simple expression	$\frac{3}{8}$ (If ξ is large and $p\simeq 0$)	Soffer (1967); Sambles & Preist (1982)

Grain boundaries (d = average grain diameter)			
A	$\frac{4}{3}\left(\frac{R}{1-R}\right)-\frac{d}{\Lambda_\infty}$	$\frac{3}{2}\left(\frac{R}{1-R}\right)$	Mayadas and Shatzkes (1970)
$A_\perp$	$\frac{2}{3}\ln(1/t)-\frac{d}{\Lambda_\infty}$	$\frac{3}{4}\ln(1/t)$	
$A_\parallel$	$\sim\frac{\frac{1}{3}\ln(1/t)}{\ln(1/v)}-\frac{d}{\Lambda_\infty}$ where $v=\frac{d}{\Lambda_\infty \ln(1/t)}$	$\frac{3}{8}\ln(1/t)$	Tellier *et al.* (1979; Pichard *et al.* (1980)

Note: See text for the meaning of the various parameters.

Both the internal boundary transmission coefficient t and external boundary specularity parameter simply give a measure of the fraction of electrons that are not diffusely scattered. As far as grain boundaries are concerned it is of no consequence whether an electron is specularly reflected back into the grain or transmitted through the boundary provided that its velocity in the direction of the current is unchanged. Stated another way, a diffuse scattering event simply returns an electron back to the equilibrium distribution f_0, regardless of whether it occurs at an external or an internal surface. It is thus not surprising that the grain boundary resistivity (5.100) can also be expressed in the form (5.114) and the corresponding values of the parameter A as a function of d/Λ_∞ are shown in Figure 5.29. In this sense it is immaterial whether t is regarded as a specular transmission or a specular reflection coefficient, and Pichard *et al.* (1981) in fact treats surface scattering in a formally identical manner to grain boundary scattering (see their equations (1) and (5) which simply exchange p for t and involve the appropriate size parameter, this being either the grain diameter or film thickness).

If the interface scattering is regarded as an intrinsic property of a phase (as it must if the dimension of that phase is $\sim\Lambda_\mathrm{I}$), equations (5.114a or b) can be used to describe the change in resistivity of that phase with a

Fig. 5.29. The parameter A as a function of d/Λ_∞ from the grain boundary scattering model (5.100) for various values of the transmission coefficient t.

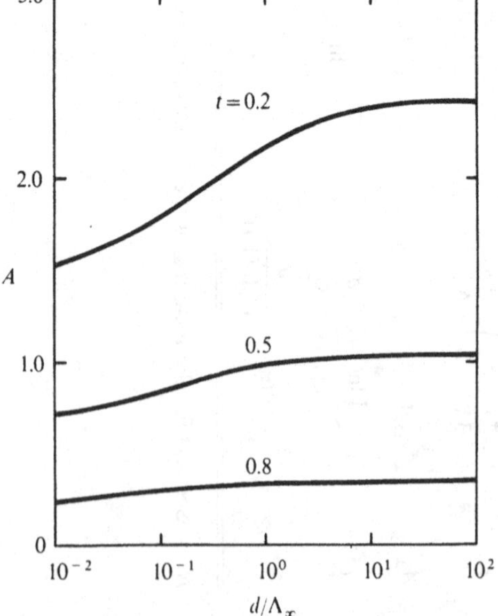

parameter A determined by any of the above models. The resistivity of the composite is then determined using one of the appropriate equations in Table 5.2 or 5.3. Alternatively, if the bulk properties of the phases are known, one would simply match the observed and calculated resistivities over a range of d or Λ_∞ (i.e. temperatures) to obtain A. Examples of such applications will be given in Section 5.7.6. The resistivity of multilayer films in which the thickness of each layer is $\sim\Lambda$ has been considered by Carcia & Suna (1983) and Dimmich (1985) using the concepts described above.

5.6 Atomic displacement effects

The diffraction effects of static and thermally induced atomic displacements will affect both the average and deviation lattice scattering terms. In general, the atomic positions are given in accordance with (2.25) and (2.29) as

$$\mathbf{R}_i = \mathbf{r}_i + \mathbf{u}_i + \boldsymbol{\delta}_i, \tag{5.115}$$

where $\mathbf{r}_i$ is the position of the undisplaced atom at site i (i.e. a perfect lattice vector), $\mathbf{u}_i$ the static displacement due to strain or size effects and $\boldsymbol{\delta}_i$ the thermally induced displacement. However, since $\mathbf{u}_i$ and $\boldsymbol{\delta}_i$ will generally be small, the exponential terms $\exp(x)$ in the structure factor can usually be expanded as $1+x+x^2/2+\cdots$, allowing the effects to be considered independently to first order, although to higher orders there will be interaction terms involving both $\mathbf{u}_i$ and $\boldsymbol{\delta}_i$. We will consider here only the first-order effects.

5.6.1 *Point defects and displacements*

We start by considering the atomic displacements associated with the relaxation of the lattice around a vacancy, substitutional impurity or interstitial in a metal or alloy. The coexisting effects of replacive and displacive correlations will be considered in the next section and we consider here just the perturbations to a lattice comprising the average potential $\bar{w}(\mathbf{r})$ at each lattice site.

(a) *Vacancy*

The introduction of a vacancy at a site $\mathbf{r}_0$ is equivalent to the addition of a potential $-\bar{w}(\mathbf{r}_0)$ at that site and introduction of the lattice distortions $\mathbf{u}_i$ leading to a total potential

$$\langle \mathbf{k}+\mathbf{q}|W(\mathbf{r})|\mathbf{k}\rangle = \frac{\bar{w}(\mathbf{q})}{N-1}\left[\sum_i^N \exp[-i\mathbf{q}\cdot(\mathbf{r}_i+\mathbf{u}_i)] - \exp(-i\mathbf{q}\cdot\mathbf{r}_0)\right]. \tag{5.116}$$

Assuming small $\mathbf{u}_i$, this becomes

$$\langle \mathbf{k}+\mathbf{q}|W|\mathbf{k}\rangle = \frac{\bar{w}(\mathbf{q})}{N-1}\left[\sum_i^N \exp(-\mathrm{i}\mathbf{q}\cdot\mathbf{r}_i) - \mathrm{i}\mathbf{q}\cdot\sum_i^N \mathbf{u}_i \exp(-\mathrm{i}\mathbf{q}\cdot\mathbf{r}_i) - \exp(-\mathrm{i}\mathbf{q}\cdot\mathbf{r}_0)\right]$$
$$= \bar{w}(\mathbf{q})S(\mathbf{q}) + \bar{w}(\mathbf{q})\,\Delta S(\mathbf{q}), \tag{5.117}$$

where the deviation in the structure factor $\Delta S(\mathbf{q})$ is given by

$$\Delta S(\mathbf{q}) = \frac{1}{N-1}\left[-\mathrm{i}\mathbf{q}\cdot\sum_i^N \mathbf{u}_i \exp(-\mathrm{i}\mathbf{q}\cdot\mathbf{r}_i) - \exp(-\mathrm{i}\mathbf{q}\cdot\mathbf{r}_0)\right]. \tag{5.118}$$

Note that we have again represented the perturbed crystal by the sum of a perfect crystal and a set of deviations. As before, the perfect crystal terms $\bar{w}(\mathbf{q})$, $S(\mathbf{q})$ cause Bragg scattering at $\mathbf{q}=\mathbf{g}$ and do not contribute to the resistivity. The deviation structure factor $\Delta S(\mathbf{q})$ exists for all $\mathbf{q}$. If distortions are neglected (i.e. $\mathbf{u}_i=0$ for all i),

$$\Delta S(\mathbf{q}) = \frac{-\exp(-\mathrm{i}\mathbf{q}\cdot\mathbf{r}_i)}{N-1}. \tag{5.119}$$

Thus

$$|\Delta S(\mathbf{q})|^2 = \frac{1}{(N-1)^2}, \tag{5.120}$$

leading to a resistivity

$$\rho = \frac{CN}{(N-1)^2}\int_0^{2k_\mathrm{F}} |\bar{w}(\mathbf{q})|^2 q^3\,\mathrm{d}q$$
$$\approx \frac{C}{N}\int |\bar{w}(\mathbf{q})|^2 q^3\,\mathrm{d}q. \tag{5.121}$$

For a small number of vacancies N_v situated at random this gives

$$\rho = Cc_\mathrm{v}\int_0^{2k_\mathrm{F}} |\bar{w}(\mathbf{q})|^2 q^3\,\mathrm{d}q, \tag{5.122}$$

where we have written the vacancy concentration $c_\mathrm{v}=N_\mathrm{v}/N$. This is the result derived by Harrison (1966) for undistorted pure metals in which case $\bar{w}(\mathbf{q})=w(\mathbf{q})$. When the distortions $\mathbf{u}_i$ are introduced it is convenient to define $\mathbf{r}_0$ as the origin so that $\exp(-\mathrm{i}\mathbf{q}\cdot\mathbf{r}_0)=1$ and the distortion are symmetrically distributed about the origin. It is then necessary to determine the displacement from some model (see discussion at end of Section 5.7.2) or experiment and average $\Delta S(\mathbf{q})$ over all directions of $\mathbf{q}$ either analytically (Harrison 1966) or by calculation of $|\Delta S(\mathbf{q})|^2$ for a number of directions of $\mathbf{q}$ and numerically averaging (Takai *et al.* 1974). We will discuss some of the results in Section 5.7.

(b) *Substitutional impurity*

Substitution of an impurity I onto one of the lattice sites $\mathbf{r}_0$ requires the removal of a host atomic potential $\bar{w}(\mathbf{r}_0)$ and addition of $w_I(\mathbf{r}_0)$. The total potential is then

$$\langle \mathbf{k}+\mathbf{q}|W(\mathbf{r})|\mathbf{k}\rangle = \frac{1}{N}\left[\bar{w}(\mathbf{q})\sum_i^N \exp[-\mathrm{i}\mathbf{q}\cdot(\mathbf{r}_i+\mathbf{u}_i)] - \bar{w}(\mathbf{q})\exp(-\mathrm{i}\mathbf{q}\cdot\mathbf{r}_0) + w_I(\mathbf{q})\exp(-\mathrm{i}\mathbf{q}\cdot\mathbf{r}_0)\right] \tag{5.123}$$

$$= \frac{1}{N}\left[\bar{w}(\mathbf{q})\sum_i^N \exp(-\mathrm{i}\mathbf{q}\cdot\mathbf{r}_i) - \bar{w}(\mathbf{q})\mathrm{i}\mathbf{q}\cdot\sum_i^N \mathbf{u}_i\exp(-\mathrm{i}\mathbf{q}\cdot\mathbf{r}_i) + [w_I(\mathbf{q})-\bar{w}(\mathbf{q})]\exp(-\mathrm{i}\mathbf{q}\cdot\mathbf{r}_0)\right]$$

$$= \bar{w}(\mathbf{q})S(\mathbf{q}) + \bar{w}(\mathbf{q})\,\Delta S(\mathbf{q}) + [w_I(\mathbf{q})-\bar{w}(\mathbf{q})]\frac{\exp(-\mathrm{i}\mathbf{q}\cdot\mathbf{r}_0)}{N}, \tag{5.124}$$

where

$$\Delta S(\mathbf{q}) = -\frac{1}{N}\mathrm{i}\mathbf{q}\cdot\sum \mathbf{u}_i\exp(-\mathrm{i}\mathbf{q}\cdot\mathbf{r}_i). \tag{5.125}$$

If the effects of the atomic displacements are ignored this gives

$$\rho = \frac{C}{N}\int_0^{2k_F} |w_I(\mathbf{q})-\bar{w}(\mathbf{q})|^2 q^3\,\mathrm{d}q. \tag{5.126}$$

For N_I impurities,

$$\rho = Cc_I\int_0^{2k_F} |w_I(\mathbf{q})-\bar{w}(\mathbf{q})|^2 q^3\,\mathrm{d}q, \tag{5.127}$$

where $c_I = N_I/N$. When the effects of the distortions are to be included, one must determine the $\mathbf{u}_i$ and perform the averaging by some of the techniques described above.

(c) *Self-interstitials*

The presence of a self-interstitial introduces an extra atomic potential $w_I(\mathbf{r}_0)$ at a position $\mathbf{r}_0$ which is no longer a lattice site:

$$\langle \mathbf{k}+\mathbf{q}|W(\mathbf{r})|\mathbf{k}\rangle$$

$$= \frac{1}{N+1}\left[\bar{w}(\mathbf{q})\sum_i^N \exp[-\mathrm{i}\mathbf{q}\cdot(\mathbf{r}_i+\mathbf{u}_i)] + w_I(\mathbf{q})\exp(-\mathrm{i}\mathbf{g}\cdot\mathbf{r}_0)\right]$$

$$= \frac{1}{N+1}\left[\bar{w}(\mathbf{q})\sum_i^N \exp(-\mathrm{i}\mathbf{q}\cdot\mathbf{r}_i) - \bar{w}(\mathbf{q})\mathrm{i}\mathbf{q}\cdot\sum_i^N \mathbf{u}_i\exp(-\mathrm{i}\mathbf{q}\cdot\mathbf{r}_i) + w_I(\mathbf{q})\exp(-\mathrm{i}\mathbf{q}\cdot\mathbf{r}_0)\right]. \tag{5.128}$$

If there is an equal probability of an A or B atom occupying an interstitial position one may assume $w_I(\mathbf{q}) = \bar{w}(\mathbf{q})$ leading to

$$\langle \mathbf{k}+\mathbf{q}|W(\mathbf{r})|\mathbf{k}\rangle = \bar{w}(\mathbf{q})S(\mathbf{q}) + \bar{w}(\mathbf{q})\Delta S(\mathbf{q}), \tag{5.129}$$

where

$$\Delta S(\mathbf{q}) = \frac{1}{N+1}\left[i\mathbf{q}\cdot\sum_i^N \mathbf{u}_i \exp(-i\mathbf{q}\cdot\mathbf{r}_i) + \exp(-i\mathbf{q}\cdot\mathbf{r}_0)\right]. \tag{5.130}$$

At $\mathbf{q}=\mathbf{g}$ the term $\exp(-i\mathbf{q}\cdot\mathbf{r}_0)$ is not in general unity, but since it is divided by $N+1$ it can be neglected leaving just the Bragg scattering due to $\bar{w}(\mathbf{q})S(\mathbf{q})$. If distortions are neglected, (5.9) then leads to

$$\rho = \frac{C}{N+1}\int_0^{2k_F} |\bar{w}(\mathbf{q})|^2 q^3 \, dq. \tag{5.131}$$

For N_I interstitials

$$\rho = Cc_I \int_0^{2k_F} |\bar{w}(\mathbf{q})|^2 q^3 \, dq, \tag{5.132}$$

where $c_I = N_I/(N+1) \approx N_I/N$, which is the same as the vacancy result (5.122) and is just the result of Harrison (1966) for self-interstitials in undistorted pure metal. If there is a preference for one type of atom (A) to occupy the interstitial position we find

$$\langle \mathbf{k}+\mathbf{q}|W(\mathbf{r})|\mathbf{k}\rangle = \bar{w}(\mathbf{q})S(\mathbf{q}) + \bar{w}(\mathbf{q})\,\Delta S(\mathbf{q}) + w_A(\mathbf{q})\exp(-i\mathbf{q}\cdot\mathbf{r}_0) \tag{5.133}$$

where $\Delta S(\mathbf{q})$ is given by equation (5.130). If the effects of distortion are neglected this gives

$$\rho = Cc_I \int_0^{2k_F} |w_A(\mathbf{q})|^2 q^3 \, dq. \tag{5.134}$$

The effects of atomic displacements again require evaluation of the $\mathbf{u}_i$ and averaging over all directions of $\mathbf{q}$.

(d) *Impurity interstitial*

The result follows directly from that of the self-interstitial (5.133) with $w_A(\mathbf{q})$ replaced by the form factor for the appropriate impurity.

5.6.2 *Thermally induced displacements*

A lattice vibration of wavenumber $\mathbf{Q}$ and mode j will give the displacements indicated in equation (2.48):

$$\boldsymbol{\delta}_i = \sum_{\mathbf{Q},j} (\mathbf{u}_{\mathbf{Q},j}\exp(\mathrm{i}\mathbf{Q}\cdot\mathbf{r}_i) + \mathbf{u}^*_{\mathbf{Q},j}\exp(-\mathrm{i}\mathbf{Q}\cdot\mathbf{r}_i)), \tag{5.135}$$

which, according to the adiabatic principle (Section 4.5.4), are static for the term of scattering an electron. Since these displacements are generally small (~0.1 Å) we can expand the exponential term in the structure factor and keep only terms to the first power in the displacements. For a particular mode $\mathbf{Q},j$ this gives

$$S(\mathbf{q})=\frac{1}{N}\sum_{i}^{N}\exp(-\mathrm{i}\mathbf{q}\cdot\mathbf{r}_i)\,[1-\mathrm{i}\mathbf{q}\cdot\mathbf{u}_{\mathbf{Q},j}\exp(\mathrm{i}\mathbf{Q}\cdot\mathbf{r}_i) \\ -\mathrm{i}\mathbf{q}\cdot\mathbf{u}^*_{\mathbf{Q},j}\exp(-\mathrm{i}\mathbf{Q}\cdot\mathbf{r}_i)]. \quad (5.136)$$

The first term is again the structure factor of a perfect lattice and does not contribute to the resistivity. The first-order terms of the electron–phonon interaction may thus be written as

$$-\mathrm{i}\mathbf{q}\cdot\mathbf{u}_{\mathbf{Q},j}\frac{1}{N}\sum_i\exp[-\mathrm{i}\mathbf{q}-\mathbf{Q})\cdot\mathbf{r}_i] \\ -\mathrm{i}\mathbf{q}\cdot\mathbf{u}^*_{\mathbf{Q},j}\frac{1}{N}\sum_i\exp[-\mathrm{i}(\mathbf{q}+\mathbf{Q})\cdot\mathbf{r}_i]. \quad (5.137)$$

The summations are over the perfect lattice and so will be equal to N when the factors $\mathbf{q}-\mathbf{Q}$ and $\mathbf{q}+\mathbf{Q}$ are equal to a reciprocal lattice vector $\mathbf{g}$, and zero otherwise. If $\mathbf{g}=0$, scattering will only occur at $\mathbf{q}=\mathbf{Q}$ and the phonon scattering matrix element is

$$\langle\mathbf{k}+\mathbf{Q}|W(\mathbf{r})|\mathbf{k}\rangle=-\mathrm{i}\mathbf{Q}\cdot\mathbf{u}_{\mathbf{Q},j}\langle\mathbf{k}+\mathbf{Q}|\bar{w}(\mathbf{r})|\mathbf{k}\rangle. \quad (5.138)$$

This represents scattering about the origin in wavenumber space and is the so-called normal (N) scattering. As the wavelength of the phonon increases, $\mathbf{Q}$ becomes smaller and so only scattering through small angles is possible at low temperatures. Note also that a purely transverse wave ($\mathbf{Q}$ perpendicular to $\mathbf{u}_{\mathbf{Q},j}$) will give no scattering. If $\mathbf{g}\neq 0$ the electron–phonon matrix elements are of the form

$$\langle\mathbf{k}+\mathbf{Q}|W(\mathbf{r})|\mathbf{k}\rangle=-\mathrm{i}(\mathbf{g}+\mathbf{Q})\cdot\mathbf{u}_{\mathbf{Q},j}\langle\mathbf{k}+\mathbf{g}+\mathbf{Q}|\bar{w}(\mathbf{r})|\mathbf{k}\rangle. \quad (5.139)$$

Thus even long wavelength phonons may give rise to large angle scattering and both purely longitudinal and transverse modes will contribute. Such scattering about the reciprocal lattice points $\mathbf{g}$ is called Umklapp (U) scattering.

We can now draw some simple conclusions about the temperature dependence of the resistivity. In the case of normal scattering, the square of the matrix element will involve terms of the form $|\mathbf{q}\cdot\mathbf{u}_{\mathbf{Q},j}|^2$. At high temperatures such terms can be evaluated with the aid of the Debye model (equation (2.47)) or the equivalent Einstein model term (2.34) to give

$$N|\mathbf{q}\cdot\mathbf{u}_{\mathbf{Q},j}|^2=\frac{2\hbar^2q_{\mathrm{D}}^2T}{Mk_{\mathrm{B}}\Theta_{\mathrm{D}}^2}. \quad (5.140)$$

This is now independent of $\mathbf{q}$ and so can be taken outside the resistivity integral (5.9). The remaining part of the integral is the effective scattering from a single average atom and so ρ is proportional to T. At low temperatures the collective motion of the atoms must be taken into account with the displacements being given approximately by equations (2.43) and (2.46)

$$N|\mathbf{q}\cdot\mathbf{u}_{Q,j}|^2 = \frac{2\hbar^2 q_D^2 T}{Mk_B\Theta_D^2}\left[\frac{\hbar\omega_{Q,j}/k_BT}{\exp(\hbar\omega_{Q,j}/k_BT)-1}\right]. \tag{5.141}$$

This leads ultimately to a T^5 dependence since the variable q in the resistivity integral (5.9) can be written in terms of $q_D T/\Theta_D$ (see Appendix D). At low temperatures the effects of Umpklapp scattering must also be included.

In addition to terms of the form described above, the squaring of the matrix elements will produce terms involving cross-products $(\mathbf{q}\cdot\mathbf{u}_{Q,j})(\mathbf{q}\cdot\mathbf{u}_{Q',j'})$ and other higher-order terms in the exponential expansion. This leads to multiphonon scattering, a Debye–Waller factor that reduces the intensity of the Bragg scattering and an associated thermal diffuse scattering, in direct analogy to the scattering of X-ray or thermal neutrons by such excitations (see e.g. Warren 1969, ch. 11; Ziman 1972, ch. 2).

The above approach to the scattering by phonons is useful in that it makes clear the nature of the N and U scattering processes. However, evaluation of the terms is not simple as it requires a full solution of the dynamical matrix and summing over modes. Furthermore, the scattering will be anisotropic and partly inelastic and so one should use a better solution of the Boltzmann equation than the simple relaxation time approximation. To this end Ziman (1960, section 9.5) has shown that the full variational expression for the resistivity together with a simple trial function of the form (5.3) leads to

$$\rho = \frac{9\pi\hbar\Omega_0}{2e^2Mk_BT}\left(\frac{1}{k_F^2S_F^2}\right) \times \int_S\int_{S'}\sum_j \frac{(\mathbf{q}\cdot\hat{\mathbf{x}})^2(\mathbf{q}\cdot\mathbf{e}_{\mathbf{q},j})^2C(|\mathbf{q}|)^2}{\{\exp(\zeta)-1\}\{1-\exp(-\zeta)\}}\frac{\mathrm{d}S\,\mathrm{d}S'}{|\mathbf{v}||\mathbf{v}'|}, \tag{5.142}$$

where $\zeta = \hbar\omega_{Q,j}/k_BT$ and $C(\mathbf{q})$ is directly related to the orthogonal planewave matrix elements $V_{Q,\mathbf{q}}$ describing the transition from $\mathbf{k}$ to $\mathbf{k}'$:

$$V_{Q,\mathbf{q}} = -i(\mathbf{e}_Q\cdot\mathbf{q})C(\mathbf{q}) \tag{5.143}$$

and $\mathbf{e}_{Q,j}$ is the polarisation vector for the phonons $\mathbf{Q},j$. This expression has been evaluated using experimentally determined dispersion relations (Darby & March 1974), a Debye dispersion relation $\omega_Q \propto \mathbf{Q}$ (Dreirach 1973) or a solution of the dynamical matrix based on known force constants (Pal 1973; Kumar 1975) in conjunction with a variety of scattering potential forms. The use of more complex trial functions has been discussed by Greene & Kohn (1965) and Ziman (1960, p. 368). It might also be noted that (5.142) leads directly to the Bloch–Guniessen formula (Ziman 1960, p. 364; Blatt 1968, p. 189). By assuming free electron values of S_F and $|\mathbf{v}_F|$ (implying a spherical Fermi surface) and taking a simpler form of the trial solution $\Phi = |\mathbf{q}|$, this expression reduces to

$$\rho = \frac{3\Omega_0 m^2}{16 k_B T \hbar k_F^6 e^2 M} \sum_j \int \frac{|\mathbf{q}||\langle \mathbf{k}+\mathbf{q}|w(\mathbf{r})|k\rangle|^2 (\mathbf{q}\cdot\mathbf{e}_{\mathbf{q},j})^2}{\{\exp(\zeta)-1\}\{1-\exp(-\zeta)\}}\,\mathrm{d}\mathbf{q}, \quad (5.144)$$

where we have substituted the pseudopotential form factors for $C(|\mathbf{q}|)$. This may be written in the form

$$\rho = CN \int_0^{2k_F} q^3 |\langle \mathbf{k}+\mathbf{q}|w(\mathbf{r})|\mathbf{k}\rangle|^2 |S(\mathbf{q})|^2\,\mathrm{d}q, \quad (5.145)$$

where the constant C is given by equation (5.10) and the square of the structure factor $|S(\mathbf{q})|^2$ is given by

$$|S(\mathbf{q})|^2 = \frac{1}{4\pi} \int \mathscr{S}(\mathbf{q})\,\mathrm{d}\mathbf{q},$$

where (5.146)

$$\mathscr{S}(\mathbf{q}) = \frac{\hbar^2}{MN} \sum_j \frac{|\mathbf{q}\cdot\mathbf{e}_{\mathbf{q},j}|^2}{k_B T\{\exp(\zeta)-1\}\{1-\exp(-\zeta)\}}.$$

Methods of evaluating the averages over the Fermi surface have been described by Bailyn (1960a) and Betts *et al.* (1956). Baym (1964) has shown that this expression can be derived from the dynamical structure factor $\mathscr{S}(\mathbf{q}, \omega)$ defined as the Fourier transform of the time-dependent pair correlation function (van Hove 1954):

$$\mathscr{S}(\mathbf{q}, \omega) = \frac{1}{2\pi N} \int_{-\infty}^{\infty} \exp(\mathrm{i}\omega t) \times \left\langle \sum_i \sum_j \exp[-\mathrm{i}\mathbf{q}\cdot\boldsymbol{\delta}_i(t)] \exp[\mathrm{i}\mathbf{q}\cdot\boldsymbol{\delta}_j(0)] \right\rangle_T \mathrm{d}t, \quad (5.147)$$

where lattice displacements are now time-dependent

$$\boldsymbol{\delta}_i(t) = \mathbf{u}_{\mathbf{Q},j} \exp[\mathrm{i}(\mathbf{Q}\cdot\mathbf{r}_i - \omega_{\mathbf{Q},j} t)]$$

and the angled brackets $\langle\ \rangle_T$ imply a thermal average at temperature T. This structure factor provides a direct measure of all density fluctuations in the lattice and so includes the Debye–Waller and all multiphonon terms. The expression for the resistivity is then (Baym 1964; Greene &

Kohn 1965)

$$\rho = C\int_0^{2k_F} q^3|\langle \mathbf{k}+\mathbf{q}|w(\mathbf{r})|\mathbf{k}\rangle|^2\,dq\int_{-\infty}^{\infty}\frac{\mathscr{S}'(\mathbf{q},\omega)\dfrac{\hbar\omega}{k_B T}\,d\omega}{\exp(\zeta)-1}, \quad (5.148)$$

where $\mathscr{S}'(\mathbf{q}, w)$ is $\mathscr{S}(\mathbf{q}, w)$ without the elastic Bragg scattering terms. This expression reduces directly to (5.144) if only the lowest-order non-vanishing terms in the displacement are retained in the expansion of the exponentials (Dynes & Carbotte 1968). As such, the Ziman form (5.145) is not expected to include any multiphonon or Debye–Waller effects. However, if a dynamical structure factor determined by neutron scattering or direct calculation is employed, then all such effects should be included. But we note also that in the high temperature limit the second integration in (5.148) becomes $\int \mathscr{S}(\mathbf{q}, \omega)\,d\omega = \mathscr{S}(\mathbf{q})$ and so, if the correct limiting form of $\mathscr{S}(\mathbf{q})$ is used with (5.145), then such effects should in fact be included. Shukla *et al.* (Shukla & Muller 1980; Shukla & VanderSchans 1980; Shukla 1980) have investigated the significance of these higher-order terms although, due to cancellation effects, it is not clear how serious their omission might be (see also Sham & Ziman 1963).

It is possible to avoid calculation of phonon frequencies and polarisations by writing the structure factor directly in terms of the dynamical matrix (Robinson & Dow 1968):

$$\mathscr{S}(\mathbf{q}) = \frac{k_B T}{M}(\mathbf{q}D^{-1}(\mathbf{q})\mathbf{q}). \quad (5.149)$$

This enables the structure factor to be determined directly from known force (or elastic) constants (see e.g. Born & Huang 1954). It might be noted that the periodicity of the dynamical matrix ensures that wavevectors outside the first Brillouin zone are properly folded back into that zone, thus ensuring that Umklapp as well as normal scattering events are automatically included.

So far we have not taken into consideration the effects of a finite conduction electron mean free path. As in the case of static disorder, only thermal disorder on a scale $\leqslant\Lambda$ will contribute to the resistivity. This means that only phonons with a wavelength $2\pi/|\mathbf{Q}| \leqslant \Lambda$ should be considered. To a first approximation this cut-off can be included by taking $|\mathbf{Q}| = 2\pi/\Lambda$ as the lower limit to the integration rather than $|\mathbf{Q}| = 0$. (The idea of a long wavelength cut-off was first suggested by Binder & Stauffer (1976) as a means of including finite mean free path effects in the static structure factor.) This will have little effect on the resistivity of good conductors ($\rho < 10\mu\Omega$ cm) but should be taken into account in studies of materials having a higher resistivity ($\rho > 50\mu\Omega$ cm). We will return to this problem in Section 8.2.2. Finally, we note that as the lattice vibrations are determined by both the atomic mass and interatomic force constants, changing the composition by alloying will in general affect the temperature dependence of the resistivity.

5.6.3 *Static atomic displacements in a concentrated alloy*

So far we have been considering the effects of atomic displacements due to impurities or due to phonons. We now turn to the effects of atomic displacements associated with different atomic sizes in a concentrated alloy. These displacements can affect both the average and deviation lattices. This means that the average lattice contributions will in general be non-zero for $\mathbf{q} \neq \mathbf{g}$ and so cross-terms will appear in the square of the pseudopotential scattering matrix element. (In an earlier study by Wells & Rossiter (1971) the effect on the average lattice and the cross-term were neglected. However, as we will show in Section 5.7.3, this omission can lead to a serious underestimation of the effect of static atomic displacements and all terms need to be included.)

As before, we write the pseudopotential as a sum over the individual ion potentials which are now located at positions $\mathbf{r}_i + \mathbf{u}_i^{\mathrm{A}}$ or $\mathbf{r}_i + \mathbf{u}_i^{\mathrm{B}}$ depending upon whether an A or B atom is at site i. From (4.73) we obtain

$$W(\mathbf{r}) = \sum_i (\sigma_i^{\mathrm{A}} w^{\mathrm{A}}(\mathbf{r} - \mathbf{r}_i - \mathbf{u}_i^{\mathrm{A}}) + \sigma_i^{\mathrm{B}} w^{\mathrm{B}}(\mathbf{r} - \mathbf{r}_i - \mathbf{u}_i^{\mathrm{B}})). \tag{5.150}$$

Following (4.74), the matrix elements of the pseudopotential are then written as

$$\begin{aligned} &\langle \mathbf{k}+\mathbf{q} | W(\mathbf{r}) | \mathbf{k} \rangle \\ &= \frac{1}{N} \sum_i [\sigma_i^{\mathrm{A}} w^{\mathrm{A}}(\mathbf{q}) \exp(-\mathrm{i}\mathbf{q} \cdot \mathbf{u}_i^{\mathrm{A}}) \\ &\qquad + \sigma_i^{\mathrm{B}} w^{\mathrm{B}}(\mathbf{q}) \exp(-\mathrm{i}\mathbf{q} \cdot \mathbf{u}_i^{\mathrm{B}})] \exp(-\mathrm{i}\mathbf{q} \cdot \mathbf{r}_i). \end{aligned} \tag{5.151}$$

The square of the matrix element is again evaluated by grouping together all terms of the same spacing $\mathbf{r}_i - \mathbf{r}_j$ and multiplying each by the number of such terms averaged over all configurations. From (2.12) and (5.151) this gives

$$\begin{aligned} &|\langle \mathbf{k}+\mathbf{q} | W(\mathbf{r}) | \mathbf{k} \rangle|^2 \\ &= \frac{1}{N} \sum_{ij} \Big[(c_{\mathrm{A}}^2 + \langle \sigma_i \sigma_j \rangle) w_{\mathrm{A}}^2(\mathbf{q}) \langle \exp[-\mathrm{i}\mathbf{q} \cdot (\mathbf{u}_i^{\mathrm{A}} - \mathbf{u}_j^{\mathrm{A}})] \rangle \\ &\qquad + (c_{\mathrm{B}}^2 + \langle \sigma_i \sigma_j \rangle) w_{\mathrm{B}}^2(\mathbf{q}) \langle \exp[-\mathrm{i}\mathbf{q} \cdot (\mathbf{u}_i^{\mathrm{B}} - \mathbf{u}_j^{\mathrm{B}})] \rangle \\ &\qquad + 2(c_{\mathrm{A}} c_{\mathrm{B}} - \langle \sigma_i \sigma_j \rangle) w_{\mathrm{A}}(\mathbf{q}) w_{\mathrm{B}}(\mathbf{q}) \langle \exp[-\mathrm{i}\mathbf{q} \cdot (\mathbf{u}_i^{\mathrm{A}} - \mathbf{u}_j^{\mathrm{B}})] \rangle \Big] \\ &\quad \times \exp[-\mathrm{i}\mathbf{q} \cdot (\mathbf{r}_i - \mathbf{r}_j)]. \end{aligned} \tag{5.152}$$

We assume that the displacements are small and so the exponentials can be expanded as a power series. The average values of the exponential terms within the large square brackets are then

$$\begin{aligned}\langle\exp[-i\mathbf{q}\cdot(\mathbf{u}_i-\mathbf{u}_j)]\rangle &= 1-i\langle\mathbf{q}\cdot(\mathbf{u}_i-\mathbf{u}_j)\rangle-\tfrac{1}{2}\langle(\mathbf{q}\cdot\mathbf{u}_i)^2\rangle\\ &\quad-\tfrac{1}{2}\langle(\mathbf{q}\cdot\mathbf{u}_j)^2\rangle+\langle(\mathbf{q}\cdot\mathbf{u}_i)(\mathbf{q}\cdot\mathbf{u}_j)\rangle+\cdots.\end{aligned}\tag{5.153}$$

We proceed by considering firstly the terms of the form $1-\frac{1}{2}\langle(\mathbf{q}\cdot\mathbf{u}_i)^2\rangle-\frac{1}{2}\langle(\mathbf{q}\cdot\mathbf{u}_j)^2\rangle$. From (5.152), these include

$$\begin{aligned}&(1-\tfrac{1}{2}\langle(\mathbf{q}\cdot\mathbf{u}_i^{A})^2\rangle-\tfrac{1}{2}\langle(\mathbf{q}\cdot\mathbf{u}_j^{A})^2\rangle)(c_A^2+\langle\sigma_i\sigma_j\rangle)w_A^2(\mathbf{q})\\ &\qquad+(1-\tfrac{1}{2}\langle(\mathbf{q}\cdot\mathbf{u}_i^{B})^2\rangle-\tfrac{1}{2}\langle(\mathbf{q}\cdot\mathbf{u}_j^{B})^2\rangle)\\ &\qquad\times(c_B^2+\langle\sigma_i\sigma_j\rangle)w_B^2(\mathbf{q})\\ &\qquad+2(1-\tfrac{1}{2}\langle(\mathbf{q}\cdot\mathbf{u}_i^{A}\rangle-\tfrac{1}{2}\langle(\mathbf{q}\cdot\mathbf{u}_j^{B}\rangle)\\ &\qquad\times(c_Ac_B-\langle\sigma_i\sigma_j\rangle)w_A(\mathbf{q})w_B(\mathbf{q}).\end{aligned}\tag{5.154}$$

If $i\neq j$, (5.154) can be written as

$$\begin{aligned}&[c_Aw_A(\mathbf{q})\exp(-M'_A)+c_Bw_B(\mathbf{q})\exp(-M'_B)]^2\\ &\qquad+\langle\sigma_i\sigma_j\rangle[w_A(\mathbf{q})\exp(-M'_A)-w_B(\mathbf{q})\exp(-M'_B)]^2,\end{aligned}\tag{5.155}$$

where we have introduced the pseudo Debye–Waller factors M'_A and M'_B defined by

$$\begin{aligned}1-\langle(\mathbf{q}\cdot\mathbf{u}_i^{A})\rangle&=\exp(-2M'_A)\\ 1-\langle(\mathbf{q}\cdot\mathbf{u}_i^{B})\rangle&=\exp(-2M'_B).\end{aligned}\tag{5.156}$$

If we assume for the moment that the atomic displacements are such that $M'_A=M'_B=M'$, the total contribution to the scattering matrix (5.152) from such terms becomes

$$\begin{aligned}&\frac{1}{N}\sum_{i\neq j}[(c_Aw_A(\mathbf{q})+c_Bw_B(\mathbf{q}))^2\exp(-2M')\exp[-i\mathbf{q}\cdot(\mathbf{r}_i-\mathbf{r}_j)]\\ &\qquad+\langle\sigma_i\sigma_j\rangle(w_A(\mathbf{q})-w_B(\mathbf{q}))^2\exp(-2M')\exp[-i\mathbf{q}\cdot(\mathbf{r}_i-\mathbf{r}_j)]].\end{aligned}\tag{5.157}$$

This summation can be taken over all i,j provided that the $i=j$ contribution is subtracted again. This gives

$$\begin{aligned}&\frac{1}{N}|\bar{w}(\mathbf{q})|^2\sum_{ij}[\exp[-i\mathbf{q}\cdot(\mathbf{r}_i-\mathbf{r}_j)]\exp(-2M')-\exp(-2M')]\\ &\qquad+\frac{1}{N}|w^{d}(\mathbf{q})|^2\left[\sum_{ij}\langle\sigma_i\sigma_j\rangle\exp(-2M')\exp[-i\mathbf{q}\cdot(\mathbf{r}_i-\mathbf{r}_j)]\right.\\ &\qquad\qquad\left.-c_Ac_B\exp(-2M')\right].\end{aligned}\tag{5.158}$$

We now add the $i=j$ contribution from the summation in (5.152) which is $1/N(c_A w_A^2(\mathbf{q})+c_B w_B^2(\mathbf{q}))$ to give, after some rearrangement,

$$|W_1(\mathbf{q})|^2=\frac{1}{N}\left[|\bar{w}(\mathbf{q})|^2\sum_{ij}\exp[-i\mathbf{q}\cdot(\mathbf{r}_i-\mathbf{r}_j)]\exp(-2M')\right.$$
$$\left.+(1-\exp(-2M'))\right]$$
$$+\frac{1}{N}|w^d(\mathbf{q})|^2\left[c_A c_B+\sum_{ij}{}'\langle\sigma_i\sigma_j\rangle\exp(-2M')\right.$$
$$\left.\times\exp[-i\mathbf{q}\cdot(\mathbf{r}_i-\mathbf{r}_j)]\right].\tag{5.159}$$

The subscript 1 on $W(\mathbf{q})$ is used to indicate that this is still only part of the total scattering matrix. Comparison with the earlier results shows that the averages $-\frac{1}{2}\langle(\mathbf{q}\cdot\mathbf{u}_i)^2\rangle$ have resulted in the introduction into the scattering matrix of the pseudo Debye–Waller factors $\exp(-2M')$ which reduce the intensities of both the Bragg peaks and the SRO diffuse scattering. In addition there is a new diffuse scattering term $1/N(1-\exp(-2M'))|\bar{w}(\mathbf{q})|^2$ which conserves the intensity lost. Averaging over all directions of $\mathbf{q}$ then gives a contribution to the scattering matrix

$$|W_1(\mathbf{q})|^2=\frac{1}{N}\left[|\bar{w}(\mathbf{q})|^2(1-\exp(-2M'))+c_A c_B|w^d(\mathbf{q})|^2\right.$$
$$\left.\times\left[1+\exp(-2M')\sum_i{}'c_i\alpha_i\frac{\sin(qr_i)}{qr_i}\right]\right],\tag{5.160}$$

where, as before, we have excluded the Bragg scattering term and adopted the shell model. If the atomic displacements are such that $M'_A\neq M'_B$ then (5.159) should be left in the more general form

$$|W_1(\mathbf{q})|^2=\frac{1}{N}\left[[c_A w_A(\mathbf{q})\exp(-M'_A)+c_B w_B(\mathbf{q})\exp(-M'_B)]^2\right.$$
$$\times\sum_{ij}\exp[-i\mathbf{q}\cdot(\mathbf{r}_i-\mathbf{r}_j)]+|\bar{w}(\mathbf{q})|^2-\bar{w}(\mathbf{q})[c_A w_A(\mathbf{q})\exp(-2M'_A)$$
$$+c_B w_B(\mathbf{q})\exp(-2M'_B)]+c_A c_B|w^d(\mathbf{q})|^2+w^d(\mathbf{q})$$
$$\times(w_A(\mathbf{q})\exp(-2M'_A)-w_B(\mathbf{q})\exp(-2M'_B))$$
$$\left.\times\sum_{ij}{}'\langle\sigma_i\sigma_j\rangle\exp[-i\mathbf{q}\cdot(\mathbf{r}_i-\mathbf{r}_j)]\right].\tag{5.161}$$

Huang (1947) has considered the next averages $\langle(\mathbf{q}\cdot\mathbf{u}_i)(\mathbf{q}\cdot\mathbf{u}_j)\rangle$ in the series (5.153) and found that they lead to an additional diffuse scattering which is highly localised about the reciprocal lattice points $\mathbf{q}=\mathbf{g}$. Since the resistivity integral is over all $|\mathbf{q}|$ only up to $2k_F$ this contribution will be small and can probably be neglected. It is important to realise that the form of scattering discussed so far arises because the spacing of a particular pair of atoms over any distance depends upon the local concentration fluctuations in the region between them.

We now turn to the second term on the right-hand side of (5.153). This is most conveniently evaluated by writing the actual static displacements in terms of the parameters suggested in (2.26) such that

$$\langle \mathbf{q}\cdot(\mathbf{u}_i^{\mathrm{A}}-\mathbf{u}_j^{\mathrm{B}})\rangle=\mathbf{q}\cdot(\mathbf{r}_i-\mathbf{r}_j)\varepsilon_{ij}^{\mathrm{AB}}, \tag{5.162}$$

and similarly for $\varepsilon_{ij}^{\mathrm{AA}}$ and $\varepsilon_{ij}^{\mathrm{BB}}$. The appropriate factors in the large square brackets of (5.152) are then of the form

$$-\mathrm{i}\mathbf{q}\cdot(\mathbf{r}_i-\mathbf{r}_j)[(c_{\mathrm{A}}^2+\langle\sigma_i\sigma_j\rangle)w_{\mathrm{A}}^2(\mathbf{q})\varepsilon_{ij}^{\mathrm{AA}}+(c_{\mathrm{B}}^2+\langle\sigma_i\sigma_j\rangle)w_{\mathrm{B}}^2(\mathbf{q})\varepsilon_{ij}^{\mathrm{BB}}$$
$$+2(c_{\mathrm{A}}c_{\mathrm{B}}-\langle\sigma_i\sigma_j\rangle)w_{\mathrm{A}}(\mathbf{q})w_{\mathrm{B}}(\mathbf{q})\varepsilon_{ij}^{\mathrm{AB}}]. \tag{5.163}$$

After some rearrangement this gives a contribution to the scattering matrix of

$$|W_2(\mathbf{q})|^2=\frac{1}{N}\sum_{ij}\Big[\exp[-\mathrm{i}\mathbf{q}\cdot(\mathbf{r}_i-\mathbf{r}_j)][-\mathrm{i}\mathbf{q}\cdot(\mathbf{r}_i-\mathbf{r}_j)]$$
$$\times(|w^{\mathrm{d}}(\mathbf{q})|^2\gamma_{ij}+\bar{w}(\mathbf{q})w^{\mathrm{d}}(\mathbf{q})\xi_{ij})\Big], \tag{5.164}$$

where

$$\gamma_{ij}=c_{\mathrm{A}}c_{\mathrm{B}}(c_{\mathrm{A}}\varepsilon_{ij}^{\mathrm{AA}}+c_{\mathrm{B}}\varepsilon_{ij}^{\mathrm{BB}})+\langle\sigma_i\sigma_j\rangle(c_{\mathrm{A}}\varepsilon_{ij}^{\mathrm{BB}}+c_{\mathrm{B}}\varepsilon_{ij}^{\mathrm{AA}}), \tag{5.165}$$

$$\xi_{ij}=(c_{\mathrm{A}}^2+\langle\sigma_i\sigma_j\rangle)\varepsilon_{ij}^{\mathrm{AA}}-(c_{\mathrm{B}}^2+\langle\sigma_i\sigma_j\rangle)\varepsilon_{ij}^{\mathrm{BB}}. \tag{5.166}$$

In deriving the above we have made use of the identity (2.28). Averaging over all directions of $\mathbf{q}$ and again employing the shell model then gives

$$|W_2(\mathbf{q})|^2=\frac{c_{\mathrm{A}}c_{\mathrm{B}}}{N}\sum_i c_i\Big[\Big(\gamma_i|w^{\mathrm{d}}(\mathbf{q})|^2+\xi_i\bar{w}(\mathbf{q})w^{\mathrm{d}}(\mathbf{q})\Big)$$
$$\times\Big(\cos(qR_i)-\frac{\sin(qR_i)}{qR_i}\Big)\Big], \tag{5.167}$$

where now

$$\gamma_i=(c_{\mathrm{A}}\varepsilon_i^{\mathrm{AA}}+c_{\mathrm{B}}\varepsilon_i^{\mathrm{BB}})+\alpha_i(c_{\mathrm{A}}\varepsilon_i^{\mathrm{BB}}+c_{\mathrm{B}}\varepsilon_i^{\mathrm{AA}}), \tag{5.168}$$

and

$$\xi_i=\Big(\frac{c_{\mathrm{A}}}{c_{\mathrm{B}}}+\alpha_i\Big)\varepsilon_i^{\mathrm{AA}}-\Big(\frac{c_{\mathrm{B}}}{c_{\mathrm{A}}}+\alpha_i\Big)\varepsilon_i^{\mathrm{BB}}. \tag{5.169}$$

Finally, the total resistivity is found by substituting $|W(\mathbf{q})|^2=|W_1(\mathbf{q})|^2+|W_2(\mathbf{q})|^2$ into the resistivity integral (5.9). The parameters M' (or $M'_{\mathrm{A}}, M'_{\mathrm{B}}$), α_i and $\varepsilon_i^{\mathrm{AA}}$, $\varepsilon_i^{\mathrm{BB}}$ may be calculated or obtained from X-ray or neutron scattering experiments (see e.g. Warren 1969). We will return to this aspect of the problem in Section 5.7.

Equation (5.167) may be written in a slightly different but directly equivalent form in terms of the pseudopotentials $w_A(\mathbf{q})$ and $w_B(\mathbf{q})$:

$$|W_2(\mathbf{q})|^2 = \frac{c_A c_B}{N} \sum_i c_i \beta_i \left(\cos(qR_i) - \frac{\sin(qR_i)}{qR_i} \right), \tag{5.170}$$

where

$$\beta_i = [w_A(\mathbf{q}) - w_B(\mathbf{q})] \left[\left(\frac{c_A}{c_B} + \alpha_i \right) \varepsilon_i^{AA} w_A(\mathbf{q}) - \left(\frac{c_B}{c_A} + \alpha_i \right) \varepsilon_i^{BB} w_B(\mathbf{q}) \right]. \tag{5.171}$$

This form is the same as that derived by Katsnel'son *et al.* (1978) and is the equivalent of the expression derived for X-ray scattering by Warren *et al.* (1951). However, by writing it in for form (5.167), the integrals over $\mathbf{q}$ for any particular system need only be evaluated once, the changes due to SRO and static displacements then being contained in the multiplying factors γ_i and ξ_i.

In summary, static atomic displacements are expected to produce the following effects: (i) a reduction in the intensity of the Bragg peaks which has no effect on the resistivity (first term in (5.159)), (ii) a reduction in the intensity of the SRO diffuse scattering (second term in (5.160)), (iii) introduction of a pseudo Debye–Waller diffuse scattering (first term in (5.160)), and (iv) introduction of a size effect diffuse scattering (5.167). The magnitudes of these effects and their dependence upon short range atomic correlations will be discussed in Section 5.7.3.

It is to be expected that there might also be a significant contribution from static atomic displacements to the resistivity of a partially long range ordered material given also by (5.167). Such a contribution will be at a maximum at T_c and must go to zero in the fully ordered state since the atoms then occupy a completely periodic lattice (neglecting here the displacements known to be associated with the antiphase domain boundaries – see e.g. Figure 2.20). As a first approximation this contribution may be expected to vary like the residual resistivity $1 - S^2$. However, there has not yet been any detailed calculation of this effect or of the scattering by the antiphase domain boundaries.

5.6.4 *Displacive transitions*

The factorisation of the potential into a structure factor and a form factor allows the resistivity due to the atomic displacements associated with a displacive phase transition to be treated in exactly the same way as a replacive phase transition leading to equations (5.9a) or (5.9b), but with the structure factor now given in terms of the appropriate displacive correlation parameters defined in Sections 2.2 and 2.7.3. In a random alloy, the form factor of interest will be that of the average lattice, $\bar{w}(\mathbf{q})$. If the alloy is non-random there will be contributions from both $\bar{w}(\mathbf{q})$ and $w^{\mathrm{d}}(\mathbf{q})$ multiplied by the appropriate structure factors. As the mathematical development is formally identical to that for replacive correlations described at some length above, we will not reproduce it here. However, with this background we know that only displacive correlations over a range $\sim\Lambda$ need be considered and that short and long range displacive correlations may cause the resistivity to either increase or decrease, depending upon the periodicity of the correlations and the influence of any associated band structure effects (cf. Sections 5.2 and 5.3). We would also expect electron scattering to occur at displacement–domain boundaries (cf. antiphase domain boundary effects). We will return to the consideration of such transitions in Section 5.7.7.

5.6.5 *Combined effects*

Some of the indirect interactions between measuring temperature and the residual resistivity of alloys with long range order (via changes in the electronic structures, Section 5.3.2) and those with inhomogeneous short range correlations or a fine dispersion of phases (via the electron mean free path, Sections 5.3.2 and 5.5.2) have already been considered. Some of the possible effects of atomic correlations on the phonon spectrum and hence upon that contribution to the resistivity have also been mentioned (Section 2.7.2). We now want to consider the direct interaction between thermally induced displacements and the resistivity due to atomic correlations and size-induced static displacements. This is done by including both the displacements $\mathbf{u}_i$ and $\boldsymbol{\delta}_i$ for each atomic species at each lattice site (Rossiter 1977). In a complete first principles calculation, the static atomic and phonon induced displacements should be determined self-consistently with respect to the local atomic environments. Here we will assume that they are already known. The square of the structure factor then contains terms of the form

$$|S(\mathbf{q})|^2 = \frac{1}{N}\sum_{ij} c_{\mathrm{B}}^2 P_{ij}^{\mathrm{AA}} \exp(-\mathrm{i}\mathbf{q}\cdot\mathbf{R}_{ij})\langle\exp[-\mathrm{i}\mathbf{q}\cdot(\mathbf{u}_i^{\mathrm{A}}-\mathbf{u}_j^{\mathrm{A}})]\rangle \times \langle\exp[-\mathrm{i}\mathbf{q}\cdot(\boldsymbol{\delta}_i^{\mathrm{A}}-\boldsymbol{\delta}_j^{\mathrm{A}})]\rangle. \tag{5.172}$$

If we assume that the lattice vibrations are harmonic (Ott 1935; Born & Sarginson 1941)

$$\langle\exp[-\mathrm{i}\mathbf{q}\cdot(\boldsymbol{\delta}_i^{\mathrm{A}}-\boldsymbol{\delta}_j^{\mathrm{B}})]\rangle = \exp[-\tfrac{1}{2}\langle[\mathbf{q}\cdot(\boldsymbol{\delta}_i^{\mathrm{A}}-\boldsymbol{\delta}_j^{\mathrm{B}})]^2\rangle]. \tag{5.173}$$

If we further assume an Einstein model for the lattice excitations and also that $|\boldsymbol{\delta}_i^{\mathrm{A}}| = |\boldsymbol{\delta}_i^{\mathrm{B}}| = |\boldsymbol{\delta}_i| = \delta$, averaging over all directions of $\mathbf{q}$ gives

$$-\tfrac{1}{2}\langle[\mathbf{q}\cdot(\boldsymbol{\delta}_i-\boldsymbol{\delta}_j)]^2\rangle = -\tfrac{1}{2}[\tfrac{1}{3}q^2|\boldsymbol{\delta}_i|^2 + \tfrac{1}{3}q^2|\boldsymbol{\delta}_j|^2]. \tag{5.174}$$

Such lattice vibrations thus introduce a Debye–Waller factor

$$\exp(-2M_E)=\exp(-\tfrac{1}{3}q^2|\delta|^2) \tag{5.175}$$

which multiplies the terms (5.161) and (5.167). The mean square displacements may be determined from equation (2.34) and will thus depend upon temperature. The effect of this extra factor may be regarded as a damping of the correlation functions α_i, γ_i and ξ_i. If we take into account the wavelike nature of the lattice vibrations, the displacements are given by (5.135). The correlated motion of atoms on neighbouring sites will then introduce a modification to this Debye–Waller factor and, for an A atom at site i and a B atom at site j, equations (5.136) and (5.173) give terms of the form

$$\exp\left\{-\sum_{\mathbf{Q},j}[\tfrac{1}{2}|\mathbf{q}\cdot\mathbf{u}^{A}_{\mathbf{Q},j}|^2+\tfrac{1}{2}|\mathbf{q}\cdot\mathbf{u}^{B}_{\mathbf{Q},j}|^2-(\mathbf{q}\cdot\mathbf{u}^{A}_{\mathbf{Q},j}) \times(\mathbf{q}\cdot\mathbf{u}^{B}_{\mathbf{Q},j})\exp[-\mathbf{Q}\cdot(\mathbf{r}^{A}_{i}-\mathbf{r}^{B}_{j})]]\right\}. \tag{5.176}$$

If we assume $\mathbf{u}^{A}_{\mathbf{Q},j}=\mathbf{u}^{B}_{\mathbf{Q},j}=\mathbf{u}_{\mathbf{Q},j}$, and average over all directions of $\mathbf{q}$ this becomes

$$\exp\left\{-\sum_{\mathbf{Q}}[q^2|\mathbf{u}_{\mathbf{Q}}|^2-q^2|\mathbf{u}_{\mathbf{Q}}|^2\exp[\mathrm{i}\mathbf{Q}\cdot(\mathbf{r}^{A}_{i}-\mathbf{r}^{B}_{j})]]\right\}, \tag{5.177}$$

since $|\mathbf{q}\cdot\mathbf{u}_{\mathbf{Q},j}|^2=\tfrac{1}{3}\mathbf{q}^2|\mathbf{u}_{\mathbf{Q},j}|^2$, but we have to sum over the three modes eliminating the factor $\tfrac{1}{3}$. The first term in (5.177) is the Debye counterpart of the (Einstein) Debye–Waller factor described above:

$$\exp(-2M_D)=\exp\left(-\sum_{\mathbf{Q}}q^2|\mathbf{u}_{\mathbf{Q}}|^2\right). \tag{5.178}$$

This may be evaluated by summing equation (2.43) over all modes (see e.g. Ziman 1972, p. 64). At high temperatures one may assume equipartition of energy and obtain $\sum_{\mathbf{Q},j}|\mathbf{u}_{\mathbf{Q},j}|^2$ from equation (2.47). The second term in (5.177) is the modification to M'_D resulting from the correlated motion of atoms on neighbouring sites. It has been evaluated for the corresponding X-ray scattering problem by Walker & Keating (1961) and decreases the magnitude of M_D for small $\mathbf{r}_i-\mathbf{r}_j$ since atoms close together vibrate more nearly in phase and so have a smaller mean square relative displacement. Following Walker & Keating we write (5.177) as

$$\exp[-2M_D\phi(r_{ij})], \tag{5.179}$$

where $\phi(r_{ij})$ is a function which increases from zero for $i=j$ and reaches unity at larger separations ($\sim$5th near neighbour) (see Warren 1969, p. 238; Moss 1964). However, as shown by Rossiter (1977), the effect of these damping terms on the resistivity is small and so one could simply take $\phi(r_{ij})=1$ without much error. Similarly, while the assumption of equal displacements for A and B atoms is not in general true, it is not felt that the additional complication of retaining different $u^{A}_{Q,j}$ and $u^{B}_{Q,j}$ leading to different Debye–Waller factors M^{A}_{D} and M^{B}_{D} would be of any benefit. We will return to this point with some numerical examples in the next section.

Finally, we want to consider an approach proposed by Bhatia & Thornton (1970) based on the thermodynamic theory of fluctuations. While this approach is only suitable for isotropic scattering in microscopically homogeneous materials, it is well suited to consideration of the resistivity of alloys around an order–disorder phase transition and liquid alloys. It also clearly introduces the effects of both atomic displacements and local concentration. Its basis lies in a generalisation of the dynamical structure factor (5.147) to include the effects of local composition fluctuations. The general scattering probability is written as

$$\Gamma(\mathbf{q},\omega)=\frac{1}{2\pi N}\int_{-\infty}^{\infty}\exp(-\mathrm{i}\omega t)\langle A^{*}(\mathbf{q},0)A(\mathbf{q},t)\rangle\,\mathrm{d}t, \tag{5.180}$$

where

$$A(\mathbf{q},t)=\sum_{i}w_{i}(\mathbf{q})\exp[\mathrm{i}\mathbf{q}\cdot\mathbf{R}_{i}(t)] \tag{5.181}$$

and $w_i(\mathbf{q})$ is the pseudopotential form factor of the ion i (see equations (4.61)–(4.63),

$$w_{i}(\mathbf{q})=\frac{1}{\Omega_{0}}\int\exp[\mathrm{i}\mathbf{q}\cdot(\mathbf{r}-\mathbf{r}_{i})w_{i}(\mathbf{r}-\mathbf{r}_{i})\,\mathrm{d}\mathbf{r}. \tag{5.182}$$

For a pure metal all $w_i(\mathbf{q})$ are equal and (5.180) becomes

$$\Gamma(\mathbf{q},\omega)=|w(\mathbf{q})|^{2}\mathscr{S}(\mathbf{q},\omega), \tag{5.183}$$

where $\mathscr{S}(\mathbf{q},\omega)$ is given by (5.147), leading directly to the expression for the resistivity (cf. (5.148)):

$$\rho=C\int_{0}^{2k_{F}}\int_{-\infty}^{\infty}\frac{\dfrac{\omega}{k_{B}T}\Gamma'(\mathbf{q},\omega)}{\exp(\zeta)-1}\,\mathrm{d}\omega q^{3}\,\mathrm{d}q, \tag{5.184}$$

where $\Gamma'(\mathbf{q},w)$ represents an average of $\Gamma(\mathbf{q},w)$ over all directions of $\mathbf{q}$ less the Bragg scattering component. Atomic displacements are considered in terms of the local number density $n_{\alpha}(\mathbf{r},t)$ defined as the number of α atoms per unit volume in a small element of volume at $\mathbf{r}$ and time t. The local deviation in the number density is

$$\delta n_{\alpha}(\mathbf{r},t)=n_{\alpha}(\mathbf{r},t)-\frac{N_{\alpha}}{V}, \tag{5.185}$$

where N_α is the number of α atoms in a volume V. Its Fourier transform is (see Section 2.4)

$$N_\alpha(\mathbf{q},t)=\int \delta n_\alpha(\mathbf{r},t)\exp(\mathrm{i}\mathbf{q}\cdot\mathbf{r})\,\mathrm{d}\mathbf{r}. \tag{5.186}$$

Similarly, the local deviation in concentration is

$$\delta c_\alpha(\mathbf{r},t)=\frac{V}{N}\,[c_\mathrm{B}\,\delta n_\mathrm{A}(\mathbf{r},t)-c_\mathrm{A}\,\delta n_\mathrm{B}(\mathbf{r},t)]. \tag{5.187}$$

Its Fourier transform is

$$c(\mathbf{q},t)=\frac{1}{V}\int \delta c(\mathbf{r},t)\exp(\mathrm{i}\mathbf{q}\cdot\mathbf{r})\,\mathrm{d}r \tag{5.188}$$

$$=\frac{1}{N}\,[c_\mathrm{B}N_\mathrm{A}(\mathbf{q},t)-c_\mathrm{A}N_\mathrm{B}(\mathbf{q},t)]. \tag{5.189}$$

The expression for $A(\mathbf{q},t)$ may thus be written as

$$\begin{aligned}A(\mathbf{q},t)&=w_\mathrm{A}(\mathbf{q})N_\mathrm{A}(\mathbf{q},t)+w_\mathrm{B}(\mathbf{q})N_\mathrm{B}(\mathbf{q},t)\\&=\bar{w}(\mathbf{q})N(\mathbf{q},t)+w^\mathrm{d}(\mathbf{q})Nc(\mathbf{q},t).\end{aligned} \tag{5.190}$$

where $N(\mathbf{q},t)$ is the number density of the average lattice. Consequently

$$\begin{aligned}\Gamma(\mathbf{q},t)=&|\bar{w}(\mathbf{q})|^2S_\mathrm{NN}(\mathbf{q},\omega)+|w^\mathrm{d}(\mathbf{q})|^2S_\mathrm{CC}(\mathbf{q},\omega)\\&+2\bar{w}(\mathbf{q})w^d(\mathbf{q})S_\mathrm{NC}(\mathbf{q},\omega),\end{aligned} \tag{5.191}$$

where

$$\left.\begin{aligned}S_\mathrm{NN}(\mathbf{q},\omega)&=\frac{1}{2\pi N}\int\exp(-\mathrm{i}\omega t)\,\mathrm{d}t\langle N^*(\mathbf{q},0)N(\mathbf{q},t)\rangle\\S_\mathrm{CC}(\mathbf{q},\omega)&=\frac{N}{2\pi}\int\exp(-\mathrm{i}\omega t)\,\mathrm{d}t\langle c^*(\mathbf{q},0)c(\mathbf{q},t)\rangle\\S_\mathrm{NC}(\mathbf{q},\omega)&=\frac{1}{2\pi}\int\exp(-\mathrm{i}\omega t)\,\mathrm{d}t\langle N^*(\mathbf{q},0)c(\mathbf{q},t)\\&\qquad\qquad+c^*(\mathbf{q},0)N(\mathbf{q},t)\rangle.\end{aligned}\right\} \tag{5.192}$$

If one now introduces

$$S_\mathrm{NN}(\mathbf{q})=\int\frac{\zeta}{\exp(\zeta)-1}\,S_\mathrm{NN}(\mathbf{q},\omega)\,\mathrm{d}\omega \tag{5.193}$$

and similarly for $S_\mathrm{CC}(\mathbf{q})$, $S_\mathrm{NC}(\mathbf{q})$ and $\Gamma(\mathbf{q},\omega)$, one finds

$$\Gamma(\mathbf{q})=|\bar{w}(\mathbf{q})|^2S_\mathrm{NN}(\mathbf{q})+|w^\mathrm{d}(\mathbf{q})|^2S_\mathrm{CC}(\mathbf{q})+2\bar{w}(\mathbf{q})w^\mathrm{d}(\mathbf{q})S_\mathrm{NC}(\mathbf{q}). \tag{5.194}$$

At temperatures above Θ_D, the scattering may be considered as elastic ($\zeta\ll1$) and so $\zeta/(\mathrm{e}^\zeta-1)\approx1$. Furthermore $\int\exp(-\mathrm{i}\omega t)\,\mathrm{d}\omega=2\pi\,\delta(t)$ and so the equations in (5.192) may be written as

$$\begin{aligned} S_{NN}(\mathbf{q}) &= 1/N\langle N^*(\mathbf{q})N(\mathbf{q})\rangle \\ S_{CC}(\mathbf{q}) &= N\langle c^*(\mathbf{q})c(\mathbf{q})\rangle \\ S_{NC}(\mathbf{q}) &= \mathrm{Re}\langle N^*(\mathbf{q})c(\mathbf{q})\rangle . \end{aligned} \tag{5.195}$$

In the long wavelength limit these partial structure factors may be roughly interpreted as the mean square fluctuation in the number of particles in a volume V: $\langle \Delta N^2\rangle/N$, the mean square fluctuation in the composition: $N\langle \Delta c^2\rangle$ where

$$\Delta c = \frac{1}{N}[c_B\,\Delta N_A - c_A\,\Delta N_B], \tag{5.196}$$

and the correlation between these two fluctuations $\langle \Delta N\,\Delta C\rangle$, respectively. Thus the effects of thermal and static atomic displacements occur in S_{NN}, atomic correlations in S_{CC} and S_{NC} is the cross-term. These structure factors may be evaluated given some suitable thermodynamic model of the solid (Bhatia & Thornton 1971; Thompson 1974.) It may be shown, for example, that for a dilute alloy ($c_A \ll 1$), $S_{CC} \approx c_A$ and the temperature independent part of $\Gamma(\mathbf{q})$ is

$$\Gamma(\mathbf{q}) = c_A\left[w^d(\mathbf{q}) - \delta\left(\frac{c_{11}+2c_{12}}{3c_{11}}\right)w_B\right]^2, \tag{5.197}$$

where δ is the fractional change in volume per unit change in composition. This is just the result deduced by Blatt (1957*a*,*b*) (see Section 6.3). Of more direct relevance to the topic of discussion of this section, Bhatia & Thornton (1971) point out that the cross-term $S_{NC}(\mathbf{q})$ which involves both the atomic displacements and the atomic correlations could be quite significant, even comparable with the phonon scattering at room temperature, so that such terms should not in general be neglected.

5.7 Some applications

Application of any of the theory presented in this chapter requires choice of a suitable form factor and evaluation of the structure factor for the particular alloy and situation of interest. We now consider some specific examples which illustrate the general principles involved.

5.7.1 *Phonon scattering*

Virtually all of the studies of phonon scattering in simple metals have been concerned with pure metals rather than alloys. We will briefly review some of these, as they give an indication of the reliability of the pseudopotential method, although the problem of determining a

Table 5.5. *Resistivities of the alkali metals at 273 K*

	$\rho_{[100]}$	$\rho_{[110]}$	$\rho_{[111]}$	ρ_{av}	ρ_{Debye}	ρ_{expt}
Li	6.77	20.33	18.50	13.62	1.09	8.55
Na	1.99	2.38	2.24	2.23	0.73	4.3
K	1.98	2.13	2.10	2.08	0.80	6.1
Rb	2.22	2.29	2.24	2.25	0.90	11.6
Cs	2.17	2.22	2.18	2.20	0.94	19.0

Source: After Robinson & Dow 1968.

suitable form factor for the alloy is not considered. Furthermore, while other materials such as Al and Pb have been considered (see e.g. Dynes & Carbotte 1968), we will consider here two particular groups of metals, the alkalis and the noble metals, as these have been subjected to considerable systematic study.

(a) *Alkali metals*

Robinson & Dow (1968) determined the resistivities of the alkali metals Li, Na, K, Rb, Cs at 273 K using the Heine–Abarenkov–Animalu form factor ((4.27), (4.28)) and structure factors determined from the dynamical matrix (5.149) and known elastic constants. They calculated the resistivity for scattering wavevectors along each of the three principle symmetry directions [100], [110] and [111] and also determined the weighted average. In the case of Li and Na they also used structure factors determined from inelastic neutron scattering data. While the calculated resistivities were not in particularly good agreement with experiment they do reveal some interesting trends as indicated in Table 5.5. Of note is the large anisotropy, particularly for Li. There was little difference between the resistivities determined from elasticity theory or neutron diffraction data, suggesting that the dynamical matrix approach is quite adequate. Use of the Debye approximation to the phonon distribution, however, gives markedly worse values for the resistivity. Dynes & Carbotte (1968) also used the Heine–Abarenkov–Animalu form factors for K and Na but took the longer path of determining the phonon frequencies and polarisations using the Born–von Karman theory to interpolate between symmetry directions for substitution into (5.146). The structure factors obtained were virtually identical to those found by Robinson & Dow and, as one might expect, the magnitude of the calculated resistivities were again not in very good agreement with experiment, although the general variation with temperature was well represented (note that their calculated resistivities should be divided by a

Table 5.6. *Core radii R_c chosen by Hayman & Carbotte to fit the resistivity to experimental data. Also shown are the Ashcroft core radii obtained by fitting to liquid metal resistivities and Fermi surface data*

	R_c (Å)	
	Hayman & Carbotte	Ashcroft
Li	—	0.561
Na	0.8282	0.878
K	1.0353	1.127
Rb	1.0422	1.127
Cs	—	1.143

Fig. 5.30. The Ashcroft form factors (solid lines) and Shaw form factors for K, Na and Heine–Abarenkov–Animalu form factor for Rb (dashed lines) (after Hayman & Carbotte 1971).

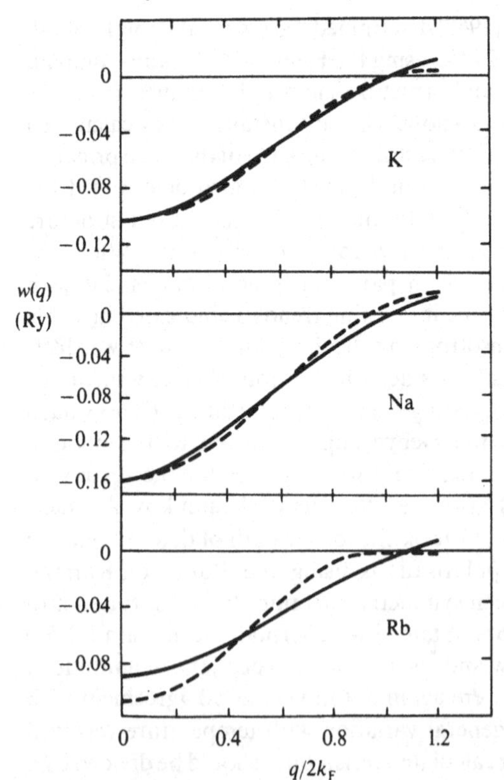

Table 5.7. *Temperature dependence of the calculated (constant volume) resistivities of Na, K, Rb, using the potentials indicated. Also shown are some experimental data*

	Potential/ experiment	Resistivity ($\mu\Omega$ cm)				
		30 K	50 K	100 K	150 K	200 K
Na	Shaw	—	0.267	0.858	1.420	1.961
	Ashcroft	—	0.357	1.093	1.792	2.467
	Experiment	—	0.314	1.108	1.857	2.566
K	Shaw	0.251	0.619	1.492	2.323	3.130
	Ashcroft	0.301	0.725	1.731	2.688	3.620
	Experiment	0.283	0.705	1.716	2.643	3.543
Rb	HAA	0.236	0.472	1.023	1.559	2.090
	Ashcroft	0.780	1.493	3.177	4.822	6.455
	Experiment	0.798	1.527	3.203	4.842	6.570

Source: After Hayman & Carbotte 1971.

factor of 2 as reported by Hayman & Carbotte (1971)). Hayman & Carbotte (1971) used essentially the same technique with the Ashcroft form factor (4.25), (4.26) and Lindhard screening function (4.36), allowing also for changes in the phonon frequencies and form factors with volume. The core radius R_c was treated as an adjustable parameter used to fit the resistivity at some particular temperature and lattice volume (usually corresponding to those of the phonon determination). The values finally chosen are given in Table 5.6. The Shaw (Na, K) and Heine–Abarenkov–Animalu (Rb) form factors were also used and are shown for comparison on Figure 5.30.

The calculated resistivities for various temperatures are given in Table 5.7 and it appears that both the Shaw and Heine–Abarenkov–Animalu form factors underestimate the resistivity. The general temperature dependence determined from the fitted Ashcroft form factor is shown in Figure 5.31. The good agreement with experiment suggests that the independent or one phonon analysis implied by the equation in Section 5.6.2 is quite adequate provided that a suitable form factor is used. It should be noted that the values of R_c used are smaller than those obtained by Ashcroft (1968) by fitting to liquid metal resistivities and Fermi surface data, which are also shown in Table 5.6. This would have the effect of pushing the first node in $w(\mathbf{q})$ to larger $\mathbf{q}$. As these nodes are near to or greater than $2k_F$ in these metals, such a shift would produce an apparent increase in the resistivity.

Hayman & Carbotte (1972) also investigated the importance of the relaxation time anisotropy by evaluating the resistivity from both $1/\langle\tau(\mathbf{k})\rangle$ and $\langle 1/\tau(\mathbf{k})\rangle$. They found a large difference (up to $\sim 40\%$) over a range of temperatures up to ~ 150 K in Li, but much smaller differences in Na, K and Rb. It should be stressed that this anisotropy results from the anisotropy in the phonon spectrum and not in the Fermi surface, which is assumed to be spherical in all of these studies. This anisotropy (also apparent in the Robinson & Dow results, Table 5.5) indicates that caution must be exercised in averaging the structure factor over all directions of $\mathbf{q}$ as implied by equation (5.146). If there is a marked anisotropy in $S(\mathbf{q})$ then one should leave the structure factor in its unaveraged form and finally obtain $\rho_p(T)$ from $1/\langle\tau(\mathbf{k})\rangle$. These authors also investigated the effects of different form factors and found that the anisotropy as well as the magnitude of $\rho_p(T)$ was dependent to some degree upon the particular form factor used, as shown in Table 5.8. Shukla & Taylor (1976) used *a-priori* form factors and phonon frequencies in their study of Na and K and found good agreement with experiment, as shown in Figure 5.32.

Fig. 5.31. Resistivity (at constant volume) as a function of temperature for Na, K and Rb. The solid circles are experimental data (after Hayman & Carbotte 1971).

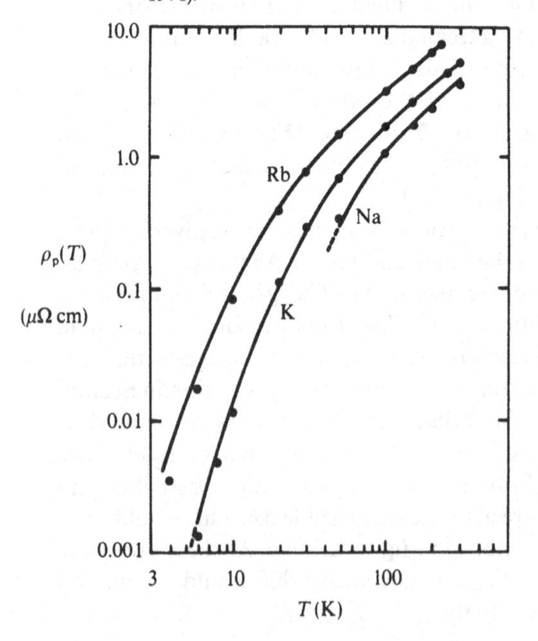

Table 5.8. *Temperature dependence of the resistivity of K using different form factors. Also shown is the ratio R of the resistivity derived from* $\langle 1/\tau(\mathbf{k})\rangle$ *to that derived from* $1/\langle\tau(\mathbf{k})\rangle$

	Form factor							
	Bardeen		Lee–Falicov		Ashcroft ($R_c = 1.0353$)		Ashcroft ($R_c = 1.2753$)	
T (K)	ρ ($\mu\Omega$ cm)	R	ρ	R	ρ	R	ρ	R
2	2.28×10^{-6}	0.891	1.52×10^{-6}	0.984	1.42×10^{-6}	0.995	5.82×10^{-6}	0.529
5	1.13×10^{-3}	0.730	5.43×10^{-4}	0.873	5.14×10^{-4}	0.935	3.19×10^{-3}	0.339
10	1.91×10^{-2}	0.888	9.96×10^{-3}	0.948	1.18×10^{-2}	0.977	3.76×10^{-2}	0.581
20	1.39×10^{-1}	0.977	8.49×10^{-2}	0.991	1.14×10^{-1}	0.997	1.84×10^{-1}	0.832

Source: After Hayman & Carbotte 1972.

Shukla & VanderSchans (1980) investigated the significance of the multiphonon and Debye–Waller terms in the resistivity of Na and K. Their results are shown in Figure 5.33 and indicate that the two effects nearly cancel, as originally proposed by Sham & Ziman (1963). Indeed, their inclusion produced a maximum change in the resistivity of Na of ~5% at the melting point, the general agreement with experiment then being within 2.5% over a wide range of temperatures, as shown in Figure 5.32.

(b) *Noble metals*

We include consideration of the noble metals in this 'simple metals' chapter as the resistivity behaviour as reasonably described in terms of a nearly free electron approximation. (This point is discussed further in the next chapter.) Pal (1973) and Kumar (1975) used the Bardeen (1937) scattering matrix elements and solution of the dynamical matrix for the phonon frequencies and polarisations. As in the case of similar studies in the alkali metals, the general variation with temperature of the resistivities of Cu, Au and Ag was in good agreement with experiment although the magnitudes were too low. Khanna & Jain (1977) used the Heine–Abarenkov–Animalu form factors and direct

Fig. 5.32. Resistivity as a function of temperature for Na and K (after Shukla & Taylor 1976). Solid line: one-phonon model: dashed line: including multiphonon and Debye–Waller effects.

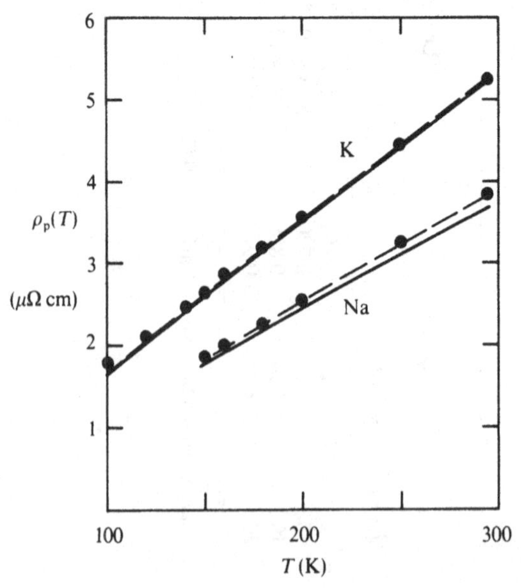

substitution of D(**q**) to obtain the structure factor (equation (5.149)). The room temperature resistivity of Cu was found to be too low while that of Ag and Au was too high, as shown in Table 5.9. They also determined the resistivities of the transition metals using this approach and, not surprisingly, the agreement with experiment was poor. Kharoo *et al.* (1978) have used the form factors of Moriarty (1970), Bardeen (1937) and Borchi & de Gennaro (1970) and a phenomenological scheme to determine the phonon frequencies and polarisations. Their results are shown in Figure 5.34 and indicate that in general the Moriarty or the Borchio–de Gennario form factors give resistivities in better agreement with experiment than the Bardeen matrix elements. Ziman (1974) has also commented upon the limitations of the Bardeen form. The effects of anharmonic processes at high temperatures has been investigated by Malinowski–Adamska & Wojtczak (1979) and found to bring the Kumar results into better agreement with experiment at high temperatures.

It thus appears that most of the lattice dynamic models can give reliable structure factors provided that the phonon spectra produced are realistic, the effects of anisotropy and temperature dependence of the

Fig. 5.33. The multiphonon (solid curve) and negative of the Debye–Waller (dashed curve) contribution to the resistivity of K and Na (after Shukla & VanderSchans 1980).

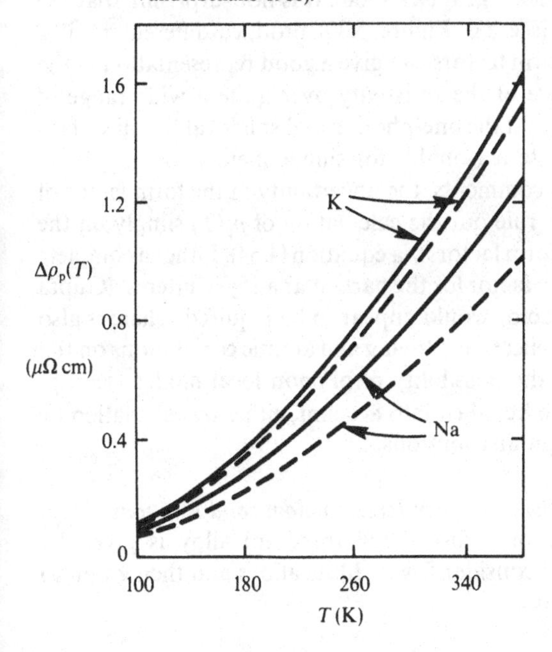

Table 5.9. *Calculated and measured resistivities of Cu, Ag and Au at room temperature*

	Resistivity at room temperature ($\mu\Omega$ cm)	
Metal	Calculated	Experimental
Cu	1.37	1.7
Ag	2.95	1.6
Au	5.90	2.2

Source: After Khanna & Jain 1977.

phonon distribution are included, and both N and U scattering are taken into account (Bailyn 1960*a*,*b*). However, the choice of the form factor is critical in obtaining agreement with experiment. This is largely because the q^3 weighting term in the resistivity expressions (5.145) or (5.148) renders the calculations of $\rho_p(T)$ particularly sensitive to the details of $w(\mathbf{q})$ (and $S(\mathbf{q})$) in the high q ($\sim 2k_F$) region. Small changes in this quantity which is usually already itself small in this region can thus produce quite large changes in $\rho_p(T)$ (see e.g. Wiser 1966). It is not surprising that the different form factors (see e.g. Figure 5.30) produce different results. However, that 'fitted' form factors can give a good representation of the temperature dependence of the resistivity over quite a wide range of temperatures indicates that the one-phonon and spherical Fermi surface approximations are quite reasonable for simple metals.

In view of the above comments, the uncertainty in the form factor of alloys would appear to rule out the calculation of $\rho_p(T)$ simply on the basis of the pure metal form factors via equation (4.64). Either a complete recalculation of the form factor for the particular alloy of interest (Gupta 1968) or a fitted form factor would appear to be required. There is also the complication of the effects of alloying and atomic correlations on the phonon spectrum and the possibility of phonon local modes (see e.g. Rowlands *et al.* 1977) to be taken into account, either by calculation or use of measured phonon distributions.

5.7.2 *Residual resistivity of disordered random solid solutions*

The resistivity of a disordered (random) alloy is given by equation (5.21). We will consider firstly dilute alloys and then go on to more concentrated alloys.

(a) *Dilute alloys*

The resistivity of dilute simple metal alloys has been considered by many workers and various general trends have emerged, all of which can be explained by (5.21). Let us consider these general trends before becoming involved in the detailed calculations.

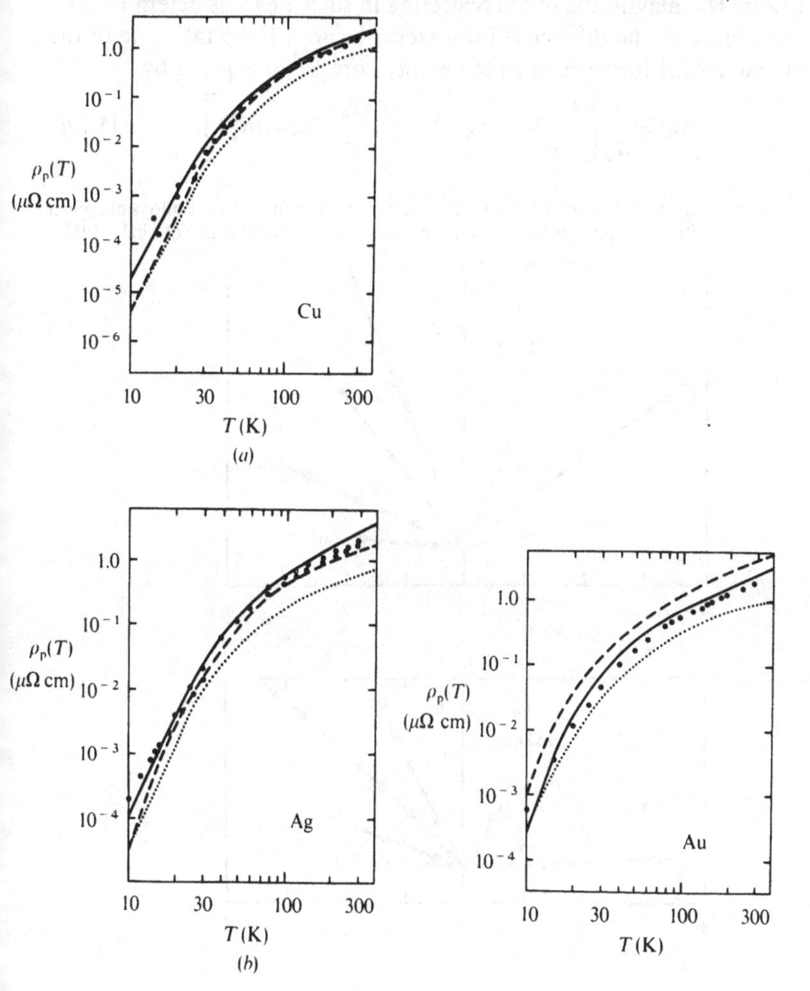

Fig. 5.34. The temperature dependence of the resistivity of Cu, Ag and Au derived from the form factors of Moriarty (1970) (dashed line); Bardeen (1937) (dotted line); and Borchi & de Gennaro (1970) (solid lines). The solid circles represent a collection of experimental data (after Kharoo *et al.* 1978).

Nordheim's rule

In dilute alloys, the resistivity depends linearly upon the concentration of the impurity. This is exemplified with the data of Linde (1932), as shown in Figure 5.35. This behaviour follows immediately from (5.21) with $c_A c_B \approx c_A$ for small c_A.

Linde–Norbury rule

The magnitude of the change in resistivity per atomic percent solute depends upon the square of the valence difference between the solute and host $(\Delta Z)^2$. This general trend is apparent in Figure 5.35. In Figure 5.36 we have plotted the change in resistivity $\Delta\rho$ as a function of $(\Delta Z)^2$. The magnitude of the scattering in such a case is determined by the square of the difference form factor $|w^d(\mathbf{q})|^2$. If we take one of the simple model forms such as the empty core, $w^d(\mathbf{q})$ is given by

$$w^d(\mathbf{q}) = \frac{1}{\Omega_0}\left[\frac{4\pi z_A e^2}{q^2}\cos(qR_c^A) - \frac{4\pi z_B e^2}{q^2}\cos(aR_c^B)\right]. \tag{5.198}$$

Fig. 5.35. Resistivity as a function of concentration for dilute alloys of copper with the impurities indicated (from Meaden 1965 using the data of Linde, 1932).

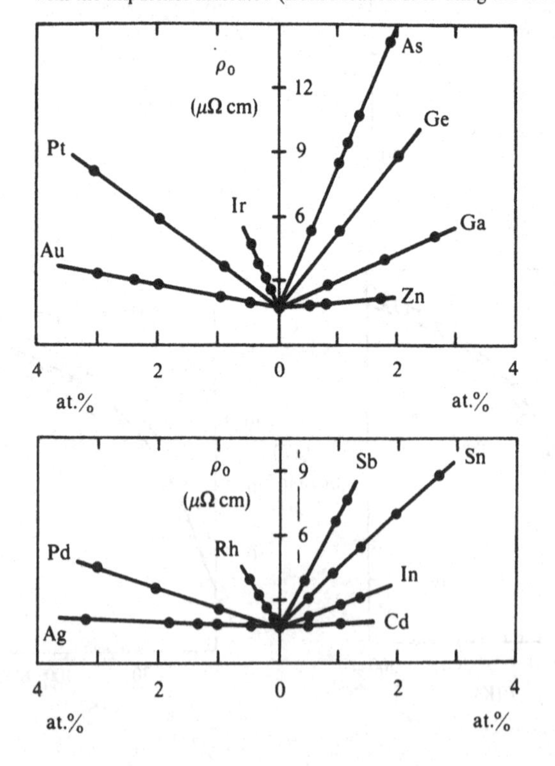

Table 5.10. *Core radii and atomic volumes Ω_0 for the Cu host and solutes in Figure 5.36*

	R_c (Å)	Ω_0 (Å^3)
Cu	0.43	11.7
Zn	0.67	15.1
Ga		19.4
Ge		22.5
As		21.5
Ag	0.55	16.8
Cd		21.4
In	0.70	25.9
Sn	0.69	26.9
Sb		30.2

Source: From Cohen & Heine 1970.

Fig. 5.36. The change in resistivity for one at.% impurity in Cu as a function of $|\Delta Z|^2$ (after Linde 1932).

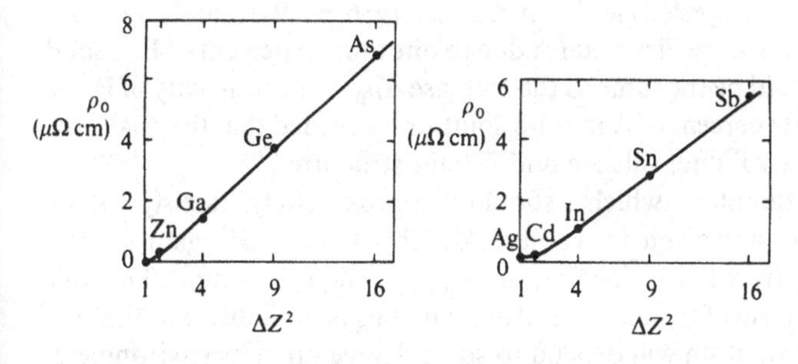

If we assume that $\Omega_0^A \approx \Omega_0^B$ so that there are not large strain effects and that $R_c^A \approx R_c^B = R_c$, this gives

$$|w^d(\mathbf{g})|^2 = \frac{4\pi(\Delta Z)^2 e^2}{q^2}\left[\frac{1}{\Omega_0}\cos^2(qR_c)\right], \tag{5.199}$$

leading to $\rho \propto \Delta Z^2$. Deviations from this rule might be expected if the atomic volumes or core radii of the solute and host differ significantly. The appropriate values for the impurities cited in Figure 5.36 are given in Table 5.10 and show that this is only approximately true so that some deviation from a pure $(\Delta Z)^2$ behaviour is expected, more so in the Ag–Sb series as observed. More significant deviations will occur if the size

Table 5.11. *Change in resistivity due to one atomic percent impurity in alloy systems that ought to obey Mott's rule*

Alloy	Ratio of atomic volumes	$\Delta\rho$ ($\mu\Omega$ cm/at %)	Alloy	$\Delta\rho$ ($\mu\Omega$ cm/at %)
Cu in Au	0.70	0.30	Au in Cu	0.49
Cu in Ag	0.70	0.07	Ag in Cu	0.07
Ag in Au	1.01	0.32	Au in Ag	0.38
Mg in Cd	1.17	0.40–0.45	Cd in Mg	0.68
Pd in Pt	0.97	0.6	Pt in Pd	0.7

Source: Data from results cited in Mott & Jones 1936; Blatt 1968; and Brown & Morgan 1971.

difference is large as lattice distortions may then play a significant effect. We will return to this later in this section.

Mott's rule

Mott suggested (see Mott & Jones 1936, p. 294) that the increase $\Delta\rho_A$ in the resistance of a metal A due to one atomic percent of B in solid solution should be the same as the increase $\Delta\rho_B$ in the resistivity of B due to one atomic percent of A in solid solution, provided that they have the same atomic volume, valence and crystal structure.

Some examples which should approximately satisfy these requirements are given in Table 5.11. This behaviour again follows immediately from (5.21) since $|w_A(\mathbf{q}) - w_B(\mathbf{q})|^2 = |w_B(\mathbf{q}) - w_A(\mathbf{q})|^2$. That this rule breaks down for non-isoelectronic metals is an indication that the potential of an atom will depend to some degree upon its environment. At the very least, the screening parameters will depend upon the electron density (via k_F) and this will depend markedly upon the number of electrons per atom of the A and B species. If the size difference is very large there will also be lattice dilation effects to be taken into account.

We turn now to some more detailed calculations of the resistivity. Harrison (1966, p. 150) performed a series of calculations of the impurity resistivity using the point-ion potential ((4.23), (4.24a) and (4.24b)) obtained by fitting the form factor to the *a-priori* calculated values (Harrison 1963) but in general the agreement was not good. As an example, the resistivities due to Zn, Mg and Ag impurities in Al determined from the point-ion and empty-core pseudopotentials using a variety of different screening and exchange and correlation corrections and equation (5.21) are given in Table 5.12, together with the

Table 5.12. *Resistivity (in μΩ cm) due to one atomic percent of Mg, Zn and Ag in Al calculated using the point-ion and empty-core potentials and Thomas–Fermi (T–F) and Lindhard (L) screening with Hubbard–Sham (H–S), Kleinmann (K) and Singwi (S) exchange and correlation corrections. Also shown are some experimental data from Blatt (1968)*

	R_c	β/Ω_{Al}	Point-ion					Empty-core					
	(Å)	(Ry)	T–F	L	L/H–S	L/K	L/S	T–F	L	L/H–S	L/K	L/S	Experiment
Al (host)	0.59	0.33											
Mg	0.73	0.375	0.64	0.78	1.35	2.75	3.56	0.19	0.22	0.33	0.49	0.63	0.45
Zn	0.67	0.24	0.07	0.08	0.11	0.15	0.18	0.10	0.11	0.15	0.20	0.25	0.22
Ag	0.55	0.04	0.48	0.59	1.04	2.52	3.24	0.21	0.24	0.36	0.67	0.83	1.1

Note: ($\Omega_{Al} = 16.7$ Å^3, $k_F = 1.75$ Å^{-1}.)

Table 5.13. *Resistivities (in μΩ cm) due to one at.% of impurity in Al and Mg hosts. See text for the significance of the columns A, B, C, D*

Host	Impurity	Point-ion	Calculations A	B	C	D	Experiment
Al	Mg	0.78	0.17	0.23	0.23	0.28	0.45
	Zn	0.08	0.13	0.14	0.14	0.17	0.72
	Ca	1.8	3.1	4.5	4.1	4.9	~4.5
Mg	Al	1.6	0.63	0.82	0.83	1.0	2.1
	Li	0.27	0.33	0.44	0.48	0.69	0.75

Source: After Gupta 1968.

experimental values. The empty-core pseudopotential parameters were determined from the liquid metal resistivities while the point-ion parameter β was determined by fitting to Harrison's *a-priori* pseudopotential. These results show that the calculated resistivity depends to some degree upon the form of screening and exchange and correlation corrections employed but there is no one particular combination that gives a significantly better agreement with experiment than any other. This is in general accord with the conclusions of the previous section. Harrison (1966, p. 152) attempted to generate a self-consistent set of point-ion parameters by fitting β to the known resistivities of impurities but did not obtain any significant improvement in the systematic agreement with experimental values. Gupta (1968) has investigated the importance of taking into account the actual environment of the impurity ion by recalculating the form factors for each situation. The changes in the form factors can change the restivities by quite a large amount, as indicated in Table 5.13. Column A refers to the recalculated bare-ion form factors, column B takes into account the non-local effects of screening and includes exchange, column C takes into account the influence of the effective mass on the screening and Fermi energy and column D incorporates the effective mass into the constant C. (The effective masses were taken from the work of Animalu & Heine (1965).) The agreement with experiment does seem to be significantly improved over the simple point-ion model by all of these corrections, although at considerable effort, and it is still not significantly better than the screened empty-core pseudopotential. This latter potential also gives moderately good agreement in a wide range of noble-metal based alloys containing impurities with valencies of 2, 3 or 4, as shown in Table 5.14. As the core radii are similar for Cu and Au, the resistivities of Cu–Au alloys are very sensitive to small differences in these parameters and therefore no results are given for this potential. The impurity resistivities determined from the Moriarty (1970) form factors for Cu, Au and Ag are also shown in this table but do not show any better general agreement with experiment. These results are also interesting in that one might expect the Born approximation to become questionable for impurity valencies greater than ~ 3 (Meyer *et al.* 1971). While the results for Ge impurities are not in good agreement with experiment being much too large (as one would expect if the Born approximation is breaking down) the results for Sn impurities are in reasonable agreement.

It has been suggested that the volume dependence of the residual resistivity might provide a more sensitive criterion for selecting among pseudopotentials (DuCharme & Edwards 1970; Foiles 1973). As shown by Foiles, the empty-core and Shaw form factors produce similar results, although it appears that when the impurity core radius is much larger than the host core radius the scattering may be sufficiently strong for the Born approximation to be suspect.

So far, we have not taken into account the possibility of effects of lattice distortion around the impurity. The effects of such distortions on the residual resistivity of dilute alkali alloys has been considered by Popovic *et al.* (1973). The atomic displacements required to evaluate the change in structure factor (5.125) were evaluated using the method of lattice statistics (Kanzaki 1957; Flocken & Hardy 1969) whereby the Fourier transforms of the atomic displacements and of the force array due to the defect:

$$\mathbf{u}(\mathbf{q}) = \sum_i \mathbf{u}_i \exp(-i\mathbf{q}\cdot\mathbf{r}_i), \tag{5.200}$$

$$\mathbf{F}(\mathbf{q}) = \sum_i \mathbf{F}_i \exp(-i\mathbf{q}\cdot\mathbf{r}_i). \tag{5.201}$$

are connected via the dynamical matrix

$$\mathbf{u}(\mathbf{q}) = [D(\mathbf{q})]^{-1}\mathbf{F}(\mathbf{q}). \tag{5.202}$$

The empty-core pseudopotential ((4.21), (4.27)) fitted to phonon spectra (Price *et al.* 1970), suitably screened by the Lindhard function (4.36) with exchange and the correlation corrections of Singwi *et al.* (1970) was used to obtain the force constants and the scattering matrix elements. As an example, the forces and displacements obtained in the first five shells about an Na impurity in Rb are shown in Table 5.15. The calculated resistivities are compared with experimental results in Table 5.16 and

Table 5.14. *Resistivity in μΩ cm due to one at.% of Cu, Ag and Au impurities in the solutes listed, determining using the empty core and Moriarty potentials. Also shown are some experimental results*

			Empty–core						Moriarty		
			Cu		Ag		Au				
Solute	R_c (Å)	Valence	Theory	Expt.	Theory	Expt.	Theory	Expt.	Cu	Ag	Au
Cu	0.430	1	—	—	0.09	0.077	—	0.45	—	1.07	1.07
Ag	0.550	1	0.10	0.14	—	—	0.09	0.36	0.53	—	0.05
Au	0.430	1	—	0.55	0.09	0.36	—	—	0.53	0.05	—
Zn	0.683	2	0.64	0.32	0.41	0.64	0.50	0.95			
Cd	0.744	2	1.13	0.30	0.52	0.38	0.82	0.63			
Hg	0.484	2	0.61	1.0	1.58	0.79	1.02	0.44			
Al	0.610	3	1.35	1.25	2.67	1.95	2.15	1.87			
Ga	0.555	3	1.96	1.42	4.25	2.36	3.35	2.2			
In	0.590	3	1.44	1.06	3.03	1.78	2.41	1.39			
Ge	0.506	4	5.66	3.79	11.0	5.5	22.4	5.2			
Sn	0.686	4	3.0	2.88	4.21	4.36	3.89	3.36			

Note: Data from DuCharme & Edwards 1970; and Blatt 1968.

Table 5.15. *Forces and displacements for the first five shells of neighbours about an Na impurity in Rb*

Neighbour	Component	Force (dyn cm^{-1}) displaced	Force (dyn cm^{-1}) non-displaced	Displacement (in units of a_0)
111	$x=y=z$	−89.3	−129.7	−0.031
200	$x(y=z=0)$	−29.4	− 17.9	0.015
220	$x=y$ $(z=0)$	5.2	6.0	−0.008
311	x	− 4.3	− 4.6	0.001
	$y=z$	− 1.5	− 1.5	−0.001
222	$x=y=z$	− 2.4	− 2.8	−0.016

Source: After Popovic *et al.* 1973.

show that the general effect of the lattice relaxation is to increase the resistivity by about 5–30%. However, the general agreement with experiment is still not very good, due no doubt to the fact that the core radius R_c was fitted to phonon spectra which involves quite different weighting factors to the resistivity calculation, allowing significant errors in the critical large q region.

In polyvalent metals application of simple pseudopotential theory results in the calculated impurity resistivities always falling below the experimental values. Fukai (1968, 1969) has suggested that the reason for this is the neglect of the mixing of the plane waves near Brillouin zone boundaries which intersect the Fermi surface in such materials. He accounted for this mixing by using a two plane wave approach with the pure metal empty-core pseudopotentials fitted to de Haas–van Alphen data. Local strain effects were also included by isotropic elasticity theory and the Boltzmann equation solved by iteration. The results of his calculations are summarised in Table 5.17. Also shown in that table are the results derived from a single plane wave model and the experimental data. It can be seen that a marked improvement has resulted from the multiple plane wave approach leading to a generally good agreement with experiment. Fukai also noted that strain and exchange and correlation effects did not produce very large corrections to the resistivity, although they are probably important for detailed quantitative analysis. He also noted that the plane wave mixing resulted in some degree of relaxation time anisotropy and that the accuracy for core radii required to reproduce the observed resistivity values to within $\pm 10\%$ varies from ± 0.005 Å for tetravalent atoms to ± 0.05 Å for monovalent atoms. Sorbello (1973) has taken this approach a stage

Table 5.16. *Calculated and experimental resistivities in μΩ cm for one at.% impurity in alkali hosts as indicated*

		Residual resistivity		
Solvent	Solute	Non-relaxed	Relaxed	Experimental
Li	vacancy	1.005	0.759	
	Na	0.426	0.458	
	K	3.084	3.288	
	Rb	4.193	4.483	
	Cs	5.529	5.958	
Na	Li	0.305	0.341	
	vacancy	1.207	0.888	
	K	1.112	1.184	
	Rb	1.903	2.021	
	Cs	3.108	3.297	
K	Li	2.288	2.914	
	Na	1.037	1.272	0.56
	vacancy	1.353	0.975	
	Rb	0.117	0.136	0.11
	Cs	0.595	0.689	1.1
Rb	Li	3.159	4.145	
	Na	1.718	2.166	
	K	0.112	0.134	0.04–0.13
	vacancy	1.434	1.050	
	Cs	0.180	0.212	
Cs	Li	4.470	6.107	
	Na	2.810	3.672	
	K	0.553	0.682	
	Rb	0.175	0.213	0.28
	vacancy	1.559	1.182	

Source: After Popovic *et al.* 1973.

further by using four OPW's. Calculations were performed with the same form factors as Fukai, as well as the Moriarty and HAA form factors using either an iterative solution to the Boltzmann equation or Ziman's approximation (equation (1.68)). The results are summarised in Table 5.18, together with the single plane wave and experimental results. As in the Fukai study it is clear that the multiple plane wave character can lead to a substantial enhancement of the residual resistivity over the nearly free electron value and improvement in agreement with experiment. The choice of different form factors has some effect on the results, although the differences are not as significant as the changes in

Table 5.17. *Experimental and theoretical residual resistivity in μΩ cm due to one at.% impurity in Al and Pb hosts using one OPW and two OPW approximations with the Gupta and Ashcroft potentials. The core radii were fitted to de Haas–van Alphen data*

			Residual resistivity				
			Gupta		Ashcroft		
Solvent	Solute	R_c (Å)	1 OPW	2 OPW	1 OPW	2 OPW	Experiment
Al	Mg	0.732	0.16	0.30	0.20	0.38	0.34–0.36
(R_c=0.610)	Si	0.522			0.35	0.58	0.56–0.60
	Cu	0.93			0.49	0.86	0.89
	Zn	0.683	0.16	0.24	0.12	0.21	0.24
	Ga	0.535			0.16	0.22	0.22
	Ge	0.506			0.51	0.77	0.79
	Ag	0.99			0.58	1.08	1.08–1.34
	Cd	0.744			0.26	0.40	0.51
Pb	Mg	0.732			1.14	3.05	3.34
(R_c=0.558)	Cd	0.744			1.23	3.25	2.52
	In	0.630			0.43	1.11	0.94–1.0
	Sn	0.591			0.11	0.28	0.25

Source: After Fukai 1969.

Table 5.18. *Experimental and theoretical residual resistivities in μΩ cm due to one at.% of the impurities indicated in an Al host. Moriarty (Cu, Ag), Heine–Abarenkov–Animalu or Ashcroft form factors were used with the one OPW and four OPW approximations. In the later case results are derived for both the full iterative solution of the Boltzmann equation and the Ziman approximation (equation (1.68))*

	Residual resistivity						
	Moriarty (Cu, Ag) or HAA			Ashcroft			
Solute	1 OPW	4 OPW Ziman	4 OPW exact	1 OPW	4 OPW Ziman	4 OPW exact	Experiment
Mg	0.21	0.37	0.38	0.18	0.30	0.31	0.34–0.36
Si	0.22	0.35	0.37	0.31	0.37	0.40	0.56–0.60
Cu	0.43	0.80	0.81	0.44	0.80	0.80	0.89
Zn	0.05	0.13	0.14	0.10	0.18	0.19	0.24
Ga	0.03	0.06	0.06	0.14	0.13	0.14	0.22
Ge	0.25	0.36	0.38	0.45	0.45	0.48	0.79
Ag	0.23	0.60	0.60	0.51	0.92	0.96	1.08–1.34

Source: After Sorbello 1973.

Table 5.19. *Experimental and theoretical residual resistivities in μΩ cm due to one at.% of the impurities indicated in Cu, Ag and Au hosts. Modified point-ion and Moriarty form factors were used for the host and Heine–Abarenkov–Animalu form factor for the impurities*

		Residual resistivity				
		Host form factor				
		Mod. point-ion		Moriarty		
Solvent	Solute	non-relaxed	relaxed	non-relaxed	relaxed	Experiment
Cu	Cu	—	—	—	—	—
	Ag	0.048	0.036	0.194	0.267	0.14
	Au	0.974	0.893	0.305	0.510	0.55
	Be	0.137	0.120	1.473	1.633	0.62
	Mg	0.248	0.309	0.771	0.613	0.65
	Zn	0.084	0.084	0.402	0.306	0.32
	Cd	0.222	0.213	0.446	0.120	0.30
	Hg	0.496	0.424	0.973	0.392	1.00
	Li	0.049	0.041	0.543	0.257	
	Na	0.272	0.254	0.687	0.774	
	K	0.715	1.092	1.868	7.714	
	Rb	3.196	2.433	8.692	14.534	
	Cs	7.270	5.190	20.192	29.228	
	Ca	0.772	1.200	2.232	3.776	
	Ba	1.529	2.645	5.279	10.923	
Ag	Cu	0.075	0.067	0.180	0.211	0.08
	Ag	—	—	—	—	—
	Au	0.814	0.815	0.079	0.077	0.36
	Mg	0.402	0.382	0.375	0.284	0.50
	Zn	0.087	0.092	0.200	0.257	0.64
	Cd	0.206	0.192	0.513	0.420	0.38
	Hg	0.433	0.391	1.428	1.107	0.79
	Li	0.014	0.011	0.318	0.187	
	Na	0.210	0.166	0.864	0.341	
	K	0.891	0.819	2.325	2.724	
	Rb	4.027	3.600	10.139	7.466	
	Cs	9.520	8.708	21.991	16.083	
	Ca	1.245	1.154	1.490	1.876	
	Ba	2.572	2.552	3.598	5.763	

going to a multiple OPW calculation. However, it should be stressed that such an improvement represents a significant increase in the complexity of the calculation. The structure factor associated with the strain around a substitutional impurity in Al has also been discussed by Werner *et al.* (1978). Singh *et al.* (1977) have determined the resistivity of impurities in noble metal alloys including the effects of lattice distortions derived from elasticity theory. They used both modified point-ion (4.24b) and Moriarty form factors for the host and Heine–Abarenkov–Animalu

Table 5.19 (*cont.*)

		Residual resistivity				
		Host form factor				
		Mod. point-ion		Moriarty		
Solvent	Solute	non-relaxed	relaxed	non-relaxed	relaxed	Experiment
Au	Cu	1.437	1.682	0.309	0.461	0.45
	Ag	0.894	0.899	0.084	0.086	0.36
	Au	—	—	—	—	—
	Mg	2.326	1.792	0.659	0.432	1.30
	Zn	1.364	1.505	0.557	0.692	0.95
	Cd	1.250	1.112	1.017	0.876	0.63
	Hg	0.686	0.393	2.252	1.692	0.44
	Li	1.035	0.803	0.758	0.475	
	Na	0.852	0.142	1.556	0.535	
	K	1.739	0.828	3.461	3.980	
	Rb	2.668	6.888	13.003	9.564	
	Cs	6.882	19.050	26.897	19.708	
	Ca	4.085	1.560	1.853	2.214	
	Ba	6.656	2.172	4.043	7.026	

Source: After Singh *et al.* 1977.

form factors for the impurities. Their results are summarised in Table 5.19. These results indicate that the maximum effect of distortion is $\sim 70\%$ for large core sized impurities and $\sim 55\%$ for small core sized impurities (except for Rb and Cs in Au which give $\sim 150\%$ changes). There is no systematic difference between the results for either host form factor and neither produces a particularly good agreement with experiment. However, once again the form factors were chosen to give a good agreement between the calculated and experimental phonon frequencies and so are not very accurate at large q.

The atomic relaxations around vacancies have been considered by a number of people (see Benedek & Baratoff 1971; Yamamoto *et al.* 1973; and Takai *et al.* 1974; Kenny *et al.* 1973; Fukai 1969 and the references cited therein). These results may be used to evaluate the appropriate structure factor (5.118). A comparison of atomic displacements using various models for the lattice distortion and various form factors for simple monovalent metals has been given by Yamamoto *et al.* (1973). The effects of plane wave mixing have been taken into account in polyvalent metals by the two OPW approaches of Fukai (1969) and Benedek & Baratoff (1971). In general, it is found that the relaxations reduce the resistivity around a vacancy, the resistivity being a minimum at the equilibrium structural disorder (Takai *et al.* 1974). The structure

Table 5.20. *Residual resistivity in μΩ cm due to one at.% vacancies in the metals indicated using a variety of calculations of atomic displacements and form factors. The values in parentheses are the values calculated when relaxation is neglected. The form factors AR, AK and AS are, respectively, the Ashcroft empty-core forms with random phase, Kleinmann and Singwi corrections for screening, exchange and correlation*

		Residual resistivity					
		Form factors					
Metal	Relaxation	Harrison	HAA	AR	AK	AS	Expt
Al	F (2 OPW)	2.12 (1.91)	1.89 (1.73)	1.72 (1.58)	—	—	2.2 ± 0.7
	BB (2 OPW)			1.02 (1.44)			1.0–3.3
	T 100 split	3.68	1.75	1.54	1.81		
	110 split	3.77	1.80	1.61	1.89		
	111 split	5.17	2.93	2.60	3.13		
	bc	3.14	1.52	1.36	1.61		
	tet	4.00	1.89	1.64	1.93		
Li	WS				1.553 (1.681)		
	JHM1				1.694		
	PLP					0.759 (1.005)	
Na	WS	0.667 (1.129)			0.702		
	JHM1	0.901			0.951		
	JHM2	0.777			0.820 (1.192)		
	WG	0.604			0.631		
	GR	0.616			0.642		
	SBF	0.768			0.811		
	FH	0.712			0.751		
	TG	0.628			0.660		
	H	0.640			0.671		
	PCP					0.888 (1.207)	

	BB (2 OPW)	0.69 (0.95)			1.9–2.1
K	WS		0.831 (1.486)		
	JHM1		0.863		
	JHM2		0.880		
	CW2		1.126		
	PCP			0.975 (1.353)	
	BB (2 OPW)	1.00 (1.36)			
Rb	WS		1.474 (2.041)		
	JHM1		1.828		
	JHM2		1.704		
	PCP			1.050 (1.434)	
CS	WS		2.489 (2.864)		
	JHM1		2.453		
	JHM2		2.570		
	PCP			1.182 (1.559)	

Notes:

WS Waseda & Suzuki (1971, 1972)
JHM1, JHM2 Johnson *et al.* (1964)
WG Wynblatt & Gjostein (1967)
GR Grimes & Rice (1968)
SBF Shyu *et al.* (1967)
FH Flocken & Hardy (1969)
TG Torrens & Gerl (1968)
H Ho (1971)
CWD Cowley *et al.* (1966)

(WS to CWD: Data from Yamamoto *et al.* (1973))

BB Benedek & Baratoff (1971)
F Fukai (1968)
PCP Popovic *et al.* (1973)
T Takai *et al.* (1972).

factor for the distortions around interstitials (5.130) has been evaluated for an Al matrix by Takai *et al.* (1972). Their results derived from a number of different form factors are shown in Table 5.20.

Hoffman & Cohen (1973) have suggested a simple calculation based on the idea that the change in resistivity due to some static displacement around an interstitial should be the same as the phonon resistivity corresponding to the same phonon-induced displacements. This theory will be discussed in more detail in Section 5.7.3 and in the next chapter. The resistivity due to various vacancy clusters and microvoids has been considered by Flynn (1962), J. W. Martin (1972), Martin & Paetsch (1973), and Hautojärvi *et al.* (1977).

As a general rule it appears that the lattice distortion around a substitutional or interstitial point defect may increase the resistivity by up to ~70% whereas that around a vacancy may decrease the resistivity by up to ~50%. It appears that strains out to at least the third nearest neighbour of the defect should be included. Since the displacement field only approaches the isotropic continuum elastic limit at some greater distance from the defect (Boyer & Hardy 1971; Kenny *et al.* 1973) it would appear that the lattice statics (Kanzaki 1957) or discrete (Shyu *et al.* 1967) methods of determining the atomic displacements should be used in preference to the elastic continuum models (Doyama & Cotteril 1967; Kanzaki 1957; Eshelby 1954, 1956). It would also appear that a multiple OPW calculation is necessary to achieve good quantitative accuracy in polyvalent metals. Form factors fitted to Fermi surface or liquid resistivity data are preferable to those fitted to atomic properties (such as phonon spectra) since they are likely to be more accurate in the critical high q region.

(b) *Concentrated alloys*

Nordheim's rule predicts that the resistivity of disordered solid solution should go as

$$\rho \propto c_A c_B.$$

Equation (5.21) is in accord with this general variation but includes a variation with $(k_F)^{-6}$ and Ω_0, both of which will in general also be composition dependent. The resistivity determined from the Moriarty form factors is compared with experimental data for disordered Cu–Au in Figure 5.37 (Wang 1970). Hillel (1970) noted the generally poor agreement between theory based on a single-OPW and experiment in Al-based alloys and constructed a simple model pseudopotential for the *difference* potential $w_d(\mathbf{q})$. He chose the point-ion form, and the values of β/Ω_0 obtained by fitting to residual resistivity data are shown in Table

5.21. Similar empirical difference pseudopotentials based on the Ashcroft empty-core form for a series of Cs–K alloys have been determined by Tanigawa & Doyama (1973).

5.7.3 *Homogeneous short range atomic correlations*

In the absence of atomic displacement effects, the resistivity of an alloy with homogeneous short range correlations is given by equation (5.18) or (5.20). The two-body correlation parameters $\langle\sigma_i\sigma_j\rangle$ or α_i are determined either theoretically from some pairwise interaction parameter (Clapp & Moss 1968; Moss & Clapp 1968) or experimentally by diffuse X-ray or neutron scattering (see e.g. Warren 1969, p. 227). The significance of correlations between various neighbours may be ascertained by interchanging the summation and integration in equation (5.20) to give

$$\rho_c = Cc_A c_B \sum_i c_i \alpha_i Y_i, \tag{5.203}$$

where

$$Y_i = \int_0^{2k_F} |\langle \mathbf{k}+\mathbf{q}|w^d(\mathbf{r})|\mathbf{k}\rangle|^2 \frac{\sin(qR_i)}{qR_i} q^3\, dq. \tag{5.204}$$

This integral (normalised to unity at $q \to 0$) is shown plotted as a function of $2k_F R_i$ in Figure 5.38 for a Cu–25%Au alloy using the

Fig. 5.37. Resistivity as a function of composition for disordered Cu–Au alloys. Solid line: values calculated from the Moriarty form factors (Wang 1970); dotted line: experimental data (Johansson & Linde 1936).

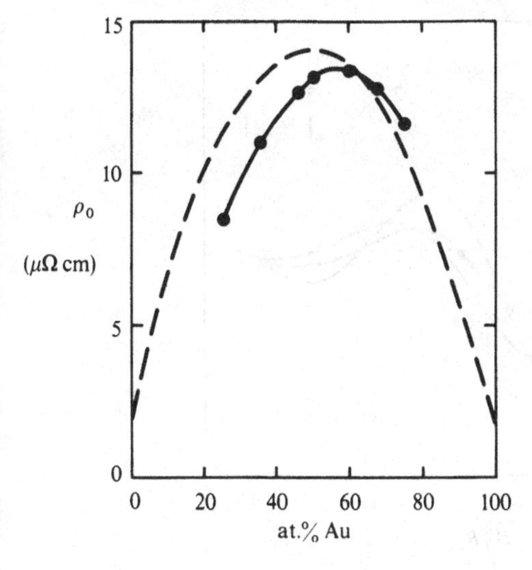

Table 5.21. *Values of β for the alloy systems indicated*

Solvent	Solute	β (Ry-Å^3)
Al	Cu	−5.2
	Ag	−5.8
	Zn	−2.8
	Mg	−3.0
	Ge	+4.0
	Si	+3.7
Cu	Ag	−0.7
	Be	+3.5
	Cd	+2.7
	In	+5.3
	Si	+9.2
Ag	Cu	+0.6
	Al	+6.7
	Zn	+4.7

Source: After Hillel 1970.

Fig. 5.38. The integrals Y_i for Cu–25%Au normalised to unity at $q \rightarrow 0$. Solid lines: (i) point-ion; (ii) empty-core; (iii) delta function. Dashed line: screened Coulomb. Dotted line: Moriarty. Dashed–dot line: empty-core. Also shown by vertical bars are the values of $2k_F R_i$ for the first four shells of neighbours.

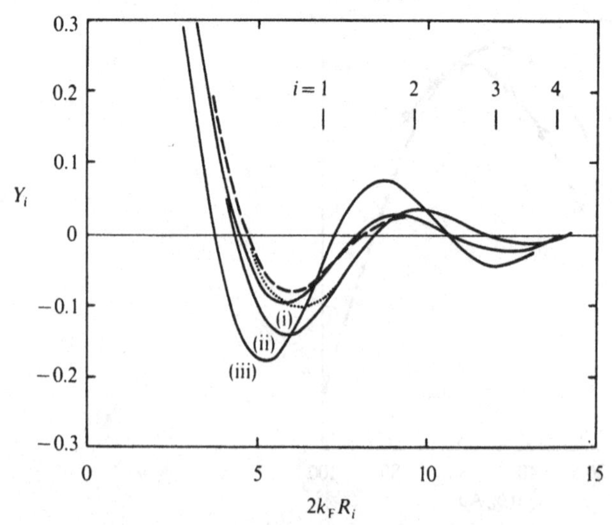

Table 5.22. *Calculated and experimental values for the resistivity (in μΩ cm) of disordered Cu–25%Au and the change caused by the reduction in SRO on heating from 405 to 450 °C (see Table 2.2). For the empty-core potential the alternative values $R_c(Cu)=0.93$ Å, $R_c(Au)=1.08$ Å quoted by Ashcroft & Langreth (1967) were used. The HAA form factors were taken from Wang & Amar (1970). The other parameters used were $k_F=1.31$ Å^{-1}, $\Omega_0=13$ Å^3*

Potential	ρ_∞	$\Delta\rho=\rho_{405}-\rho_{450}$	$\Delta\rho/\rho_\infty$
Moriarty	10.96	0.28	0.025
Empty-core	2.22	0.091	0.040
HAA	2.61	0.076	0.031
Expt	11.35	0.045	0.004

Moriarty, empty-core, point-ion, simple delta function (i.e. $z=0$ in the point-ion potential) and screened Coulomb ($\beta=0$ in the point-ion potential) scattering potentials. The resistivity due to some particular set of SRO parameters (α_i) is then determined by finding the value of Y_i at each value of $2k_F R_i$ for the particular lattice concerned (see Appendix E). The decaying nature of these curves and of the spatial variation of the pairwise correlation parameters indicates that the resistivity should be determined largely by the first few SRO parameters. The resistivities calculated for a random alloy using the Moriarty, HAA and empty-core form factors are shown in Table 5.22 together with the change in resistivity and fractional change in resistivity resulting from the different degree of SRO at 405 °C and 450 °C (see Table 2.2). These results show that provided the $2k_F R_i$ do not fall too near the zeros in Y_i, the actual form of the potential is not very critical in determining the fractional change in resistivity due to SRO, although it will affect the magnitude of the resistivities calculated. However, as a contra-example, if the same potentials are used for Au–Ag alloys, the Moriarty form factors give a positive Y_1 whereas all others give a negative Y_1. Only the latter predict an increase in ρ with increased SRO as observed experimentally. It might be noted that the Moriarty form factors also give a poor prediction of the disordered resistivity in this system (0.05μΩ cm per at.% compared to the experimental value of 0.36μΩ cm per at.%). The effects of different screening and exchange and correlation corrections on the integral Y_i for

Table 5.23. *Calculated and observed direction of the change in residual resistivity of a number of short range ordering alloy systems (D: decrease; I: increase)*

Alloy	Change in resistivity with increased SRO Empty-core	Experiment	Reference
Cu–30%Zn	D	D	Damask 1956*b*
Cu–50%Zn	D	D	Honeycombe & Boas 1948
Cu–25%Au	I	I	Damask 1956*a*
Cu–75%Au	I	I	Chardon & Radelaar 1969
Ag–30%Zn	D	D	Chardon & Radelaar 1969
Cu–10%Al	D	D	Meisterle & Pfeiler 1983*a*
Cu–20%Al	D	I(?)	Trieb & Veith 1978
Ag–10%Al	D	D	Pfeiler *et al.* 1984
Au–35%Ag	I	I	Lücke & Haas 1973

Cu–30%Zn with an empty-core pseudopotential are shown in Figure 5.39. While these different forms are known to give different residual resistivities of the disordered alloy, they all give virtually identical values of Y_1 and Y_2 and so the fractional change in resistivity with short range atomic correlations is again not very sensitive to the form used.

The predicted direction of variation of ρ with increased SRO for a number of different alloy systems is compared with the observed behaviour in Table 5.23. In deriving these results the empty-core pseudopotential was used and it was assumed that $\alpha_1 > 0$ and $\alpha_2 < 0$. In all cases except Cu–20%Al the correction variation is predicted. However, in this latter case the ordering is known to be very inhomogeneous (Trieb & Veith 1978) and there is a large atomic size difference in this system. We discuss this effect later on. As a general rule, either the point-ion or empty-core model pseudopotentials appear to give a reasonable prediction of the dependence of the resistivity upon short range atomic order, particularly if fitted to the disordered alloy resistivity. Nevertheless, when exploring new alloy systems it would always be advisable to examine the sensitivity of the results to the form of the scattering potential, particularly if the first one or two zeros in Y_i occur near values of $2k_F R_i$ corresponding to atomic positions in the lattice. Exactly the same arguments apply to the early stages of clustering except that, of course, α_1 will then be positive. However, it appears that special care is required for polyvalent materials such as the Al-based GP zone forming alloys (Al–Zn, Al–Ag, Al–Mg, etc.), where construction of $w^d(\mathbf{q})$ from the point-ion or empty-core pseudopotentials of the constituents can often lead to the wrong results in a single OPW calculation. As an example, the integrals Y_i obtained from the point-ion and empty-core potentials for Al and Zn are shown in Figure 5.40. It can be seen that the resultant values of Y_1 are then negative for nearest neighbours, wrongly predicting a decrease in ρ_0 with increased clustering. Also shown in that diagram are the integrals corresponding to the point-ion difference pseudopotential of Hillel (see Table 5.21) and the Shaw (1973) optimised pseudopotential. Both of these predict the correct trend with clustering since they both give a positive Y_1. Hillel (1970) has drawn attention to this problem, particularly noting the discrepancy between the point-ion difference potential and the empty-core potentials in Al–Zn. The corresponding curves for Al–Ag are shown in Figure 5.41. Again it is clear that the scattering potential obtained for the point-ion or empty-core potentials of Al and Ag are not correct ($Y_1 \approx 0$), while the point-ion difference potentials of Hillel corectly

Fig. 5.39. The integral Y_i as a function of $2k_F R_i$ for Cu–30%Zn using an empty-core pseudopotential and different screening and exchange and correlation corrections: (*a*) Thomas–Fermi; (*b*) Lindhard with no exchange and correlation; (*c*) Lindhard with Hubbard–Sham exchange and correlation; (*d*) Lindhard with Kleinman exchange and correlation; (*e*) Lindhard with Singwi exchange and correlation. Also shown are the values of $2k_F R_i$ for the first two shells of neighbours.

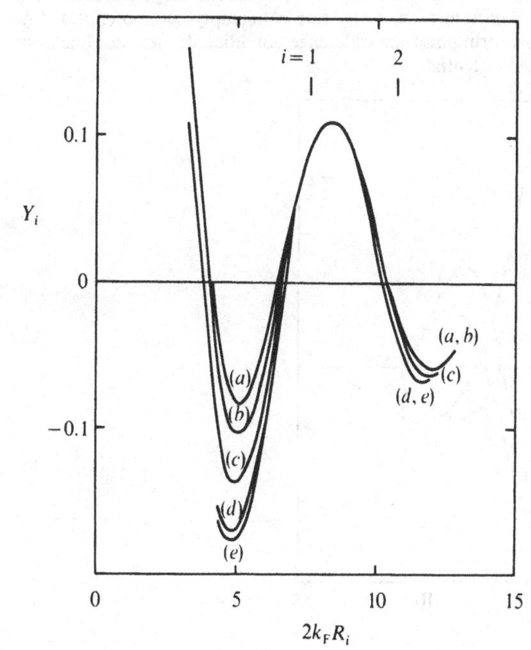

predicts a positive Y_1. In such systems it thus appears that the safest choice for a single-OPW calculation is the point-ion difference potential with β adjusted to give the correct disordered alloy resistivity. The empty-core pseudopotential appears to give reasonable results in a multiple–OPW calculation (Yonemitsu & Matsuda 1976).

However we must also consider the effects of atomic displacements. Borie (1957) has shown that the terms $\langle(\mathbf{q}\cdot\mathbf{u}_i)^2\rangle$ may be evaluated by assuming that each atom causes a spherically symmetric distortion which may be one of two kinds depending upon whether an A or B atom occupies the site in question. The deviation of an atom from its undisplaced position is given by the vector sum of the individual displacements caused by each atom in the crystal. If the lattice is assumed to be an isotropic elastic medium, the distortion from each centre will decay as $1/r^2$, and so at a site n

$$\mathbf{u}_n = \sum_{j \neq n} d_j \frac{\mathbf{r}_{nj}}{|\mathbf{r}_{nj}|^3}, \tag{5.205}$$

where the distortion parameter d_j is either d_A or d_B depending upon the occupancy of site j. Since the total change in volume due to all distortion

Fig. 5.40. The integral Y_i as a function of $2k_F R_i$ for Al–Zn alloys: solid line from point-ion potentials of Al and Zn; dotted line from empty-core potentials of Al and Zn; dashed line form point-ion difference potential; dashed–dot line from the Shaw optimised potential.

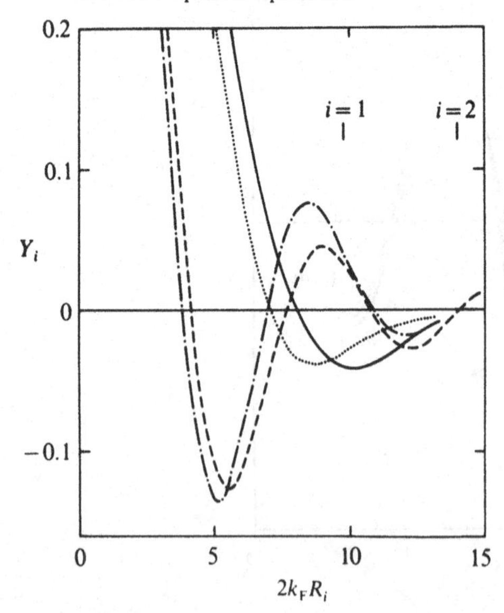

centres is zero on the average lattice,

$$c_A d_A + c_B d_B = 0. \tag{5.206}$$

We can then write (Borie's equations 20, 21)

$$2M'_A = \langle(\mathbf{q}\cdot\mathbf{u}_i^A)^2\rangle$$

$$= \sum_{i\neq j} d_A^2(1-\alpha_{ij})\left(\frac{c_A}{c_B}+\alpha_{ij}\right)(\mathbf{q}\cdot\mathbf{r}_{ij})^2|\mathbf{r}_{ij}|^6, \tag{5.207}$$

and similarly for $2M'_B$. If all α_{ij} are small, these become (Born & Misra 1940)

$$2M' = \tfrac{1}{2}\langle(\mathbf{q}\cdot\mathbf{u}_i^A)^2\rangle + \tfrac{1}{2}\langle(\mathbf{q}\cdot\mathbf{u}_j^B)^2\rangle. \tag{5.208}$$

The distortion parameters d_A, d_B are related to the strain parameters $\varepsilon_i^{AA}, \varepsilon_i^{BB}$ in this model since, by the definition of d,

$$\frac{\Delta r}{r} = \frac{d}{r^3}. \tag{5.209}$$

If an A atom is the cause of the distortion, for example, a nearest neighbour A atom will suffer a strain,

$$\frac{\Delta r}{r} = \frac{1}{2}\left[\frac{r_1^{AA} - \bar{r}_1}{\bar{r}_1}\right] = \frac{\varepsilon_1^{AA}}{2}, \tag{5.210}$$

Fig. 5.41. The integral Y_i as a function of $2k_F R_i$ for Al–Ag alloys: solid line from point-ion potentials of Al and Ag; dotted line from empty-core potentials of Al and Ag; dashed line from point-ion difference potential.

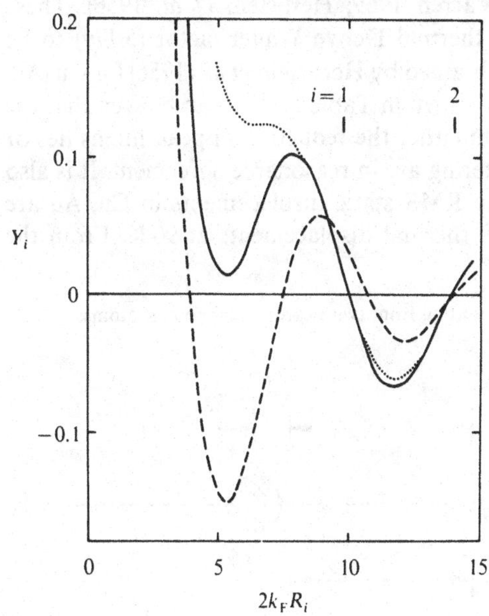

where the distances are as shown in Figure 5.42. For nearest neighbours $\bar{r}_1 = a_0\sqrt{2}/2$ and so

$$\varepsilon_1^{AA} = \frac{d_A 4\sqrt{2}}{a_0^3}. \tag{5.211}$$

Thus from (5.208):

$$2M' = 1.05a_0^2 \frac{c_A}{c_B} (\varepsilon_1^{AA})^2 |\mathbf{q}|^2. \tag{5.212}$$

From (5.206) we also have

$$c_A \varepsilon_1^{AA} + c_B \varepsilon_1^{BB} = 0. \tag{5.213}$$

As noted by Borie, these results should be regarded as approximate since the model assumes isotropic elasticity and it has been assumed that the atomic displacement for an atom at j introduced by an atom at i does not depend upon the kind of neighbouring atoms around j. Cowley (1968) has given a calculation of the X-ray scattered intensities which does not make such an assumption, although this involves higher order correlation parameters which are generally not available. The quantities M (or M_A, M_B) or ε_i^{AA}, ε_i^{BB} may be obtained from diffuse X-ray or neutron scattering experiments. The pseudo Debye–Waller factors give a reduction in Bragg peak height as the scattering angle increases, whereas the parameters ε_i^{AA}, ε_i^{BB} give a modulated diffuse scattering as discussed in Section (5.6.3) (see also Warren 1969; Herbstein *et al.* 1956). These techniques also allow the thermal Debye Waller factor (5.179) to be determined. Some results obtained by Herbstein *et al.* (1956) for Cu_3Au, CoPt, NiAu and LiMg are shown in Table 5.24. It can be seen that the values obtained for M' from either the reduction in peak intensities or the modulated diffuse scattering are in reasonable agreement. It is also interesting to note that the RMS static displacements in Cu, Au are roughly equal to the RMS thermal displacements at 90 K. From the

Fig. 5.42. The strain resulting from two nearest neighbour A atoms.

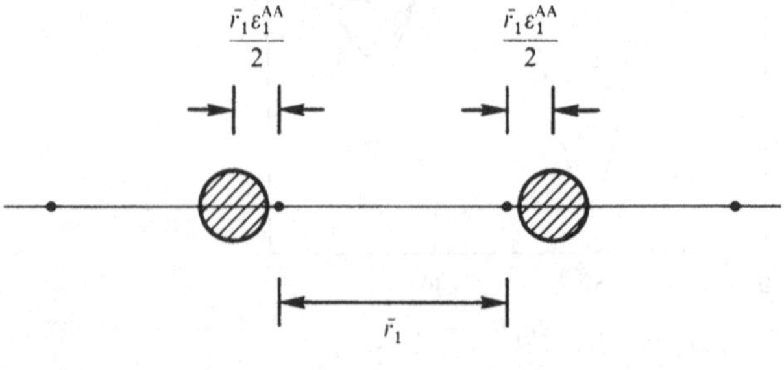

Table 5.24. *Static and dynamic Debye–Waller factors and atomic displacements for the alloys indicated*

	M'/q^2 (Å²)		RMS static displacements	M_D/q^2 (Å²)		RMS dynamic displacements (Å)	
Alloy	D–W	Diffuse	Å	295 K	90 K	295 K	90 K
Cu–25%Au	0.0011	0.0010	0.08	0.0033	0.0012	0.14	0.09
Co–50%Pt	0.000 76	0.000 44	0.07	0.0022	0.000 82	0.12	0.07
Ni–48%Au	0.0020	0.0022	0.11	0.0042	0.0026 (180 K)	0.16	0.13 (180 K)
Li–49%Mg	0.023	—	0.11	0.010	0.0042	0.25	0.16

Source: After Herbstein *et al.* 1956.

value of M'/q^2 for Cu, Au and equations (5.212) and (5.213) we find

$$\begin{aligned}\frac{M'}{q^2} &= 0.0011\\ \varepsilon_1^{\mathrm{AA}} &= -0.007 \\ \varepsilon_1^{\mathrm{BB}} &= 0.021.\end{aligned} \tag{5.214}$$

(The signs of $\varepsilon_1^{\mathrm{AA}}$ and $\varepsilon_1^{\mathrm{BB}}$ are inferred from the atomic sizes). The variation of M' with SRO is approximately obtained from (5.207), i.e.

$$\begin{aligned}M'_{\mathrm{A}} &= 0.0011\frac{c_{\mathrm{B}}}{c_{\mathrm{A}}}(1-\alpha_i)\left(\frac{c_{\mathrm{A}}}{c_{\mathrm{B}}}+\alpha_i\right),\\ M'_{\mathrm{B}} &= 0.0011\frac{c_{\mathrm{A}}}{c_{\mathrm{B}}}(1-\alpha_i)\left(\frac{c_{\mathrm{B}}}{c_{\mathrm{A}}}+\alpha_i\right).\end{aligned} \tag{5.215}$$

(The problem of largely different $M'_{\mathrm{A}}, M'_{\mathrm{B}}$ will be discussed below.)

From (5.214), (5.215), (5.160) (or (5.161)) (5.167) and (5.9), and by analogy with the SRO case, the effects of neighbour distance may be found by writing the total contribution due to size effects as

$$\begin{aligned}\rho_{\mathrm{PDW}}+\rho_{\mathrm{SE}} = C\int|\bar{w}(\mathbf{q})|^2[1-\exp(-2M')]q^3\,\mathrm{d}q\\ +Cc_{\mathrm{A}}c_{\mathrm{A}}c_{\mathrm{B}}\sum[c_i\gamma_iY_i+c_i\xi_iX_i],\end{aligned} \tag{5.216}$$

where ρ_{PDW} is the pseudo Debye–Waller term (first term on the right hand side of (5.216), ρ_{SE} is the size effect modulated diffuse scattering due to the different atomic sizes (second term on the right hand side) and

$$X_i = \int_0^{2k_{\mathrm{F}}} \bar{w}(\mathbf{q})w^{\mathrm{d}}(\mathbf{q})\left(\cos(qR_i)-\frac{\sin(qR_i)}{qR_i}\right)q^3\,\mathrm{d}q. \tag{5.217}$$

The extension of this to different pseudo Debye–Waller factors $M'_{\mathrm{A}}, M'_{\mathrm{B}}$ is straightforward (see equation (5.161)). The total residual resistivity can then be written as

$$\rho_{\mathrm{T}} = \rho_0+\rho_{\mathrm{SRO}}+\rho_{\mathrm{SE}}+\rho_{\mathrm{PDW}} \tag{5.218}$$

The resistivity of Cu–25%Au calculated from the Moriarty, Heine–Abarenkov–Animalu and empty-core form factors is shown in Table 5.25. The first five SRO parameters of Moss corresponding to quench temperatures 405 °C and 450 °C (Table 2.2) have been used to determine the SRO contribution, but only the first parameter was used for the size effect and pseudo Debye–Waller contributions. The results in this table reveal some interesting differences. For example, the Moriarty form factors give the largest disordered and SRO contributions but a similar size effect contribution to the empty-core potential which gives the lowest disordered and SRO contributions. The HAA form factors give

Table 5.25. *Resistivity contribution in μΩ cm from the disordered lattice, short range order (SRO), static atomic displacements (SE) and pseudo Debye–Waller factor (PDW) for Cu–25%Au using the form factors indicated*

Potential	Quench temperature (°C)	Resistivity contribution Disordered	SRO	SE	PDW	Total
Moriarty	405	10.96	2.83	0.78	2.67	17.24
	450	10.96	2.55	0.79	2.73	17.03
	$\gg T_c$	10.96	0	1.08	3.04	15.09
HAA	405	2.61	0.49	−0.10	0.43	3.43
	450	2.61	0.41	−0.09	0.44	3.37
	$\gg T_c$	2.61	0	−0.05	0.49	3.05
Empty-core	405	2.22	0.67	0.80	0.46	4.15
	450	2.22	0.58	0.84	0.48	4.12
	$\gg T_c$	2.22	0	1.10	0.56	3.88

intermediate values of the disordered and SRO contributions but a small and negative size effect term. The empty-core form factors give a size effect contribution that is larger than the SRO term. These differences result from the different shapes of the form factors, the different contributions depending upon the magnitudes of either $\bar{w}(\mathbf{q})$ or $w^d(\mathbf{q})$, or both. In all three cases the nett effect of SRO is still to cause an increase in ρ but because SRO causes a decrease in the size effect term for the Moriarty and empty-core potentials, the magnitude of the overall increase is reduced for these potentials, as indicated in Table 5.26.

It is also apparent that while the pseudo Debye–Waller diffuse scattering may produce quite a large resistivity, it is not very dependent upon the degree of SRO (within the limitations of (5.215)). However, the incorporation of different values of M'_A and M'_B, even in the disordered state (see Walker & Keating 1961; Chipman 1956; Krivoglaz 1969, p. 158), could produce large changes to this contribution. In this particular example both the static displacements and short range order have resulted in an increase in resistivity (except with the HAA form factor). However, for different alloys it is quite possible that either or both contributions could be of the opposite sign. It is also quite clear that unless the atoms in an alloy are of very similar size, the effects of static atomic displacements are just as important as those of short range atomic correlations and should not be neglected in a calculation of the resistivity. It is thus not surprising that theories which neglect this size effect cannot explain the observed behaviour in systems with large

Table 5.26. *Change in resistivity of Cu–25%Au in μΩ cm if only the short range order contribution is considered and if the size effect contribution is included*

	Fractional change in resistivity $(\rho_{405}-\rho_{450})/\rho_{\infty}$		
Potential	only SRO	with SE+PDW	Experiment
Moriarty	0.025	0.124	
HAA	0.031	0.023	0.004
Empty-core	0.040	0.013	

atomic size differences. This provides a possible explanation for the discrepancy in the Cu–Al alloys noted in Table 5.23. However, it is also clear that this makes the choice of suitable form factors even more critical.

We can also estimate the effects of non-zero measuring temperature on the resistivity due to atomic correlations and size effects. As shown in Section 5.6.4 this results in a thermal diffuse scattering characterised by a Debye–Waller term $\exp(-2M_D\phi(r_{ij}))$ (equation (5.179)). As suggested in that section we take $\phi(r_{ij})=1$. The parameters M_D or M_E could be evaluated from the Debye or Einstein models but we have an experimentally derived value given in Table 5.24. Using the value of $M/q^2=0.0012$ then results in a reduction of the SRO and size effect terms of less than 1%. We do not include the effects of thermal vibrations on the pseudo Debye–Waller term as these will be included in the overall phonon scattering. Compared to the other uncertainties in the calculation this is a small effect, although it does indicate that the change in ρ with SRO will be slightly less at higher measuring temperatures (assuming that the degree of SRO remains constant). This is in qualitative agreement with the observation by Damask (1956*a*) that the slope of the resistivity against quench temperature decreases by about 2% for a change in measuring temperature from 4 to 77 K. In general, however, it is felt that this effect will be negligible. As such there appears to be no need to consider the effects of different amplitudes of *thermal* vibration of the A or B atoms.

Hoffman & Cohen (1973) have proposed a somewhat more empirical means of determining the atomic size effect resistivity. In their model it is assumed that a particular mean square static displacement caused by lattice strain will cause the same resistivity as that produced by an identical mean square thermal displacement. Implicit in this model is the unlikely assumption that the spatial correlation of atomic displacements produced by strain or thermal excitation is the same. Nevertheless they do find reasonable agreement between prediction and experiment in some Fe–C and Fe–Ni–C martensites. We can apply their model to the Cu–25%Au data given in Tables 5.24 and 5.25. The RMS static and dynamic displacements are roughly equal at 90 K at which temperature the thermal component of the resistivity is $\sim 1\mu\Omega$ cm (Rossiter & Bykovec 1978). The calculated resistivity contribution from static atomic displacements depends upon the form factors used and is $\sim 4\mu\Omega$ cm (Moriarty), $1.5\mu\Omega$ cm (empty-core) and $0.5\mu\Omega$ cm (Heine–Abarenkov–Animalu). As the discrepancies here are not much larger than the discrepancies in the total resistivities, this simple approach may give a useful check of the magnitudes of calculated atomic displacement resistivities.

In view of the above discussions we come to the following conclusions regarding the selection of pseudopotential form factors for use with concentrated alloys. The safest approach is to recalculate the pseudopotentials for each particular alloy of interest. The effects of lattice distortion and scattering anisotropy should be taken into account and in polyvalent metals the mixing of plane wave states at the Brillouin zone boundaries should be included by using multiple OPWs. If this sort of facility is not available then the following general guidelines will assist in getting the most out of a simpler one-OPW calculation. If the atomic size difference is not large (i.e. less than about 20%, say) the form factors that give a disordered alloy resistivity in best agreement with experiment should be chosen. The effects of atomic correlations should then be reasonably well reproduced since both these and the disordered resistivity depend upon the differences in the atomic form factors. If the atomic size difference are large (>20%) then it will be necessary to include the scattering due to size effects. The pseudopotential form factors chosen should then give a reasonable value for the temperature-dependent resistivity as well as the residual resistivity since both the size effect and phonon scattering depend upon the average pseudopotential. In polyvalent metals particularly, it may be found that no potentials constructed from the form factors of the alloy constituents meet these requirements, in which case it may be necessary to model the average and difference pseudopotential form factors directly, fitting the difference form factor parameters to the disordered residual resistivity and, if necessary, the average form factor parameters to the temperature-dependent resistivity. The effects of exchange and correlation do not seem to be particularly important (in view of the other uncertainties), and so one might just as well neglect these and use the simple Lindhard (RPA) screening formula. The size effect parameters may be calculated using the lattice statics or discrete methods, or obtained from diffuse X-

Table 5.27. *Some non-simple binary alloy systems in which the dependence of the resistivity on short range atomic distribution has been investigated*

Alloy system	Reference
AgMg	Grayevskaya *et al.* (1975)
AgMn	Thomas (1951); Koster & Rave (1961)
AlCr	Kovacs–Csetenyi *et al.* (1973)
AlFe	Davies (1963), Thomas (1951)
AlNi	Starke *et al.* (1965)
AlMn	Kedves *et al.* (1976)
AuCo	Hurd & McAlister (1980); Henger & Korn (1981)
AuNi	Hurd *et al.* (1981)
CuCo	Servi & Turnbull (1966); Haller *et al.* (1981)
CuCr	Szenes *et al.* (1971); Kovacs & Szenes (1971); Williams (1960)
CuFe	Boltan (1960)
CuNi	Rossiter (1981*b*) and references therein; Wagner *et al.* (1981); Thomas (1951)
CuMn	Segenreich *et al.* (1978); Köster & Rave (1961)
PdAu	Lücke *et al.* (1976); Haas & Lücke (1972)
PdCo	Katsnel'son *et al.* (1968)
PdFe (Au)	Kuranov (1982)
PdPt	Katsnel'son & Alimov (1967)
NiCr	Heidsiek *et al.* (1977)
NiW	Katsnel'son *et al.* (1968)
FeCr	Richter *et al.* (1974)
FeCo	Pureur *et al.* (1982)
Steels	Wilson & Pickering (1968); Raynor & Silcock (1968, 1970); Silcock (1970*a*,*b*); Jones *et al.* (1971)

ray or neutron scattering data. Finally one should also check that the values of $2k_F R_i$ for the material being investigated do not lie close to the zeros of the SRO Y_i integrals (5.204) and the corresponding size effect integrals Y_i and X_i (5.204), (5.217). If this is not the case then the sensitivity of the predicted resistivity due to short range atomic correlations upon the details of the form factors should be investigated to establish a suitable confidence level.

It might be noted in passing that there are many non-simple metal alloys that also show a marked dependence of the resistivity upon the state of atomic correlations, although in most cases it is likely that the behaviour is complicated by band structure and/or magnetic effects (see Chapter 6). Some of the systems investigated are indicated in Table 5.27 and include the Cu–Ni system which has been investigated in some detail by Wagner *et al.* (1981). Using the Warren–Cowley parameters determined from diffuse neutron scattering for various clustering states corresponding to temperatures between 380 K and 730 K, the resistivity contribution due to short range clustering in Ni–41.4%Cu was calculated according to (5.20) using a simple delta function potential (i.e. a point-ion potential with $\Delta Z = 0$). Good quantitative agreement was found between the calculated and experimentally determined relative resistivity changes, as shown in Figure 5.43. It is evident that both the sign and the magnitude of the calculated resistivity changes are in excellent agreement with the experimentally measured data. It was also found that the behaviour was mainly determined by the first shell parameter, the higher shell contributions being smaller in magnitude and largely cancelling each other. The good agreement in this ferromagnetic alloy indicates a relatively small and constant magnetic scattering at the temperature of measurement. However, while much of the non-simple metal behaviour follows the same trends as in the simple metals there is really no justification for applying the nearly free electron theory to such cases and the form factors, etc., used should be regarded simply as adjustable scattering parameters.

Because of the linear relationship between ρ and the SRO parameters α_i (5.20), resistivity studies are often used to obtain information about the energies and kinetics of ordering or clustering processes and the effects of cold work and particle irradiation. In some cases the process can be described in terms of a single activation energy, as in the studies of Ag–(3–

Fig. 5.43. Relative resistivity change of Ni–41.4%Cu as a function of temperature. The calculations were done using the first seven α_i (●) and only α_1 (○) and the solid curve represents the experimental result (after Wagner *et al.* 1980).

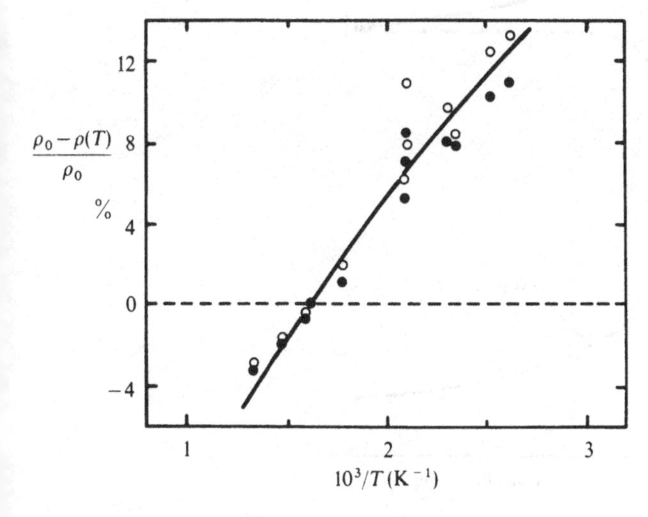

15)%Al by Meisterle & Pfeiler (1983*a*) (see also Pfeiler *et al.* 1984) or Cu–Ni by a number of different groups (Köster & Schüle 1957; Schüle & Kehrer 1961; van Royen *et al.* 1973; Bronsveld & Radelaar 1974; Poerschke & Wollenberger 1976, 1980; Poerschke *et al.* 1980; Charsley & Shimmin 1979). However, in other cases it is evident that either two or more coexisting processes are occurring or there is a distribution of relaxation times. For example, multiple relaxation times have been found necessary to explain the ordering kinetics in αCu–Zn (Trattner & Pfeiler 1983*a*, 1983*b*; Pfeiler *et al.* 1985), Cu–(5–20)%Mn (Segenreich *et al.* 1978; Reihsner & Pfeiler 1985); Cu–(5–20)%Al (Trieb & Veith 1978) and Ag–20%Al (Meisterle & Pfeiler 1983*b*). Kohl *et al.* (1983) found that the two relaxation times deemed necessary to fit the resistivity behaviour of Au–15%Ag could be attributed to thermal vacancy assisted diffusion promoting short range ordering and vacancy annihilation by migration to fixed sinks at low vacancy concentrations and by formation of di-vacancies at higher vacancy saturations. The resultant variation of resistivity with annealing time is shown schematically in Figure 5.44. Similar behaviour was found in Au–Pd (Lücke *et al.* 1976) and Ni–11% Cr (Heidsick *et al.* 1977), although, in the case of the Au–Pd alloys, and in contrast to Au–Ag, the excess vacancies were found to be annealed out

Fig. 5.44. Schematic variation of resistivity due to SRO and vacancy annihilation as a function of annealing time.

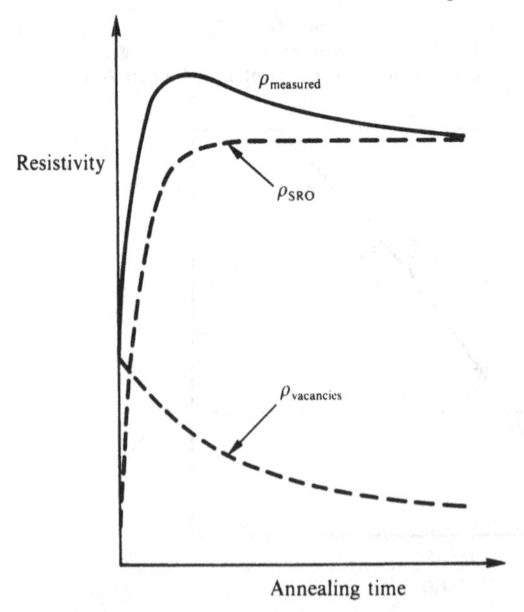

long before the equilibrium degree of SRO was attained. The importance of vacancy formation and annihilation in short range ordering processes was also noted in the resistivity study of αCu–Al by Eguchi *et al.* (1978). Other systems investigated in this way include Au–(1–25)%Cu (Korevaar 1961) and Ag–31%Zn (Chardon & Radelaar 1969). The complications associated with multiple relaxation times that can arise when particle irradiation is used to enhance diffusion has been discussed by Bartels *et al.* (1985) in relation to Au–15 at.%Ag alloys.

5.7.4 *Long range ordering*

The effects of long range atomic order on the resistivity have been most intensively studied in the Cu–Au system. However even in this case there has not been a comprehensive, systematic investigation of the variation of the long and short range atomic correlations with composition and temperature. As a first approximation we may use the Bragg–Williams model to determine S. The resistivity is then given by equation (5.59), (5.61) and (5.70). If it is assumed that the alloy has the maximum degree of order at all compositions (see Figure 5.11), the residual resistivity varies with composition as shown on the right hand side of Figure 5.45. The experimental results of Johansson & Linde (1936) are shown on the left hand side of this diagram. Note that the theoretical result is symmetrical about the equiatomic composition but the high ageing temperature used in the experimental study (200 °C) would have prevented the Au_3Cu ordered phase being noticed. (At lower ageing temperatures the limited atomic diffusion means that the Au_3Cu ordered phase may only be revealed by particle irradiation or some other diffusion-enhancing technique.)

The Bragg–Williams model could also be used to determine the variation of S with temperature. However, Anquetil (1962) has determined the equilibrium (i.e. at-temperature) variation of resistivity $\rho(S, T)$ and degree of long range ordered S with temperature. His resistivity results averaged with those of Sykes & Evans (1936), Nix & Shockley (1938), and Siegel (1940), and corrected for a residual resistivity in the completely ordered state of $0.7\mu\Omega$ cm are shown as the dotted curve labelled (a) in Figure 5.46. The variation of residual resistivity $\rho_0(S)$ deduced from these results has been averaged with the room temperature results of Wright & Thomas (1958) and Sykes & Evans (1936), scaled to give the same values of resistivity $\rho_0(0)$ for a specimen quenched from T_c and again corrected for the residual resistivity of $0.7\mu\Omega$ cm. This is shown as the dotted curve labelled (b) in Figure 5.46. In both cases the error bars give an indication of the actual spread in data.

In order to determine the resistivity from (5.74) it is necessary to deduce appropriate values for A, B/n_0 and $\rho_0(0)$. The value of $\rho_0(0)$ may be obtained directly from curve (b) and is 9.2$\mu\Omega$ cm. Coles (1960) has suggested that A should be ~0.5. From the general results given in

Fig. 5.45. Variation of resistivity with composition for long range ordered Cu–Au alloys. The experimental results of Johansson & Linde (1936) are shown in the full curve on the left, whereas the theoretical curve is shown on the right-hand side. The resistivity of a disordered alloy is shown as the dotted curve. The theoretical curve is scaled so that the disordered resistivities are the same at Cu–50%Au. The experimental data have been reduced by 1.7$\mu\Omega$ cm to give $\rho=0$ for pure Cu.

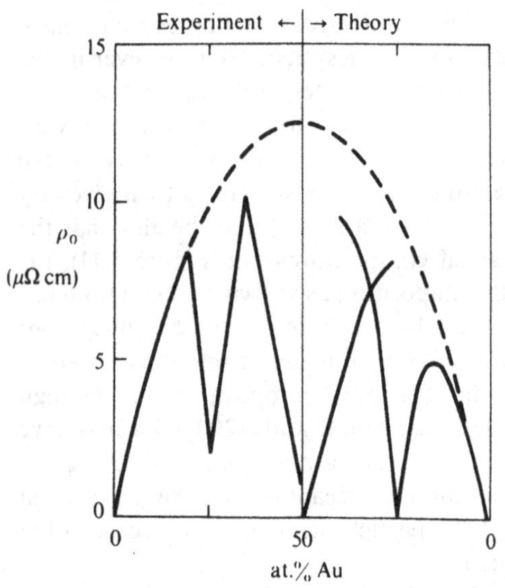

Fig. 5.46. Experimental (solid points) and theoretical (solid curve) variation of ρ with temperature for Cu_3Au: (*a*) at-temperature; (*b*) quenched.

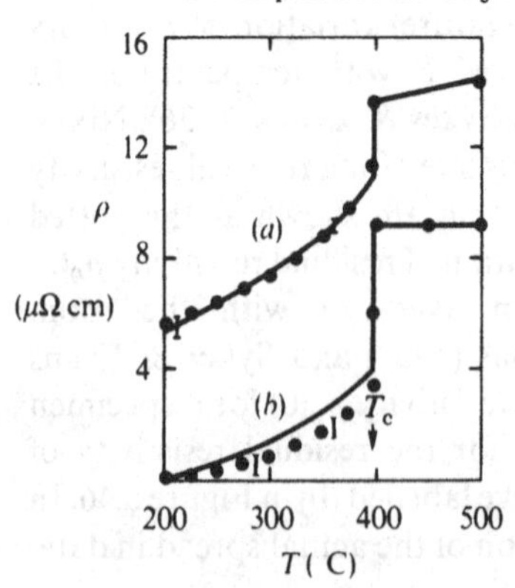

Figure 5.12 it is clear that the values of A largely determines the height of the discontinuity at T_c and for this study a value $A = 0.4$ will be used. The value of B/n_0 is then determined from the experimental temperature coefficient of resistivity and is $B/n_0 = 6.8 \times 10^{-3} \mu\Omega\,\text{cm}\,\text{K}^{-1}$. Using the variation of S with temperature determined by Anquetil, together with these values, leads to results shown as the solid lines in Figure 5.46. While there is a small discrepancy in the quenched results, the general shape of the curve is reproduced quite well and the at-temperature results are in almost complete agreement.

One of the few other materials that has been studied experimentally in sufficient detail so far is Fe_3Al. However, since Fe is not a simple metal the temperature dependence is complicated by the presence of the unfilled d-band which in effect gives rise to an increase in n_{eff} with increasing temperature. There is also the complication of the effects of magnetic ordering below the ferromagnetic Curie temperature. (These effects are discussed in more detail in the next chapter.) Nevertheless, Cahn & Feder (1960) have deduced the variation of resistance R with temperature expected if these complications did not arise. Their results are shown as the dotted curve in Figure 5.47. The variation of S with temperature may be deduced from their calculated values of the atomic scattering probability (see their fig. 7) and the assumption of a $1 - S^2$ dependence in accordance with (5.59) or (5.69). It is then possible to determine a value of A from (5.68), (5.71), (5.72) and their 'freedom of conduction electrons' result (see again their fig. 7). At low temperatures

Fig. 5.47. The experimental (dotted curve) and theoretical (solid curve) variation of resistance of Fe_3Al as a function of temperature: (*a*) at-temperature; (*b*) quenched.

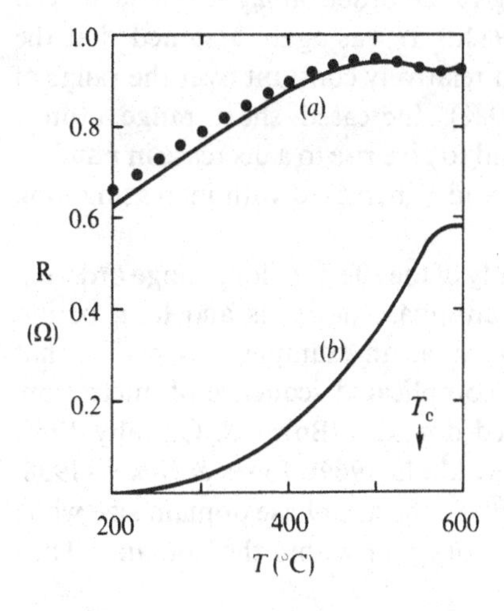

(i.e. below 450 °C) this gives $A \approx 0.7$. At higher temperatures the aforementioned effects of s-d scattering would lead to lower values of A. Having found A, the value of B/n_0 may then be determined from the temperature coefficient of resistivity which, again at lower temperatures, given $B/n_0 \approx 4 \times 10^{-4} \Omega\,K^{-1}$. The value of $R_0(0) = 0.58\Omega$ was taken from their experimentally determined value at 77 K (they quote resistances rather than resistivities). The predicted behaviour of R is then indicated by the full curves in Figure 5.47. The estimated at-temperature behaviour found by Cahn & Feder is reproduced remarkably well considering the approximations involved and the fact that there has been no attempt to optimise the parameters involved. The quenched behaviour is shown as curve (b) and although Cahn & Feder did not determine the $T \to 0$ K behaviour experimentally, it is of the same form as that found at 77 K. Use of an experimentally determined variation of S with temperature which was reported in a later study (Lawley & Cahn 1961) leads essentially to the same result. The observed change in behaviour from a monotonically decreasing ρ with decreasing temperature in quenched specimens to an increasing ρ with decreasing temperature just below T_c when determined at-temperature is properly reproduced by equation (5.74). The similar resistivity behaviour found in Ni_3Al by Corey & Lisowsky (1967) could presumably be explained in the same way.

In these discussions we have not included the effects of short range order which would modify the behaviour near T_c along the lines discussed in 5.3.1(b), resulting in the presence of a small cusp in the resistivity of Cu_3Au at T_c, for example. However, these corrections will be small and are not expected to change the general form of the results. The effects of both short and long range order on n_{eff} and τ have been investigated in FeCo (Rossiter 1981*a*). It was again assumed that the magnetic scattering was small and relatively constant over the range of measuring temperatures (4.2–250 K). Increased short range atomic correlations near T_c were then found to give rise to a decrease in τ and an increase in n_{eff}. Below T_c both n_{eff} and τ increased with increasing long range order.

Further complications in the study of the effects of long range ordering can arise with the formation of antiphase domains and long period superlattices. If we again take Cu_3Au as an example, it is known that ordering below T_c proceeds by a complicated sequence of nucleation, growth and coalescence of ordered domains (Burns & Quimby 1955; Nagy & Nagy 1962; Poquette & Mikkola 1969). Jones & Sykes (1938) investigated the dependence of ρ upon the antiphase domain size while maintaining an equilibrium degree of order within the domains. They found that ρ decreases with increasing domain size εa_0 for $\varepsilon a_0 > 100$ Å. The resistivity varies linearly with $1/(\varepsilon a_0)$ in this size range, suggesting that this contribution to ρ is due to a scattering of electrons from the domain boundaries. For $\varepsilon a_0 < 100$ Å the resistivity becomes progressively independent of ε, approaching the disordered value as $\varepsilon \to 0$. As suggested by Jones & Sykes this transition in behaviour occurs when the domain size is comparable to Λ. Burns & Quimby (1955) measured the variation of ρ with ageing time at different ageing temperatures and found that, for ageing temperatures in the range T_c to 360 °C, ρ passes through a maximum as ageing proceeds and that the position of the maximum moves to shorter ageing times at lower ageing temperatures. The presence of such a peak may now be attributable to an increase in ρ_0 as the domains grow to a size comparable to Λ_l (and hence the degree of SRO increases) and then a decrease in ρ_0 as they continue to grow and the electrons encounter larger and larger regions of long range order as discussed at the end of Section 5.4. Rossiter & Bykovic (1978) found a contribution to $d\rho/dT$ in Cu_3Au that depended upon the *deviation* in ρ from the value expected of a homogeneous LRO state. It is again likely that this deviation is the result of scattering from antiphase domain boundaries and so the additional contribution to $d\rho/dT$ may come from the effects of finite Λ discussed at the end of Section 5.4. However, as this particular study did not include a determination of the size of the antiphase domains, no detailed comparison with theory is possible. A review of ordering in Cu_3Au and some of the earlier resistivity measurements is also given in that paper. It is also likely that the phonon spectrum will be somewhat order dependent (see e.g. Gilat 1970; Evseev & Dergachev 1976; Martin, D. L. 1972; Hartmann 1968; Synecek 1962; Towers 1972) leading to an additional dependence of $d\rho/dT$ on the state of order.

5.7.5 *Precipitation*

There has not as yet been any complete calculation of the resistivity during precipitation which properly encompasses both the early nucleation and growth and subsequent coarsening stages. Such a calculation would need to be based on the iterative solution of the Boltzmann equation described in Section 5.2.2 in order to account for the mean free path effect at intermediate zone sizes, and use some suitable model to determine the resistivity of the multiphase solid at large zone sizes, as discussed in Section 5.5. Nevertheless some aspects of the problem have received detailed attention. For example, *ab-initio* calculations of the effects of anisotropy in Al–Zn, Al–Ag, Al–Cu and Cu–Be have been reported by Edwards & Hillel (1977) using fitted point-ion pseudopotentials and the Ziman approximation (1.68) to take account of relaxation time anisotropy. Some of these results were applied in detail to ageing in the Al–Zn system by Hillel & Edwards (1977) with the aid of the Hillel–Edwards–Wilkes (1975) phenomenological model. The effects of relaxation time anisotropy have also been investigated using one-OPW models by Guyot & Simon (1977) and Luiggi (1984). Yonemitsu & Matsuda (1976) used a 4-OPW model to investigate anisotropy effects in Al–Ag. In general it has been confirmed that an increased anisotropy of scattering does lead to a reduction in the resistivity below the isotropic value as the zone size increases. At large zone sizes $R \gg \Lambda$ it can also lead to a decrease in ρ with increasing zone size. The effects of the three different atomic environments at larger zone sizes, i.e. the matrix, the zone and the matrix–zone boundary, have been considered by Merlin & Vigier (1980) and Glässer & Löffler (1984). Merlin & Vigier assumed a simple law of mixtures for these three components (i.e. the lower bound for conductivity in (5.78) or $T11$ in Table 5.3), whereas Glässer & Löffler employed an effective medium approach ($T1$) and emphasised the importance of the local electron mean free path. In another study, Glässer (1984) employed the t-matrix instead of the pseudopotential scattering matrix (see Section 6.4) but the results are only expected to be valid in the limit of small zone sizes due to the omission of finite mean free path and scattering anisotropy effects.

The influence of the measuring temperatures has been investigated by Perrin & Rossiter (1977), Vigier & Merlin (1983) and Vigier *et al.* (1983). Some typical results are shown in Figures 5.48 and 5.49, confirming the

Fig. 5.48. Electrical resistivity of Al–5.3at.%Zn as a function of time of ageing at 45 °C. Solid curve: $T_m = 318$ K; dashed curve: $T_m = 77$ K (after Vigier & Merlin 1983).

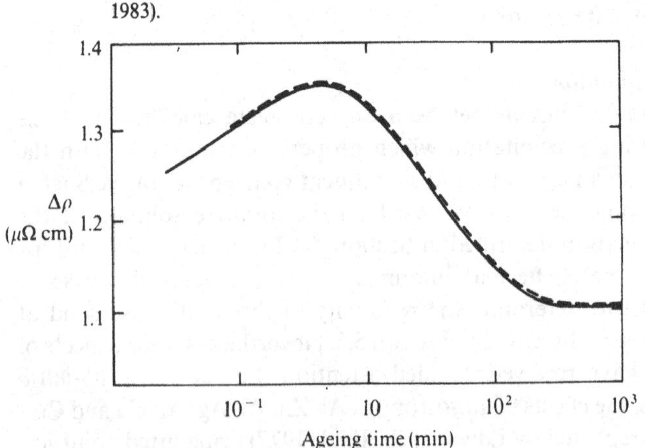

conclusion of Section 5.2.2 that T_m has little effect on the resistivity peak if the zones are spherical but that the peak shifts to higher ageing times (i.e. larger zone sizes) with an increase in T_m if the zones are more two-dimensional. A complete investigation of this effect must include the determination of Λ_l as well as the zone size distribution and the effects of the strain fields around the 'platelet' zones.

Mimault *et al.* (1978, 1981) Radomsky & Löffler (1978, 1979*a*,*b*) and Löffler & Radomsky (1981) have proposed methods of determining the resistivity of alloys during GP zone formation and spinodal decomposition using experimental or theoretical small angle X-ray scattering data to determine the structure factor. However, again such approaches are strictly only valid in the small zone limit since they do not take proper account of the effects of finite electron mean free path or scattering anisotropy.

Finally, we mention some results that relate to the size-induced deviations from Matthiessen's rule discussed at the end of Section 5.2.2. As noted there, this is just another way of looking at the variation in the resistivity peak with measuring temperature. Perrin & Rossiter (1977) determined the temperature dependence of the resistivity of an Al–6.8 at.%Zn alloy at various stages of decomposition. While there was no detectable variation in ageing time (i.e. particle size) corresponding to the resistivity peak, the temperature coefficient of resistivity was observed to change by a small amount over a limited temperature range. Values of $d\rho/dT$ at 40 K for various ageing times are shown in Figure 5.50. While the lack of any determination of particle size precludes a

Fig. 5.49. Electrical resistivity of Al–1.72 at.%Cu as a function of time of ageing at 40 °C. Solid curve: $T_m = 40$ °C; dashed curve: $T_m = 77$ K (after Vigier *et al.* 1983).

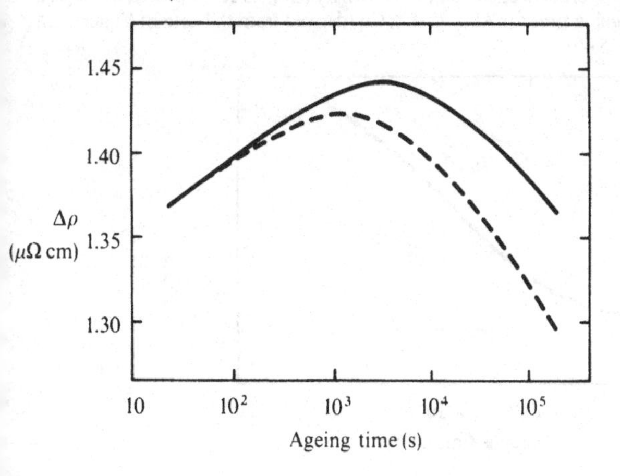

detailed comparison with the model calculations reported in Section 5.2.2, the general form of the result is similar to those given by the solid lines in Figure 5.9(*a*). This behaviour was observed over the temperature range 30 K–50 K where the rate of change of ρ with T was largest and where there is likely to be the greatest sensitivity of the resistivity to such microstructural effects. However, the change is not large and there is a need to obtain more data on this and other systems over the low temperature region. By contrast the Al–1.72 at.%Cu results of Vigier *et al.* (1983) are replotted into this format in Figure 5.51. Here the change in $d\rho/dT$ corresponds to the dashed line in Figure 5.9(*a*) indicating that scattering anisotropy is playing an important role.

Similar effects have been observed in alloys containing transition metals (see e.g. Kovacs & Szenes 1971; Szenes *et al.* 1971; Jones *et al.*

Fig. 5.50. Temperature coefficient of resistivity $d\rho/dT$ measured at 40 K as a function of ageing time for Al–6.8at.%Zn (after Aubauer & Rossiter 1981).

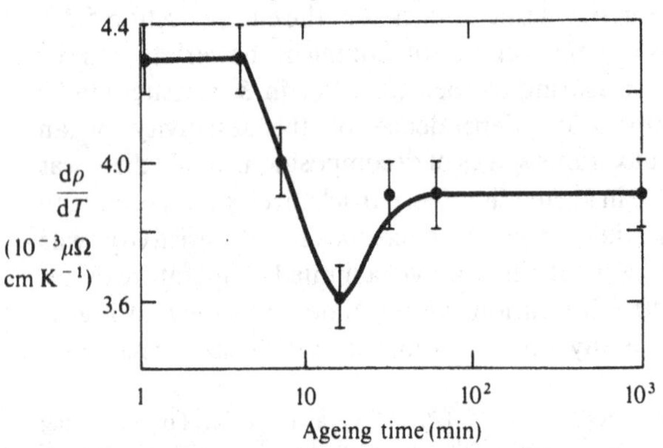

Fig. 5.51. Temperature coefficient of resistivity ($\Delta\rho$(313 K – $\Delta\rho$(77 K))/236 as a function of ageing time for Al–1.72at.%Cu (derived from the data of Vigier *et al.* 1983, Figure 5.49).

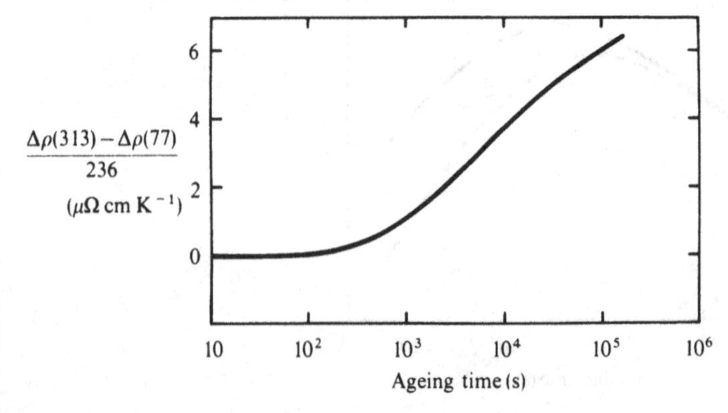

1971; Silcock *1970a,b*). However, as noted in Section 5.2.2, the uncertainty about the intrinsic temperature coefficients of resistivity of the zones and matrix, coupled with possible magnetic effects (see Chapter 7), makes a direct comparison of dubious value.

Similar resistivity peaks have also been found in systems undergoing spinodal decomposition (see e.g. the results of Larsson 1967; Mimault *et al.* 1978) and it is suggested that the origin of these peaks again lies also in mean free path and anisotropy effects. However, few results are available precluding any detailed analysis at this stage.

5.7.6 *Long range phase separation*

While the variation of resistivity with composition has been determined all the way across the phase diagrams of many alloys (see e.g. Schröder 1983), there are few instances where the metallurgy of the specimens has been characterised in sufficient detail to allow an investigation of the merits of the various formulae for two phase materials described in Section 5.5. Nevertheless, eutectic alloys have come under some scrutiny in this regard, presumably because of the comparatively simple geometries of the phase distributions involved. We will start by considering the case of phase distributions that are large on a scale of the mean free path so that size effects can be ignored.

(a) *Scale of phase separation* $\gg \Lambda$

Watson *et al.* (1975) fabricated Bi–Cu composite alloys with specific volume fractions and spatial distributions of the minor phase (Cu). A comparison of the resistivities measured perpendicular to the fibres at 77 K and those calculated from the Raleigh formula (5.95) is given in Figure 5.52. As the data were obtained from specimens having two

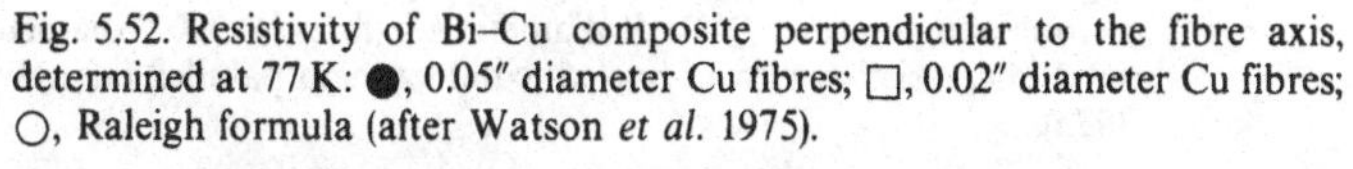

Fig. 5.52. Resistivity of Bi–Cu composite perpendicular to the fibre axis, determined at 77 K: ●, 0.05″ diameter Cu fibres; □, 0.02″ diameter Cu fibres; ○, Raleigh formula (after Watson *et al.* 1975).

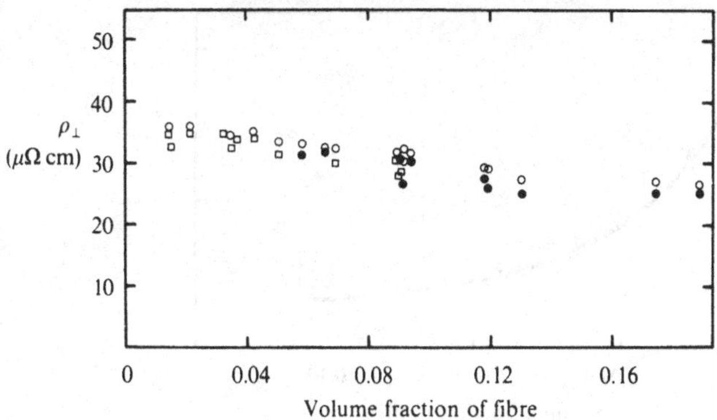

different fibre diameters (both of which were much larger than Λ) there appears to be no discernible variation that could be ascribed to fibre–matrix interface scattering effects.

The resistivity determined with the current parallel to the fibres is shown in Figure 5.53 and is fitted to the lower bound (i.e. the upper conductivity bound) of (5.78) representing the phases in parallel. Given the apparent reliability of these two results, Watson *et al.* devised some electrical analogue models for a complex three-dimensional fibre array having discontinuous fibres (see Section 5.5.1) and which appear to give a good fit to data obtained from both artificially prepared specimens and a series of InSb–Sb eutectic alloys investigated previously by Liebmann & Miller (1963). In another study, Landauer (1952) has compared the effective medium result for random spheroids (equation *T*3 in Table 5.2) with experimental data for a wide range of different two-phase mixtures. In many cases ($Bi–Bi_2Tb$, Cd–Pb, Cu_2Sb–Sb, Cu–Fe, Pb–Sb) excellent agreement was obtained, as indicated in Figure 5.54 and 5.55.

In other cases (Mg_2Pb–Pb, Bi–Sn) the agreement was not quite as good, although still quite acceptable, whereas in some others (Sn–Te, Ag–Bi, Mg_2Sn–Sn, $MgZn_2$–Mg) the reported agreement (usually at one end of the composition range), was poor, possibly because of changes to the electronic structure. Hashin & Shtrikman (1962) have reconsidered the Mg_2Pb–Pb data and found that it lies within the bounds 5.79(a) and 5.80(a), as shown in Figure 5.56.

The electrical resistivity of some directionally solidified Al–Al_3Ni, Al–Al_6Fe and Al–Al_9Co eutectics has been measured as a function of

Fig. 5.53. Resistivity of Bi–Cu composite parallel to the fibre axis, determined at room temperature: ○, 0.05″ diameter Cu fibres; ●, 0.02″ diameter Cu fibres; solid line, resistivity lower bound (conductors in parallel) (after Watson *et al.* 1975).

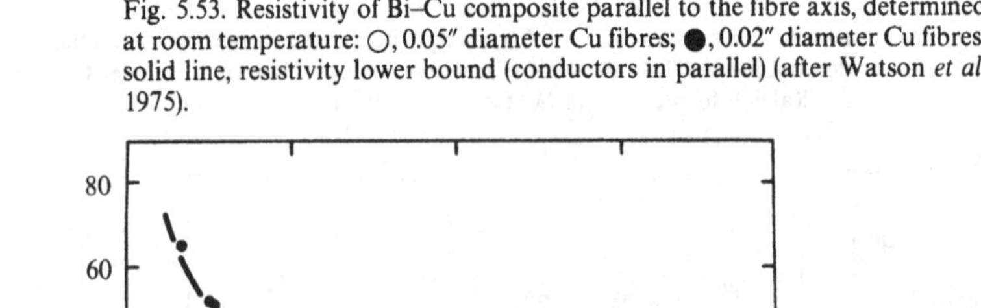

temperature over the range 77–300 K by Jenkins & Arajs (1983). The measurements were made with the current parallel ($\rho_{\parallel}$) and perpendicular ($\rho_{\perp}$) to the fibre axis. Various different formulae were used to obtain the resistivities of the components. In general it was found that

Fig. 5.54. Resistivity of two-phase Cd–Pb mixture as a function of volume fraction of Pb: continuous line, theory; points, experimental values (after Landauer 1952).

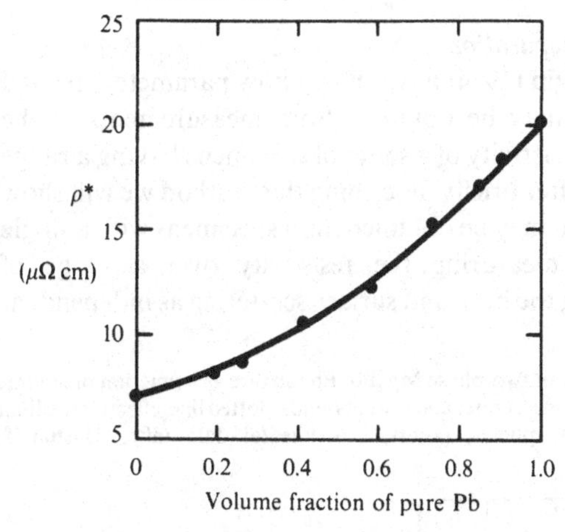

Fig. 5.55. Resistivity of two-phase Cu_2Sb–Sb mixture as a function of volume fraction of Sb; continuous line, theory; points, experimental values (after Landauer 1952).

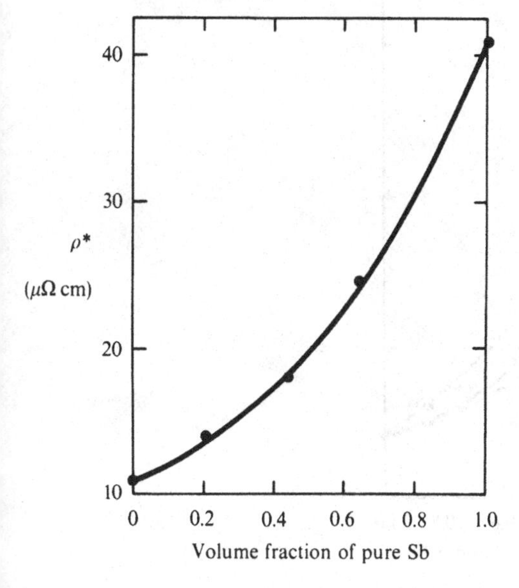

the resistivity of the dispersed phase so determined was very sensitive to the accuracy of the measurements. Nevertheless, their results show that the effective medium theory for aligned fibres (Table 5.3, equation $T10$) gives apparently reliable data. In all cases the resistivity anisotropy $\rho_{\perp}/\rho_{\parallel}$ was found to increase markedly at low temperatures, indicating the increasing importance of size effect scattering from the interfaces. It is to this problem that we now turn.

(b) *Scale of phase separation* $\leqslant \Lambda$

Simoneau & Bégin (1974) have shown how parameters related to size effect scattering may be obtained from measurements of the parallel and transverse resistivity of a series of specimens having a range of interphase spacings. After briefly describing this method we will show how the same information may be obtained from specimens with a single interphase spacing by measuring the resistivity over a range of temperatures. By treating the bulk and surface scattering as independent

Fig. 5.56. Resistivity of two-phase Mg_2Pb–Pb mixture as a function of volume fraction of Pb: solid lines, upper and lower bounds; dotted line, effective medium theory for random spheres; points, experimental data (after Hashin & Shtrikman 1962).

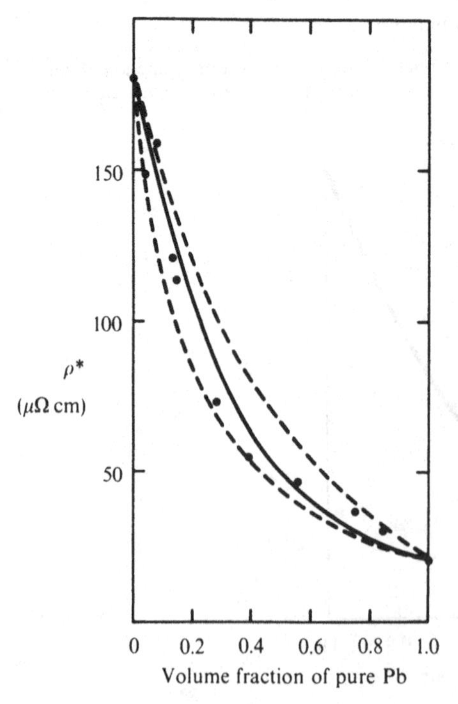

processes (i.e. neglecting cross-terms in the structure factor) the resistivity of a material with a mean spacing d between surfaces may be written in the form (5.114a). For a well-aligned array of scattering rods in an isotropic matrix, Simoneau & Bégin take the view that, if the rods are aligned in the z direction, scattering in the xy plane is always diffuse because of the isotropic distribution of curvatives of these surfaces. When the current is parallel to the rods they assume that a fraction of the scattering α will be diffuse. This point of view is at variance with the physical interpretation of the specularity parameters in terms of phase shifts induced by surface irregularities. In fact the angular dependence of the specularity parameter must be included at an earlier stage of the calculation (see e.g. Ziman 1960, ch. XI; Soffer 1967; Sambles & Preist 1982). Nevertheless they write (5.114a) as

$$\frac{\rho(d,t)}{\rho_\infty}=1+\frac{\alpha\Lambda_\infty}{d}. \tag{5.219}$$

It is thus suggested that their parameter α should be interpreted as a measure of the effective number of small angle scattering events that cause diffuse scattering at the boundary, i.e. it is a measure of the angle dependence of the specularity parameter p, in particular its rate of decrease as the angle of incidence decreases from $\theta=90°$ (grazing incidence). In a two-phase solid, the resistivities in the directions parallel and transverse to the fibres are given by equations of the form $T13$–$T15$ and $T9$, $T10$ shown in Table 5.3. Simoneau & Bégin write these in the general form (assuming that the matrix is the more conducting phase)

$$\rho_\parallel(D)=\frac{\rho_\mathrm{m}(D)}{\chi_\parallel}, \tag{5.220}$$

$$\rho_\perp(D)=\frac{\rho_\mathrm{m}(D)}{\chi_\perp}, \tag{5.221}$$

where the subscript m denotes the bulk matrix value and D is the mean fibre spacing. The proportionality constants $\chi_\parallel$, $\chi_\perp$ are related to the geometrical arrangement of the fibres as well as the resistivity of the matrix and the fibres. For a small volume fraction v_m of high resistivity fibres, $\chi_\parallel\approx v_\mathrm{m}$. By using (5.219) and the assumption that the mean free path should be proportional to the width of the rods d and their density ($\propto 1/D^2$), these equations become

$$\rho_\parallel(D)=\frac{\rho_\mathrm{m}}{\chi_\parallel}+\alpha\frac{\rho_\mathrm{m}\Lambda_\mathrm{m}d}{\chi_\parallel D^2}, \tag{5.222}$$

$$\rho_\perp(D)=\frac{\rho_\mathrm{m}}{\chi_\perp}+\frac{\rho_\mathrm{m}\Lambda_\mathrm{m}d}{\chi_\perp D^2}. \tag{5.223}$$

For a series of fibre composites with fixed volume fraction of fibre v_f, the ratio D/d is a constant and equations (5.222) and (5.223) predict that $\rho_{\parallel}(D)$ and $\rho_{\perp}(D)$ should vary as $1/D$. The quantities ρ_m, $\chi_{\perp}$ and $\rho_m\Lambda_m$ can then be obtained from plots of $\rho_{\parallel}(D)$ and $\rho_{\perp}(D)$ as a function of $1/D$, as shown in Figure 5.57, with the $1/D \to 0$ intercepts and experimental value of v_f giving ρ_m and $\chi_{\perp}$ and the ratio of slopes giving α. In the study of these parameters in a Al–Al_3Ni eutectic, they found that both α and $\rho_m\Lambda_m$ increased with temperature. This is contrary to *a-priori* expectations since p is determined by the surface roughness and $\rho_m\Lambda_m$ is determined simply by the free area of the Fermi surface. However, it could be explained by an angle-dependent specularity parameter, as discussed in Section 5.2.2.

If a number of specimens with widely varying lamellar or rod spacing D are not available, an alternative analysis is possible by determining the resistivity as a function of temperature over the range from 4.2 K to above Θ_D. Equations (5.222) and (5.223) may be written as

$$\rho_{\parallel}(D,T) = \frac{\rho_m(T)}{\chi_{\parallel}} + \alpha\frac{\rho_m\Lambda_m d}{\chi_{\parallel}D^2}, \tag{5.224}$$

$$\rho_{\perp}(D,T) = \frac{\rho_m(T)}{\chi_{\perp}} + \frac{\rho_m\Lambda_m d}{\chi_{\perp}D^2}, \tag{5.225}$$

Fig. 5.57. Parallel, ●, and perpendicular, ○, resistivities of a well-aligned Al–Al_3Ni eutectic as a function of inter-rod distance D (after Simoneau & Bégin 1974).

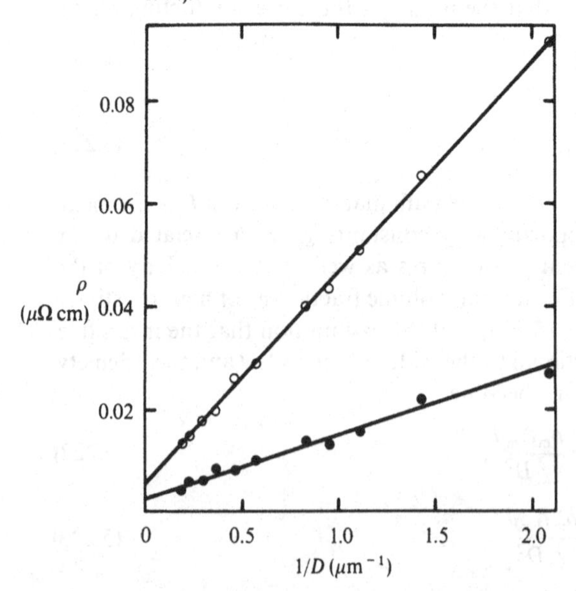

where it has been assumed that $\chi_{\|}$ and $\chi_{\perp}$ are temperature independent. This is a reasonable assumption provided that the matrix constitutes the larger volume fraction and that the included phase is a reasonably good conductor. The latter requirement excludes composites which incorporate semiconducting phases from consideration. We could add to these equations the enhanced surface scattering $\rho_s(D, T)$ of Olsen (equaton (5.113)). However, in many eutectics this turns out to be some orders of magnitude smaller than the observed resistivities and so will be neglected for the moment. From equations (5.224) and (5.225) we find

$T > \Theta_D$

$$\frac{\dfrac{\partial \rho_{\|}(D, T)}{\partial T}}{\dfrac{\partial \rho_m(T)}{\partial T}} = \frac{1}{\chi_{\|}}, \tag{5.226}$$

$$\frac{\dfrac{\partial \rho_{\perp}(D, T)}{\partial T}}{\dfrac{\partial \rho_m(T)}{\partial T}} = \frac{1}{\chi_{\perp}}. \tag{5.227}$$

$T \to 0$

$$\rho_{\|}(D, 0) = \alpha \frac{\rho_m \Lambda_m d}{\chi_{\|} D^2}, \tag{5.228}$$

$$\rho_{\perp}(D, 0) = \frac{\rho_m \Lambda_m d}{\chi_{\perp} D^2}. \tag{5.229}$$

Provided that $\rho_m(T)$ is known, all constants can be obtained. The above equations are based on the validity of Matthiessen's rule and it is known that there are significant size-dependent deviations from this rule. In fact, by comparing the experimentally determined $\rho(D, T)$ curves with those calculated from (5.224) and (5.225) over all temperatures between $\sim$4.2 K and Θ_D reveals deviations of the form shown in Figure 5.58. These are of exactly the same form as found in pure metal fibres and thin wires (Holwech & Jeppesen 1967; von Bassewitz & Mitchell 1969) and are expected from either the Fuchs–Sondheimer angle-independent p models or from the Ziman–Soffer model which includes an angle dependence of p, as discussed in Section 5.5.2 (see also Sambles & Elsom 1980; Sambles & Preist 1982). However, as these deviations exist only over a limited temperature range it is expected that the analysis described above (which relies on the high and low temperature limits) should remain valid. It is a relatively straightforward matter to incorporate an angle-dependent specularity using Soffer's equations. In the limit $D \gg \Lambda_m$, the appropriate expression for A in (5.114a) may be taken from Table 5.4. One may then employ either form of the analyses described above to find the parameter ξ ($\propto h/\lambda_k$). For smaller D there is no simple analytic form of results that would lead to simple expressions of the form (5.224) or (5.225). Nevertheless, it would be quite feasible to perform the necessary numerical integrations and fit the data with an appropriate choice of the parameter h/λ_k. This done, neither $\rho_m \Lambda_m$ nor the specularity parameter should need to be temperature dependent.

By way of example, the temperature dependences of the transverse and parallel resistivities of an Al–Al_6Fe eutectic are shown in Figure 5.59. These are shown fitted to the resistivity for pure Al (assumed for $\rho_m(T)$) multiplied by a constant thereby determining the parameters $\chi_\parallel$ and $\chi_\perp$. Note that the discrepancy between $\rho_\perp(D, T)$ and $\rho_{Al}(T)/\chi_\perp$ or $\rho_\parallel(D, T)$ and $\rho_{Al}(T)/\chi_\parallel$ is indeed of the form shown in Figure 5.58. From the ratio of $\rho_\parallel(D, 0)\chi_\parallel$ to $\rho_\perp(D, 0)\chi_\perp$ at $T \to 0$, one obtains directly the value for the parameter α (equations (5.228) and (5.229)). From Figure 5.59 this is ~ 0.5. The parameter $\chi_\parallel$ should be given by the reciprocal of the volume fraction of matrix. Results for a series of Al–Al_6Fe specimens having a range of D spacings from 0.53 to 0.98 are shown in Table 5.28. These give an average volume fraction of matrix determined from $\chi_\parallel$ of 0.89 ± 0.03 and may be compared with the volume fraction obtained by metallographic analysis of 0.920 ± 0.005. Where such a range of D spacings are available, further checks are possible from plots of $\rho_\parallel(D, 0)\chi_\parallel$ and $\rho_\perp(D, 0)\chi_\perp$ against $1/D$. If the matrix contains significant levels of impurity, the additional resistivities $\rho'/\chi_\parallel$ and $\rho'/\chi_\perp$ should be added to (5.224) and (5.225) respectively. However, values of α may still be

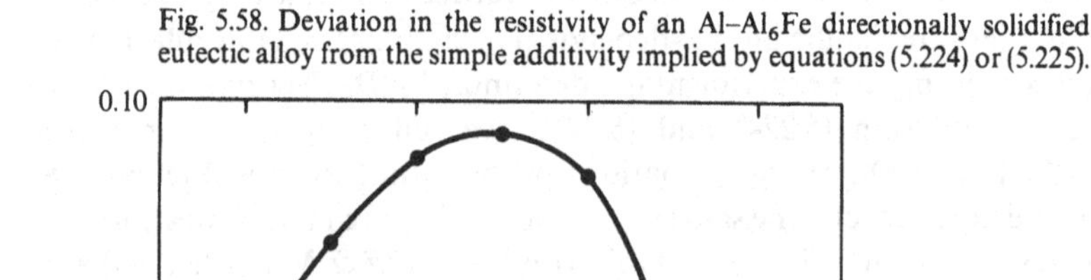
Fig. 5.58. Deviation in the resistivity of an Al–Al_6Fe directionally solidified eutectic alloy from the simple additivity implied by equations (5.224) or (5.225).

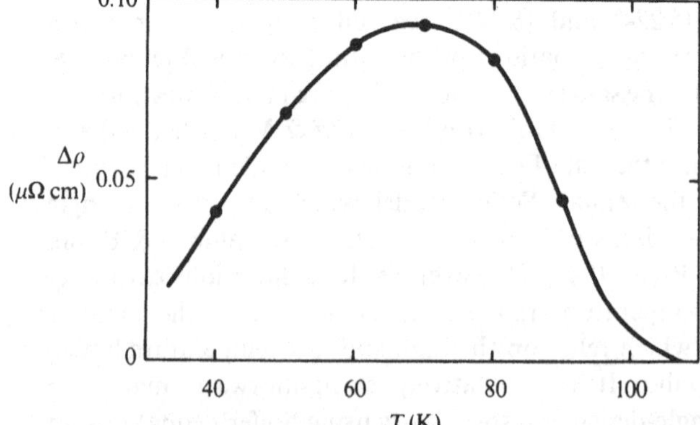

Table 5.28. *Values of the parameter* $\chi_\parallel$ *(proportional to the reciprocal of the volume fraction of matrix) for Al–Al_6Fe specimen produced at different growth rates*

Growth rate (mm s^{-1})	Fibre spacing (μm)	$\chi_\parallel$
218	0.98	1.15
218	0.95	1.11
765	0.72	1.12
765	0.63	1.16
896	0.58	1.17
896	0.53	1.09

Fig. 5.59. Temperature dependence of the perpendicular and parallel resistivities of an Al–Al_6Fe directionally solidified eutectic. The points are experimental data and the solid lines are $\rho_{Al}(T)/\chi_\perp$ and $\rho_{Al}(T)/\chi_\parallel$ for the perpendicular and parallel resistivities respectively (G. V. Thomson, unpublished).

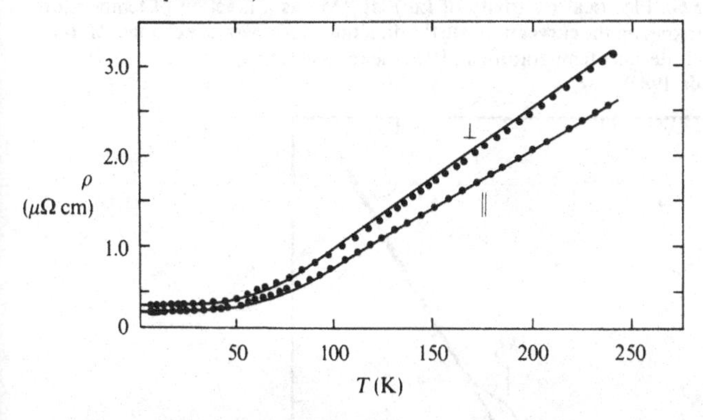

obtained from the gradients of such plots and known values of $\rho_\infty\Lambda_\infty$. The resistivity of the included phase may be obtained from some suitable expression for the composite resistivity (see Tables 5.2, 5.3) in the size or temperature range where the effects of interface scattering are much reduced (i.e. D, $d > \Lambda$).

Should the rods or lamellae be the more highly conducting phase, equations (5.222) and (5.223) can simply be rewritten as

$$\rho_\parallel(d, T) = \frac{\rho_f(d, T)}{\chi_\parallel}, \tag{5.230}$$

$$\rho_\perp(d, T) = \frac{\rho_f(d, T)}{\chi_\perp}, \tag{5.231}$$

Table 5.29. *Summary of some of the studies of the electrical resistivity of eutectics*

System	T_m (K)	Reference	Comments
Bi–MnBi	77–30	Yim & Stofko (1967)	Single (d (~4 μm), anisotropic matrix
Bi–Ag	77–300	Digges & Tauber (1970)	anisotropic matrix
Al–Ni (2–28wt.%Ni)	42, 77, 297	Simoneau & Bégin (1973)	includes hypo- and hyper-eutectics
Al–Al_3Ni	4.2, 77	Simoneau & Bégin (1974)	$D = 0.5$–3.5 μm
Al–Al_3Ni	77–300	Jenkins & Arajs (1983)	Single *d*
Al–Al_6Fe	77–300	Jenkins & Arajs (1983)	Single *d*
Al–Al_9Co	77–300	Jenkins & Arajs (1983)	Single *d*
Ag–Ca	17, 77, 298	Frommeyer & Brion (1981)	$d = 100$ Å–1 μm

Fig. 5.60. Electrical resistivity of Li–10at.%Mg as a function of temperature. The arrows on the curves indicate the direction of temperature change, M_s is the martensite start temperature and M_r the reversion temperature (after Oomi & Woods 1985).

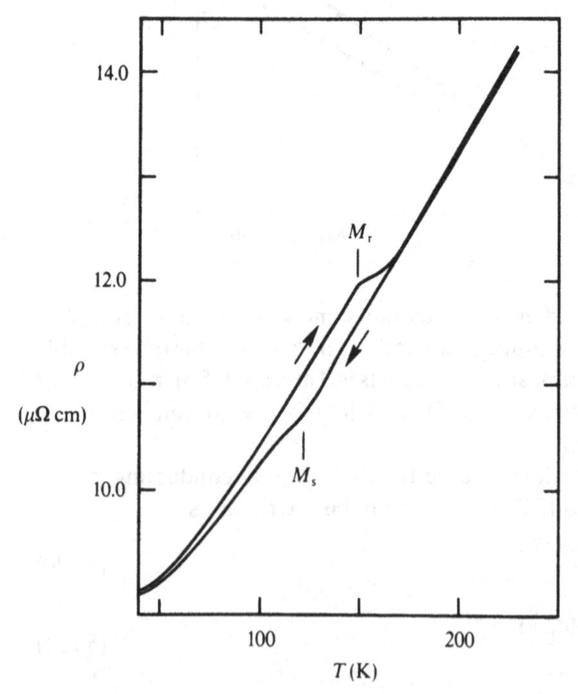

where d is the appropriate thickness. The usual thin film or thin wire expressions can be used to determine ρ_f. For example, in the case of conducting rods with diameters $d \gg \Lambda_1$,

$$\rho_{\parallel}(d, T) = \frac{\rho_{\infty}(T)}{\chi_{\parallel}} + \alpha \frac{\rho_{\infty}\Lambda_{\infty}}{\chi_{\parallel} d}, \tag{5.232}$$

$$\rho_{\perp}(d, T) = \frac{\rho_{\infty}(T)}{\chi_{\perp}} + \frac{\rho_{\infty}\Lambda_{\infty}}{\chi_{\perp} d}, \tag{5.233}$$

where the subscript ∞ refers to the bulk properties of the fibre. Thus $\rho_{\parallel}$ and $\rho_{\perp}$ should vary as $1/d$ and this variation or the variation with T can be used to extract the required information (although if v_f is small and the resistivity of the matrix high, the temperature dependence of the matrix in the transverse resistivity would need to be taken into account). For smaller d a full solution involving the angle-dependent specularity is again required. Some details of experimental studies of electrical properties of eutectic alloys are summarised in Table 5.29.

5.7.7 *Displacive transitions*

As noted in Section 5.6.4, displacive phase transitions are expected to influence the electrical resistivity in much the same way as replacive effects. As an example of the type of behaviour observed, the temperature dependence of the resistivity during the formation and reversion of martensite in a Li–10%Mg alloy is shown in Figure 5.60. However, detailed analysis of such data is again dependent upon a proper characterisation of the structure in terms of the short and long range correlation parameters, the size and size distribution of the domains, etc., and there is still much work to be done in this area.

6 Non-simple, non-magnetic metals and alloys

The previous chapter was concerned with the electrical resistivity of simple metals and alloys. The result was a quantitative understanding of much of the behaviour associated with dilute and concentrated simple-metal alloys in terms of the electronic scattering processes and microscope phase distribution. Unfortunately there has been less progress in the understanding of non-simple metals and alloys. This is not especially due to any lack of a basic understanding of the processes concerned, but rather to the complicated nature of the techniques that are currently employed. In the case of simple metals the analysis was enormously simplified by the introduction of a weak scattering formalism. This was manifest at many different stages of the calculation: the approximate solution of the Boltzmann equation, the use of perturbation theory, the assumption of nearly free electrons and linear screening, for example. In alloys containing transition metals, however, the distinction between the valence and core states is no longer so clear-cut. The d-states in the transition metals cannot simply be bundled into the core region and incorporated into some smoothly varying lattice pseudopotential. The wavefunctions associated with these states have significant amplitude outside the atomic core region and play an important role either by overlapping with those from adjacent atoms to form a narrow d-band or via the effects of hybridisation with the sp conduction bands. As a further consequence the Fermi surface may be quite irregular and have some parts that are s-like and others that are d-like. The complications introduced by these factors lead to a considerable increase in the complexity of the calculations so that more sophisticated methods of analysis are involved. As a result, the research papers concerned are often presented in rather abstract notation, to the extent that an intelligent though non-expert worker in the field may find it difficult if not impossible to 'spot the physics'. Quite often it may not be clear what approximations have been introduced, and results which may allegedly be appropriate to some real alloy system may have in fact been obtained from some quite unrealistic model density of states, for example. It is thus quite difficult for a non-expert to know what, if anything, has been achieved by these more elaborate theories and when, if at all, he or she can have confidence in the results produced. The purpose of this chapter is to give a brief overview of some of these techniques and indicate the current state of progress in relation to determining the electrical resistivity of real, non-simple metals and alloys. In passing we will also gain further insight into the nature and validity of the pseudopotential approximation.

6.1 Band structure and the electrical resistivity

Before we consider some of the theories appropriate to electron scattering in non-simple metals, we want to consider the general importance of the effects of a changing band structure on the electrical resistivity. Here we will continue to use the terminology of 's-bands' and 'd-bands', although it is recognised that s–d hybridisation may render some parts of the Fermi surface s-like and other parts d-like even though these regions may lie within the same Brillouin zone, in which case the distinction is quite artificial. The significance of alloying effects on the band structure, and hence the electrical resistivity, was initially recognised by Mott (see Mott & Jones 1936) who stressed the importance of the high density of states in the d-band (or d-like region of the Fermi surface). He assumed that the current was largely carried by the s electrons since these have an effective mass similar to that of the conduction electrons in copper, while the d electrons have a much higher effective mass and therefore a much lower mobility. Impurities, phonons and electron–electron interactions will cause a scattering of these s electrons into vacant s- and d-states but, since the scattering probability depends upon the density of states into which the electrons are scattered (equation (1.2)), s–d scattering occurs much more frequently than s–s scattering. Electrons scattered into the d-band do not contribute significantly to the conduction because of their increased effective mass and so the s–d scattering mechanism could expain the high resistivities of the transition metals and also some of the qualitative effects of alloying, as discussed in Section 6.7. A rapid change in $N_d(E_F)$ with increasing energy can also lead to a modification of the temperature dependence of resistivity. This is because a thermal broadening of the Fermi surface of $\sim k_B T$ can then produce a significant change in $N_d(E_F)$. Jones (1956) has shown that such an effect would lead to an additional temperature-dependent term of the form

$$\rho(T)=\rho_0(1-AT^2), \tag{6.1}$$

where

$$A=\frac{\pi^2 k_F^2}{6}\left[3\left(\frac{1}{N(E_F)}\,\frac{dN(E_F)}{dE}\right)^2-\frac{1}{N(E_F)}\,\frac{d^2N(E_F)}{dE^2}\right]. \tag{6.2}$$

Note that this term arises from *impurity* scattering into the d-band and is in addition to the usual phonon terms. Such a mechanism can explain why the resistivity of transition metals falls below the linear variation with temperature expected in simple metals at high temperatures. However, a calculation of the electronic structures and the s–s and s–d scattering rates in non-simple alloys is a difficult problem, particularly if the alloy is concentrated. Furthermore, recent calculations are beginning to throw some doubt on the basic assumptions of this simple model (see Section 6.7.2). There is an additional temperature dependence of the resistivity that results from the presence of a partially filled d-band. As we have discussed previously, the d electrons are largely localised at the transition metal sites and are screened by conduction electrons. However, the conduction electrons are scattered by the screened Coulomb electron–electron interaction. Because of the higher effective mass of the d electrons, the s electrons may suffer a large change in velocity (it is assumed that the d electrons lose their momentum to the lattice). As both the d and s electrons are subject to Fermi–Dirac statistics, the maximum energy an s electron can lose is $\sim k_B T$, all lower energy states being filled. The number of vacant states within an energy range of $k_B T$ of the Fermi energy is $\sim N_d(E_F)k_B T$. However, the d electrons that take part in the scattering process must also lie within an energy range $k_B T$ of the Fermi level and so the probability of finding such an electron is also $N_d(E_F)k_B T$. The total scattering probability is thus proportional to $N_d(E_F)^2(k_B T)^2$ and the electrical resistivity due to such an electron–electron scattering process should go as

$$\rho = \rho_0 + BT^2. \tag{6.3}$$

Electron–electron interactions between electrons in the s-band will also produce extra electron scattering, as discussed briefly in Chapter 1, but the effect will be much smaller because of the lower density of states in the s-band $N_s(E_F)$. Some typical values of the coefficient of the T^2 term in the resistivity are given in Table 6.1 and can be seen to be quite small so that any such contributions are likely to be observed only in pure metals at low temperatures where the phonon and residual impurity scattering are negligible.

We will show in the next chapter that spin density fluctuations and spin waves can also give rise to a T^2 term in the low temperature resistivity of magnetic and nearly magnetic materials, and it is often difficult in practice to distinguish between the different mechanisms. In the meantime let us consider some of the theoretical approaches relevant to the study of such non-simple metals and alloys.

6.2 Model and pseudopotentials in non-simple metals

The OPW pseudopotential approximation discussed in the previous chapters relies on the distinction between the core and valence

Table 6.1. *Coefficient of the T^2 term in the resistivity of some metals*

Metal	Coefficient of T^2 term ($\times 10^{-6} \mu\Omega$ cm K^{-2})
Mn	0.15×10^6
Fe	13
Co	13
Ni	16
Y	~ 300
Zr	~ 80
Mo	7.9
Pd	33
Pt	14

Source: After Dugdale 1977.

states and its success depends upon the rapid convergence of the OPW's onto the actual wavefunctions. In non-simple metals the d-states are not core-states and so are not included in the basis set for the OPW expansion. However, as the d-states are strongly localised their representation by an OPW expansion will only be slowly convergent. Thus the essential simplification offered by the pseudopotential method is lost. A method which augments the basis sets of OPWs by including functions which vanish in the interstitial regions of the crystal but give a good description of the outer core functions and d-band states near the nucleus has been described by Deegan & Twose (1967). Use of such an overcomplete set restores the rapid convergence and has been used by Harrison (1970, p. 201) to define a transition metal pseudopotential. Unlike the simple metal pseudopotential, this form contains a term that diverges at a particular 'resonance' energy E_r. This term is responsible for the hybridisation of the d-band with the sp conduction band, as illustrated in Figure 6.1. Screening is simply incorporated by independently screening the additional 'hybridisation' form factor and adding this to the normal screened pseudopotential form factor, as shown in Figure 6.2. Such a transition metal pseudopotential could only be used in a perturbation expansion at energies well-removed from the resonance, otherwise the calculations become very complex. However, such a situation should obtain in the noble metals, justifying use of the pseudopotential approximation for such metals.

Moriarty (1970, 1972) has determined the form factors of Cu, Ag and Au allowing for such hybridisation effects and his 1970 form factors are shown in Figure 6.3. While the minima at small q may in fact be spurious (Moriarty 1972) the values in the critical $q \sim 2k_F$ region appear to give reasonable results in Cu–Au alloys, but not in Au–Ag, as discussed in the previous chapter. Moriarty noted that the effects of hybridisation were quite significant, as evidenced by his calculated form factors for Cu with and without hybridisation which are shown in Figure 6.4. The neglect of hybridisation would reduce the calculated resistivity of liquid copper by a factor of 4.

In a later work, Moriarty (1974) extended the calculation to include the heavy alkali and alkaline–earth metals K, Rb, Cs, Ca, Sr, Ba (which have empty d-bands above the Fermi level) and the group IIB (Zn, Cd,

Fig. 6.1. (*a*) d-band crossing s-band; (*b*) the effects of s–d hybridisation.

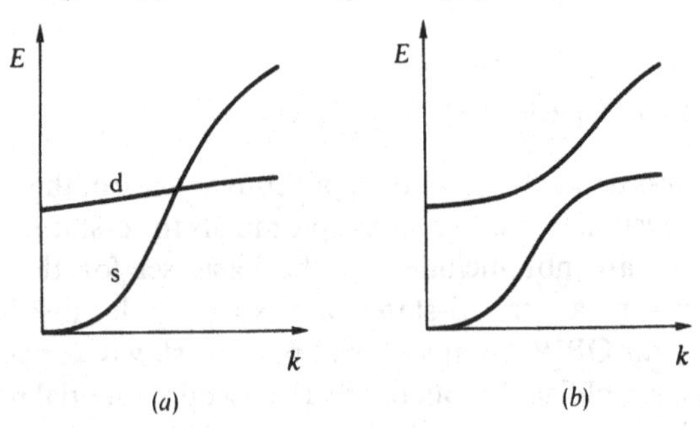

Fig. 6.2. Computed pseudopotential form factor (*a*) and screened hybridisation factor (*b*) for copper, and the sum of the two (*c*) (after W. A. Harrison 1969).

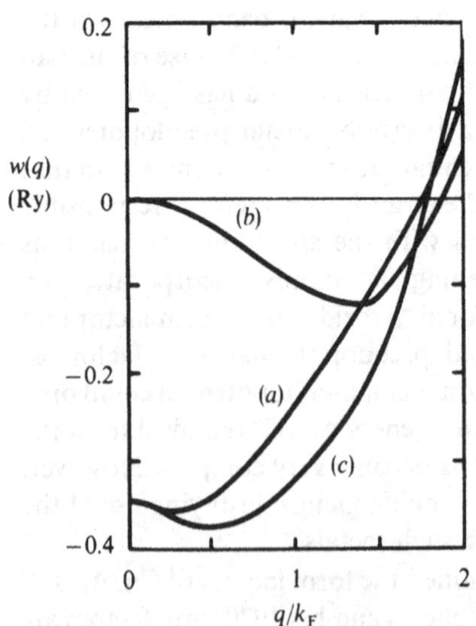

Hg) and group IIIA (Ga, In, Th) metals where d-bands are filled. The effects of hybridisation in these metals are much smaller as expected, although still quite significant in the alkaline–earth and group IIB metals. Some of the computed form factors are shown in Figure 6.5.

Animalu (1973) has considered extensions of the model potential concept to transition metals by effectively including a resonance term

Fig. 6.3. Form factor of Cu, Ag and Au (due to Moriarty 1970).

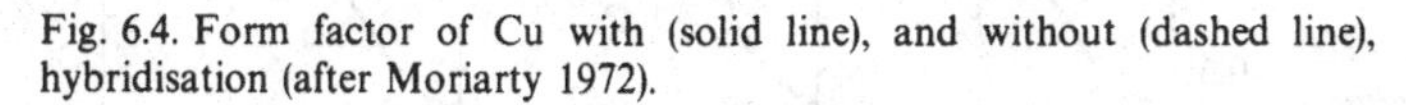

Fig. 6.4. Form factor of Cu with (solid line), and without (dashed line), hybridisation (after Moriarty 1972).

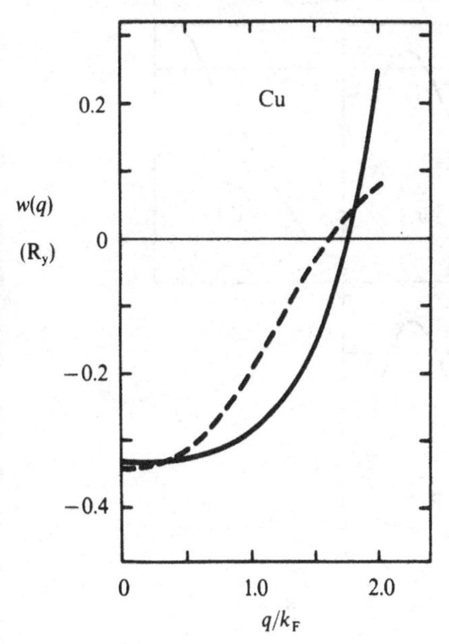

which varies rapidly with energy. Some model potential parameters for a number of metals are given in Table 6.2 and have been fitted to various aspects of their electronic structure such as phonon spectra, liquid metal resistivity and energy bands. Dagens (1976) has also considered a model potential including a resonance term. His use of perturbation theory restricts its application to completely full or empty d-band metals (i.e. E_F far removed from E_d) and results are given for Cu, Ag and Ca in Figure 6.6.

However, there is a different formalism of electron scattering which has more general application for the transition metals. This is called the phase shift method and will now be considered in some detail.

Fig. 6.5. Some computed form factors for K, Ca, Rb, Sr, Cs, Ba, Zn, Cd (after Moriarty 1974).

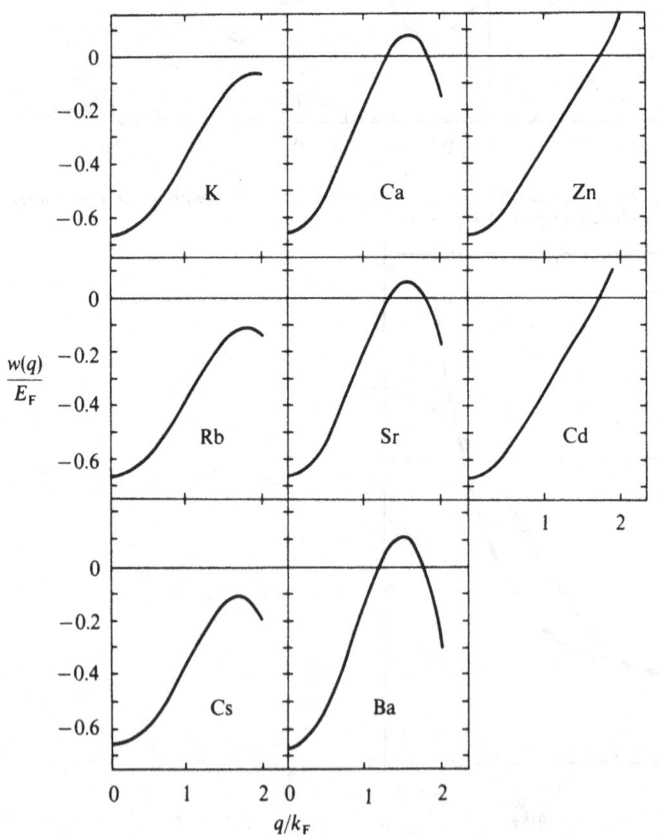

Table 6.2. *Some parameters for the Animalu transition metal model potential*

Metal	Well depth for each angular momentum component (Ry) A_0	A_1	A_2	R_H (Å^2)	Ω_0 (Å^3)
Cu	0.50	0.80	0.43	1.16	11.7
Ag	0.45	0.80	0.43	1.37	17.1
Au	0.30	1.00	0.42	1.37	17.0
Ti	4.60	5.00	4.20	1.06	17.6
V	6.50	7.00	5.80	0.85	13.9
Cr	3.20	2.94	2.80	1.32	11.9
Mn	1.78	1.96	1.74	1.16	12.1
Fe	3.20	3.30	2.80	1.06	11.8
Co	1.98	2.10	1.96	1.16	11.1
Ni	1.98	2.10	1.96	1.16	10.9
Pd	1.78	1.50	1.74	1.37	14.7
Pt	1.94	2.22	1.70	1.37	15.0
Mo	4.60	5.86	5.00	1.06	15.6
W	4.60	5.70	5.00	1.06	15.8

Source: See Animalu (1973) for a more extensive list.

Fig. 6.6. Resonant model potential form factors for Cu, Ag and Ca (after Dagens 1976).

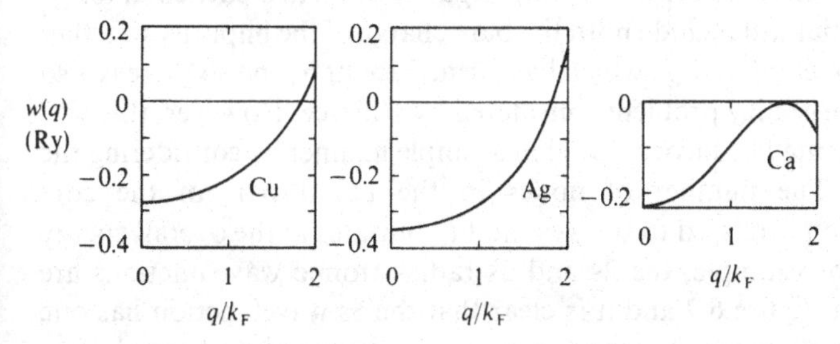

6.3 The phase shift method

The use of the Born approximation and perturbation theory in the pseudopotential methods rests on the assumption that the scattering potential V (or W) is small. An exact treatment of the scattering of plane waves from any arbitrary spherically symmetric potential $V(r)$ is given by the phase shift method described in quantum mechanics texts (see e.g. Schiff 1968, ch. 5: Messiah 1961, ch. 10). As a consequence of the spherical symmetry of the scattering potential, it is possible to write a solution of the Schrödinger equation in terms of spherical harmonics (or partial waves).

$$\Psi(r,\theta)=\sum_l \frac{y_l(r)}{r} P_l(\cos\theta), \tag{6.4}$$

where $y_l(r)$ is a solution of the radial equation

$$\left[\frac{\hbar^2}{2m}\frac{d^2}{dr^2}+\left(E-V(r)-\frac{\hbar^2 l(l+1)}{2mr^2}\right)\right]y_l(r)=0, \tag{6.5}$$

and $P_l(\cos\theta)$ is a Legendre polynomial. Assuming that $V(r)\to 0$ as $r\to\infty$, the asymptotic form of y_l is

$$y_l \sim a_l \sin(kr-\tfrac{1}{2}l\pi+\eta_l). \tag{6.6}$$

The parameters η_l are the phase shifts which provide all of the information we need about the scattering from such a centre. The scattering amplitude $f(\theta)$ is then given by

$$f(\theta)=\frac{1}{k}\sum_{l=0}^{\infty}(2l+1)\exp(i\eta_l(k))\sin\eta_l(k)P_l(\cos\theta). \tag{6.7}$$

The indices $l=0,1,2,\ldots$ correspond to the angular quantum numbers of the s,p,d,... states and the scattering probability $P(\theta)$ (or differential cross-section) is just $|f(\theta)|^2$. The problem now is to calculate the phase shifts for some given scattering potential but we still have the problem of choosing a suitable potential for an impurity atom in a particular alloy. This potential will include both the bare charge of the impurity together with the screening charge which has formed about it, and so there is also the self-consistency problem considered by Hartree. However, this self-consistency may be incorporated in a simple manner by considering the following. The number of nodes in the radial part of the core wavefunction is related to the availability of states in the overall energy scheme. For example, the 2s and 3s radial atomic wavefunctions are sketched in Figure 6.7 and it is clear that the 3s wavefunction has one more node than the 2s wavefunction. Each time the phase shift η_l increases by π one l-state is lost. Thus the phase shift is directly related to the number of states available below the Fermi energy. Each l-state adds $2(2l+1)$ independent angular momentum electron states, the factor of 2 allowing for electrons of opposite spin.

When an impurity is added to a host, the number of new states below E_F must equal the number z of excess electrons introduced by the impurity since E_F must be uniform (it is a chemical potential) and remain unchanged from that of the pure host. The phase shifts must thus be directly related to z. This relationship is known as the Friedel sum rule

(Harrison 1970, p. 182; Dugdale 1977, p. 144):

$$z=\frac{2}{\pi}\sum_{l}(2l+1)\eta_l. \tag{6.8}$$

Note that its derivation relies on the assumption that the electronic charge of an impurity must be screened within a finite distance from the impurity and that the Fermi wavevector at large distances from the impurity must be the same as in a pure crystal. The latter assumption restricts its applicability to dilute alloys. Nevertheless, in such cases it can lead to a simplification in determination of the scattering potential, particularly when a single phase shift component dominates the scattering. For example, the screened scattering potential may be written in some parametric form such as the screened Coulomb potential $[(\Delta ze^2/4\pi\varepsilon_0 r)\exp(-\lambda r)]$, square well, muffin-tin or some other form. The phase shifts are then determined within the atomic cell, usually by numerical integration of the Schrödinger equation, and the parameters

Fig. 6.7. The radial part of atomic 2s and 3s wavefunctions. Note that the introduction of an extra node in the 3s wavefunction may be brought about by introducing a phase shift of π in the 2s wavefunction (i.e. by compressing the 2s curve along the r axis by an amount π/k).

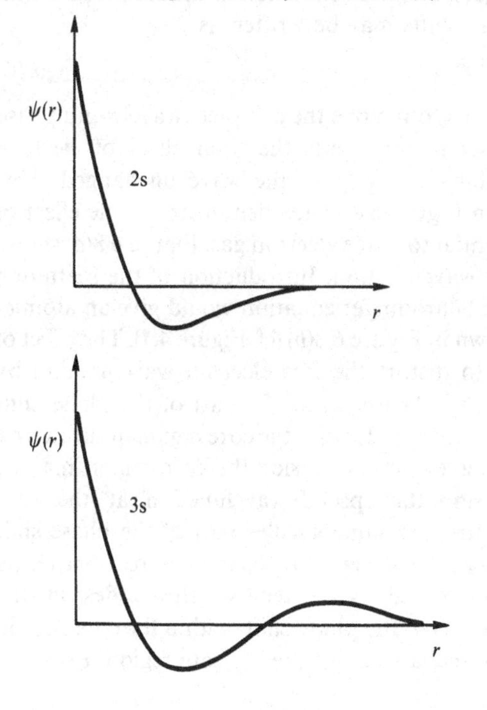

adjusted to produce phase shifts that satisfy the Friedel sum rule. If a single phase shift dominates the scattering it may be found *directly* from equation (6.8), representing a considerable simplification of the problem. These phase shifts, evaluated at E_F, may be used to determine the scattering amplitude and hence the electrical resistivity. However, as pointed out by Ziman (1964, 1969, p. 250), the sum only fixes the scattering potential at $q=0$ whereas the resistivity integral involves all q up to $q=2k_F$, and is weighted in favour of the large q values. Satisfying the Friedel sum rule (6.8) is thus no guarantee of the accuracy of the calculated resistivity.

Phase shifts for a variety of metals and alloys have also been obtained by numerical integration of the Schrödinger equation with a muffin-tin potential that has been fitted to known band structures (Khanna & Jain 1974). Later in this chapter when some applications of this technique are discussed we will make use of these results.

If the phase shifts are small compared to π (i.e. if the scattering is weak) the Born approximation may be used giving (Harrison 1970, p. 186)

$$\frac{m\Omega_0}{2\pi\hbar^2}\langle\mathbf{k}+\mathbf{q}|w(\mathbf{r})|\mathbf{k}\rangle = -\frac{1}{k}\sum_l (2l+1)\eta_l P_l(\cos\theta). \tag{6.9}$$

In fact this equation suggests an alternative view of the pseudopotential approximation. The phase shifts may be written as

$$\eta_l = 2p_l\pi + \delta_l, \tag{6.10}$$

where p_l is an integer chosen to remove the complete wavelength phase shifts (2π). The remainder δ_l represents the total effect of electron scattering since phase shifts of 2π leave the wave unchanged. This behaviour is illustrated in Figure 6.8 which demonstrates the effect of adding a perturbing potential to a free electron gas. Figure 6.8(*a*) shows an unperturbed electron wavefunction. Introduction of the scattering potential and solving the Schrödinger equation would give an atomic-like wavefunction, as shown in Figure 6.8(*b*) (cf. Figure 4.1). The effect of this potential has been to distort the free electron wavefunction by introduction of a phase shift $2\pi+\delta_0/k$, the 2π part of the phase shift result in the introduction of two nodes into the core region (indicated by r_c). However, the same wavefunction outside the core region may be obtained by fitting a smoother pseudowavefunction at the core boundary while maintaining the same smaller part of the phase shift δ_0/k. Thus we can see that the effect of replacing the real scattering potential by a pseudopotential is to remove the nodes in the wavefunction (or 2π multiples of the phase shift) within the core region, leaving the wavefunction unchanged outside the core region. Provided that the remaining phase shift δ_l is small, perturbation theory may be applied to determine the behaviour outside the core region. Note however, that even though δ_l is less than 2π, it may still be quite large in terms of scattering theory ($\delta_l \sim \pi/2$ producing the *maximum* scattering amplitude), invalidating the use of the Born approximation. We return to this matter briefly at the end of Section 6.4. When a solute atom has the same valency as the matrix $\Delta z = 0$. However, there will still be scattering because of detailed differences in the atomic potentials which appear in the screening radius or atomic volume. For example, Blatt (1957*a*,*b*) and Harrison (1958) have suggested that the lattice distortions arising from atomic size difference lead to an effective impurity valency:

$$z' = z_B - z_A\left(1 + \frac{\delta v}{\Omega}\right), \tag{6.11}$$

where δv is the local dilation about the impurity and z_A, z_B the valencies of the host and impurity respectively. A more formal derivation has been given by Ziman (1969, p. 261). Similarly, Stern (1967) has considered the effects of an impurity with a different screening radius. If these effects are not large, the scattering will be weak allowing use of the Born approximation and direct calculation of the phase shifts from equation (6.9). However, Sorbello (1980) has cautioned that the distortion field

Fig. 6.8. (*a*) Unperturbed free electron wavefunction. (*b*) Atomic-like wavefunction due to the presence of the impurity atom. (Note the phase shift δ_0/k.) (*c*) Pseudowavefunction which is equivalent to (*b*) outside the core region r_c but which lacks the nodes of the true wavefunction within r_c.

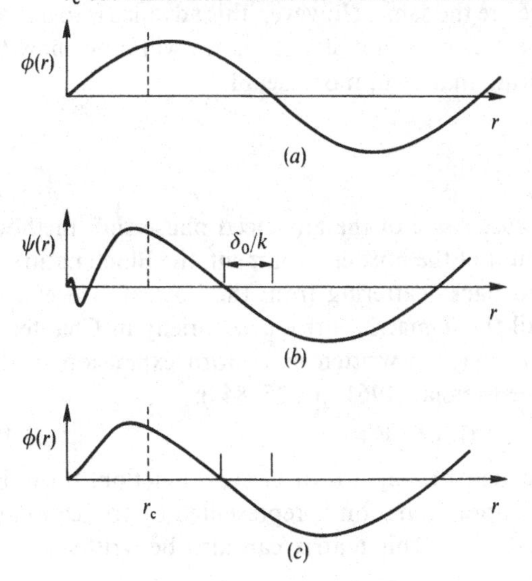

surrounding an impurity may in fact cause significant backscattering of the conduction electrons which is neglected in the Born approximation. We will discuss these effects further in Section 6.6.2 with reference to some particular systems. It is interesting to note that the Friedel oscillations in charge density around an impurity also follow directly from the phase shift method (see e.g. Ziman, p. 159; Harrison, p. 184).

In a metal containing impurities or in an alloy there is still, however, the general problem of knowing what potential to use in order to determine the phase shifts. Ideally one should calculate a new potential for each composition and state of order by fitting to known Fermi surface or band structure information, and use this to determine the phase shifts at E_F by numerical integration of the Schrödinger equation in the atomic cell. In the simplest approximation due to Friedel the host environment is replaced by a constant potential which is taken to be zero. The integration over the scattering angle θ in the resistivity integral may then be carried out over a spherical Fermi surface using the properties of Legendre polynomials to give (de Faget *et al.* 1956; Huang 1948)

$$\int_0^\pi \left| \sum_{l=0}^{\infty} (2l+1) \exp(i\eta_l) \sin \eta_l P_l(\cos\theta) \right|^2 (1-\cos\theta) \sin\theta \, d\theta$$

$$= 2 \sum_{l=0}^{\infty} (l+1) \sin^2(\eta_l - \eta_{l+1}). \qquad (6.12)$$

It may appear that this equation represents an advance over the pseudopotential method in that the scattering is treated exactly. Otherwise the nearly free electron assumptions of plane waves and a spherical Fermi surface are the same. However, this advance is to a large extent illusory if the scattering is not strong, as we will now show by considering the scattering matrix in more detail.

6.4 The *T*-matrix

Before we discuss some of the advanced phase shift methods which take better account of the host environment, the Bloch nature of the electron states and back-scattering from the host, it is useful to consider in more detail the *T*-matrix introduced briefly in Chapter 3. Formally, the *T*-matrix may be written as a Born expansion in the scattering potential (see Messiah 1961, p. 825, 849):

$$T = V + VG_0V + VG_0VG_0V + \cdots, \qquad (6.13)$$

where G_0 is the free particle propagator or Green's function. It can be seen that the first Born approximation is represented by truncation of this series after the first term. This matrix can also be written in the multiple scattering formalism as (Beeby & Edwards 1963)

$$T=\sum_i t_i+\sum_{i\neq j} t_iG_0t_j+\sum_{i\neq j\neq k} t_iG_0t_jG_0t_k+\cdots, \tag{6.14}$$

where t_i is now the transition matrix of a single scattering site i giving the probability of scattering an electron from state $\mathbf{k}$ to $\mathbf{k}'$ by a single scattering event (Greenwood 1966). The double scattering terms $t_iG_0t_j$ represent a scattering of the electron by atom j, then propagation of the scattered electron from j to i, and finally the scattering of the electron by atom i into the final state $\mathbf{k}'$. For the case of plane wave electron states lying on a spherical Fermi surface, the t-matrix is related to the phase shifts by (Messiah 1961)

$$\begin{aligned}\langle\mathbf{k}'|t|\mathbf{k}\rangle &= t(\mathbf{k},\mathbf{k}')\\ &= -\frac{2\pi\hbar^2}{mk_F\Omega_0}\sum_l(2l+1)\sin(\eta_l(E_F))\exp(i\eta_l(E_F))P_l(\cos\theta),\end{aligned} \tag{6.15}$$

where $\eta_l(E_F)$ are the phase shifts evaluated at E_F ('scattering on the energy shell'). The Ziman expression for ρ in terms of this matrix (neglecting the higher order multiple scattering terms) can then be written as (Greene & Kohn 1965; Dreirach 1973)

$$\rho=\frac{12\pi\Omega_0}{16\hbar e^2v_F^2k_F^4}\int_0^{2k_F}|S(q)|^2|t(\mathbf{k},\mathbf{k}')|^2q^3\,dq, \tag{6.16}$$

where v_F is the Fermi velocity. This is sometimes known as the extended Ziman formula since it is the same as the Ziman equation (5.9b) with the scattering probability $\langle\mathbf{k}'|w(\mathbf{r})|\mathbf{k}\rangle$ replaced by the corresponding t-matrix.

We can now picture the type of error that occurs when the Born approximation fails. If the scattering potential is progressively increased in strength, the t-matrix does not continue to increase but fluctuates as bound states are introduced as discussed in Section 3.2.1. However, in the Born approximation the scattering is proportional to the first term in equation (6.13), i.e. $|\langle\mathbf{k}'|V(\mathbf{r})|\mathbf{k}\rangle|^2$, and simply increases steadily as $V(\mathbf{r})$ is increased, predicting resistivities that are often much too large. An extension of the above to include some of the multiple scattering terms has been given by Dunleavy & Jones (1978). However, before pursuing these refinements we should note that in going to a more general description of the scattering process it is again possible for the scattering rate to vary over the Fermi surface. This means that, as in the case of multiple-OPWs, the Boltzmann equation must be solved at each specific location on the Fermi surface requiring accurate Fermi surface and electron wavefunction data as well as a suitable form for the scattering potential. Some of these methods used have been briefly discussed in Section 1.8. As a general guideline to the type of calculation involved, the transition probability $Q_{\mathbf{kk}'}$ is given by

$$Q_{\mathbf{kk}'} = \frac{2\pi}{\hbar} |T_{\mathbf{kk}'}|^2 \, \delta(E_{\mathbf{k}} - E_{\mathbf{k}'}). \tag{6.17}$$

This may be evaluated using equation (6.15) (or some more exact form discussed in the next section) and phase shifts derived at the Fermi surface. It is then substituted into the appropriate form of the Boltzmann equation (see Section 1.8) which is solved to obtain the resistivity. It is useful to note that the Ziman approximation (1.68) may also be written in terms of the t-matrix. For example, for a concentration c_{I} of independent scatterers:

$$\frac{1}{\tau(\mathbf{k})} = \frac{2\pi c_{\mathrm{I}}}{\hbar} \sum_{\mathbf{k}'} |t(\mathbf{k}, \mathbf{k}')|^2 (1 - \cos(\mathbf{v}_{\mathbf{k}}, \mathbf{v}_{\mathbf{k}'})). \tag{6.18}$$

Coleridge (1972) and Sorbello (1974b) have found that expressions of this form gave a very good estimate of the resistivity of dilute Cu and Au-based alloys with results usually within a few percent of those obtained by an iterative solution of the Boltzmann equation over the Fermi surface, although the details of the anisotropy were often in error (see also the comments in Sections 1.8.5 and 5.7.2, and Leavens & Laubitz 1976). In the Friedel approximation the states $\mathbf{k}, \mathbf{k}'$ are plane waves and $\mathbf{v}_{\mathbf{k}}$ is parallel to $\mathbf{k}$. Equation (6.18) then effectively reduces to the corresponding extended Ziman formula (equation (6.16)) for a dilute alloy.

Finally, we return to consideration of the validity of perturbation methods in metallic solids. Firstly, the phase shifts for a pseudopotential are given correctly by equation (6.9) since the pseudopotential is chosen so that its matrix elements give the correct scattering *within* the Born approximation; i.e. they approximate the matrix elements of the single ion t-matrix (Sorbello 1977; Greenwood 1966). It would thus be quite wrong to treat a linearly screened pseudopotential as a real potential and integrate over the atomic sphere to determine 'better' phase shift values (see e.g. Harrison 1970, p. 187). Secondly, even though the phase shifts of a single ion may be $\sim\pi/2$ (i.e. the scattering is not weak in the perturbation sense) perturbation methods may still be valid for a metallic solid provided that the ionic potentials add together so that the total pseudopotential $V(r) = \sum_i V(r_i)$ is smooth, i.e. the scattering perturbation is weak. Such a situation should occur in an alloy provided that its constituents have a reasonably similar electronic structure. However, as noted by Heine & Weaire (1970), a spatially random arrangement of atoms or a vacancy in an otherwise perfect lattice might not lead to such a condition.

6.5 Advanced phase shift methods: the KKR–Green's function method

We are now in a position to return to the application of phase shift methods to the calculation of resistivity of non-simple alloys. This requires some means of determining the phase shifts which takes proper account of their complicated electronic structure (which, as discussed in Chapter 4, can be determined by APW and KKR methods). All the required information about the bands is contained in the logarithmic derivatives of the wavefunctions at the surfaces of the core spheres which can in turn be written in terms of phase shifts. One way to proceed is to describe the s and p phase shifts in terms of some simple-metal potential while the $l=2$ (i.e. d) phase shift could be taken to be a resonant form such as (see e.g. Messiah 1961, p. 399)

$$2\tan\eta_2=\frac{\Gamma}{E_r-E}, \tag{6.19}$$

where Γ is the width of the resonance. In a calculation Γ, E_r, η_0 and η_1, could then be regarded as parameters which are adjusted to fit the electron energies to known band structures and then used directly in a resistivity calculation, as described above. This is probably a better alternative to the transition metal model- or pseudopotnetial in such cases since the effects of the narrow d-band enter in a natural way through the d phase shifts. However, it is possible to perform *ab-initio* calculations and avoid any assumption about the form of the resonance. For example, the KKR method has been extended by Morgan (1966) to include the Bloch nature of the conduction electrons (see also Blaker & Harris 1971; Harris 1973; Podloucky *et al.* 1980). This technique assumes that the impurity is localised at a single site in a lattice of muffin-tin potentials and the Schrödinger equation is solved exactly using Green's function techniques. So while the perturbing potential is localised, the full spatial extent of the wavefunctions is taken into account and the calculation effectively incorporates details of the band structure into the resistivity calculation. The scattering may be treated with the *t*-matrix to include multiple scattering effects and so include the effects of back-scattering from the host atoms. Approaches such as these are quite acceptable in a perfectly periodic (or nearly periodic) system since the Bloch theorem allows solution of the Schrödinger equation in a single unit cell. However, concentrated alloys will certainly lack such translational periodicity so that the Bloch theorem fails and the

Table 6.3. *Theoretical and experimental resistivity in μΩ cm due to one at.% solute in Cu, Au and Ag solvents*

Solvent	Solute	Equation (6.22)	Resistivity Coleridge *et al.*	Experiment
Cu	Ni	1.10	1.11	1.11
	Ge	3.76	3.79	3.79
	Au	0.5	0.55	0.55
	Fe	12.2	12.0	11.5
	Al	1.44	1.6	1.1
Au	Ag	0.34	0.36	0.36
	Cu	0.48	0.46	0.45
	Zn	0.76	1.00	0.95
	Ga	2.15	—	2.09
Ag	Au	0.38	0.39	0.38
	Sn	4.37	4.30	4.3

Source: After Gupta & Benedek 1979.

Schrödinger equation must be solved with a potential of infinite extent. For the moment let us restrict consideration to pure or nearly pure metals (i.e. the dilute alloy limit) and then return to the discussion of concentrated alloys later in Section 6.7. Coleridge *et al.* (1974) have used this KKR–Green's function technique with the inclusion of multiple scattering effects and shown that the summation in (6.15) reduces to an integration of the anisotropic mean free path over the Fermi surface which has to be evaluated numerically. In this calculation the phase shifts become

$$\phi_l = \Delta\eta_l + \theta_l, \tag{6.20}$$

where $\Delta\eta_l$ is the difference between the impurity and host phase shifts

$$\Delta\eta_l = \eta_l^{\mathrm{i}} - \eta_l^{\mathrm{h}}, \tag{6.21}$$

and θ_l is the phase shift due to back scattering from the host. Gupta & Benedek (1979) and Schöpke & Mrosan (1978), have also considered Bloch states and multiple scattering effects and shown that the resistivity can be written in a form analogous to that deduced from equations (1.3), (6.7) and (6.12):

$$\rho = \frac{4\pi c_{\mathrm{I}}\hbar}{z^{\mathrm{h}}e^2k_{\mathrm{F}}} \sum_{l=0} (l+1)\sin^2(\Phi_{l+1} - \Phi_l), \tag{6.22}$$

thereby avoiding the integration over the Fermi surface (z^{h} is the valence of the host). This simple equation was found to produce good agreement with the numerical integration technique of Coleridge *et al.*, as evidenced by the results quoted in Table 6.3.

One problem with the simple muffin-tin model is that the effects of the impurity must be restricted to a single atomic (Wigner–Seitz) cell. This makes it difficult to account for charge transfer and lattice distortion effects. Lodder *et al.* (Lodder 1976, 1977; Lodder & Braspenning 1980; Braspenning & Lodder 1981; Lodder *et al.* 1982; Boerrigter *et al.* 1983; Braspenning *et al.* 1984) included such effects by determining the scattering of a cluster consisting of the impurity atom surrounded by one or two shells of relaxed host atoms. The alloy is thus represented by an extended impurity cluster in free space, although the states **k** are taken to be not plane waves but states corresponding to a perfect host cluster. The resulting $\tau^{-1}(\mathbf{k})$ is again found to be anisotropic. A somewhat different cluster approach has been described by Johnson *et al.* (1979) in which an impurity is embedded in a 19-atom cluster representing the local crystalline environment up to second nearest neighbours. The calculation is based on Slater's $X\alpha$ approximation for exchange and correlation (see Chapter 4) and a multiple scattered wave formalism (Johnson 1973). Before we extend the discussion to include problems associated with concentrated alloys, let us consider applications of these techniques and compare some of the results obtained.

6.6 Some applications

6.6.1 *Pure noble and transition metals*

Dreirach (1973) has determined the phonon-limited resistivities of Cu, Ag, Au and Ni. Potentials derived from self-consistent band structure calculations were used for Cu and Ni, while for Ag and Au suitable muffin-tin potentials were constructed (see e.g. Loucks 1967). The phase shifts were then determined by numerically integrating the Schrödinger equation inside the muffin-tin sphere and the t-matrix evaluated at the Fermi energy. Even with the further simplifying assumptions of a spherical Fermi surface, plane wave conduction states and the Debye model for the lattice vibrations, the results were found to be in quite good agreement with experiment, as shown in Table 6.4. Also shown in that table are some results from Khanna & Jain (1974) for Fe, Cu, W, Nb, Ta, Mo and Pd. The phase shifts for Fe and Cu were obtained from published muffin-tin potentials. For W, Ta, Mo, Pd and Nb the phase shifts at E_F tabulated by Evans *et al.* (1973*a*) were used. In this study a more accurate description of the thermal motion of the ions was obtained by solving the dynamical matrix leading to reasonable agreement with experimental results over wide temperature range in all cases except Fe (in which case it was not clear whether spin–disorder scattering was included). Some of these results are shown in Figure 6.9.

Table 6.4 *Resistivity at 295 K (in μΩ cm) determined by Dreirach (Cu to Ni) and Khanna & Jain (Fe to Pd) compared with experimental values*

Metal	Resistivity at 295 K Theory	Experiment
Cu	1.9	1.7
Ag	1.5	1.61
Au	1.9	2.2
Ni	5.2	7.0
Fe	11.4	9.8
W	7.2	5.3
Nb	18.4	14.5
Ta	13.2	13.1
Mo	6.1	5.3
Pd	5.2	10.5

Source: After Dreirach 1973; Khanna & Jain 1974.

Fig. 6.9. Calculated (solid lines) and experimental (dashed lines) temperature dependence of the electrical resistivity of Cu, Ta, Mo and Fe (after Khanna & Jain 1974).

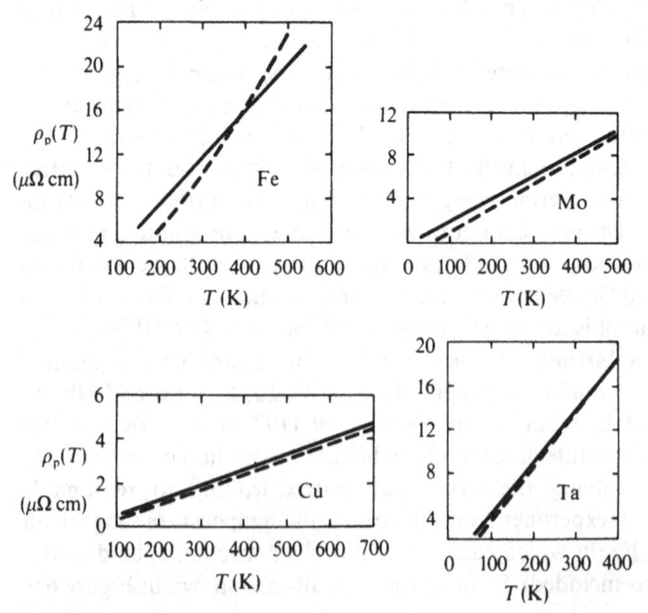

The resonant model potential of Dagens has been used to determine the electron–phonon matrix elements by Kus (1978). By using a numerical solution of the Boltzmann equation he obtained the variation of electrical resistivity with temperature for Cu and Ag shown in Figure 6.10 together with results obtained by the lowest order variational solution and experimental data. These results are seen to be in better agreement with experiment than those based on the Moriarty pseudopotential form factors which predict a much stronger electron–phonon coupling (Black 1974).

The rigid muffin-tin model (analogous to the rigid-ion model discussed in Chapter 4) has also been used to determine the resistivity of 3d and 4d transition metals (see e.g. Butler 1981, 1982). Some results from Mazin *et al.* (1984) are shown in Table 6.5.

6.6.2 *Dilute alloys: bound and virtual bound states*

If the phase shift of an impurity atom passes through π, the nature of the electron states associated with that atom changes rapidly

Fig. 6.10. Temperature dependence of the resistivity of Cu and Ag determined using the Dagens resonant model potential. Solid lines: numerical solution of Boltzmann equation; dashed lines: lowest order variational solution (after Kus 1978). Also shown is the experimental variation (●) (White & Woods 1959).

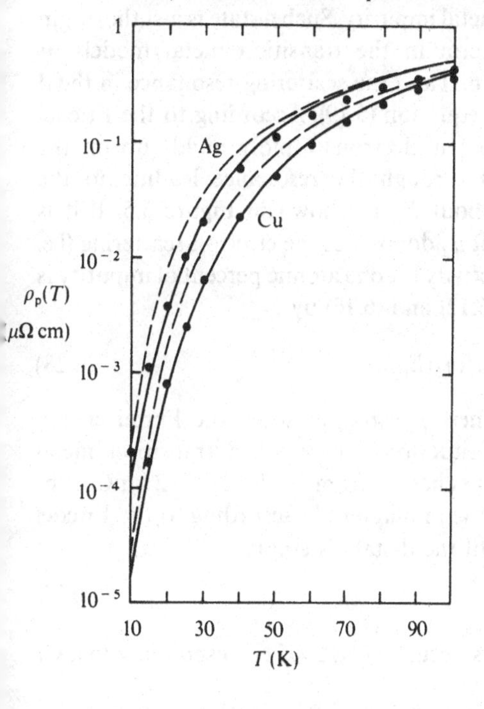

Table 6.5. *Theoretical and experimental resistivities (in μΩ cm) of some 4d transition metals at 273 K*

	Resistivity at 273 K	
Metal	Theory	Experiment
Nb	14.3	13.5
Mo	5.8	4.9
Rh	4.1	4.3
Pd	10.4	9.7
Ag	1.8	1.7

Source: After Mazin *et al.* 1984.

and becomes an exponentially decaying localised or 'bound' state, as discussed in Section 3.2 and later in Section 7.3. In simple metals the screening effects of the mobile electrons would preclude the formation of such states, but with transition metal hosts or impurities the effects of unfilled d-levels must be considered. These can lie at energies within the conduction band and so there is a finite probability that an electron will momentarily occupy them leading to a virtual bound state localised in the vicinity of the transition metal impurity. Such a state is just the origin of the divergent resonant term in the transition metal model- or pseudopotential (see Section 6.1) and the scattering resonance in the d partial wave as embodied in equation (6.19). According to the Friedel sum rule, a charge of up to ten electrons rapidly builds up on the impurity as E_F is increased through the resonance leading to the increased density of states about E_r, as shown in Figure 3.5. If it is assumed that the *d* phase shift η_2 dominates the electron scattering (i.e. $\eta_0=\eta_1=0$), the electrical resistivity for one atomic percent of impurity is given from equations (6.7), (6.12) and (6.16) by

$$\rho=\frac{30\pi^3\hbar^3\times 10^{-2}}{me^2\Omega_0 k_F^2 E_F}\sin^2(\eta_2(E_F)). \tag{6.23}$$

This reaches a maximum when $\eta_2=\pi/2$. i.e. when the Fermi energy coincides with E_r. Such a situation occurs when transition metal impurities are added to Al. As z changes from -1 for Ni, -2 for Co etc., the *d* phase shifts must increase in magnitude according to the Friedel sum rule (equation (6.5)) until the d-state is empty:

$$\eta_2=\frac{z\pi}{10}. \tag{6.24}$$

The maximum scattering thus occurs when $z=5$ corresponding to a Cr impurity. By using this simple approach the predicted resistivity is then $\sim 5\mu\Omega\,\text{cm/at.}\%$ compared to the experimental value of $8\mu\Omega\,\text{cm/at.}\%$. The resistivity should decrease on either side of Cr as found experimentally. The values calculated on the basis of this simple Friedel model are shown in Figure 6.11 and Figure 6.12 together with the

Fig. 6.11. Resistivity per atomic % impurity for 3d transition metals in Al host. Solid circles are the experimental results of Babic *et al.* (1972, 1973); dotted line is the Friedel model; dashed line is the single scattering model of Schöpke & Mrosan (1978); dot–dash line is the local spin density result of Nieminen & Puska (1980); and solid line is the multiple scattering model of Schöpke & Mrosan (1978).

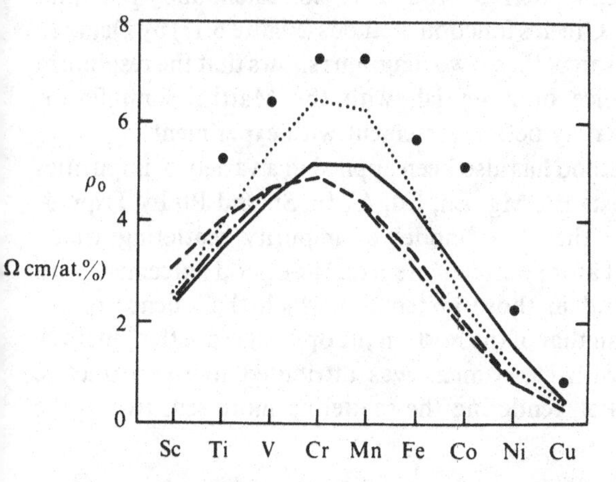

Fig. 6.12. Resistivity per atomic % impurity for 3d transition metals in Al host. Solid circles are experimental results of Babic *et al.* (1972, 1973); dotted line is Friedel model; dashed line is cluster model; solid line is Gupta–Benedek model.

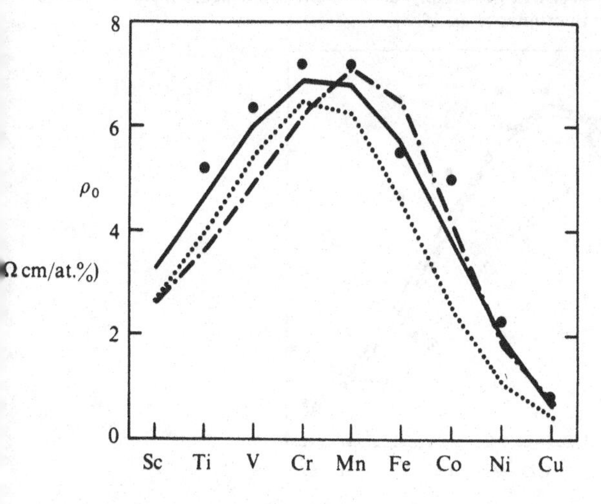

experimental values of Babic *et al.* (1972, 1973), the single and multiple scattering results of Schöpke & Mrosan (1978) and the local spin density approximation results of Nieminen & Puska (1980), the Gupta–Benedek multiple scattering model (1979) and the cluster model of Boerrigter *et al.* (1983).

The importance of the choice of the scattering potential in determining the phase shifts was investigated by Boerrigter *et al.* (1983) who determined the impurity resistivities from the Friedel, Gupta–Benedek and cluster models using both a simple muffin-tin potential according to the Mattheiss prescription (Mattheiss 1964) (see Figure 6.12) and the phase shifts derived from a self-consistent alloy potential determined by KKR–Green's function methods (Figure 6.13) by Deutz *et al.* (1981). A comparison of these two diagrams shows that the results are sensitive to the choice of potential, with the Mattheiss muffin-tin potential giving generally better agreement with experiment.

The phase shift method has also been applied to a variety of impurities in the polyvalent hosts Be, Mg, Zn, Cd, Al, In, Sn and Pb by Tripp & Farrell (1973) using the Blatt model of impurity scattering which includes the effects of atomic size differences. Here good agreement with experiment was found in those systems in which the valence of the impurity was less than that of the host. In the opposite case the generally poorer agreement with experiment was attributed to the attractive nature of the potential, rendering the scattering more sensitive to the

Fig. 6.13. Residual resistivities for 3d impurities in Al. Dotted curve is Friedel model, solid line is Gupta–Benedek model and dashed curve is cluster model using a self-consistent potential. The solid circles are again the experimental results (after Boerrigter *et al.* 1983).

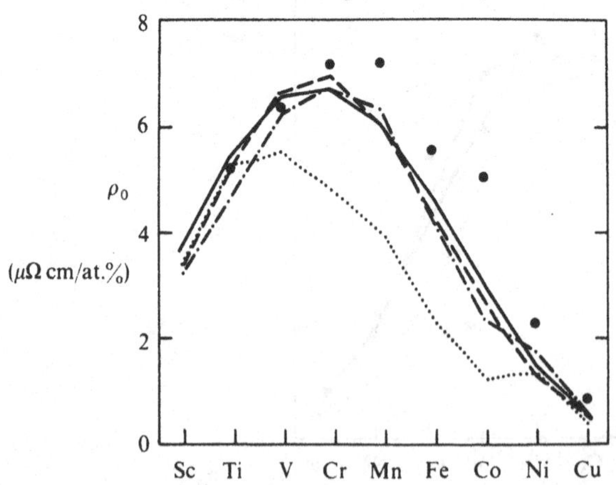

short range details of the potential in the core. The pseudopotential results presented in the previous section support this view, showing an increased sensitivity to changes in the ion-core radius for such a situation. Deutz *et al.* (1981) have also determined the scattering phase shifts for some sp (Si, P, S, Cl, Ga, Ge, As) impurities in Al using the KKR–Green's function method.

From the results quoted above, it appears that with 3d transition metal impurities in Al the effects of backscattering are important, as evidenced by the difference between the Friedel or single scattering Schöpke–Mrosan models and the Gupta–Benedek and Lodder *et al.* multiple scattering cluster models, with the latter giving better agreement with experiment. Furthermore, the impurities Cr, Mn and Fe carry a magnetic moment in Al (Deutz *et al.* 1981) and so will give additional spin-fluctuation scattering effects of the kind to be considered in the next chapter. It also appears that once again the choice of scattering potential is important. Boerrigter *et al.* (1983) could not find any significant effects due to charge transfer or lattice distortion and adopted the suggestion of Tripp & Farrell (1973) that Fermi surface effects are important in leading to a reduction in the number of conduction electrons.

Many other dilute alloy systems have also been considered, notably transition and non-transition metal impurities in noble metal hosts. In dilute alloys of copper with 3d transition metal impurities, at moderate temperatures (>4 K) there are two peaks in the resistivity curve, one at Fe and another at Ti. This is because the d-shell is divided into two spin–split subshells of different energy (see e.g. Heeger 1969). As the different impurities from Ti to Ni are added, the electrons first fill one d subshell which can hold five electrons and then fill the other subshell with the opposite spin direction. The results obtained from a KKR–Green's function calculation for Cu and Ag hosts by Podloucky *et al.* (1980) are shown together with the results from the cluster model of Johnson *et al.* (1979) and some experimental results in Figure 6.14(*a*). The KKR–Green's function results for an Ag host are shown in Figure 6.14(*b*). In the study by Podloucky *et al.* it was found that V, Cr, Mn and Fe impurities in Cu and Ti, V, Cr, Mn, Fe and Co impurities in Ag suffered such spin splitting. The scattering within each subshell is analogous to that from a ferromagnetic above the critical transition temperature (i.e. there is complete spin mixing), and so the total resistivity is given by an average of the spin 'up' and spin 'down' resistivities. The mechanism of two-band conduction is discussed in more detail in Section 7.1.2. However, at very low temperatures the magnetic moments of the impurity are effectively quenched and the curve exhibits only a single peak. This behaviour is shown in Figure 6.15, together with the results of the Friedel model, the theory of Gupta & Benedek which includes backscattering, the Coleridge *et al.* theory which takes into account both backscattering and band structure effects, and the cluster model of Lodder *et al.* which includes local distortion effects.

The results from a cluster calculation of Johnson *et al.* (1979) derived from a self-consistent field calculation are shown compared with some experimental results in Figure 6.16. Also shown in this diagram are the results of Mrosan & Lehmann (1976*a*,*b*), calculated from equation (6.18),

Fig. 6.14. (*a*) Residual resistivity of 1at.%3d impurity in Cu. The filled circles are experimental results; the solid curve is the Podloucky *et al.* KKR–Green's function result; and the dashed curve is the cluster model result of Johnson *et al.* 1979. (*b*) Residual resistivity of 1at.%3d impurity in Ag. The solid line is the KKR–Green's function result and the solid circles are the experimental data (after Podloucky *et al.* 1980).

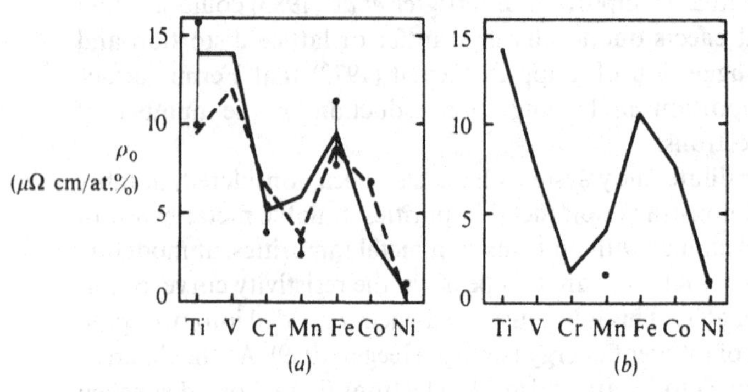

Fig. 6.15. The residual resistivity due to lat.%3d transition metal impurity in copper at low temperatures. Dotted curve – Friedel model; dashed line – Gupta–Benedek; solid line – Lodder *et al.*; solid circles – experimental data (after Lodder *et al.* 1982).

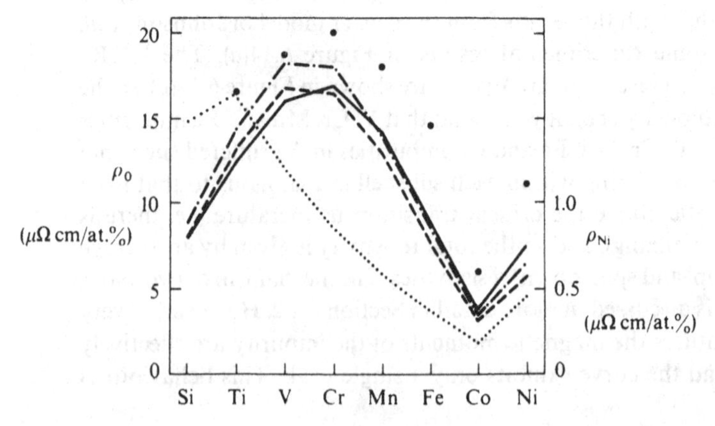

and phase shifts derived from a muffin-tin potential constructed according to the Mattheiss (1964) prescription. Inclusion of multiple scattering effects and Fermi surface anisotropy led to similar results (Mertig *et al.* 1982). While the Johnson *et al.* cluster model gives as good an agreement with experiment as the other theories, the exchange splittings predicted do not appear to be in such good agreement (Podloucky *et al.* 1980; Braspenning *et al.* 1984).

In other studies, Coleridge (1972) has determined the anisotropic relaxation times of Ni, Fe and Al impurities in Cu. The calculation of resistivity was found to be very sensitive to the addition of extra phase shifts, indicating the need for caution in neglecting the higher order ($l > 3$) phase shifts. The Cu–Ni system indicated a scattering dominated by the d phase shifts but with appreciable s and sp contributions, the Cu–Fe system exhibited essentially pure d scattering and the Cu–Al system was dominated by p scattering but with appreciable s and d contributions. Some other calculations of the phase shifts and/or resistivities of impurities in noble metals may be found in Blatt (1957*a*,*b*): Cu, Zn, Ga, Ge, As, Ag, Cd, In, Sn, Sb in Cu and Ag; Lasseter & Soven (1973): Al, Zn, Ga, Ge, Ag, Cd, Sn, Au in Cu.

The resistivity of 4d transition metal impurities in Cu has also been investigated by Lodder *et al.* (1982), again using a variety of different theoretical models. Their results are shown in Figure 6.17. These authors

Fig. 6.16. Residual resistivity due to 3d transition metal impurities in Cu without spin splitting. Filled circles are experimental data; dashed line is the Johnson *et al.* $X\alpha$ cluster calculation; solid line is the multiple scattering calculation of Mrosan & Lehmann.

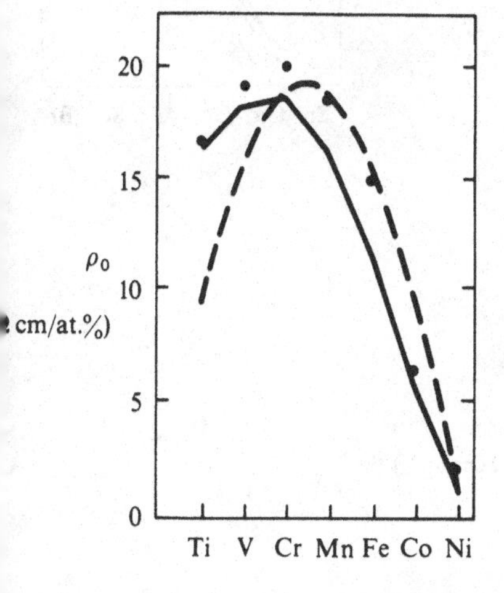

Fig. 6.17. The residual resistivity due to 1at.%4d transition metal impurity in copper. The different curves are as for Figure 6.15 (after Lodder *et al.* 1982).

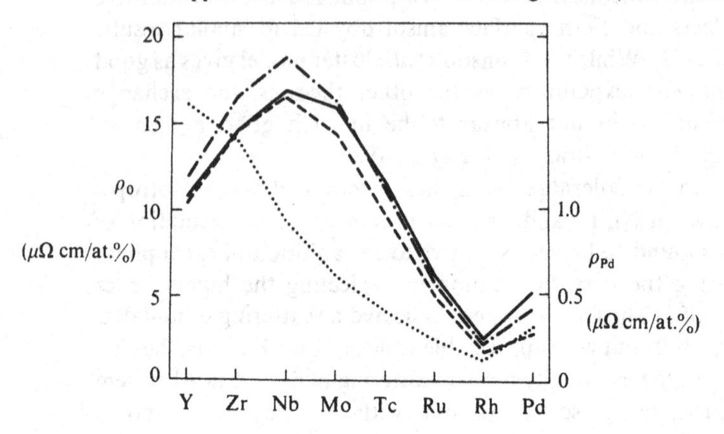

Fig. 6.18. Residual resistivities of copper with the impurities indicated; curves as for Figure 6.15 (after Lodder *et al.* 1982).

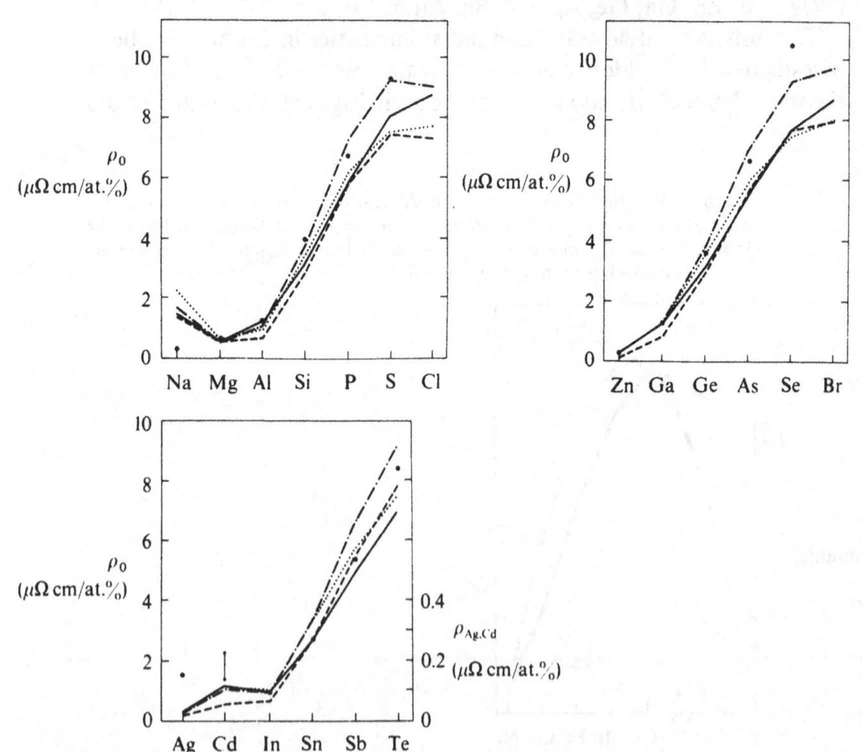

Table 6.6. *Theoretical and experimental residual resistivities in μΩ cm due to one at.% Cu, Ag and Au impurities in Cu, Ag and Au solvents. Both non-relativistic (Dickey et al. 1967) and relativistic (Brown & Morgan 1971) theoretical results are shown. The experimental results are also taken from Brown & Morgan*

		Resistivity		
Solvent	Solute	Non-relativistic	Relativistic	Experiment
Cu	Ag	0.06		0.07
	Au	0.11		0.49
Ag	Au	0.01	0.86	0.38
	Cu	0.06	0.02	0.07
Au	Cu	0.09		0.30
	Ag	0.01		0.32

have considered a range of other impurities in copper leading to the results summarised in Figure 6.18.

The residual resistivities of some 4d and 5d transition metal impurities in Au have been considered by Julianus *et al.* (Julianus & de Chatel 1984: Julianus *et al.* 1984).

Relativistic effects have been investigated by Brown & Morgan (1971) using the formalism developed by Morgan (1966). These results are particularly interesting in that they provide a possible explanation for the much larger resistivities of noble metal–noble metal alloys that contain Au compared to those that do not. The difference is attributed to relativistic effects which are expected to be important in all alloys containing atoms having an atomic number greater than ~70. Their results are shown in Table 6.6., together with the non-relativistic phase shift calculations of Dickey *et al.* (1967). The latter authors also considered a variety of other monovalent impurities in monovalent hosts including Li, Ni, K, Rb, Cs and vacancies. They generally found good agreement with experimental values in the alkali–alkali systems and alkalis with vacancies, although they have emphasised the need for accurate phase shift data. Their results for the alkali–noble metals are less convincing because of the neglect of the effects of bound and virtual bound states in the calculation.

As a more general observation it appears that multiple scattering effects are important with transition metal impurities in both Al and noble metal hosts, but less so with sp impurities. Both the cluster model of Lodder *et al.* and the KKR–Green's function calculation of Coleridge *et al.* adequately describe this effect, although the latter approach has the advantage of also including band structure effects and, perhaps for this reason, often gives a slightly better agreement with experiment. On the other hand, the cluster models have the advantage of being able to include charge transfer and lattice distortion effects in a direct manner. The simple form given by equation (6.18) also gives good results when used with muffin-tin phase shifts derived from the Mattheiss prescription (solid line in Figure 6.13(*b*) and dot–dash line in Figure 6.16). We will return to the scattering caused by magnetic (or nearly magnetic) impurities in Chapter 7.

6.7 Concentrated alloys

In Section 6.5 we noted that the Bloch theorem is generally not valid in concentrated alloys, due to the lack of translational periodicity. Furthermore, in concentrated alloys the impurity d-states may overlap on adjacent transition metal ions forming a more extended cluster d-band and ultimately an extended crystal d-band. The details of these electronic changes and hence the effects on electron scattering are likely to depend upon the detailed atomic arrangements. It should thus come as no surprise that the determination of the electrical resistivity of complicated concentrated alloys which may be quite inhomogeneous represents a formidable task.

One of the main problems in determining the properties of a concentrated alloy (even if completely homogeneous) thus lies in the averaging that must be performed over lattice sites or electron states in order to allow the calculations to proceed. We discuss firstly the simplest forms of averaging as embodied in the virtual crystal and rigid band approximations.

6.7.1 *First order theories: the virtual crystal and rigid band approximations*

In these (and most other) approximations the alloy is considered to be completely homogeneous. The virtual crystal approximation (briefly introduced in Chapter 4) then constructs an average lattice potential from the constituent ion potentials

$$\langle V \rangle = c_A V_A + c_B V_B. \tag{6.25}$$

However, such a prescription is of limited use for alloys containing transition metals as it will always predict a single d-band whereas many such alloys have two distinct d-bands well-separated in energy and each associated mainly with one of the alloy constituents. Some typical examples of this type of behaviour are found in the Cu–Zn and Cu–Ni systems (see e.g. Ehrenreich & Schwartz 1976). An alternative approach is the rigid band model discussed in Chapter 3. This is an entirely empirical approach whereby a fixed density of states $N(E)$ appropriate to the alloy of interest is assumed and is filled to the appropriate Fermi level E_F determined by the concentration n of valence electrons:

$$n = \int_{-\infty}^{E_F} N(E)\,dE. \tag{6.26}$$

This density of states may be based on a model calculation or experimental data and may or may not have any precise relationship to those of the alloy components. For example, the density of states for a typical transition metal alloy determined by Slater (1936) is shown in Figure 6.19. As the total number of valence electrons is increased by alloying, E_F shifts up the energy axis as indicated.

However, in many systems the density of states is far from constant. For example, the experimentally determined density of states for Ag–15%Pd and Ag–80%Pd are shown in Figure 6.20 and it is clear that their shapes are very composition dependent. In fact Stern (1967) has argued that the rigid band approximation is only expected to be valid in dilute alloys containing isoelectronic components.

Despite the obvious limitations of these approximations they can give a qualitative understanding of the resistivity of some concentrated alloys. The most celebrated example is perhaps, surprisingly (in view of the foregoing), again the Ag–Pd system. It is assumed that Pd contains an s-band similar to that of Ag, and a narrow d-band which overlaps the s-band, and with the Fermi level near the top of the d-band. Within the general spirit of the rigid band model, as Ag is added to Pd the s electrons

Fig. 6.19. Density of states for 3d transition metal alloys determined by Slater 1936. The location of E_F for various concentrations of valence electrons is indicated on the energy axis (after Slater 1936).

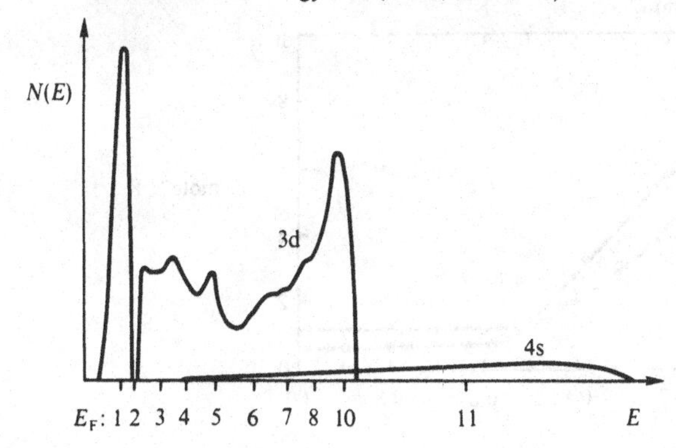

from the Ag atoms are supposed to fill the d-band leading to a decrease in $N(E_F)$ until the d-band is filled, whereupon it reaches a constant value typical of the s-band. The electronic specific heat coefficient γ and the magnetic susceptibility χ are both directly related to the density of states and so reflect this behaviour, as shown in Figure 6.21, with the d-band becoming completely filled at $\sim 60\%$Ag. On the basis of these results Mott assumed that there were about 0.6 holes per atom in the d-band and 0.6 electrons per atom in the s-band of pure Pd. As shown by Mott & Jones (1936), both the specific heat coefficient and susceptibility curves lead to the approximate description for the density of states in the unfilled d-band:

$$N_d(E_F) \sim (0.6 - c_A)^2, \quad c_A \leqslant 0.6, \tag{6.27}$$

where c_A is the atomic faction of Ag. In a later study Coles & Taylor (1962) used the electronic specific heat data in its raw form to give

$$N_d(E_F) \propto \frac{(\gamma_c - \gamma_{0.65})}{\gamma_{0.65}}, \tag{6.28}$$

Fig. 6.20. Experimentally determined density of states for Ag–15%Pd and Ag–80%Pd (after Norris & Myers 1971).

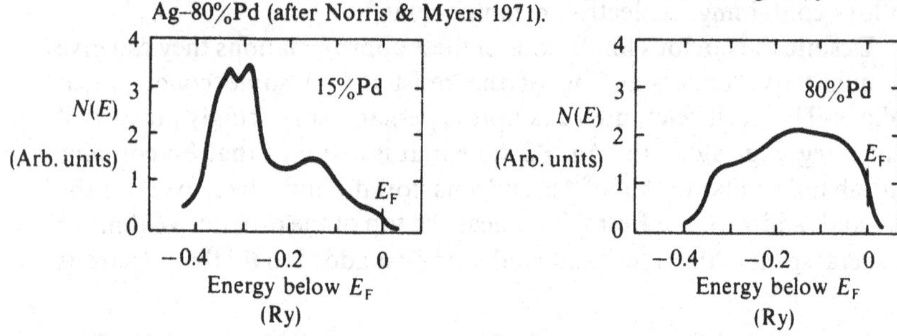

Fig. 6.21. Electronic specific heat coefficient γ (Montgomery *et al.* 1967) and magnetic susceptibility χ (Hoare *et al.* 1953) of Ag–Pd alloys as a function of composition.

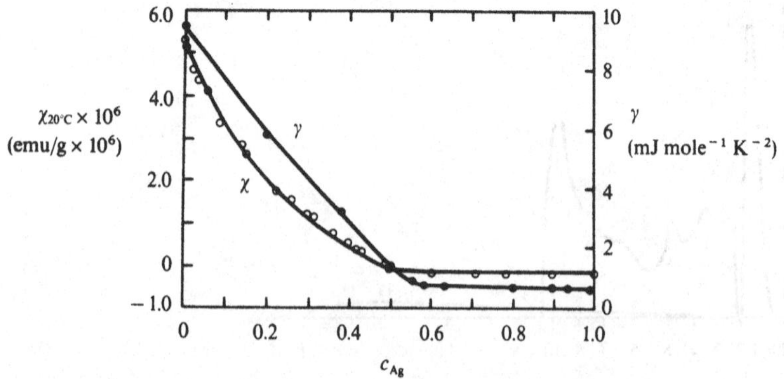

where γ_c is the value of γ at a composition c. If the atoms are randomly distributed the probability of scattering from an s-state to a d-state is given by

$$\int |\psi_d V^d \psi_s \, dS|^2 N_d(E_F), \tag{6.29}$$

where the virtual crystal approximation has been used to define a difference scattering potential V^d by analogy to equation (4.66). The magnitude of the scattering potential at a Pd site is then $c_A V^d$ (cf. equation (4.67)) and the fraction of such sites is $(1-c_A)=c_B$. The s–d scattering probability is thus proportional to

$$(0.6-c_A)^2 c_A^2 c_B, \tag{6.30}$$

where equation (6.27) has been used for $N_d(E_F)$. Since the s–s scattering probability is proportional to $c_A c_B$ (equation (5.20)), the total resistivity may be written as

$$\begin{aligned} \rho_{tot} &= \rho_{ss} + \rho_{sd} \\ &= A c_A c_B + |B(c_0 - c_A)^2 c_A^2 c_B|_{c_A \leqslant c_0}, \end{aligned} \tag{6.31}$$

where c_0 is the concentration of A atoms at which the d-band just becomes filled and A and B are constants. This simple function is shown fitted to the experimental data for Pd–Ag and Pd–Au alloys in Figure 6.22. The general agreement with experiment appears to support the

Fig. 6.22. Experimental and theoretical composition dependence of the residual resistivity of Pd–Ag (*a*) and Pd–Au (*b*): solid lines, experimental results (after Coles & Taylor 1962 and Kim & Flanagan 1967); dashed lines, fitted theoretical curves (equation (6.31)) with $c_0 = 0.65$ for Ag–Pd, 0.6 for Au–Pd.

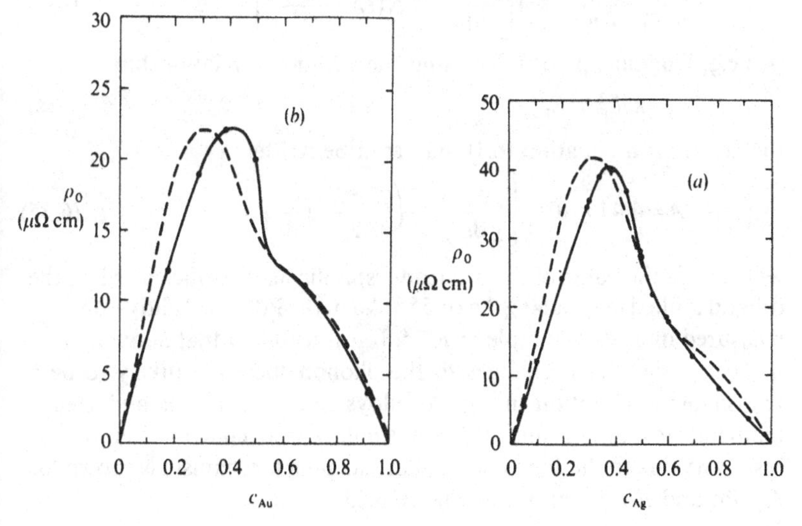

basic concepts of Mott's s–d scattering mechanism. Dugdale & Guénault (1966) have refined this approach to relax some of the restrictions of the rigid band model and allow a relative shift in the energies of the s- and the d-bands with composition. They also found a good agreement between the experimental and theoretical resistivity curves while allowing for the somewhat lower number of 0.35 electrons per atom in the Pd s-band, which is in better agreement with Fermi surface data (Vuillemin & Priestly 1965).

Note that small amounts of Ag or Au in Pd do not give a significant s–d scattering since the scattering potential is only present at the impurity atom sites. If any d-states of the impurity atoms lie well below the Pd d-band, the wavefunction of the empty d-states will have very small amplitude *at the site of the impurity atom* and so the scattering probability $|\langle\psi_d|V^d|\psi_s\rangle|^2$ will be small. At higher concentrations both the Pd and Ag sites act as scattering centres and so the s–d scattering rate will be appreciable. If the d-states of the impurity atoms are not well-separated from the host d-band, virtual bound states may cause a large scattering amplitude, as discussed in the previous section. Coles & Taylor have also shown how the thermal broadening mechanism of Jones (equations (6.1), (6.2)) may be related to the electronic specific heat via the rigid band model since

$$\frac{1}{N(E)}\frac{dN(E)}{dE}=2\frac{dN(E)}{dc}, \tag{6.32}$$

and

$$\frac{1}{N(E)}\frac{d^2N(E)}{dE^2}=4\left[\frac{dN(E)}{dc}+N(E)\frac{d^2N(E)}{dc^2}\right] \tag{6.33}$$

(see e.g. Dugdale, p. 278). By using the additional relationship

$$\gamma=\tfrac{2}{3}\pi^2k_F^2N(E_F), \tag{6.34}$$

the factor A in equation (6.1) may then be written as

$$A=4.41\times10^{-3}\left[\left(\frac{d\gamma_c}{dc}\right)^2-\left(\frac{\gamma_c-\gamma_{c_0}}{2}\right)\frac{d^2\gamma_c}{dc^2}\right], \tag{6.35}$$

where γ_{c_0} is the value of the electronic specific heat coefficient when the d-band is filled (e.g. at 60 %Ag or 55 %Au in the Pd-based alloys) and γ is measured in units of J mole^{-1} K^{-2}. If it is assumed that alloying does not cause significant changes to the phonon spectrum (likely to be a reasonable assumption in Ag–Pd alloys since the masses and Debye temperatures of Ag and Pd are similar), the contribution to the resistivity due to the Jones thermal broadening mechanism is shown for Ag–Pd and Au–Pd alloys in Figure 6.23.

While the good agreement between experiment and theory (even including the reproduction of the small bump in the thermal contribution to the resistivity of Ag–Pd alloys at ~50%Ag) cannot be taken as a justification of the rigid band model, it again supports the general features of the Mott s–d scattering model and emphasises the importance of band structure effects on the electrical resistivity of non-simple alloys (but see also the discussion in Section 6.7.2). At higher temperatures it might be expected that the $(1-AT^2)$ term could swamp the usual phonon scattering term leading to a decrease in the resistivity with increasing *T*. Indeed, Coles & Taylor did find a maximum in the resistivity of a Pd–39%Ag alloy at ~350 K which was attributed to such an effect. Similar behaviour was found by Ricker & Pflüger (1966) and Ahmad & Greig (1974) in Pd–40%Ag. However, Arajs *et al.* (1977) could find no such maximum in well-annealed Pd–Ag alloys, although some incipient behaviour was found when a Pd–40%Ag alloy was cold-worked, leading to the suggestion that strain and/or short range ordering effects might be the cause of the maximum, when observed. Rowland *et al.* (1974) sought similar behaviour in Pd–Au alloys and did find a dip in $d\rho/dT$ at ~700 K in alloys containing 40–70%Au but admitted that it was difficult to see how an s–d scattering mechanism could be involved at such high Au compositions. As a consequence, they also suggested that SRO effects might be the cause. Ricker & Pflüger

Fig. 6.23. The experimental and theoretical thermal contribution to the resistivity: (*a*) Ag–Pd alloys at 293 K (after Coles & Taylor 1962); (*b*) Au–Pd alloys at 195 K (after Kim & Flanagan 1967).

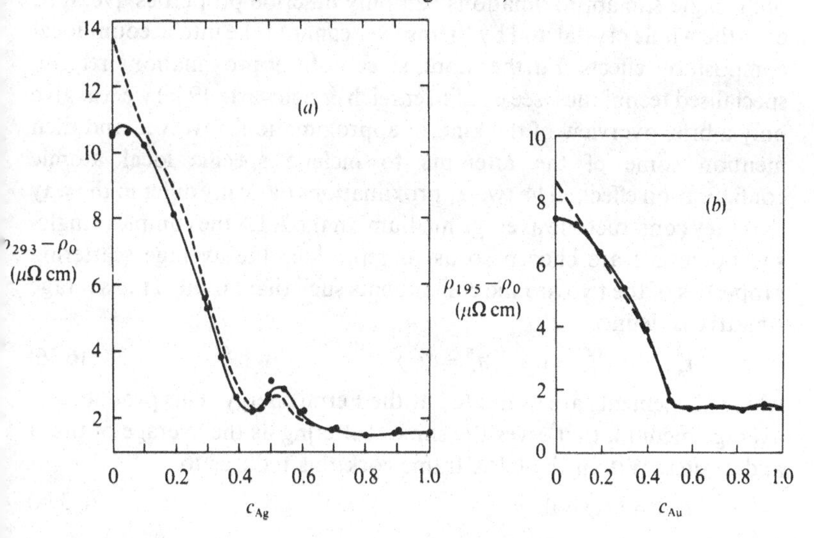

(1966) could find no maximum in Pd–Rh alloys up to ~15%Rh, these being prime candidates for such an effect. Reproducible maxima are found in a range of paramagnetic Cu–Ni alloys, but this behaviour has been attributed to magnetic clustering effects and will be discussed more fully in the next chapter. The elusive nature of such a maximum and its apparent sensitivity to specimen preparation would seem to argue against its being caused predominantly by a thermal broadening mechanism, although it may be important in reducing $d\rho/dT$ so that other effects such a short range ordering could be observed. Many of these alloys also have a maximum in the resistivity at low temperatures, which has also been attributed to a thermal broadening effect (Rowlands *et al.* 1971; Greig & Rowlands 1974). However, it now appears that such behaviour is more likely to result from local spin fluctuations in these enhanced paramagnets and this will also be discussed in the next chapter.

6.7.2 *Advanced theories: the average t-matrix approximation (ATA) and coherent potential approximation (CPA)*

We return now to the problem of selecting a suitable means of averaging over lattice sites but seek an improvement over the virtual crystal and rigid band approximations. Two approaches that are in current use are the average t-matrix approximation (ATA) and coherent potential approximation (CPA). In their basic form, both of these methods effectively decouple a single site which is then embedded in some average or effective medium. However, as we found in Chapter 2, such single-site approximations can only describe properties averaged over the whole crystal and by themselves cannot take into account local composition effects. Furthermore, since both approximations rely on specialised techniques (see e.g. Ehrenreich & Schwartz 1976) we will give only a brief overview of the kind of approximations involved and then mention some of the attempts to include specific local atomic configuration effects. The two approximations basically differ in the way that they construct the average medium. In the ATA the complex single-site potentials are chosen so as to reproduce the average scattering properties of the two atomic constituents such that an site n the average t-matrix is defined by

$$t_n^{\text{ATA}} = c_A t_n^A + (1 - c_A) t_n^B = \langle t_n \rangle, \tag{6.36}$$

where all elements are evaluated at the Fermi energy. This produces an average medium that gives the same scattering as the average of the A and B sites. Within the CPA framework this reduces to

$$t_n^{\text{CPA}} = \langle t_n \rangle = 0 \tag{6.37a}$$

or

$$\langle t_\alpha \rangle = \langle t_\beta \rangle = 0 \tag{6.37b}$$

in a two-sublattice model of long range order (Brauwers *et al.* 1975). The average medium is now constructed so that replacement of the effective potential at any site by the true potential of an A or B atom would, on average, produce no change in the scattering. In this approximation the potential of the effective medium (usually taken to have a muffin-tin form) must be obtained self-consistently, whereas in the ATA it is usually constructed from the constituent atom potentials. The ATA potential may thus be regarded as the starting point in the CPA iteration towards self-consistency. Being self-consistent the CPA generally leads to a better description of the effects of alloying on the electronic structure. A pictorial representation of these two approximations is given in Figure 6.24. The angled brackets $\langle\ \rangle$ indicate an average over all configurations (and hence there is no information relating to the detailed atomic distributions). In the ATA, the scattering is the same at every site of the effective medium (Figure 6.24(*a*), equation (6.36)). In the CPA, the potential associated with each effective atom is the coherent potential and is chosen so that the scattering introduced by replacing an effective atom by a real A or B atom is zero (Figure 6.24(*b*), equation (6.37)). Note that the CPA is directly analogous to the effective medium approach described in Section 5.5.1 in relation to the resistivity of multiphase materials. The atomic potentials used in these constructions must yield the correct band structures for the pure A and pure B components and be correctly placed on the same absolute energy scale. There should also be some means of incorporating charge transfer effects if there is any difference in valency.

Fig. 6.24. Pictorial representation of the definition of the effective medium for (*a*) ATA and (*b*) CPA.

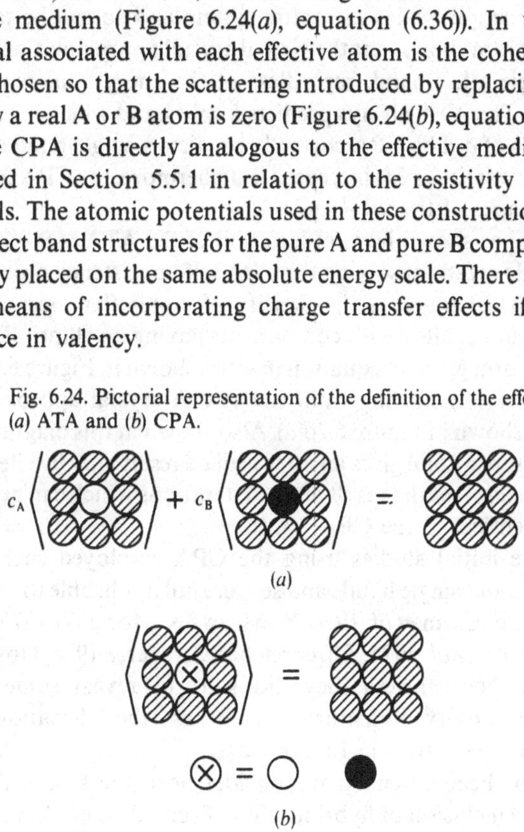

c_A + c_B =

(*a*)

=

=

(*b*)

An interesting comparison between the predictions of the virtual crystal, ATA and CPA techniques is given in Figure 6.25 which shows the predicted density of states for a single band A–50%B alloy in which the pure components have a density of states of the Hubbard form:

$$\begin{aligned} N(E) &= \left(\frac{2}{\pi W^2}\right)(W^2 - E^2)^{1/2}, \quad &|E| \leqslant W, \\ N(E) &= 0, &|E| \geqslant W, \end{aligned} \tag{6.38}$$

and the energy W is defined by

$$\left. \begin{aligned} \varepsilon^{\mathrm{A}} &= \tfrac{1}{2} W \delta, \\ \varepsilon^{\mathrm{B}} &= -\tfrac{1}{2} W \delta, \end{aligned} \right\} \tag{6.39}$$

where ε^{A} and ε^{B} are the one-electron energy levels of the pure components, such that δ is a dimensionless parameter

$$\delta = \frac{(\varepsilon^{\mathrm{A}} - \varepsilon^{\mathrm{B}})}{W}. \tag{6.40}$$

The energy W simply scales the entire system and so can be taken as unity. This diagram shows that the virtual crystal approximation always predicts a single alloy band, regardless of the energy difference δ or concentration. The ATA always predicts a split band which is incorrect for small values of δ. The CPA correctly predicts a single band at small values of δ which splits as δ is increased. Furthermore, the CPA correctly predicts the shapes of the sub-bands.

Chen (1973) has shown how an iteration of the ATA may provide a relatively simple means of solving the CPA self-consistency problem. As an example, the changes in the density of states that occur during iteration for a model alloy with components having an elliptical density of states of the form given in equation (6.40) is shown in Figure 6.26. The first iteration ($N = 1$) starting from the virtual crystal approximation gives the ATA shown in Figure 6.26(*a*). Also shown in this diagram is the CPA result. Figure 6.26(*b*) gives an intermediate result after five iterations, and Figure 6.26(*c*) gives the result after 30 iterations which can be seen to be virtually identical to the CPA result.

Many of the initial studies using the CPA employed such model density of states and single bands and so were not applicable to real alloy situations (see e.g. Levin *et al.* 1970; Yonezawa & Morigaki 1973; Elliott *et al.* 1974; Morita *et al.* 1975; Ehrenreich & Schwartz 1976; Hoshino & Watabe 1975). Nevertheless they did start to reveal some of the experimentally observed features, such as the deviation from Nordheim's rule, as shown in Figure 6.27.

The CPA has been extended to consider both the s- and d-bands allowing for the inclusion of hybridisation effects (Brouers & Vedyayev

Fig. 6.25. Comparison of the density of states calculated in (*a*) the virtual crystal approximation; (*b*) ATA; and (*c*) CPA for a model A–15%B alloy (after Velicky *et al.* 1968).

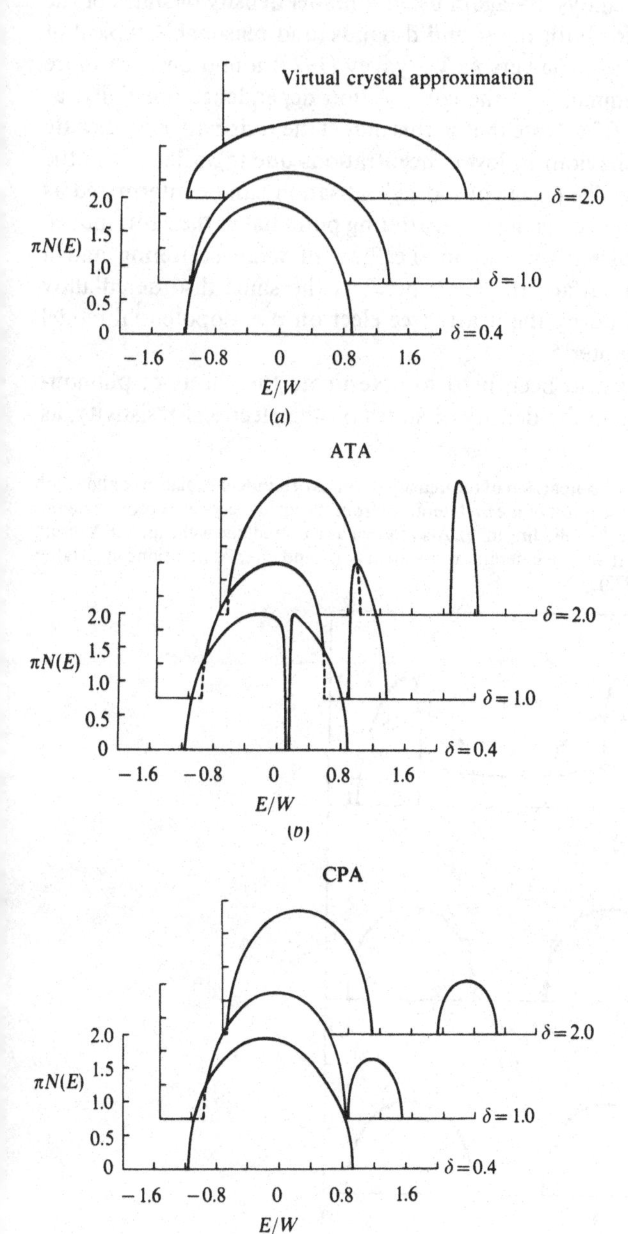

1972; Brouers & Brauwers 1975; Ehrenreich & Schwartz 1976; Elk *et al.* 1978; Paja 1981), essential to the consideration of noble metal and transition metal alloys. By again using a model density of states of the Hubbard form for both the s- and d-bands (and reasonably typical of transition metals), Brouwers & Vedyayev (1972) found an even more pronounced asymmetry in the composition dependence resistivity, as shown in Figure 6.28. Note that in this model the resistivity may deviate from a linear behaviour at low concentrations due to variations in the density of states. The effects of s–d hybridisation can be interpreted as having the same effect as the s–d scattering potential in the Mott model. Wang (1973) has shown that in the limit of weak scattering and a spherical Fermi surface the CPA predicts the same disordered-alloy resistivity as found by the nearly free electron pseudopotential model discussed in Chapter 5.

The CPA has also been used to investigate the effects of phonon-induced changes in the density of states on the electrical resistivity, as

Fig. 6.26. Comparison of the density of states of a model equiatomic alloy with $\delta = 2.0$. The result of the first iteration from the virtual crystal approximation is shown as the solid line in (*a*). Also shown as a dotted line is the full CPA result. The result after five iterations is shown in (*b*) and after 30 iterations in (*c*) (after Chen 1973).

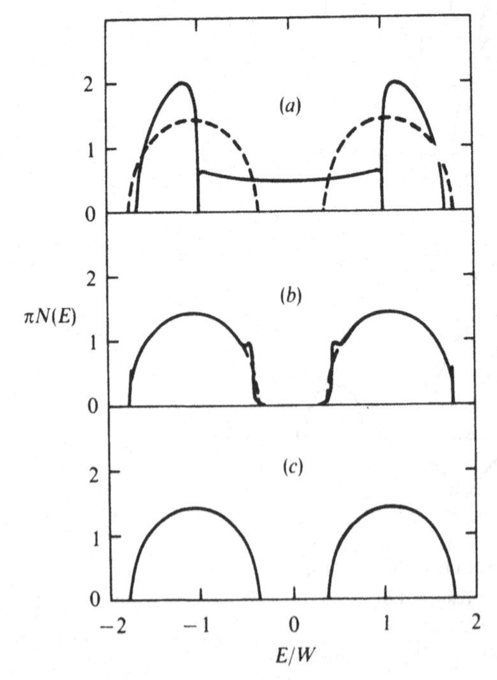

discussed in Section 6.1. Chen *et al.* (1972) have considered the effects of thermal disorder on a model alloy using the single band CPA and found that it causes a broadening and smearing of the density of states, as shown in Figure 6.29. In the weak scattering limit (small δ) the thermal

Fig. 6.27. Resistivity as a function of composition for a model A–B alloy with $\delta = 0.5$. The dotted line is the Nordheim ($c_A c_B$) approximation. The solid curves refer to the CPA results for electron concentrations per atom of $x = 0.2$ and $x = 0.5$ as indicated (after Levin *et al.* 1970).

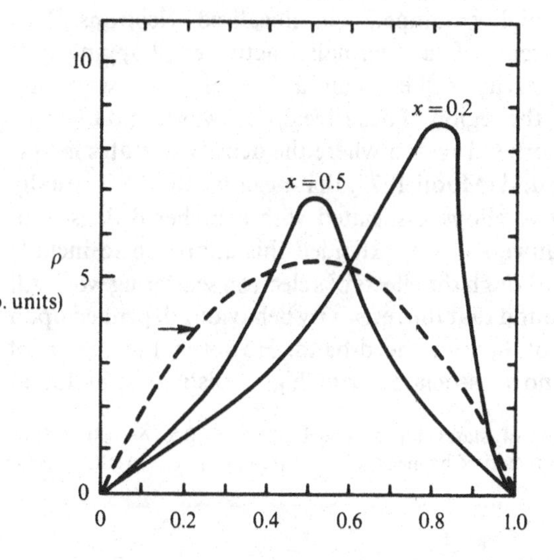

Fig. 6.28. Total residual resistivity as a function of concentration in the CPA two-band model (after Brouers & Vedyayev 1972).

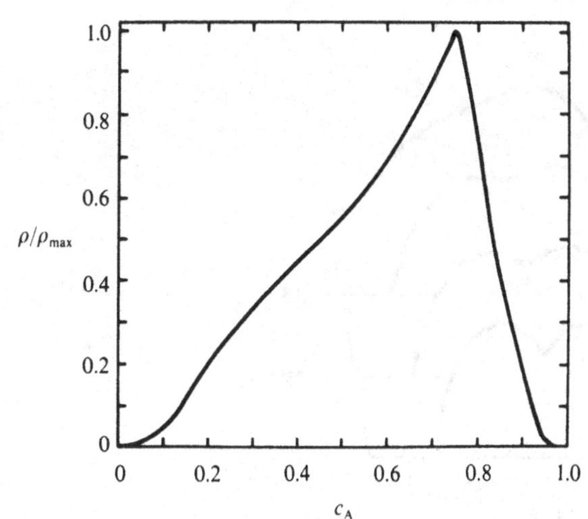

disorder effects cause an increase in the resistivity. However, in the strong scattering case (large δ) it was found that the resistivity could increase, decrease or remain constant with temperature, depending upon the location of E_F with respect to the energy band. While these authors have cautioned that for a concentrated, strongly scattering alloy, the contributing factors to the conductivity (as well as the formula itself) tend to lose their simple meaning, the resistivity decrease with increasing temperature is interpreted as a regime in which the thermal fluctuations assist the motion of highly damped 'quasilocalised' electrons. This behaviour is reminiscent of a thermally activated hopping-type mechanism of conductivity rather than a free electron scattering mechanism. However, the region of negative $d\rho/dT$ was found to occur when E_F lies in the interband region where the density of states is low. Experimentally it is found (Mooij 1973) that negative $d\rho/dT$ is usually found in highly resistive alloys associated with a higher d density of states. Brouers & Brauwers (1975) extended this approach to include both s- and d-bands (although the effects of s electron scattering were still neglected) and again found that the resistivity behaviour depended upon the relative position of E_F and the d-band. However, the region of negative $d\rho/dT$ was now associated with high resistivities, as found

Fig. 6.29. Density of states for a model alloy ($c=0.5$, $\delta=0.8$) at four temperatures characterised by the parameter α ($\propto T/T_{\text{melting point}}$) (after Chen *et al.* 1972).

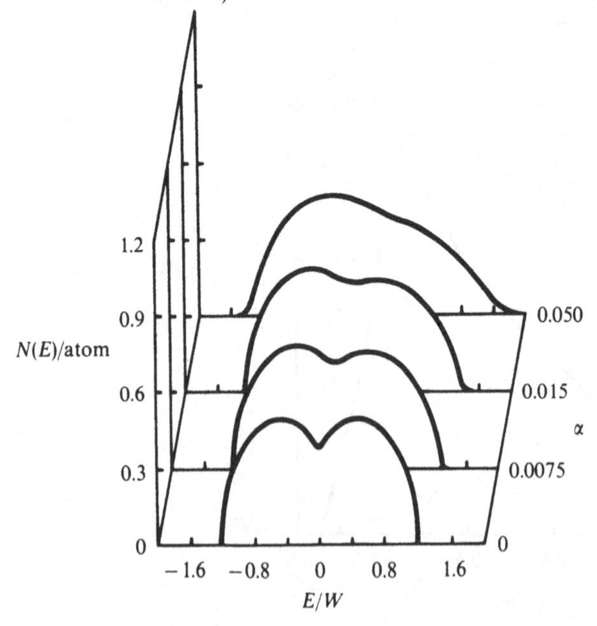

experimentally. Elk *et al.* (1979) also considered the problem using the two-band CPA but included the effects of s-electron scattering, the thermal smearing of the Fermi surface and strong electron correlations. They also found that a negative $d\rho/dT$ could be associated with high resistivity and that the magnitude of the variation of ρ with T was increased when compared to the uncorrelated case (Elk *et al.* 1978). However, in view of the model nature of the alloys studied, no detailed comparison with experiment is yet possible. Richter & Schiller (1979) used a modified version of this model and managed to show that the mixing of static atomic disorder and dynamic temperature-dependent scattering would result in a negative $d\rho/dT$ when the static disorder scattering was sufficiently large, corresponding to high residual resistivities. The problem of highly resistive alloys will be discussed further in Chapter 8.

The order-dependence of the resistivity of a one-orbital model alloy has been determined by Brauwers *et al.* (1975) using the CPA and the same elliptical form of the density of states decribed above. They determined the temperature variation of the resistivity from the completely ordered state ($S=1$ at $T=0$ K) to the order–disorder transition temperature shown in Figure 6.30, together with the $1-S^2$ variation predicted for the relaxation time by equations (5.59) and (5.60).

Paja (1976,) has also considered the effects of LRO in a one-orbital model alloy and found that ordering could produce either an increase or decrease in resistivity depending upon the alloy composition (see also

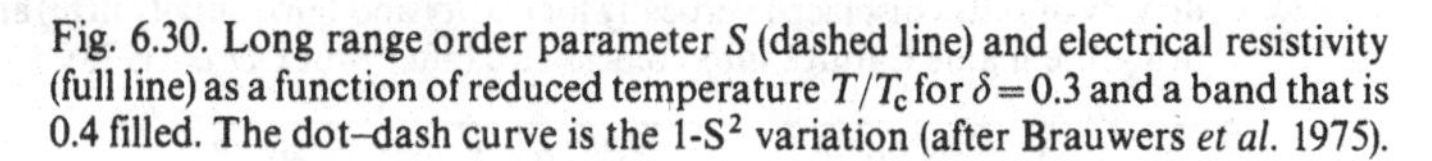
Fig. 6.30. Long range order parameter S (dashed line) and electrical resistivity (full line) as a function of reduced temperature T/T_c for $\delta=0.3$ and a band that is 0.4 filled. The dot–dash curve is the $1-S^2$ variation (after Brauwers *et al.* 1975).

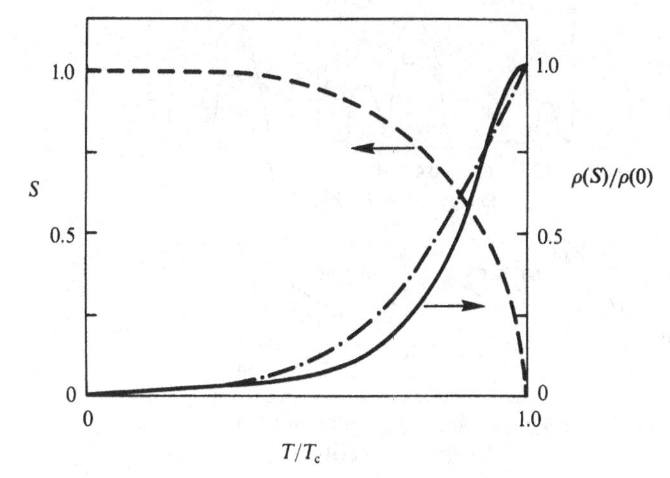

Borodachev *et al.* 1976). However, again such model studies should be generalised to take into account the s electrons in the conduction band before any meaningful comparison with experiment is possible.

Other studies using the CPA have included a calculation of the electronic structure of disordered alloys using KKR–Green's function techniques (see e.g. Ehrenreich & Schwartz 1976; Pindor *et al.* 1980; Gordon *et al.* 1981; Temmerman *et al.* 1978; Ebert *et al.* 1985). These have produced electronic structures which compare quite favourably with those determined experimentally, as shown in Figure 6.31. The general effects of ordering on the electronic structure have been considered by Plischke & Mattis (1973), Brouers *et al.* (1974*a*), Maksymowicz (1976), Kudrnovský & Velický (1977), Goedsche *et al.* (1978), and Paja (1981). The CPA has also been used to determine the electronic structure of an ordered-disordered interface by Parent *et al.* (1982), paving the way for a calculation of the electronic scattering from such interfaces. We might note that the optical properties of concentrated alloys have also been considered with reference to the single-band and two-band CPA by determining the frequency dependence of conductivity (Brouers *et al.* 1974*c*; Velický & Levin 1970).

A variation of the CPA using pseudopotentials has been used by Clark & Dawber (1972) to determine the residual resistivity of disordered Au–

Fig. 6.31. Density of states of a series of random Ag–Pd alloys: (*a*) 15%Pd; (*b*) 30%Pd; (*c*) 40%Pd; (*d*) 60%Pd; (*e*) 80%Pd. The full line gives the calculated density of states and the broken line gives the experimentally determined density of states displaced vertically for clarity and horizontally in (*a*) and (*b*) to bring the major features into coincidence (after Stock *et al.* 1973).

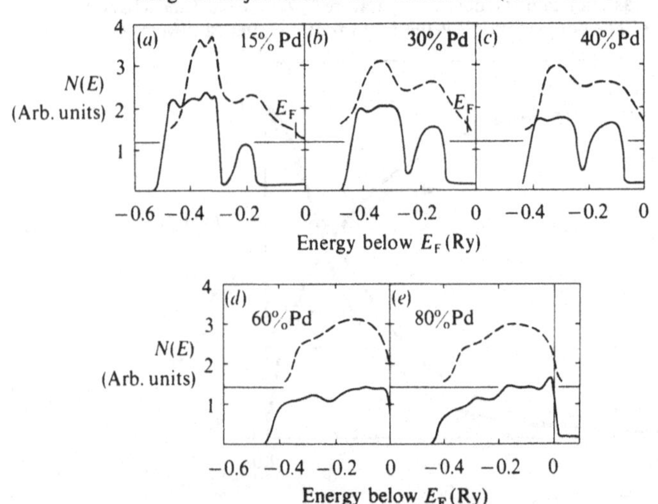

Ag alloys. They found essentially the same variation with composition as predicted by Nordheim's rule and also noted that consideration of both the s- and d-bands produced results that were virtually the same as obtained if only the s-band were used, again reinforcing the supposition that the noble metal alloys can be regarded as simple metals, at least as far as the resistivity is concerned. A further discussion of the coherent pseudopotential approximation has been given by Drchal & Kudrnovský (1976).

More recently the CPA calculations have been extended to include complete self-consistency so that charge transfer between the alloy components can be taken into account (Winter & Stocks 1983). As an example, the resistivity of disordered Ag–Pd alloys obtained by Butler & Stocks (1984) and Stocks & Butler (1981) is shown in Figure 6.32. As the only inputs to the calculation were the atomic numbers of Ag and Pd, the agreement with experiment is encouraging. Furthermore, they have noted that their results could not be interpreted strictly in terms of the simple s–d model of Mott. For example, the majority current carrying electrons (i.e. the s electrons in the Mott model) had predominantly d character in the Pd-rich alloys. The asymmetry in the composition dependence of the resistivity then comes largely from the higher Fermi velocity of the s electrons in silver, the average *scattering* rate being roughly the same in Ag–20%Pd as in Ag–80%Pd. It is hoped that this

Fig. 6.32. Calculated (dashed line) and measured residual (solid circles) resistivities of Ag–Pd alloys (after Butler & Stocks 1984).

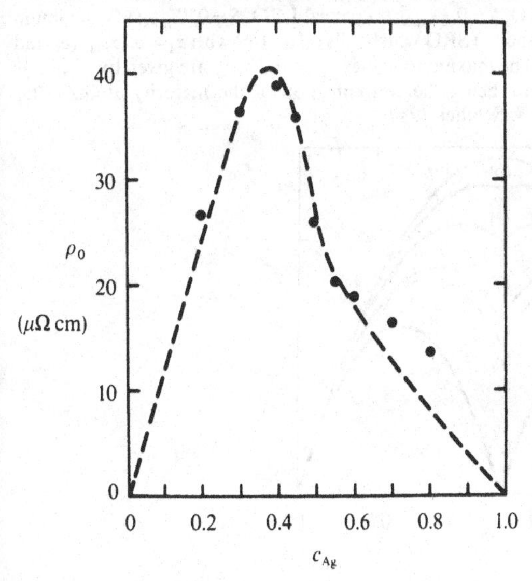

technique will be applied to obtain meaningful first-principles calculations of the resistivity of other disordered alloys containing transition metals.

Stern (1971) has noted that the self-consistency constraint on the potential demands that the perturbation of an impurity be more extended than allowed by the single-site CPA models described above. Attempts have been made to extend the theory to include scattering from small clusters embedded in the effective medium. In this way the effects of specific short range atomic configurations may also be taken into account (Tsukada 1972; Brouers *et al.* 1974*b*; House *et al.* 1974; Moorjani *et al.* 1974; Czycholl & Zittartz 1978; Batirev *et al.* 1980; Gonis & Freeman 1984; Gonis *et al.* 1984; Zin & Stern 1985). These studies have concentrated mainly on the electronic structure, although the transport properties study of Czycholl (1978), based on a model which included the effects of clustering and s–d hybridisation, do indicate significant differences to results obtained from the single site CPA. However, as is generally the case with these non-simple metals and alloys, these techniques are quite specialised and require powerful computing facilities. As such, they are still very much in the domain of the specialist rather than providing a general purpose tool. Furthermore, it is doubtful whether the current approaches can be extended beyond the first few shells of neighbours, leaving them far short of the 10–100 Å

Fig. 6.33. Dependence of the residual resistivity ρ/ρ_{max} upon alloy concentration for different degrees of SRO and LRO: (*a*) complete disorder, $S=0$; (*b*) partial LRO, $S=0.3S_{max}$; (*c*) partial LRO, $S=0.7S_{max}$; (*d*) maximum LRO, $S=S_{max}$; (*e*) and (*f*) SRO which falls off as $1/r^3$ with $\alpha_1=0.2\alpha_{max}$ (*e*), and $\alpha_1=-0.4\alpha_{max}$ (*f*). The maximum values S_{max} and α_{max} are given by $S_{max}=\frac{1}{2}c$ and $\alpha_{max}=(1-c)/c$, c being the concentration of the majority atoms (after Christoph, Richter & Schiller 1980).

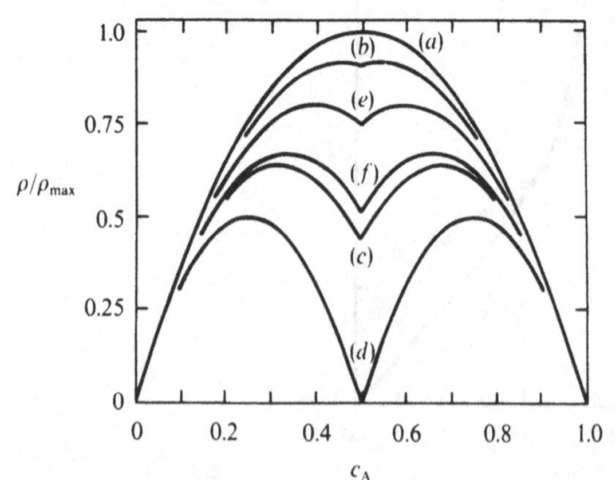

cluster sizes where the interesting resistivity behaviour described in Chapter 5 is observed. Nevertheless, it is only through such studies that techniques and databases suitable for 'non-expert' users may hopefully be developed.

Finally, the force–force correlation formalism briefly described at the end of Chapter 1 has been used to investigate the effects of long- and short-range order (Christoph *et al.* 1980; Richter & Schubert 1983). Some results are shown in Figure 6.33. This technique has also been applied to precipitation in Al–Zn (see also Schubert & Richter 1984) and a maximum found in the resistivity as a function of GP zone radius. However, this maximum occurs at a small cluster radius of about 1.5 times the nearest neighbour separation, which is much smaller than that found experimentally. This effect appears to be related to a scattering resonance occurring when $k_F \approx a_0$, much along the lines originally suggested by Mott (1937) but not currently in favour for explaining the observed behaviour. The electrical resistivity and thermoelectric power of spinodally decomposing alloys has also been considered using this formalism (Schubert & Löser 1985).

7 Magnetic and nearly magnetic alloys

The structure of magnetic material was discussed in some detail in Chapter 3 and we now want to consider how this magnetic structure effects the motion of conduction electrons. However, the purpose of this chapter is not so much to give a complete theory of the resistivity of magnetic materials but, rather, to alert the reader to some of the complications that can arise when the alloy being investigated is magnetic or nearly magnetic. We will thus concentrate mainly on those aspects of the resistivity of magnetic materials that have a dependence upon composition and atomic correlations. Furthermore, as much of the theory of the magnetic state is still being developed, some of the analysis will of necessity be qualitative rather than quantitative.

7.1 Magnetic materials with long range magnetic order

7.1.1 *Overview*

As noted in Section 3.1, there are still some aspects of the nature of the magnetic state at finite temperatures in transition metals that are uncertain. As such we cannot as yet give a comprehensive discussion about the variation of resistivity with temperature and the effects of alloying. References to some of the more general theories of the resistivity of magnetic materials are given in Table 7.1. We can, however, give a qualitative description by adopting the view that a clear separation is possible between the conduction and magnetic electrons and that they are scattered by deviations from perfect ordering of the magnetic moments. This allows a direct analogy between the magnetic case and the effects of atomic correlation discussed at some length in the previous two chapters. The dependence of the resistivity upon long and short range magnetic order can then be deduced from the long range and pairwise spin–spin correlation functions defined in equations (3.7) and (3.8). In a concentrated alloy the resistivity is then determined by the combined effects of the atomic scattering and the spin–disorder scattering. We will assume in this section that these scattering processes are independent and that the atomic arrangement remains unchanged over the temperature range of interest. However, it should be noted that the magnetic correlation parameters may not be independent of the atomic correlation parameters if the magnetic moment at a site depends (as is likely) upon the atom at that site and the nature of its immediate

Table 7.1. *General theories of resistivity of magnetic materials*

Magnetic state	Material	Reference
Ferromagnet	Pure metal	Kasuya (1956)
	Pure metal	de Gennes & Friedel (1958)
	Pure metal	Mills & Lederer (1966)
	Alloy	Masharov (1966)
	Dilute alloy	Mills *et al.* (1971)
	Alloy	Brouers *et al.* (1973)
	Alloy	Akai (1977)
	Pure metal	Joynt (1984)
Weak ferromagnet	Dilute alloy	Long & Turner (1970)
	Pure metal	Ueda & Morija (1975)
	Alloy	Goedsche *et al.* (1975, 1979)
Simple anti-ferromagnet	Pure metal	Yosida (1957)
Rare-earth	Pure metal	Elliott & Wedgwood (1963)
	Pure metal	Miwa (1963)

environment. We will further assume that there is complete spin mixing so that there is no distinction between the spin-up and spin-down sub-bands. Furthermore, as noted in Section 1.7, the change in energy caused by the scattering is small and so we will treat the scattering here as quasielastic.

The short-range interaction between the conduction electron and localised spin may be approximated by a delta-function scattering potential: $w(\mathbf{q})=\text{const.}$ By following the theoretical development given in Section 5.3.1 for atomic correlations, the resistivity due to long-range spin correlations may then be written as

$$\rho_{\mathrm{m}}=\rho_{\infty}\left[1-\frac{\langle \mathbf{S}\rangle^2}{S(S+1)}\right], \tag{7.1}$$

where

$$\rho_{\infty}=\text{const.}\ S(S+1), \tag{7.2}$$

and const. $=30\times10^{-6}\,\Omega$ cm for transition metals and $7.5\times10^{-6}\,\Omega$ cm for rare-earth metals (Weiss & Marrota 1959). By analogy with the Bragg–Williams theory, the variation of $\langle\mathbf{S}\rangle$ with temperature may be derived from the equivalent (single-site) molecular field approximation to give a resulting variation of ρ with T of the form shown in Figure 7.1. Also shown is the experimental data for the magnetic contribution to the resistivity of Gd.

As in the case of long range atomic ordering, long range antiferromagnetic ordering may introduce new energy gaps at superlattice Brillouin zone boundaries, which will modify this behaviour and possibly result in an increase in ρ just below T_N (cf. Section 5.3.2) (see e.g. Miwa 1962, 1963; Elliott & Wedgwood 1963; Ausloos 1976). There is in fact much variation in the observed temperature dependence of $\rho_m(T)$ as will be evident from a perusal of the experimental data given in Meaden (1965) and Schröder (1983). Probably the only reliable generalisation that can be made is that there will usually be a change in gradient of $\rho_m(T)$ at or near the magnetic transition temperature. This can often be used to gain information about the presence and composition of a magnetic phase.

The effects of short range magnetic correlations may be found by substitution of m_i for $\langle\sigma_i\sigma_j\rangle$ in (5.19) leading to (cf. (5.203)):

$$\rho_m = AS(S+1)\sum_i c_i m_i Y_i, \tag{7.3}$$

where A is a constant. As in the case of atomic SRO, magnetic SRO may cause an increase or a decrease in ρ_m depending upon the signs of m_i and Y_i. On the basis of the assumed delta-function scattering potential, a cusp-like anomaly at T_c (i.e. $d\rho_m/dT<0$ above T_c) is expected to result from magnetic SRO in a ferromagnetic system ($m_1>0$) only if the number of conduction electrons per atom $n\sim2$. If $n\sim1$ then $d\rho_m/dT$ is expected to be positive above T_c and so ρ should decrease smoothly through the transition region. Similarly, in a simple two sub-lattice antiferromagnet ($m_i<0$), $d\rho_m/dT$ will be positive if $n\sim2$ and negative if

Fig. 7.1. Reduced resistivity ρ_m/ρ_∞ as a function of reduced temperature T/T_c given by the molecular field approximation for $S=2$, $S=\frac{7}{2}$ and $S=\infty$. The points are experimental data for Gd corrected for the phonon scattering contribution (after de Gennes & Friedel (1958).

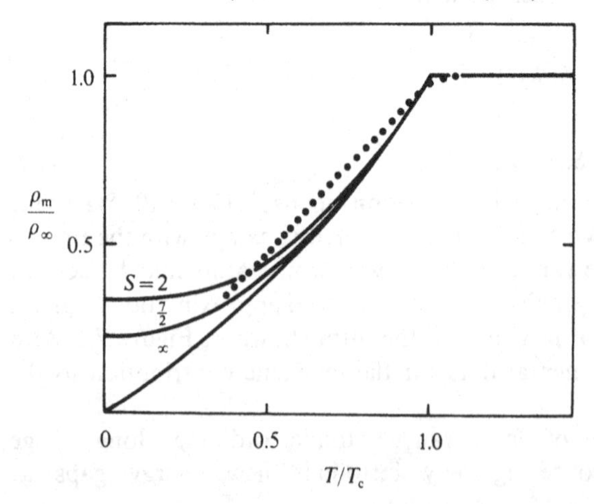

Table 7.2. *Magnetic short-range order parameters for Fe at the temperatures indicated. Also shown are values of c_i and Y_i, assuming a delta-function scattering potential and $n=0.7$*

Shell	m_i					
i	754 °C	790 °C	836 °C	854 °C	c_i	Y_i
0	1	1	1	1	1	1
1	0.201	0.149	0.111	0.111	8	−0.117
2	1.184	0.141	0.102	0.102	6	−0.051
3	0.158	0.107	0.081	0.081	12	0.033
4	0.141	0.090	0.064	0.064	24	−0.020
5	0.137	0.090	0.060	0.060	8	−0.026
6	0.115	0.068	0.051	0.047	6	−0.002
7	0.111	0.055	0.038	0.034	24	0.051
8	0.107	0.055	0.038	0.034	24	0.016
9	0.102	0.043	0.030	0.026	24	0.002
10	0.094	0.038	0.026	0.021	32	−0.009

Source: After Gersch *et al.* 1956.

Table 7.3. *Calculated and observed values of ρ_T/ρ_∞ for Fe*

Temperature	ρ_T/ρ_∞	
°C	Calculated	Experimental
754	0.78	0.78
790	0.83	0.84
836	0.88	0.87
854	0.88	0.88

Source: Experimental data from Arajs & Colvin 1964.

$n \sim 1$. A more detailed evaluation of (7.3) requires the parameters m_i which may be obtained from diffuse neutron scattering data. As an example, the correlation parameters obtained by Gersch *et al.* (1956) for Fe at various temperatures near T_c are shown in Table 7.2 together with the values of Y_i corresponding to $n=0.7$. Equation (7.3) then leads to the calculated values of ρ_m/ρ_∞ shown in Table 7.3, together with the experimental values of Arajs & Colvin (1964) which were corrected for a phonon contribution of $3.5\times10^{-2}\mu\Omega$ cm K^{-1} and $\rho_\infty=82.5\mu\Omega$ cm. These values were obtained from an extrapolation of the experimental data to high temperatures where it is assumed that the magnetic contribution to the resistivity is constant. The good agreement evident in

Table 7.2 must be regarded as fortuitous because of the approximations involved, but it does serve to illustrate the effects of spin–disorder scattering.

The discussion above has assumed that the magnetic structure is homogeneous. But, as discussed in Section 3.2.3, we know that this is not the case on a large scale due to the formation of magnetic domains. However, as also noted in that section, the domain walls are comparatively thick (~100–1000 Å) and, as this thickness is generally much greater than the electronic mean free path in most alloys, they simply represent a magnetically disordered phase of rather small volume fraction as far as the resistivity is concerned. In very pure materials which can have a much longer mean free path at low temperatures, this may not be the case and the domain walls can then make a significant contribution to the resistivity (see e.g. Cabrera 1977; Nascimento & Cabrera 1981). The magnitude of this contribution is $\sim 10^{-2} \mu\Omega$ cm (Ramanan & Berger 1981), again clearly indicating that it will be a negligible effect in alloys which usually have much larger residual resistivities. Nevertheless, the effects of domains may be apparent in the resistivity just at the Curie point. We will discuss this aspect further in Chapter 8.

Simple models such as those described above can also be used to gain specific information about the low temperature behaviour of the electrical resistivity. As discussed in Chapter 3, the low temperature excitations of the spin system are spin waves. In the absence of other scattering mechanisms, the resistivity due to coherent scattering from spin waves varies at T^2 at moderately low temperatures (see e.g. Mills & Lederer 1966). In the presence of other scattering mechanisms the coherency of the magnon scattering may be destroyed leaving an incoherent contribution which goes as $T^{3/2}$ (see e.g. Masharov 1966; Long & Turner 1970; Goedsche *et al.* 1979). However, there are other contributions to the temperature dependence of the resistivity that should also be considered. Firstly, there will be a contribution which arises from the interference between the atomic potential and magnetic scattering (see e.g. Yosida 1957; Koon *et al.* 1972; Levin & Mills 1974; Troper & Gomes 1975). Secondly, magnetic materials by definition have a complicated band structure and this must also be taken into account. As foreshadowed in Section 1.7.3, one of the lowest order effects of the band structure is the possibility of a different scattering rate for electrons having their spins aligned parallel to the magnetic moment in comparison to those with opposite spin direction. We will discuss this aspect of the problem in detail in the next section. There is also the possibility of thermal broadening and other band structure related effects, as discussed throughout Chapter 6.

Finally, from the brief discussion of the effects of alloying on the magnetic structure given in Chapter 3, it is clear that changes in the composition and atomic configuration of an alloy are likely to have a marked effect on the magnetic contribution to the resistivity as well as on the atomic contribution. Even though the details of these changes may not be clearly understood, it is still possible to obtain useful information about the state of the alloy. For example, in a two-phase system having only one magnetic phase, it is likely that the total magnetic contribution to the resistivity depends upon the volume fraction of that phase (provided of course that it is distributed in regions large compared to Λ_1). Resistivity studies can then be used to obtain information about the onset and rate of decomposition (see e.g. Schlawne 1983). Furthermore, as the Curie temperature is usually composition dependent, information about the composition of a phase can also be obtained by locating the changes in gradient of the resistivity–temperature curves at T_c. If the material contains two magnetic phases, this type of analysis may still be possible provided that their Curie points occur at sufficiently different temperatures (see Figure 3.16).

7.1.2 *Two-sub-band model*

As originally suggested by Mott (1935, 1964), the conduction electrons in a ferromagnetic material may be classified into two groups having spin directions either parallel or antiparallel to the magnetisation. Those electrons in the spin-up (↑) sub-band are conventionally regarded as having a spin parallel to the majority spin band (i.e. the direction of magnetisation) within a domain. Those in the spin-down (↓) sub-band have a spin parallel to the minority spin band. Note that this separation implies that the conduction electron mean free path Λ is much shorter than the size of the magnetic domains. This is generally true except for very pure materials at low temperatures. Conventions for assigning the spin direction from experimental results have been discussed by Dorleijn (1976). Elastic scattering does not change the population of the sub-bands, but inelastic (spin–flip) scattering will result in transfer of momentum between the two groups of electrons. In the simplest approximation it is assumed that the conduction bands contain nearly free electrons. The current is then carried by the nearly free spin-up and spin-down bands in parallel. This two-sub-band (or two-current) model has been described in detail by Fert & Campbell (1976) and Dorleijn (1976). The variational solution of the Boltzmann equation proceeds as described in Section 1.8.5, but now the trial function $\Phi_{\mathbf{k}}$ is given by a linear combination of two functions $\Phi_{\mathbf{k},\sigma}$ with $\sigma = \uparrow, \downarrow$, so that (1.49) becomes

$$\Phi_{\mathbf{k}} = \eta\uparrow\Phi_{\mathbf{k}\uparrow} + \eta_{\downarrow}\Phi_{\mathbf{k}\downarrow}. \tag{7.4}$$

As before, the coefficients $\eta_\uparrow$ and $\eta_\downarrow$ are adjusted to minimise equation (1.53). Fert & Campbell show that this leads to the expression

$$\rho = \frac{\rho_\uparrow\rho_\downarrow + \rho_{\uparrow\downarrow}(\rho_\uparrow + \rho_\downarrow)}{\rho_\uparrow + \rho_\downarrow + 4\rho_{\uparrow\downarrow}}, \tag{7.5}$$

where $\rho_{\uparrow\downarrow}$ describes the contribution of the spin–flip processes and each resistivity ρ_σ is given by the sum of the phonon and impurity scattering contributions:

$$\rho_\sigma = \rho_{0\sigma} + \rho_{p\sigma}(T), \tag{7.6}$$

i.e. Matthiessen's rule is assumed valid within each sub-band.

The three parameters which characterise the temperature dependent part of the resistivity viz, $\rho_{p\uparrow}(T)$, $\rho_{p\downarrow}(T)$ and $\rho_{\uparrow\downarrow}(T)$, are assumed to be independent of the type of impurity and so are characteristic of the host only. In general, the impurity scattering will contribute little to the $\rho_{\uparrow\downarrow}$ term (Monod 1968). As discussed in Section 3.2.5, the excitations within the spin system at low temperatures are spin waves (magnons), which give rise to an inelastic electron scattering. This mechanism introduces a temperature dependent resistivity contribution $\rho_{\downarrow\uparrow}(T)$ which must 'freeze out' as $T \to 0$ K. Another possible contribution to $\rho_{\uparrow\downarrow}$ comes from spin ↑–spin ↓ electron–electron interactions, but this is thought to be small (Bourquart *et al.* 1968). A final contribution to $\rho_{\uparrow\downarrow}$ can come from the spin–orbit interaction. This is independent of the impurity concentration and so is likely to be most noticeable in alloys having a low residual resistivity ($< 1\mu\Omega$ cm). Such a 'zero temperature' term $\rho_{\uparrow\downarrow}(0)$ was found necessary to give a phenomenological explanation of the residual resistivity of dilute magnetic alloys (Jaoul & Campbell 1975). In most reasonably concentrated alloys, this contribution can also be neglected. Thus, as $T \to 0$ K, $\rho_{\uparrow\downarrow}(T) \to 0$ and the resistivity is given by

$$\rho_0 = \frac{\rho_{0\uparrow}\rho_{0\downarrow}}{\rho_{0\uparrow} + \rho_{0\downarrow}}. \tag{7.7}$$

Note that even though the sub-bands are assumed to obey Matthiessen's rule (equation (7.6)), the total resistivity does not. At high temperatures the spins will be completely mixed so that the resistivity for each spin direction will be given by the simple average of the spin-up and spin-down resistivities:

$$\rho_\uparrow(T) = \tfrac{1}{2}[\rho_{0\uparrow} + \rho_{0\downarrow} + \rho_{p\uparrow}(T) + \rho_{p\downarrow}(T)], \tag{7.8}$$

and similarly for $\rho_\downarrow(T)$. Since there are two sub-bands in parallel, the measured resistivity is

$$\rho(T) = \tfrac{1}{4}[\rho_{0\uparrow} + \rho_{0\downarrow} + \rho_{p\uparrow}(T) + \rho_{p\downarrow}(T)]. \tag{7.9}$$

Table 7.4. *Values of the parameters used to interpret the sub-band resistivities of Ni-based alloys*

Temperature (K)	$\rho_{\uparrow\downarrow}(T)$ ($\mu\Omega$ cm)	$\rho_0(T)$ ($\mu\Omega$ cm)	$\rho_{p\uparrow}(T)$ ($\mu\Omega$ cm)	$\rho_{p\downarrow}(T)$ ($\mu\Omega$ cm)	μ
77	0.9 ± 0.3	0.32	0.38	1.9	5 ± 1
200	5 ± 2	2.5	3	15	5 ± 2
300	11 ± 4	5.4	6.7	27	4 ± 2

Source: After Fert & Campbell 1976.

The apparent residual resistivity from a high temperature measurement is then

$$\rho_0(\mathrm{HT}) = \tfrac{1}{4}[\rho_{0\uparrow} + \rho_{0\downarrow}]. \tag{7.10}$$

Thus if ρ_0 is obtained from both high and low temperature measurements, $\rho_{0\uparrow}$ and $\rho_{0\downarrow}$ can be found from (7.7) and (7.10) (see e.g. Farrel & Greig 1968). However, to obtain $\rho_0(HT)$ from (7.9) requires knowledge of $\rho_{p\uparrow}(T)$ and $\rho_{p\downarrow}(T)$. Fert & Campbell (1976) have cautioned that $\rho_{p\sigma}(T)$ may differ from the pure host $\rho(T)$ and have given the values appropriate to Ni-based alloys shown in Table 7.4. Values of $\rho_p(T) = \frac{1}{4}[\rho_{p\uparrow}(T) + \rho_{p\downarrow}(T)]$ for some Fe based alloys are shown in Figure 7.2. On the basis of the known band structure of Fe (i.e. a similar density of states for the $\uparrow$ and $\downarrow$ bands at the Fermi energy), they suggest that $\mu = \rho_{p\uparrow}(T)/\rho_{p\downarrow}(T) = 1$. At intermediate temperatures the resistivity will change because of the increasing thermal contribution (equation (7.6)), as well as the increased spin mixing due to magnon scattering ($\rho_{\uparrow\downarrow}(T)$). This gives rise to large deviations from Matthiessen's rule such as have been observed in dilute alloys of Fe, Ni and Co with transition metal impurities over the temperature range 1–300 K. Some typical results for two Fe–Cr alloys are shown in Figure 7.3.

From (7.5) and (7.6),

$$\rho(T) = \frac{[\rho_{0\uparrow} + \rho_{p\uparrow}(T)][\rho_{0\downarrow} + \rho_{p\downarrow}(T)] + \rho_{\uparrow\downarrow}(T)[\rho_{0\uparrow} + \rho_{p\uparrow}(T) + \rho_{0\downarrow} + \rho_{p\downarrow}(T)]}{\rho_{0\uparrow} + \rho_{p\uparrow}(T) + \rho_{0\downarrow} + \rho_{p\downarrow}(T) + 4\rho_{\uparrow\downarrow}(T)} \tag{7.11}$$

Using different alloys of different compositions, the different parameters can in principle be evaluated at each temperature (see Durand & Gautier 1970; Loegel & Gautier 1971 and appendix in Fert & Campbell 1976).

The variation of $\rho_{\uparrow\downarrow}(T)/T^2$ with temperature for Ni derived by Fert & Campbell with the aid of these techniques is shown in

Figure 7.4. This variation is as expected for scattering by magnons (Bourquart *et al.* 1968) since at moderate temperatures, $\rho_{\uparrow\downarrow}(T) \propto T^2$, as noted in the previous section. The dip in $\rho_{\uparrow\downarrow}(T)/T^2$ below 25 K can be attributed to a separation of the spin-up and spin-down Fermi surfaces. When the magnon wavevectors **q** can no longer span this gap, $\rho_{\uparrow\downarrow}(T)$

Fig. 7.2. Temperature-dependent part of the resistivity $\rho_p(T)$ as a function of T for Fe–Cr (●), Fe–Mn (○), Fe–Ni (□) and pure Fe(■) (after Fert & Campbell 1976).

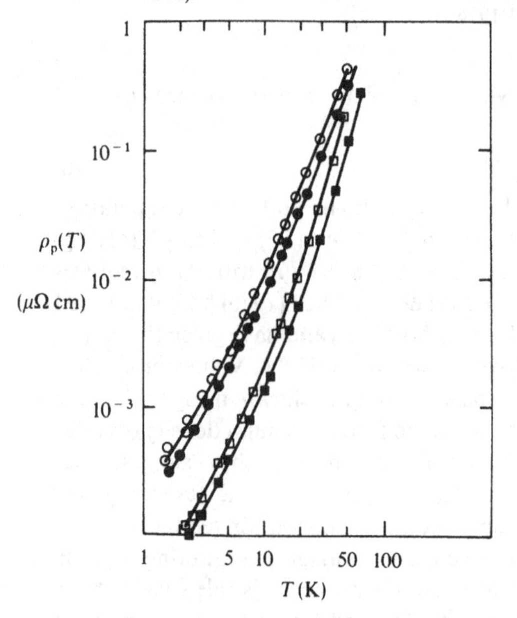

Fig. 7.3. Deviation from Matthiessen's rule for two Fe–Cr alloys; ○, Fe–0.6% Cr; ●, Fe–0.3%Cr (after Fert & Campbell 1976).

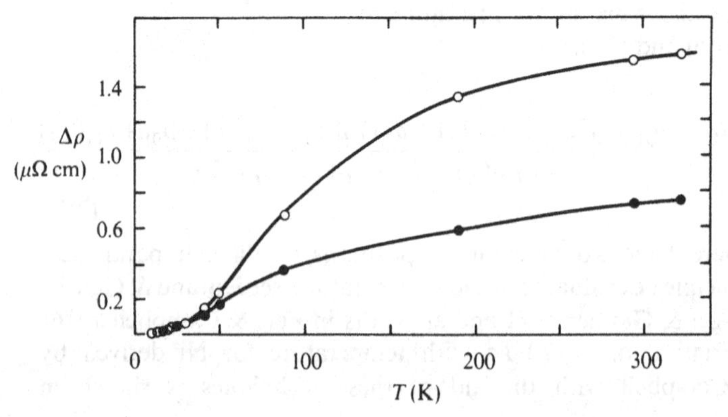

should decrease exponentially with T. The actual form of this contribution may be written as (Fert 1969)

$$\rho_{\uparrow\downarrow}(T) = \text{const. } T^2 F\left(\frac{\theta}{T}\right), \tag{7.12}$$

where

$$F(x) = x\left[\frac{x}{1-e^{-x}} - \log(e^x - 1)\right], \tag{7.13}$$

and θ is a characteristic temperature of the magnons which incorporates the size of the gap between the $\uparrow$ and $\downarrow$ Fermi surfaces and spin wave stiffness. The function $F(x)$ is shown plotted in Figure 7.5 and is indeed very similar to the higher temperature part of the experimentally determined variation (Figure 7.4). At low temperatures the magnon scattering may become incoherent as described in the previous section leading to a $T^{3/2}$ power law. The upturn at $T \leqslant 10$ K in Figure 7.4 is just of this form. If one assumes that for $T \geqslant 100$ K, $\rho_{\uparrow\downarrow}(T) = AT^2$ (Dorleijn 1976; Kaul 1977; Farrell & Grieg 1968), it is possible to derive the variations of $\rho_{p\uparrow}(T)$ and $\rho_{p\downarrow}(T)$ with temperature. Some results for pure Ni are shown in Figure 7.6 and indicate a decreasing ratio μ with increasing temperature. At low temperatures μ should be approximately constant since $\rho_{\uparrow\downarrow}(T)$ then tends to zero. The general form of this behaviour is in accord with the Mott s–d scattering model. As the temperature is increased, the probability for $s^\uparrow$ to $d^\uparrow$ scattering increases as the density of states in the band increases with decreasing spontaneous magnetisation. At the Curie temperature both the d-subbands are equally populated so that $\rho_\uparrow \approx \rho_\downarrow$.

Another way of investigating the sub-band resistivities which avoids the uncertainties in the temperature-dependent terms involves

Fig. 7.4. The variation of $\rho_{\uparrow\downarrow}(T)/T^2$ with T for Ni (after Fert & Campbell 1976).

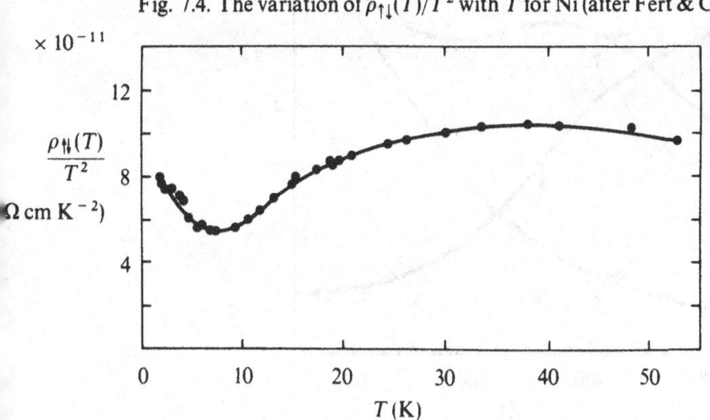

measurement of the low temperature resistivities of samples containing two different impurities. The low temperature resistivity for a dilute ternary alloy containing A and B impurities is then

$$\rho = \frac{(\rho_{A\uparrow} + \rho_{B\uparrow})(\rho_{A\downarrow} + \rho_{B\downarrow})}{\rho_{A\uparrow} + \rho_{B\uparrow} + \rho_{A\downarrow} + \rho_{B\downarrow}}. \tag{7.14}$$

This implies a deviation from Matthiessen's rule:

$$\begin{aligned}\Delta\rho_0 &= \rho_0(AB) - \rho_0(A) - \rho_0(B) \\ &= \frac{(\alpha_A - \alpha_B)^2 \rho_0(A)\rho_0(B)}{(1+\alpha_A)^2 \alpha_B \rho_0(A) + (1+\alpha_B)^2 \alpha_A \rho_0(B)},\end{aligned} \tag{7.15}$$

Fig. 7.5. The magnon scattering function $F(\theta/T)$ as a function of reduced temperature T/θ.

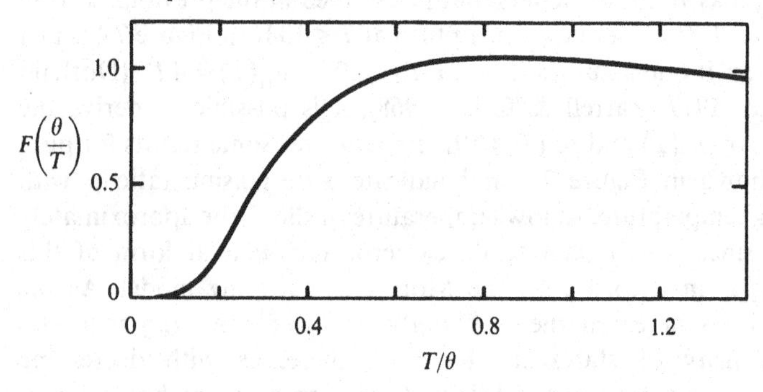

Fig. 7.6. Temperature dependence of the spin-up and spin-down resistivity and the ratio $\mu = \rho_{p\downarrow}(T)/\rho_{p\uparrow}(T)$ for pure Ni (after Kaul 1977).

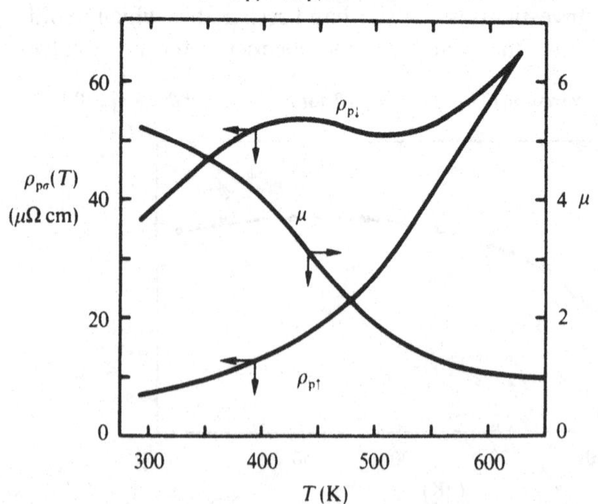

where $\alpha_A = \rho_{A\downarrow}/\rho_{A\uparrow}$, $\alpha_B = \rho_{B\downarrow}/\rho_{B\uparrow}$. The residual resistivities of ternary Fe- or Ni-based alloys containing various pairs of impurities do indeed show a large DMR which can be fitted to (7.15) to obtain values of α_A, α_B and hence $\rho_{A\uparrow}, \rho_{A\downarrow}, \rho_{B\uparrow}, \rho_{B\downarrow}$ from

$$\rho_{A\uparrow} = \frac{1}{1+\alpha_A}\rho_\uparrow, \qquad \rho_{A\downarrow} = \frac{\alpha_A}{1+\alpha_A}\rho_\downarrow, \tag{7.16}$$

and similarly for ρ_B (Fert & Campbell 1971, 1976). The results for some 3d impurities in Ni are shown in Figure 7.7. Dorleijn (1976) has shown that the values of $\rho_{\sigma A}$ or $\rho_{\sigma B}$ obtained by this method are not dependent on the particular partner *B* chosen. These results can be interpreted in terms of the Friedel virtual bound state model discussed in the previous chapter. For Co, Fe and Mn impurities, the virtual bound state resonance lies below the Fermi level. The resistivity $\rho_{0\uparrow}$ is then due only to the weaker s–s scattering and thus is smaller than $\rho_{0\downarrow}$. For Cr, V and Ti impurities, the virtual bound state is pushed through the Fermi level leading to a peak in $\rho_{0\uparrow}$ at Cr. Similar results have been obtained for 3d impurities in Fe (Fert & Campbell 1976) and Co (Durand & Gautier 1970), as shown in Figure 7.8, and for heavier transition impurities in Ni

Fig. 7.7. The residual resistivity of 3d impurities in Ni for each spin direction: solid lines, Fert & Campbell (1976); dotted lines, Dorleijn & Miedema (1975*a*).

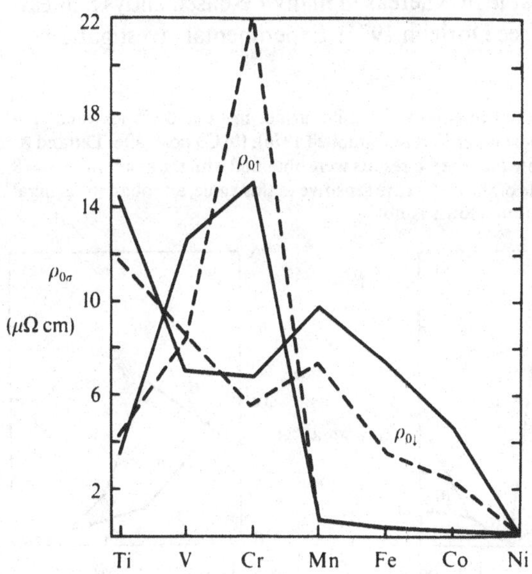

and Co (Durand & Gautier 1970; Loegel & Gautier 1971; Dorleijn & Miedema 1975*a*), as shown in Figures 7.9 and 7.10. These results again indicate the existence of virtual bound states and the existence of a period effect (i.e. a dependence upon the position of the impurity in the periodic table).

Different sub-band resistivities will lead to an anisotropy that appears as a dependence of ρ upon the direction of the current in relation to the direction of magnetisation. The magnitude of this effect is expressed as the ratio

$$\frac{\Delta\rho}{\rho}=\frac{\rho_{\parallel}-\rho_{\perp}}{\rho_{\parallel}}, \tag{7.17}$$

where $\rho_{\parallel}$ and $\rho_{\perp}$ are the resistivities for the current parallel and perpendicular to the saturation magnetisation, respectively, and extrapolated to zero magnetic induction **B**. In the actual measurement of this quantity an external field is required to saturate the specimen in the required direction. An extrapolation is then required over the demagnetising field of $\sim\mu_0 M_s$ to correct for the Kohler (1938) magnetoresistance:

$$\frac{\Delta\rho(\mathbf{B})}{\rho_0}=f\left(\frac{\mathbf{B}}{\rho_0}\right), \tag{7.18}$$

where f is a function characteristic of the material which is independent of ρ and T. For most Ni-based alloys (except Ni–Cu) a quadratic function gives a reasonable fit whereas in many Fe-based alloys a linear extrapolation is better (see Dorleijn 1976). Experimental investigation of

Fig. 7.8. The residual resistivity of 3d impurities in Fe and Co for each spin direction: (*a*) Fe host (after Fert & Campbell 1976); (*b*) Co host (after Durand & Gautier 1970). Note that the Co results were obtained with the assumption $\mu=8$ and the magnitude of the results are sensitive to this value, although the general form of the variation shown is not.

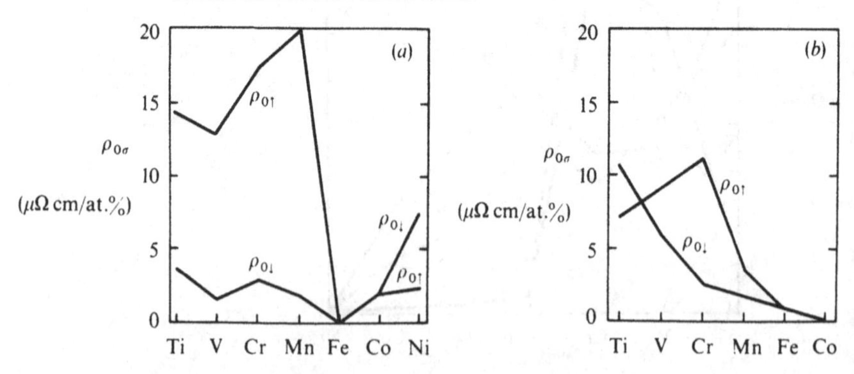

the function in (7.18) will indicate whether any correction to these forms is necessary. Campbell *et al.* (1970) have discussed this anisotropy in terms of the two-band model and a spin–orbit mechanism proposed by Smit (1951) and derived the formula

$$\frac{\Delta\rho}{\rho}=\frac{\gamma(\rho_{\downarrow}-\rho_{\uparrow})\rho_{\downarrow}}{\rho_{\uparrow}\rho_{\downarrow}+\rho_{\uparrow\downarrow}(\rho_{\uparrow}+\rho_{\downarrow})}, \tag{7.19}$$

where γ is a constant that is dependent upon the strength of the spin–orbit iteration. However, while this expression has met with some success (Kaul 1977; Campbell *et al.* 1970), it has been found somewhat lacking in other studies (Dorleijn & Miedema 1975*b*; Dorleijn 1976). Muth & Christoph (1981) have derived an alternative expression using

Fig. 7.9. Sub-band resistivities of 4d (*a*) and 5d (*b*) transition metal impurities in Ni (after Dorleijn & Miedema 1975*a*).

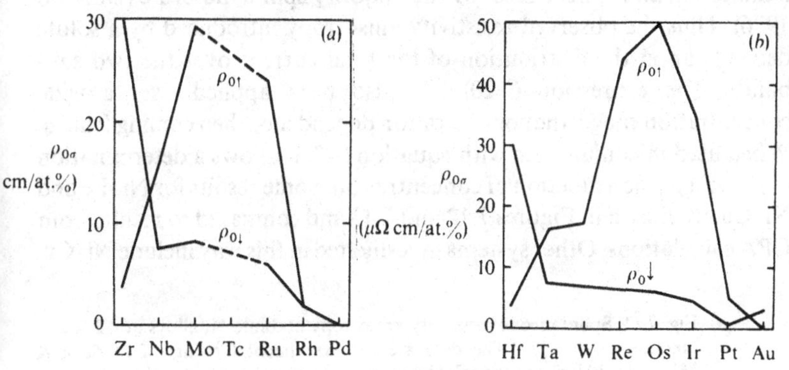

Fig. 7.10. Sub-band resistivities of 4d transition metal impurities in Co (after Durand & Gautier 1970).

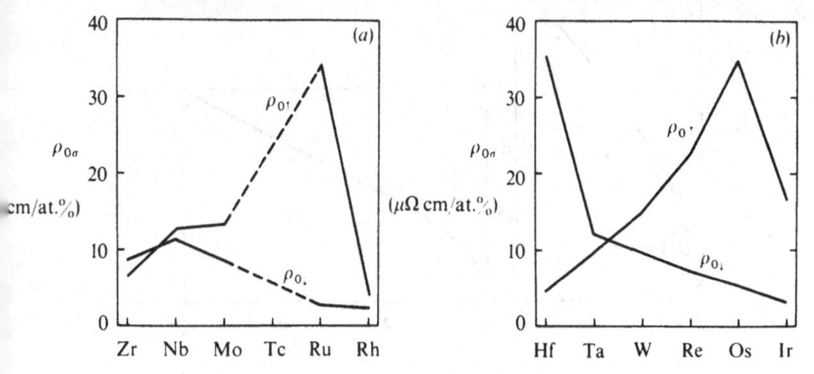

the force–force correlation function method (see Section 1.8.6). At low temperatures this is

$$\frac{\Delta\rho}{\rho}=\left(\frac{\gamma_1}{\alpha}-\gamma_2\right)\frac{\mu^2-\alpha^2}{1+\alpha}, \tag{7.20}$$

where, in the dilute (independent scattering) limit, γ_1, γ_2 and μ do not depend upon the concentration of the scatterers. Experiments confirm that this is true up to an impurity concentration of about 6% (Campbell *et al.* 1970; Dorleijn 1976), beyond which the impurities presumably start to strongly interact. Figure 7.11 shows this expression fitted to the dilute Ni alloy results of Dorleijn & Miedema (1975*b*). The parameters used were $\gamma_1=-0.0425$, $\gamma_2=0.01$ and $\mu=0.22$. As implied by (7.20) the resistivity anisotropy is different for different solutes. This might be expected since the anisotropy is positive for the majority spin bands of Ni and the minority spin band of Fe, is negative for the minority spin band of Ni, and is near zero for the majority spin band of Fe (Dorleijn 1976). Thus the observed resistivity anisotropy introduced by a solute depends upon the distribution of the total current over the two sub-bands. The expression (7.20) has also been applied over a wide concentration range, the concentration dependence then coming from α. When used in conjunction with equation (7.7) it allows a determination of $\rho_{0\uparrow}$ and $\rho_{0\downarrow}$ as a function of concentration. Some results for Ni–Fe and Ni–Cu are shown in Figures 7.12 and 7.13 and compared to results from CPA calculations. Other systems investigated in this way include Ni–Cu,

Fig. 7.11. Spontaneous resistivity anisotropy of dilute Ni alloys at 4.2 K as a function of $\alpha=\rho\downarrow/\rho\uparrow$. The dots are the experimental points of Dorleijn & Miedema (1975*b*) and the solid lines are (7.18) with the values of the parameters given in the text: (*a*) $\alpha<3$; (*b*) $\alpha>3$ (after Muth & Christoph 1981).

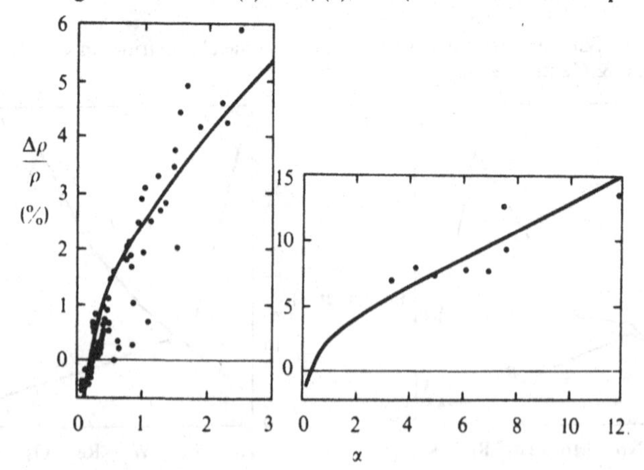

Fig. 7.12. Sub-band resistivities of Ni–Fe alloys at 4.2 K (after Muth & Christoph 1981). The dashed curves are from the CPA calculations of Akai (1977).

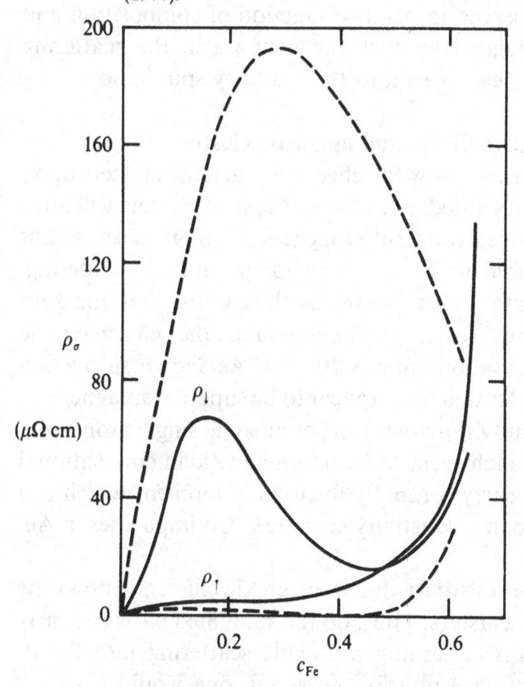

Fig. 7.13. Sub-band resistivities of Ni–Cu alloys at 4.2 K (after Muth & Christoph 1981). The dashed curves are from the CPA calculations of Brouers *et al.* (1973).

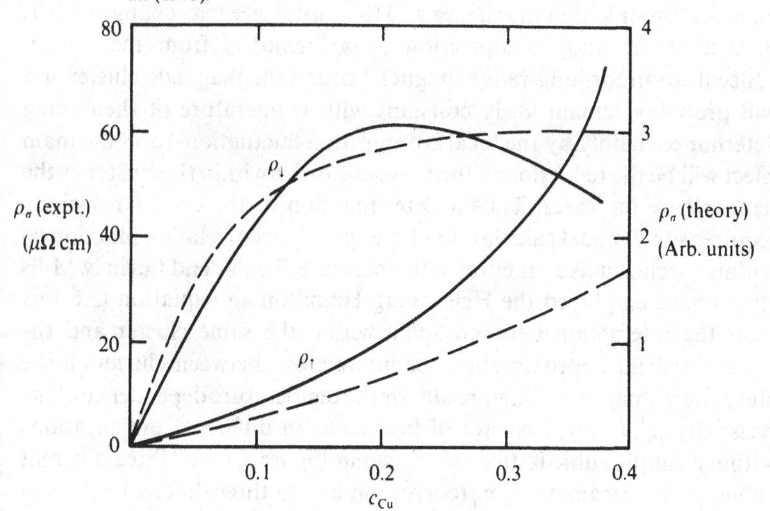

Ni–Co (Muth & Christoph 1981), Ni–Mn (Muth 1983*a*), Ni–Cr, Ni–Mo, Ni–W, Ni–Ru, Ni–V (Muth 1983*b*). Muth (1985) has shown that a maximum is usually observed in $\rho_{0\uparrow}$ as a function of composition and suggested that this is related to a strong increase in the scattering probability of the spin $\downarrow$ electrons into the minority spin band.

7.2 Local environment effects and magnetic clusters

We want to consider now the effects of magnetic clusters upon the electrical resistivity. As noted in Section 3.2.2, such clusters will form in alloys containing a concentration of a magnetic component somewhat smaller than that required to sustain long range magnetic ordering. Implicit in the discussion which follows is the notion that the spin fluctuation time of the cluster is longer than the characteristic conduction electron relaxation time $\sim 10^{-14}$ s. As shown in Section 3.2.2, this includes particles that may appear to be superparamagnetic in a bulk magnetisation study. Furthermore, even some single atoms in a non-magnetic matrix, which seem to be non-magnetic in conventional bulk studies, do in fact carry a rapidly fluctuating moment which can appear to be quite static in a resistivity study (e.g. Co impurities in Au; Henger & Korn 1981).

The calculation of the resistivity due to magnetic clusters follows the same path as for atomic clusters. Thus, on the simplified basis of nearly free conduction electrons (assuming negligible scattering into the d-band) and a delta function scattering potential, one would expect a magnetic cluster to cause an initial increase in ρ_m if the cluster size is less than Λ_l and if $n \sim 2$. However, in the magnetic case both the degree of magnetic order within the cluster and the cluster size may change with T. In fact, if the alloy composition is far removed from the critical concentration for long range magnetic order, the magnetic cluster size will probably remain fairly constant with temperature (it then being determined mainly by the local composition fluctuations) and the main effect will be the reduction in the magnetic order within the cluster as the temperature increases. Thus a determination of the cluster resistivity requires some model calculation of the spin–spin correlation function m_i within the cluster as a function of temperature. To this end Levin & Mills (1974) have employed the Heisenberg Hamiltonian (equation (3.6)) to treat the interaction between spins within the same cluster and the molecular field approximation for interactions between clusters if the alloy is ferromagnetic. Their result for the temperature dependence of the resistivity $\rho_m^{cl}(T)$ for a cluster of four spins in different configurations within a simple cubic lattice are shown in Figure 7.14 for three different values of the parameter $k_F a_0$ (corresponding to three different values of

n). These results were obtained using the first Born approximation. Inclusion of the next higher terms in the perturbation expansion can lead to a Kondo effect at low temperatures (see Section 7.4). There are two main features that we want to point out. Firstly, the general form of the behaviour is not particularly sensitive to the cluster topology and, secondly, whether the cluster causes an increase or decrease in ρ (as evidenced by the value $\rho_{\mathrm{m}}^{\mathrm{cl}}(0)$ at $T \rightarrow 0\,\mathrm{K}$ compared to the high temperature (spin-disordered) value $\rho_{\mathrm{m}}^{\mathrm{cl}}(\infty)$) depends upon $k_{\mathrm{F}}a_0$ (i.e. n) as indicated above. It might also be noted that the cluster resistivity does not reach the high temperature value until temperatures well above $T \approx 5J^{\mathrm{dd}}/k_{\mathrm{B}}$. Increasing the size of the cluster was found to increase the magnitude of the deviation in $\rho_{\mathrm{m}}^{\mathrm{cl}}(0)$ from the high temperature value.

The effects of a finite mean free path were not considered by Levin & Mills but can be considered along the same lines as for atomic clusters presented in Section 5.2.2. The behaviour shown in Figure 7.14 effectively assumes that the cluster size d is smaller than Λ. If $d > \Lambda$ the cluster resistivity $\rho_{\mathrm{m}}^{\mathrm{cl}}(0)$ should always lie below the high temperature value $\rho_{\mathrm{m}}^{\mathrm{cl}}(\infty)$ since a conduction electron within the cluster is then effectively inside a body with long range magnetic order. The magnetic contribution to the resistivity will thus pass through a maximum with

Fig. 7.14. The magnetic contribution of the resistivity of a cluster of four atoms as a function of reduced temperature $k_{\mathrm{B}}T/J^{\mathrm{dd}}$ for three different configurations (I, II, III) and three values of $k_{\mathrm{F}}a_0$ as indicated. The resistivity has been normalised to that of a single isolated spin (after Levin & Mills 1974).

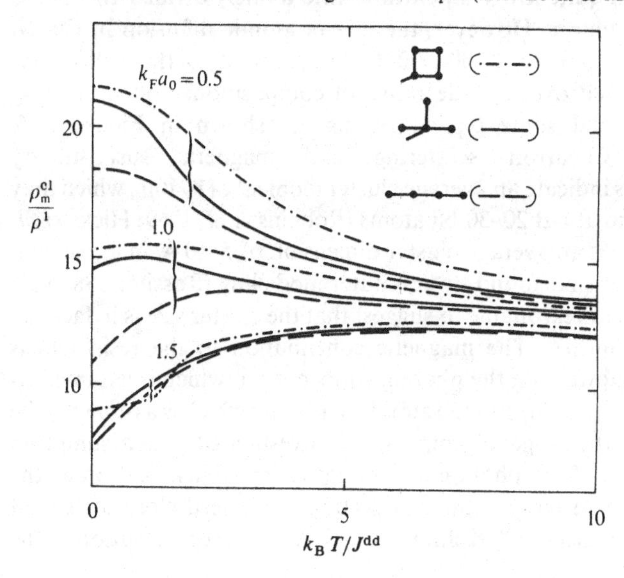

increasing cluster size at $d \sim \Lambda_l$ provided that $k_F a_0$ (or n) is such that the cluster initially causes an increase in ρ_m. However, in this case there is the additional complication that the magnitude of the mean magnetic moment per atom (and hence the effective spin) may depend upon the number of other magnetic atoms and degree of atomic correlation within the cluster, as discussed in Section 3.4.2. The details of this effect are not well understood but it is likely to be due to a change in the spin fluctuation rate, as discussed in that section and to be considered further in Section 7.3. Nevertheless, the magnitude of the effect on a timescale of $\sim 10^{-14}$ s is not really known. Unfortunately, there are few systems that permit an experimental verification of this behaviour. One requires that the alloy can be prepared in an initially fairly random state, that magnetic clusters associated with composition fluctuations can be formed by suitable heat treatment and that the temperature dependence of the resistivity can be determined over a temperature range at least up to $\sim J^{dd}/k_B$ without introducing changes to the atomic configuration. Many potentially suitable alloys involving a non-magnetic host with Fe, Co or Ni solutes have been investigated (e.g. Au or Cu solvents with Fe, Co, Ni solutes). The precipitate formed is usually fairly pure Fe, Co or Ni and so J^{dd}/k_B is given roughly by their bulk Curie temperatures which are 770 °C, 1131 °C and 358 °C respectively (see Cullity 1972, appendix 5). This effectively rules out systems containing Fe or Co precipitates, as significant atomic diffusion will occur before the variation of ρ_m^{cl} with temperature due to magnetic disordering can be determined. Au–Ni decomposes at quite low temperatures into a finely divided two-phase modulated structure. However, the rate of atomic diffusion in Cu–Ni alloys is quite low up to $\sim$200–300 °C. The resistivity of these alloys has been investigated over a wide range of compositions and measuring temperatures and some typical results are shown in Figure 7.15. Furthermore, neutron scattering and magnetic susceptibility measurements indicate an average cluster moment of 8–10μ_B which may be attributed to about 20–30 Ni atoms (Robbins *et al.* 1969; Hicks *et al.* 1969). This gives an average cluster dimension of 5–10 Å, in agreement with the cluster sizes found by computer modelling (Rossiter 1981*b*). It does not seem unreasonable to suggest that the cluster size is in fact less than Λ, as required. The magnetic contribution to the resistivity is obtained by subtracting the phonon contribution (which is assumed to be the same as pure Cu) and the atomic scattering which is assumed to be constant over the range of temperatures investigated. The assumption about the form of the phonon scattering has some justification as the behaviour at high temperatures ($T > 600$ K) is indeed close to that of pure Cu (see e.g. Schüle & Kehrer 1961). The atomic contribution to the resistivity is (somewhat arbitrarily) taken as the remaining resistivity at 700 K at which temperature ρ_m appears to have been much reduced. The resultant curves are shown in Figure 7.16. No data is shown below 100 K as the behaviour becomes complicated by resistivity minima (see Section 7.3.4). The data of Schüle & Kehrer (1961) leads to a similar set of results and the notion of a significant 'magnetic' contribution to the resistivity even at high Cu concentrations has been supported by Ikeda & Tanosaki (1984).

The calculated curves of Levin & Mills for four Ni concentrations are shown in Figure 7.17, the behaviour below T_c in the ferromagnetic alloys (46, 50%Ni) having been omitted. These curves were calculated with the assumption the $k_F a_0 = 1$. This is considerably less than the value of 2.28

Fig. 7.15. The resistivity of Cu–Ni alloys as a function of temperature. The Ni contents are as indicated on the curves (after Houghton *et al.* 1970*a*).

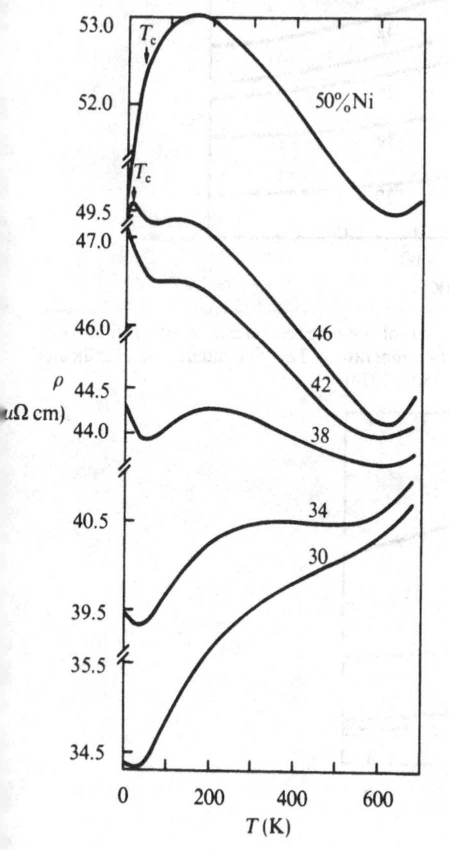

obtained by a nearly free electron approximation, but is necessary in order to reproduce the increase in ρ_m with magnetic clustering observed. In view of the errors likely to have already resulted from the assumptions of nearly free electrons and localised moments, it is not thought that this discrepancy is of any particular significance. While the magnitudes of the calculated resistivities are generally too low, the general form of the results are in reasonable agreement. The experimentally determined

Fig. 7.16. The magnetic cluster $\rho_m^{cl}(T)$ and atomic disorder ρ_0 contributions to the resistivity of Cu–Ni as a function of temperature. The Ni contents are as indicated on the curves (after Rossiter 1981*b*).

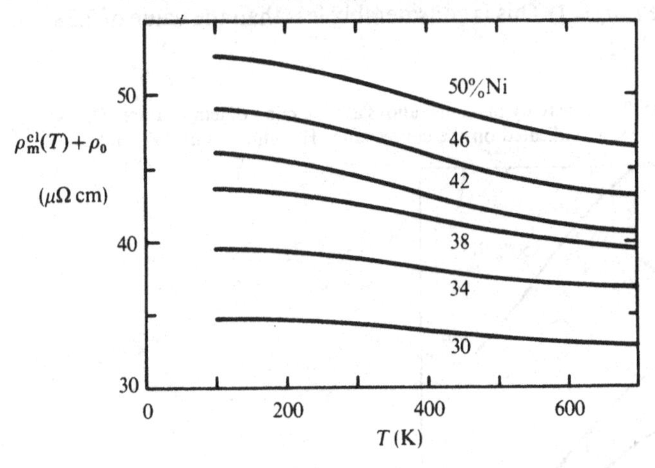

Fig. 7.17. The predicted variation of the magnetic cluster contribution to the resistivity $\rho_m^{cl}(T)$ of Cu–Ni with temperature. The Ni contents are as indicated on the curves (after Levin & Mills 1974).

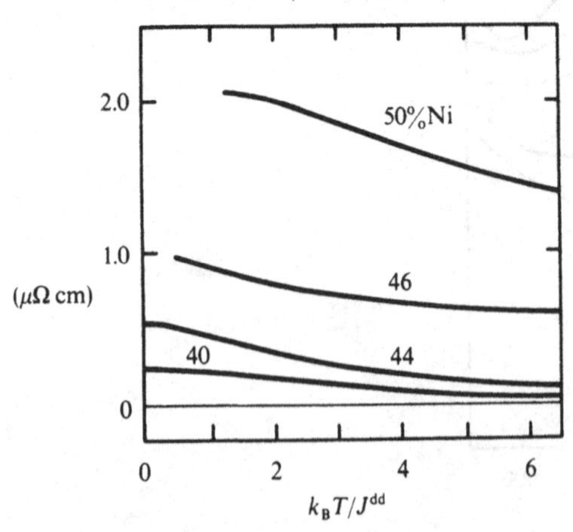

contribution ρ_m^{cl} at ~ 100 K is quite sensitive to composition. This indicates that the magnetic nature of the clusters is indeed very sensitive to the local environment. This fact has been used to provide a possible explanation for the observation that increased clustering causes a decrease in ρ in alloys containing more than 25%Ni but an increase in ρ in alloys having a lower Ni content (Rossiter 1981*b*).

7.3 Nearly magnetic systems: local spin fluctuations

The concept of spin fluctuations was discussed in some length in Section 3.3. The resistivity resulting from these fluctuations has been the subject of a number of investigations (Schindler & Rice 1967; Lederer & Mills 1968; Kaiser & Doniach 1970; Jullien *et al.* 1974; Fischer 1974, 1978, 1984; Ueda & Morija 1975; Riseborough 1984). However Rivier & Zlatic (1972*a*,*b*) have presented a formalism (albeit based on some rather drastic approximations) which treats both simple and transition metal hosts having resonant or exchange enhancement effects and we will use their results here.

7.3.1 *Kondo alloys*

We start by considering alloys that have a simple metal host and transition metal impurity. Here the behaviour results from the presence of a virtual bound state and is described by the Anderson model (see Section 3.2.1). The resistivity in this case results from the scattering of conduction electrons by the localised spin fluctuation. However, the electrons being scattered are those which have already formed the resonant virtual bound state from the conduction bands by scattering from the impurity potential, and thus have already experienced the resonant phase shift of $\pi/2$ (i.e. they are at the 'unitary limit'). The additional scattering by the thermal excitations of the spin fluctuation drives them off-resonance and the resistivity decreases as the temperature increases. This can only occur if the spin fluctuation exists for a time that is long compared to their thermal fluctuations. The transition from the unitary limit to zero thus occurs through the temperature range $T<T_{sf}$ to $T>T_{sf}$. The universal behaviour calculated by Rivier & Zlatic (1972*a*) is shown in Figure 7.18. This curve has a number of distinct regimes:

(i) $T \to 0$ ($X < 0.4$)

The resistivity is a parabolic function of T:

$$\rho = 1 - \frac{X^2}{8}$$
$$= 1 - \left(\frac{T}{\theta_1}\right)^2, \qquad (7.21)$$

where

$$X = 2\pi U N_d(E_F)T/T_{sf},$$

or

$$\theta_1 = \frac{\sqrt{2}\,T_{sf}}{\pi U N_d(E_F)}, \tag{7.22}$$

and U is the Coulomb repulsion between d electrons of opposite spin at the impurity. The result shown in Figure 7.18 was calculated with the assumption that the Fermi surface was infinitely sharp at E_F. Inclusion of a thermal broadening leads to a slightly different characteristic temperature $\theta_1' = \theta_1\sqrt{2}$ (Rivier 1968).

(ii) *Intermediate temperatures* $(0.9 < X < 2.1)$
The resistivity is linear

$$\rho = \alpha(1 - \beta X)$$
$$= \alpha\left(1 - \frac{T}{\theta_2}\right), \tag{7.23}$$

where

$$\alpha = 1.0742$$
$$\beta = 0.1766$$
$$\theta_2 = 2\theta_1.$$

Fig. 7.18. The reduced resistivity as a function of the reduced temperature $X = 2\pi U N_d(E_F)T/T_{sf}$. The position of T_{sf} for $UN_d(E_F)^d \approx 1$ is as indicated (after Rivier & Zlatic 1972*a*).

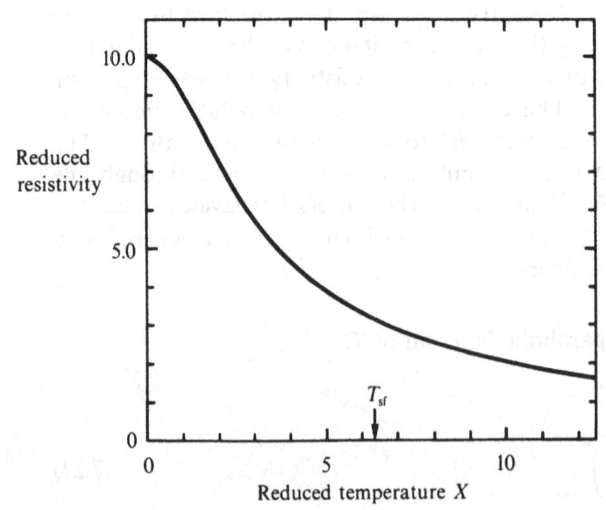

(iii) *Above the 'Kondo' (or spin fluctuation) temperature* $T > T_{sf}$ $(6.5 < X < 9)$

The resistivity exhibits the familiar logarithmic dependence characteristic of magnetic impurities

$$\begin{aligned}\rho &= A - B \ln X \\ &= C - B \ln T,\end{aligned} \tag{7.24}$$

where

$$A = 0.757,$$
$$B = 0.565,$$

(iv) *High temperature* $(X > 9)$

The resistivity goes to zero as $1/T$,

$$\rho = \frac{\theta_1'}{T}, \tag{7.25}$$

Most of these regions have been found experimentally (see e.g. Daybell & Steyert 1968; van Dam & van den Berg 1970; Star 1972; Star *et al.* 1972*a,b*; Greig & Rowlands 1974). Rizzuto *et al.* (1973) have collected together some data and plotted them in the above form. Their results are shown in Figure 7.19. The normalising temperature θ_R was determined by a fit to the quadratic behaviour found at low temperatures

$$\rho(T) = \rho_0\left(1 - \left(\frac{T}{\theta_R}\right)^2\right), \tag{7.26}$$

and so, from (7.22),

$$\theta_R = \frac{\sqrt{2}\, T_{sf}}{\pi U N_d(E_F)}. \tag{7.27}$$

Whether or not this behaviour appears as the uniquitous minimum in the total resistivity as a function of temperature as shown in Figure 7.20, for example, depends upon the magnitude of the localised spin fluctuation effect in relation to the phonon contribution, a combination of the two being responsible for the observed behaviour. Transition metal impurities *except* Fe, Ni, Co or Mn in Pd, Pt or Rh also exhibit the same form of behaviour, as indicated by the Pt–Cr results in Figure 7.19. The origin of the effect in these materials also lies in spin fluctuations but here the resonance is due mainly to the host rather than the impurity. We will discuss this in the next section.

7.3.2 *Exchange-enhanced alloys*

Dilute transition metal impurities Mn, Fe, Co or Ni in Pd, Pt or Rh hosts form a class of exchange enhanced alloys (see Section 3.3). The resistivity due to spin fluctuations in such materials has also been determined by Rivier and Zlatic (1972*b*). Again if the temperature is above T_{sf}, the conduction electrons are scattered by the thermal fluctuations of the localised spin on a timescale short compared to the lifetime of the spin fluctuation. However, in this case the conduction electron is simply scattered by the sum of the impurity potential and the scattering potential due to the localised spin fluctuation. (In the 'Kondo' case the conduction electron had to first scatter into the extra localised orbital before seeing the spin fluctuation.) The high temperature resistivity is thus characteristic of the scattering by a disordered array of spins, as in the case of a ferromagnetic material above T_c, and reaches a plateau at the spin disorder limit (cf. Section 7.1.1). This will correspond to the unitary limit with a phase shift of $\pi/2$. Such materials are consequently better described by the Wolff rather than the Anderson model (see Section 3.2.1). Using a similar analysis to the Kondo case,

Fig. 7.19. Normalised resistivity as a function of normalised temperature T/θ_R. The experimental points for each system were obtained from a number of specimens having a range of compositions (after Rizzuto *et al.* 1973).

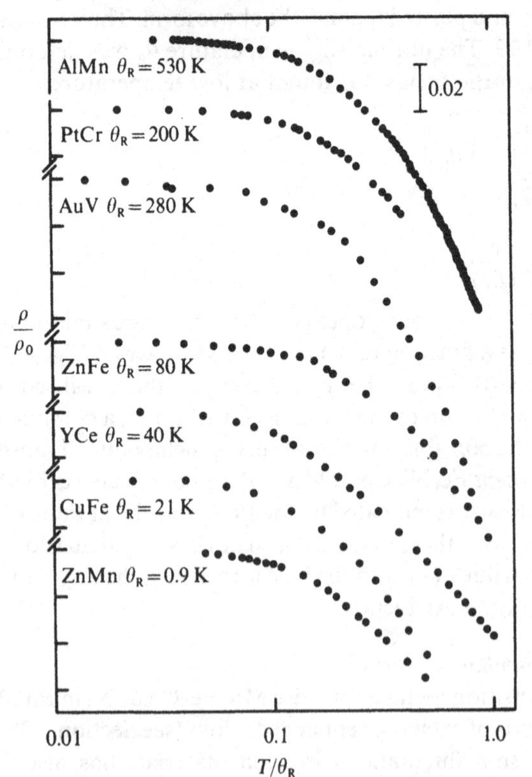

Rivier & Zlatic again determined a universal curve describing the variation of the reduced resistivity with reduced temperature. This is shown in Figure 7.21. This curve is in fact a mirror image of that shown in Figure 7.18, i.e. ρ (exch. enhanced) $= 1 - \rho$ (Kondo), and can again be divided into a number of characteristic regimes:

(i) *Low temperatures* ($T \to 0$)
The resistivity increases as

$$\rho = \tfrac{1}{2}\pi^2 \left(\frac{T}{T_{sf}}\right)^2. \tag{7.28}$$

As in the case of the Kondo alloys, including the effects of thermal smearing of the Fermi surface gives a slight modification:

$$\rho = \frac{3\pi^2}{4}\left(\frac{T}{T_{sf}}\right)^2. \tag{7.29}$$

(ii) *Intermediate temperatures* ($T \sim T_{sf}$)
The resistivity becomes linear,

$$\rho = \gamma(T - \theta), \tag{7.30}$$

where

$$\gamma = 1.12/T_{sf},$$
$$\theta = 0.67\,T_{sf}.$$

Fig. 7.20. Electrical resistivity of some dilute Cu–Fe alloys as a function of temperature (after Franck *et al.* 1961).

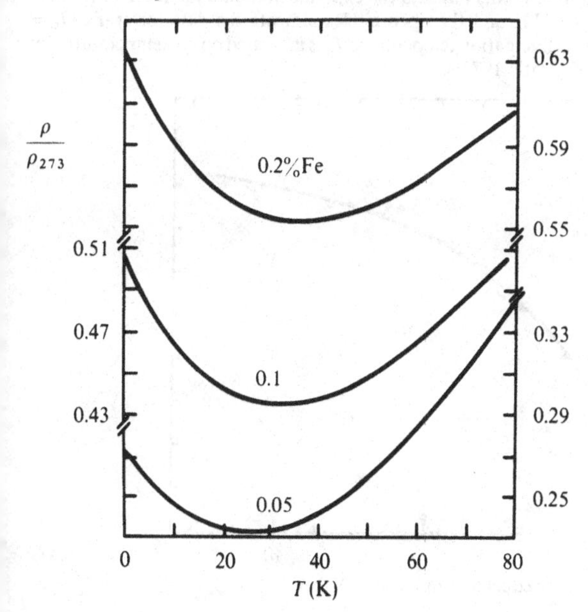

(iii) *Above the spin fluctuation temperature* ($T > T_{sf}$)
The impurity now has a magnetic moment and the resistivity is logarithmic:

$$\rho = C + B \ln(T/T_{sf}), \tag{7.31}$$

where

$$B = 0.24,$$
$$C = 0.68.$$

(iv) *High temperatures* ($T \to \infty$)
The resistivity approaches the spin disorder limit as

$$\rho = 1 - \frac{T_{sf}}{T}. \tag{7.32}$$

Again many of these regions have been found experimentally. Some results for dilute Rh–Fe and Ir–Fe alloys are also shown in Figure 7.21.

The nature of the localised spin fluctuations are the same in the Kondo or exchange-enhanced alloys. Whether the resistivity increases or decreases with increasing temperature depends upon their electronic structures and in particular whether the host conduction states are orthogonal to the impurity states. If they are, the conduction electron must first leak into the extra orbital to see the localised spin fluctuation,

Fig. 7.21. The normalised resistivity as a function of reduced temperature $X = 2\pi T/T_{sf}$. The solid points indicate the experimental data for Rh–Fe (less than 0.1at.%Fe; $T_{sf} = 12$ K and the open circles indicate the data for Ir–Fe ($T_{sf} = 225$ K). The spin fluctuation temperature T_{sf} is indicated on the temperature axis (after Rivier & Zlatic 1972*b*).

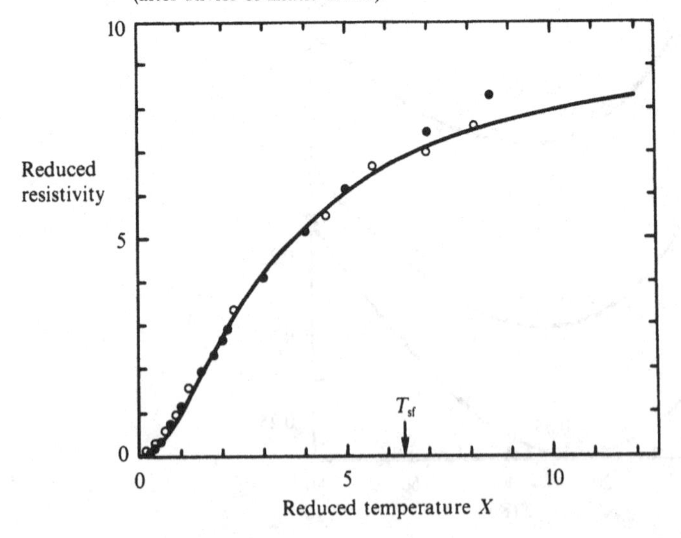

and the behaviour is given by the Anderson model. If they are not (i.e. the extra 'orbital' is simply a linear combination of host d-states without additional orbitals), then the fluctuations are seen by the conduction electrons in the conduction band and the Wolff model applies. The resonance occurs in the Wolff model if the phase shift due to the *host* potential rises rapidly through $\pi/2$, as is more likely to occur in transition hosts. The change in the orthogonality is illustrated by the series of Cr, Mn, Fe or Co impurities in Pd, Pt or Rh. The orthogonality decreases from Cr to Co. Thus dilute alloys of Pd, Pt or Rh with Cr act as Anderson alloys and exhibit resistivity minima at low temperatures, whereas the impurity states of Mn, Fe, Co in Pd, Pt or Rh are sufficiently non-orthogonal for them to act progressively (from Mn to Co) more like Wolff alloys, with a resistivity that increases monotonically with temperature. The effects of this orthogonality have been discussed in detail by Rivier & Zitkova (1971). A summary of the type of behaviour found in a variety of alloys is given in Table 7.5.

The above results are based on the assumption of a single conduction band, that there is zero host enhancement, that the virtual bound state (determined by the impurity potential scattering) lies at E_F and consists of a single orbital in the Anderson model and of zero atomic scattering potential in the Wolff model. It appears that some of these assumptions are not trivial. For example, Fischer (1974) has shown that in alloys with a small enhancement of the host susceptibility, the resistivity due to the localised spin fluctuation in the Wolff model is modified by the host scattering potential (measured in terms of the shift V of the resonance with respect to the Fermi surface), as shown in Figure 7.22. These results show that both the magnitude and sign of T^2 coefficient are dependent upon the degree of atomic scattering. In fact for very large V one obtains the mirror image of the curve for $V=0$. This corresponds to the Anderson model with $V=0$ and reflects the changes expected with differing degrees of orthogonality between the host and impurity states described above.

7.3.3 *Composition dependence*

So far the impurities have been treated as independent. As such, the magnitude of the resistivity contribution from the localised spin fluctuations should be proportional to the concentration of impurity c_I, while the characteristic temperature should be independent of c_I. This behaviour is generally found, as may be verified from the results given in the references cited in Section 7.3.1). However, careful analysis (Star 1972; Star *et al.* 1972*a*,*b*) has revealed in some cases an apparent dependence of T_{sf} upon c_I, as shown, for example, in Figure 7.23. This

Table 7.5. *Nature of impurity states in a number of alloy systems*

	Solute	
Solvent	Anderson	Wolff
Al	Cr	
	Mn	
Zn	Mn	
	Fe	
Cu	Ti	
	V?	
	Cr	
	Mn	
	Fe	
	Co	
	Ni (concentrated)	
Rh	Cr	Mn
		Fe
		Co
		Ni
Pd	V	Mn?
	Cr	Fe
	Mo	Co
	Ru	Ni
	Rh	
	Ag (concentrated)	
	Pt (concentrated)	
	Au (concentrated)	
	U	
	Np	
Pt	Cr	Mn
		Fe
		Co
		Ni
Au	Ti	
	V	
	Cr	
	Mn	
	Fe	
	Co	
	Ni	

behaviour could result from an increased contribution from Fe–Fe pairs, which have a much lower T_{sf} (see Section 3.3), as c_1 increases. Such an effect would be manifest as a deviation above the T^2 dependence with decreasing temperature. This deviation may be reduced by application of a magnetic field, partly quenching the Kondo (or localised spin

Fig. 7.22. Normalised resistivity due to localised spin fluctuations as a function of reduced temperature for different atomic scattering strength (measured in terms of the shift V of the resonance with respect to the Fermi surface in units of k_BT) (after Fischer 1974).

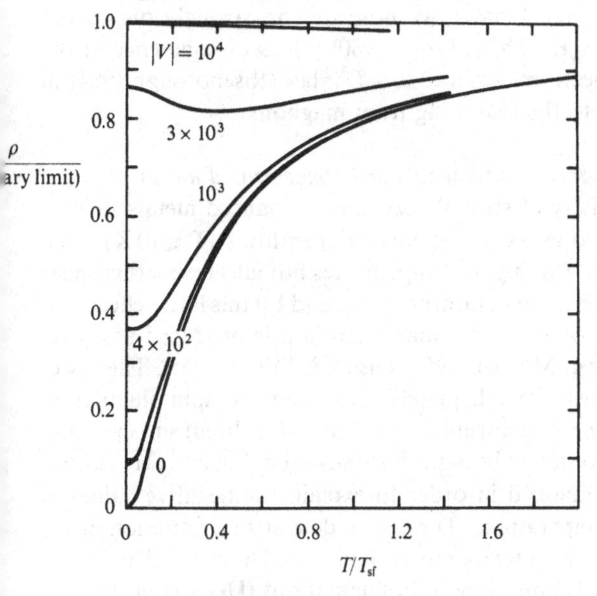

Fig. 7.23. Characteristic Kondo temperature T_K (αT_{sf}) of dilute alloys as a function of Fe concentration (after Star *et al.* 1972*a*).

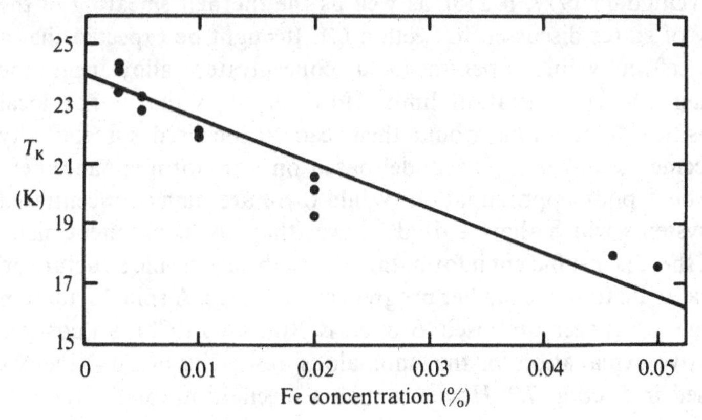

fluctuation) state due to impurity atom pairs etc. which have a lower T_{sf} than the isolated moments. If this is the cause of the composition dependence, it indicates that the true single isolated moment behaviour could only be observed in very dilute alloys. The magnitude of the resistivity contribution due to pairs of atoms may be expected to vary as c_I^2, triples as c_I^3, and so on. In this way, for example, the minimum in resistivity of rapidly quenched dilute Cu–Co and Au–Co alloys (produced by the sum of the localised spin fluctuation and phonon scattering) is attributed mainly to pairs of Co atoms (Haller *et al.* 1981; Henger & Korn 1981). Finally, we note that in strongly disordered systems (e.g. $Ce_{0.9-x}La_xTh_{0.1}$, Grier 1980) a loss of coherence in the spin–fluctuation spectrum may lead to a $T^{3/2}$ law (Riseborough 1984), in the direct analogy to the scattering from magnons.

7.3.4 *Nearly magnetic pure metals and concentrated alloys*

The resistivity of strongly exchange enhanced metals such as Pd, Np and Pu varies as T^2 at low temperatures ($T \leqslant 10$ K), then becomes linear with T at higher temperatures but falls below this linear variation at still higher temperatures. In Np and Pu this latter effect is so strong that saturation or even a maximum in ρ is produced at higher temperatures (see e.g. Meaden 1966; Olsen & Elliott 1965). The lower temperature regimes are adequately described by spin fluctuation theories by assuming a uniform (as opposed to localised) susceptibility enhancement and so should be of the form shown in Figure 7.21. Various factors have been invoked in order to explain the negative values of $d\rho/dT$ at higher temperatures. These include a strong variation of the enhancement factor with temperature (Kaiser & Doniach 1970; Jullien *et al.* 1974), a suppression of the spin fluctuations (Ueda *et al.* 1975), a temperature dependent hybridisation (Arko *et al.* 1972), mean free path effects (Dugdale 1977, p. 215), as well as the thermal smearing of the density of states discussed in Section 6.1. It might be expected that a similar effect would operate in a concentrated alloy near the ferromagnetic concentration limit. However, it is likely that local composition fluctuations would then lead to localised susceptibility enhancements and that any model based on a uniform enhancement would be a poor approximation. While there are many concentrated alloy systems which show a $d\rho/dT$ lower than the component metal values, there is insufficient information about the electronic structure on the local scale to make further progress at this stage. A spin fluctuation mechanism has been proposed (Brouers & Brauwers 1978) as a possible alternative explanation for the 'anomalous' resistivity of Cu–Ni alloys discussed in Section 7.2. However such a mechanism (or the thermal smearing mechanisms) could not explain the relatively sharp transition back to 'normal' behaviour at $T \approx 600$ K and so is not thought to be the dominant mechanism in that system.

As also noted in Section 6.7.1 some concentrated alloy systems exhibit resistance minima at low temperatures (see Table 7.5), the Pd–Ag, Pd–Au, Pt–Au, Pt–Pd and Cu–Ni systems have been studied in greatest detail. In the Pd based alloys, the depth of the minimum is greatest at Pd–50%Au, Pd–40%Ag (Edwards *et al.* 1970), as shown in Figure 7.24. Murani (1974) found a large positive T^2 coefficient with Ag concentrations greater than ~60%.

In the investigation of Pt–Au and Pt–Pd alloys by March (1978), only the resistivity between 1.6 and 4.2 K was determined. However, the slope $\Delta\rho/\Delta T$ may be taken as a measure of the magnitude of the spin fluctuation effect, and is plotted in Figure 7.25.

As suggested by Edwards *et al.* and Murani, the Pd–Au and Pd–Ag results can be understood on the basis of localised d states associated with statistical clusters of Pd atoms. In Ag- (or Au-) rich alloys, these lie below the Fermi level and so a Wolff-type scattering results, with ρ increasing with T (Figure 7.21). At Pd concentrations above ~60%, the s–d scattering rate increases with the appearance of holes in the d-band (see Sections 6.1 and 6.7). The scattering phase shifts associated with the localised pockets of holes then become larger, the coefficient of the T^2

Fig. 7.24. Depth of resistivity minima as a function of concentration for: Pd–Ag (○), Pd–Au (●) (after Edwards *et al.* 1970).

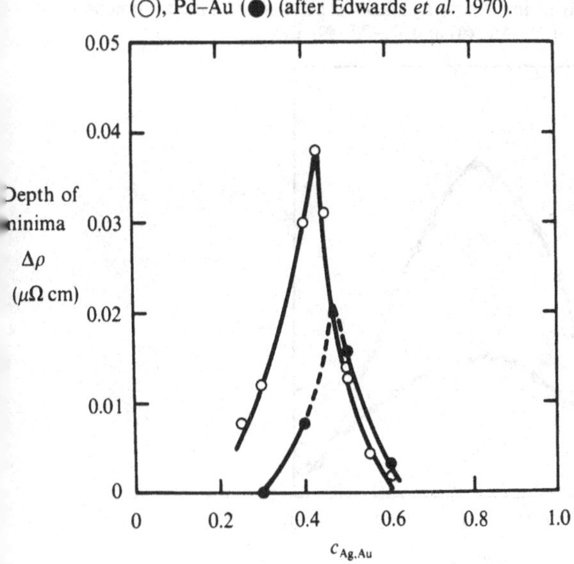

term becomes negative and the behaviour more Anderson-like (Figure 7.19). At higher Pd concentrations the hole regions form a band and the behaviour is suppressed. This qualitative explanation is in general agreement with the experimentally determined variation of electronic structure with composition (Myers *et al.* 1968; Norris & Nilsson 1968; Norris & Myers 1971). The change in sign of the T^2 coefficient with the position of the impurity level with respect to the Fermi level is also expected theoretically (Fischer 1974), as shown in Figure 7.22. It seems likely that a similar qualitative explanation can be applied to the Pt-based alloys. Concentrated paramagnetic Cu–Ni alloys also exhibit low temperature resistivity minima, as shown in Figure 7.26. These have been attributed to the presence of magnetic or nearly magnetic Ni clusters (Houghton *et al.* 1970*b*; Crangle & Butcher 1970; Legvold *et al.* 1974) and are characteristic of a Kondo–Anderson type of behaviour.

The low temperature properties of a range of other Ni-based alloys (Mo, Cu, V, Ru, Rh, Ir, Pd) in the nearly ferromagnetic composition range have been investigated by Amamou *et al.* (1974, 1975) and interpreted on the basis of localised spin fluctuations associated with nearly magnetic regions as well as magnetic regions with reduced characteristic temperatures. In these cases however, the coefficient of the T^2 term remained positive, characteristic of a Wolff model. As in the case of the Pt- and Pd-based alloys, a better understanding of these

Fig. 7.25. Change in resistivity $\Delta\rho = \rho(1.6) - \rho(4.2)$ as a function of concentration for Pt–Au (●) and Pt–Pd (■) alloys.

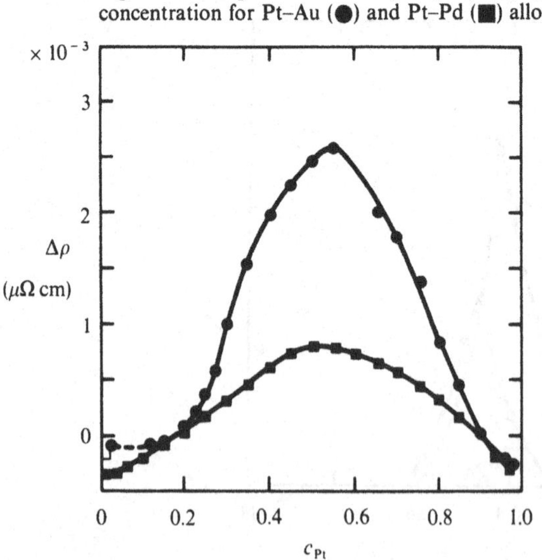

phenomena must await a more detailed knowledge of the electronic structure of the materials on a local scale.

7.4 Spin glasses

The magnetic structure of spin glasses was discussed in Section 3.2.2. In order to calculate the resistivity of these materials, some model of the excitation spectrum and dispersion of the diffuse spin waves is required. Rivier & Adkins (1975) regarded the spin waves as completely diffusive (i.e. overdamped) and found a $T^{3/2}$ behaviour in the resistivity

Fig. 7.26. Resistivity of Cu–Ni alloys as a function of temperature. The Ni concentrations are as indicated on the curves which have been offset vertically to allow them all to appear in detail on the same graph (after Houghton *et al.* 1970*b*).

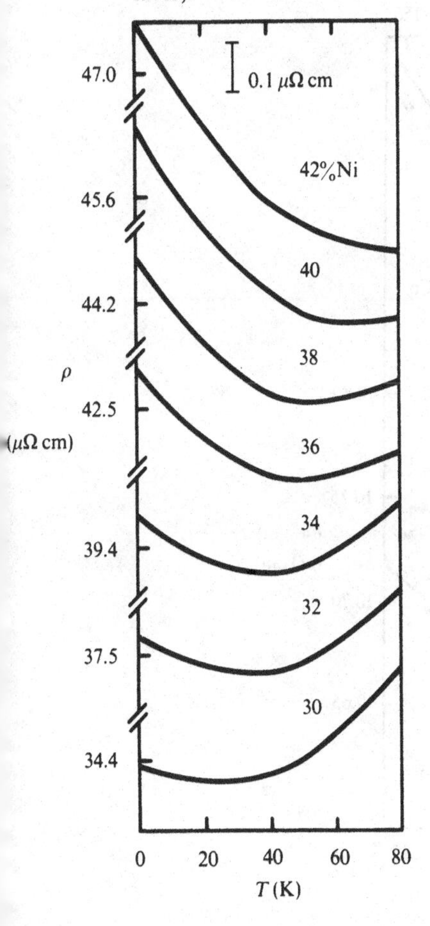

at low temperatures and a flattening-off or saturation at higher temperatures (the effects of localised spin fluctuation (or Kondo effect) was not included in the calculation). The magnitude of the $T^{3/2}$ term is weakly concentration dependent, becoming smaller with increased concentration of magnetic impurities. This $T^{3/2}$ behaviour at very low temperatures relies on the presence of very long wavelength diffuse modes. Conduction electron scattering by other imperfections will render these modes inoperative and a T^2 law will result. The point of transition between these two regimes and the slope of the T^2 behaviour

Fig. 7.27. Resistivity (corrected for the phonon scattering) as a function of $T^{3/2}$ for Au–Cr, Au–Mn and Ag–Mn (after Ford & Mydosh 1974).

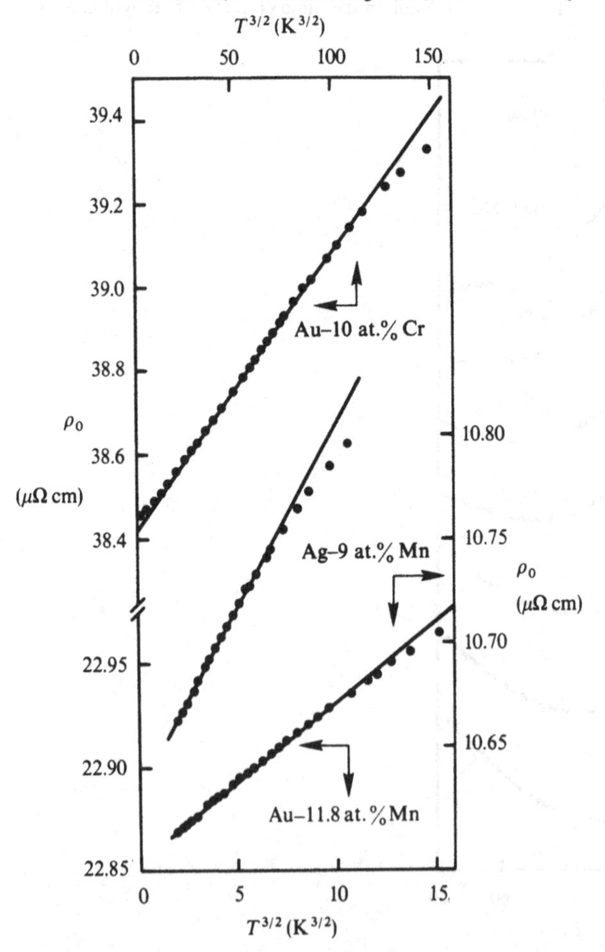

should thus be sensitive to the level of non-magnetic impurities and local structure. These results are in qualitative agreement with the observed low temperature behaviour, as shown in Figure 7.27.

The low temperature behaviour has also been explained on the basis of damped ferromagnetic modes ($\omega \propto q^2$), which lead to a $T^{3/2}$ law changing to T^2 law if the modes are overdamped (Fischer 1979), and a discrete excitation spectrum model (Campbell *et al.* 1982).

At higher temperatures the effects of the spin fluctuations (i.e. the Kondo effect) must be taken into account. If it is assumed for simplicity that T_{sf} remains fixed and that the RKKY interactions quench the resonant spin fluctuation state below T_0, the combined effects may be represented schematically as in Figure 7.28. A maximum in the low temperature resistivity at T_{max} thus results from a competition between the spin glass RKKY interaction and the Kondo effect. Such a maximum is a characteristic of the resistivity of many spin glasses based on noble metal hosts, as shown, for example, in Figure 7.29. Note that there is no obvious anomaly at T_0, as expected, since on the characteristic timescale of a resistivity measurement ($\sim 10^{-14}$ s) the spin glass structure is well formed at temperatures much higher than T_0. The conceptual extension of this simple model to the more realistic case where the characteristic temperature T_{sf} steadily decreases towards zero with decreasing temperature is straightforward. This maximum tends to disappear as the concentration increases, leaving only an inflection point in the temperature dependence of the resistivity. Fischer (1981) has considered the effects of different spin-wave modes while including the Kondo effect

Fig. 7.28. Schematic representation of the resistivity due to isolated impurities ρ_{sf}, interacting impurities ρ_{RKKY} and phonons ρ_p as a function of log T. The combined effect is shown as the solid line (after Schilling *et al.* 1976).

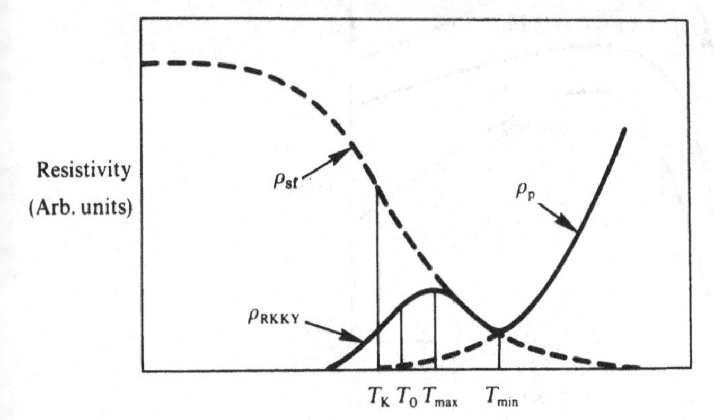

and confirmed the occurrence of such a maximum (see also Larsen 1976, 1978). However, a complete calculation which self-consistently includes the effects of both the spin fluctuations and the co-operative excitations of the spin systems via the RKKY interaction has yet to be performed. Variations in the position of the resistivity maximum and T_0 following heat treatment have been attributed to the effects of an increased mean free path Λ supporting the concept of a damping factor $e^{-r/\Lambda}$ discussed in Section 3.2.2 (Buchmann *et al.* 1977; Zibold 1979). This effect has also been investigated by alloying with a third element by Srivastava *et al.* (1981*a*,*b*). We might also note that a finite conduction electron mean free path is an integral part of the theories of Kinzel & Fischer (1977) and Larsen (1977). Spin glass behaviour is also observed in Wolff-type alloys such as Pd–Mn or Rh–Fe (Rusby 1974; Sarkissian & Taylor 1974), again characterised by a $T^{3/2}$ law at low temperatures. However, in these cases

Fig. 7.29. Temperature dependence of the resistivity (in $\mu\Omega$ cm and corrected for the phonon contribution) of Au–Cr, Au–Mn, Au–Fe and Ag–Mn. The freezing temperatures T_0 are as indicated by the arrows on each curve (after Ford & Mydosh 1974).

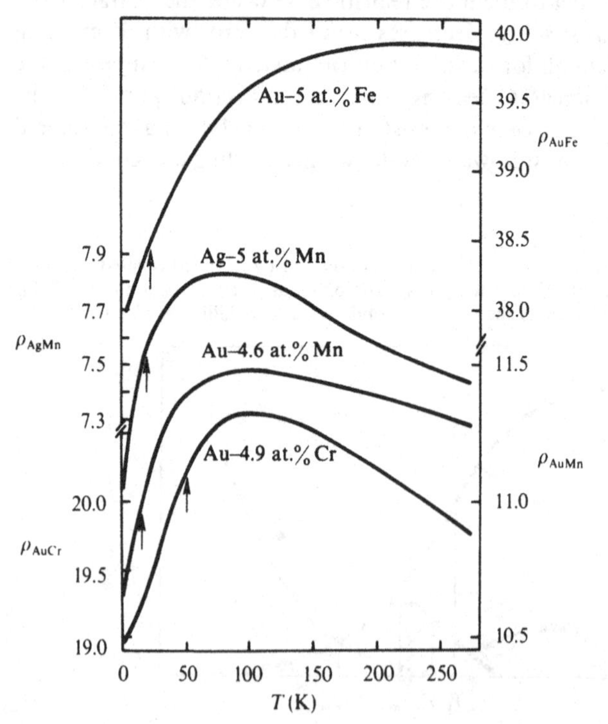

the coefficient of the $T^{3/2}$ term may be negative if the magnitude of the atomic scattering potential is sufficiently large (Rivier 1974). The general condition for such a spin glass to exhibit a resistance *minimum* appears to be

$$\rho(T \to 0) > \tfrac{1}{2}\rho_{\infty}, \tag{7.33}$$

where ρ_{∞} is the high-temperature limiting value of the spin glass contribution. This inversion in slope with increased atomic scattering is of the same nature as discussed in relation to Figure 7.22.

8 Other phenomena

In this final chapter, three other aspects of the resistivity of metals and alloys will be discussed. As with many of the topics included in this book, each of these could easily justify a complete text in its own right and so the discussions here will be of necessity rather brief. It is hoped, however, that this introduction to these topics, all of which are still in a state of development, will serve to assist the interested reader in pursuing the matters in more detail.

8.1 Resistivity at the critical point

8.1.1 *Some general comments*

As noted in Chapter 2, a second order phase transition is characterised by a long range order parameter that continuously decays to zero as the critical transition temperature T_c is approached. Any of the physical properties of such a system which depend upon the derivative of S (or M) with respect to temperature, dS/dT (or dM/dT), such as the specific heat or temperature coefficient of resistivity $d\rho/dT$, will diverge at T_c. However, this gives a static picture of the transition which is somewhat misleading. Near T_c, small fluctuations in temperature produce fluctuations in the correlation parameter that are large in both magnitude and spatial extent. Since the resistivity is proportional to the magnitude of such correlations (the subject of Chapters 5 and 6), it is expected that the detailed behaviour at the critical point will be determined by these fluctuations. Nevertheless, the behavour of such properties at the critical point is not very system dependent. For example, the variation of $d\rho/dT$ with temperature is similar in many magnetic and atomic order-disorder systems as T approaches T_c, provided that they have the same dimensionality and symmetry of the ordered phase. This is known as the 'universality hypothesis' and the materials are said to belong to the same 'universality class'. We thus define a reduced temperature ε to give a general measure of the difference between the system temperature and T_c (or T_N):

$$\varepsilon = \left|\frac{T-T_c}{T_c}\right| \quad \left(\text{or} \left|\frac{T-T_N}{T_N}\right|\right). \tag{8.1}$$

We are interested in the asymptotic laws governing the variation of a property P as $T \to T_c$ (i.e. $\varepsilon \to 0$). This asymptotic behaviour is described

Table 8.1. *Definitions of critical exponents for magnetic systems*

Exponent	Definition	Conditions H	Conditions M	Quantity
α'	$C_H \sim \varepsilon^{-\alpha'}$	0	0	specific heat at constant magnetic field
α	$C_H \sim \varepsilon^{-\alpha}$	0	0	
β	$M \sim \varepsilon^{\beta}$	0	$\neq 0$	zero-field magnetisation
γ'	$\chi_T \sim \varepsilon^{-\gamma'}$	0	$\neq 0$	zero-field isothermal susceptibility
γ	$\chi_T \sim \varepsilon^{-\gamma}$	0	0	
δ	$H \sim \lvert M \rvert^{\delta} \operatorname{sign}(M)$	$\neq 0$	$\neq 0$	critical isotherm
ν'	$\xi \sim \varepsilon^{-\nu'}$	0	$\neq 0$	correlation length
ν	$\xi \sim \varepsilon^{-\nu}$	0	0	
η	$\Gamma(r) \sim \lvert r \rvert^{-(d-2+\eta)}$	0	0	pair correlation function (d = dimensionality)

Source: From Stanley 1971.

in terms of the critical exponents, but these may depend upon whether we approach T_c from temperatures above or below T_c, written as $T \rightarrow T_c^+$ or $T \rightarrow T_c^-$ (or $\varepsilon \rightarrow 0^+$; $\varepsilon \rightarrow 0^-$) respectively. These exponents are defined by

$$P(\varepsilon) \sim \varepsilon^{\lambda};\ \varepsilon \rightarrow 0^+,$$
$$\sim \varepsilon^{\lambda'};\ \varepsilon \rightarrow 0^-. \tag{8.2}$$

By this it is meant that

$$\lim_{\varepsilon \rightarrow 0^+} \frac{\ln P(\varepsilon)}{\ln \varepsilon} = \lambda, \tag{8.3}$$

and similarly for λ'. Note that this does not strictly mean that

$$P(\varepsilon) = A\varepsilon^{\lambda}, \tag{8.4}$$

where A is a constant, as there will usually be some correction terms over a finite range of ε giving something like

$$P(\varepsilon) = A\varepsilon^{\lambda}(1 + a\varepsilon + \cdots), \tag{8.5}$$

although for $\varepsilon \rightarrow 0$, (8.4) is often employed. Furthermore, for small ε it is difficult to distinguish between $A\varepsilon^{\lambda}$ and $A \ln \varepsilon$, and such a singularity is sometimes referred to as a logarithmic divergence. There are a range of critical exponents appropriate to different physical quantities, as shown, for example, in Table 8.1 (see e.g. Stanley 1971). The exponents are not all independent and several relations have been established between them. The scaling hypothesis (Kanadoff *et al.* 1967) is concerned with determining the minimum number of quantities required to define all of the critical exponents and is framed in terms of the correlation length ξ. This length was introduced conceptually in Section 2.2 and gives a measure of the rate of decay in the atomic or magnetic correlation with distance from some arbitrary site. The scaling hypothesis states that the correlation length ξ should be the longest and only relevant length in determining critical phenomena; i.e. near the critical point ξ is much longer than interatomic distances and so any characteristic dimensions on a scale of a few angstroms (such as lattice parameters, for example) are unimportant. The dominant temperature dependence of all physical quantities near T_c is then assumed to be determined by ξ which diverges with a positive exponent ν:

$$\xi \sim \varepsilon^{-\nu}. \tag{8.6}$$

Thus all other physical quantities depend upon ε only through their dependence upon ξ. This concept is of particular interest with regard to the resistivity since a change in temperature near T_c is likely to cause a transition from the short ($\xi \ll \Lambda$) to the long ($\xi \gg \Lambda$) range correlation regimes or vice-versa.

Application of the renormalisation group (Wilson 1972, 1974; Wilson & Fisher 1972; Wilson & Kogat 1974; Fisher 1974; Ma 1973; Wallace & Zia 1978) has led to considerable clarification of the concepts of scaling and universality. However, the field of critical behaviour, whether experimental, computer-simulation or theoretical, is a vast subject and the reader is referred to some of the excellent reviews for more detailed information (Fisher 1967, 1982; Heller 1967; Kanadoff *et al.* 1967; Mouritson 1984).

8.1.2 *The electrical resistivity near T_c*

It must be stated from the outset that a rigorous determination of the behaviour at the critical point depends upon a knowledge of the temperature and composition dependence of the local electronic structure, including the effects of fluctuations. As should be evident from discussions throughout the rest of this book, this problem is far from being solved. Experimentally the situation is not much better. The analysis of data around the critical point is very sensitive to the choice of the actual value of T_c (or T_N) (see e.g. Heller 1967). In fact Ausloos & Durczewski (1980) have remarked that even the sign of the critical exponent cannot be determined from strict data analysis alone. Nevertheless, in order to give the reader some feeling for the sort of analysis involved let us consider a somewhat simplified approach to the problem. As in Chapters 6 and 7 we will assume a model in which the scattering centres are localised (whether atoms or spins) and in which there is complete spin mixing near the critical point so that there is no distinction between the spin-up and spin-down sub-bands in the magnetic case. In order to determine the electrical resistivity from (5.9b) we need some suitable model for the pairwise or spin–spin correlation functions. In the spirit of the universality hypothesis we need make no distinction between spin direction or atom type and use a 'universal' correlation function which is written† as $\Gamma(\mathbf{R}, T)$, $(\equiv\langle\sigma_0\sigma_{\mathbf{R}}\rangle$ or $\langle\mathbf{S}_0\mathbf{S}_{\mathbf{R}}\rangle)$. The classical Ornstein–Zernike form may be written as

$$\Gamma(\mathbf{R}, T)=C_0\left(\frac{a}{R}\right)\exp(-R/\xi). \tag{8.7}$$

Note that this decreases as either R or T (via (8.6)) increases, as required. This correlation function describes only fluctuations in the long range order (see e.g. Fisher 1964) and leads to a temperature coefficient of resistivity which varies with temperature as (see Craig *et al.* 1967)

$$\frac{\mathrm{d}\rho(T)}{\mathrm{d}T}\sim\frac{T_c-T}{|T_c-T|}b\varepsilon^{2\nu-1}\ln\varepsilon, \tag{8.8}$$

where b is a positive constant. The classical value of ν is $\frac{1}{2}$ giving

$$\frac{\mathrm{d}\rho(T)}{\mathrm{d}T}\sim b\ln\varepsilon. \tag{8.9}$$

Note that $\mathrm{d}\rho(T)/\mathrm{d}T$ changes sign from positive below T_c to negative above T_c and so (8.9) always predicts an upward-pointing cusp in ρ at T_c (de Gennes & Friedel 1958; Kim 1964). However, as discussed at length in Chapter 5, the conduction electrons are only affected by fluctuations over a range $R\leqslant\Lambda_l$. As Λ_l will be smaller than ξ near T_c, it is only fluctuations in the *short-range* part of the correlation function that need be considered at such temperatures. (This was originally pointed out in relation to behaviour at the critical point by Fisher (1968).) The Ornstein–Zernike correlation function is thus expected to be valid only at temperatures farther removed from T_c. A more suitable correlation parameter for the range $R/\xi\ll 1$ is given by scaling theory (Fisher 1964):

$$\Gamma(\mathbf{R}, T)=D(R/\xi)(a/R)^{1+\eta}. \tag{8.10}$$

For small R/ξ the function $D(R/\xi)$ may be expanded as (Fisher & Langer 1968):

$$D(x)=D_0+D_1\frac{T_c-T}{|T_c-T|}x^{(1-\alpha)/\nu}-D_2x^{1/\nu}+\cdots. \tag{8.11}$$

Quantitative information about the coefficients $D_0, D_1, D_2, \ldots$ may be

† As in Section 5.2, the vector $\mathbf{R}$ joins pairs of lattice sites.

obtained by renormalisation group methods. However, if the same expansion is used to derive the singularity in the magnetic specific heat, we have the constraint $D_1 > 0$. Substitution of this correlation function for $\langle \sigma_i \sigma_j \rangle$ in (5.18) leads to

$$\frac{d\rho(T)}{dT} \sim \varepsilon^{-\alpha}. \tag{8.12}$$

The qualitative variation with temperature of the structure factors that result from these correlation functions (see e.g. (5.17), (5.146)):

$$\mathscr{S}(\mathbf{q}, T) = \sum_{\mathbf{R}} \Gamma(\mathbf{R}, T)\, e^{-i\mathbf{q}\cdot\mathbf{R}}\, e^{-|\mathbf{R}|/\Lambda}, \tag{8.13}$$

are shown in Figure 8.1.

As the q^3 weighting factor in the resistivity integral leads to a dominant contribution for $q \approx 2k_F$, these results can roughly be pictured as the variation of ρ with T. Note that they represent a completely different behaviour at T_c: the Ornstein–Zernike model gives a maximum at T_c, whereas the *slope* of the scaling result is a maximum at T_c. To this behaviour must be added the effects of phonon scattering to obtain the total resistivity, possibly masking the broad hump just above T_c in the scaling result. The differences are more apparent if we plot $d\rho(T)/dT \approx d\mathscr{S}(\mathbf{q}, T)/dT|_{qa_0=1}$, as shown in Figure 8.2. In view of the variation of ξ with ε, it is expected that the resistivity may initially follow the Ornstein–Zernike result for temperatures above T_c but switch to the scaling result when $\xi \sim \Lambda_1$ (see also Schwerer 1971). This change in behaviour has been investigated theoretically and experimentally in some detail and we will return to it later in this section.

Fig. 8.1. Temperature dependence of $\mathscr{S}(\mathbf{q}, T)$ for various values of $|\mathbf{q}|$. (*a*) Ornstein–Zernike approximation. (*b*) Scaling theory (after Parks 1972; Fisher & Burford 1967).

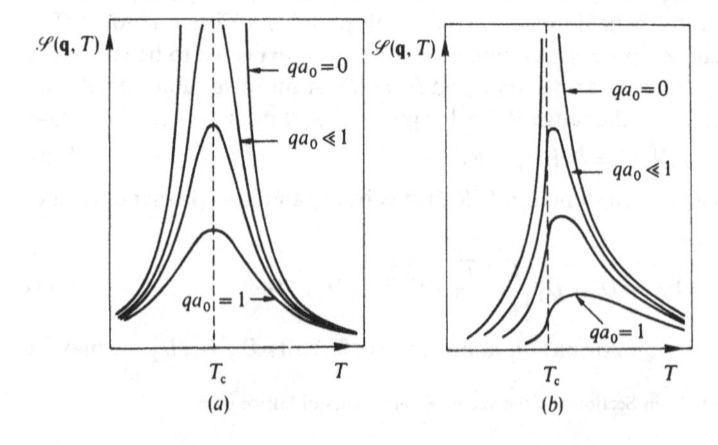

The discussion above has concentrated on the singularity and the critical exponents which relate to the correlation length. The actual change in the resistivity with temperature in the vicinity of T_c will depend also upon the nature of the fluctuations: whether short range atomic ordering or clustering in the atomic case or antiferromagnetic or ferromagnetic ordering in magnetic systems. This information is contained in the periodicity of the structure factor, as described at some length in the earlier chapters of this book. This periodicity can be taken into account in a straightforward way with the method of Richard and Geldart (1977). For clustering or ferromagnetic ordering, the correlation function is positive for all pairs of atoms or spins and given directly by (8.7) or (8.8). For atomic ordering or antiferromagnetic spin alignments, $\Gamma(\mathbf{R}, \mathrm{T})$ will be negative or positive depending upon the distance separating the atoms or spins. This is simply taken into account by including a modulating term:

$$\Gamma(\mathbf{R}, T) = \cos(\mathbf{Q}\cdot\mathbf{R})\Gamma_0(\mathbf{R}, T), \tag{8.14}$$

where $\mathbf{Q}$ is determined by the periodicity of the atomic or magnetic structure, so that $\cos(\mathbf{Q}\cdot\mathbf{R}) = \pm 1$ at the appropriate sites. Such an approach allows for the consideration of simple ordered structures (e.g. AB, A_3B or cubic two-sublattice antiferromagnets) as well as the more complicated spiral spin structures discussed in Chapter 3. For example, in the case of helical rare-earth antiferromagnets (helimagnet), the axis of the helix is along the *c* axis of the *hcp* structure and

$$\mathbf{Q} = \frac{2\pi}{P}\hat{\mathbf{c}}, \tag{8.15}$$

where $\hat{\mathbf{c}}$ is a unit vector along that axis and the period P is approximately $9c$ and $4c$ for terbium and dysprosium respectively near T_N (Koehler 1972). However, if the ordered structure has a periodicity that differs from the crystal lattice, the presence of new superlattice Brillouin zone boundaries may also affect the resistivity, as discussed in Section 5.3.2.

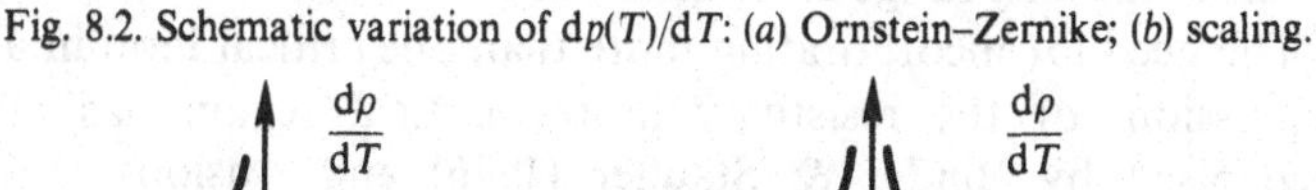

Fig. 8.2. Schematic variation of $\mathrm{d}\rho(T)/\mathrm{d}T$: (*a*) Ornstein–Zernike; (*b*) scaling.

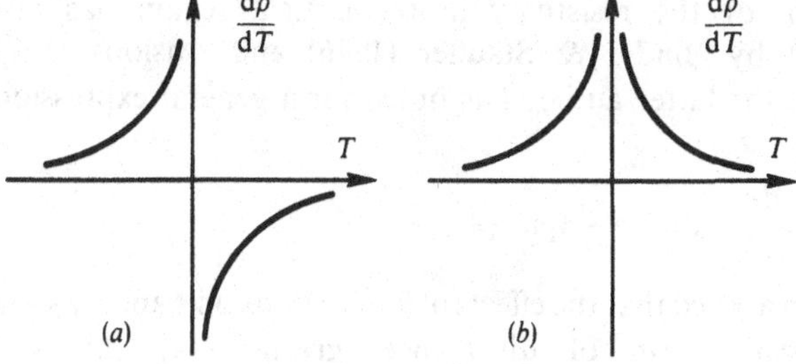

The energy gaps introduced at these boundaries are of course just another aspect of the scattering process and so will also be affected by fluctuations in the correlation parameter at the critical point. However, it is suggested that they are determined mainly by the long range correlations ($n_{\text{eff}} \propto S^2$ or M^2: see equation (5.68)) and so lead to a singularity in n_{eff} of the form (see e.g. Fisher & Langer 1968; Takada 1971; Ausloos 1976)

$$\frac{\mathrm{d}n_{\text{eff}}}{\mathrm{d}T} \sim \varepsilon^{-(\alpha+\gamma-1)}. \tag{8.16}$$

Note that this effect will exist just above T_c as well as below T_c since there will be significant long range spin fluctuation over this whole temperature range. From 1.5:

$$\frac{1}{\rho}\frac{\mathrm{d}\rho}{\mathrm{d}T} = -\frac{1}{n_{\text{eff}}}\frac{\partial n_{\text{eff}}}{\partial T} - \frac{1}{\tau^{-1}}\frac{\partial \tau^{-1}}{\partial T}, \tag{8.17}$$

and so

$$\frac{1}{\rho}\frac{\mathrm{d}\rho}{\mathrm{d}T} = c_1^- \varepsilon^{2\beta-1} + c_2^- \varepsilon^{-\alpha'}; \quad T \to T_c^-, \tag{8.18a}$$

$$\frac{1}{\rho}\frac{\mathrm{d}\rho}{\mathrm{d}T} = c_1^+ \varepsilon^{2\beta-1} + c_2^+ \varepsilon^{-\alpha'}; \quad T \to T_c^+, \tag{8.18b}$$

where we have used the scaling result

$$\alpha + 2\beta + \tau = 2, \tag{8.19}$$

and defined a parameter β' by

$$\alpha' + 2\beta' + \tau' = 2. \tag{8.20}$$

At temperatures further above T_c, the Ornstein–Zernike behaviour (equation (8.9)) should replace the second term on the right hand side of (8.18b) when $\xi \sim \Lambda_1$. At all temperatures below T_c, ξ remains longer than Λ and so there is no deviation in the critical behaviour from (8.18a). The total resistivity well below T_c will of course be determined mainly by the effects of the long range correlations as discussed in Sections 5.3 and 7.1.

The need for incorporating more than one critical term in a general expression for the resistivity in the critical region has also been recognised by Binder & Stauffer (1976) and Ausloos (1977*b*). In particular, the latter author has opted for a general expression of the form

$$\frac{1}{\rho}\frac{\mathrm{d}\rho}{\mathrm{d}T} = c_1\varepsilon^{-\lambda_1} + c_2\varepsilon^{-\lambda_2} + \cdots. \tag{8.21}$$

It was then argued that the effects of fluctuations on τ and n_{eff} should lead to the *same* form of divergence, giving $-\lambda_1 = 2\beta - 1$ for an antiferromagnet both above and below T_N. The second term was found to be $\lambda_2 = \alpha$ for $T < T_N$ leading to (8.18a). On the other hand Malmström & Geldart (1980) have argued that the effects of long range fluctuations disappear in both the τ and n_{eff} contributions, so that α and α' should be the leading critical exponents in antiferromagnets as well as ferromagnets. However, comments made at the beginning of this section regarding our current lack of knowledge of the true local electronic structure cannot be stressed too highly.

The magnitudes and signs of the coefficients C_1^-, C_1^+, C_2^- and C_2^+ in (8.18a and b) depend upon the variation of both the structure factor and form factor with q, in the same way as was discussed in relation to the effects of atomic correlations in Chapter 5. However, there are some constraints:

(i) They may be positive or negative but each pair C_1^-, C_1^+ and C_2^-, C_2^+ must have the same sign.
(ii) If increased short range correlations lead to an increase in ρ then C_2^-, C_2^+ should be negative.
(iii) If the change in n_{eff} with the onset of long range ordering (atomic or antiferromagnetic) leads to an increase in ρ, C_1^-, C_1^+ should be negative. In case of phase separation or ferromagnetic ordering $C_1^- = C_1^+ \approx 0$.
(iv) The ratio C_1^+/C_1^- and C_2^+/C_2^- are expected to lie in the range $\sim$0.5 to 0.6 (Fisher 1967; Ausloos & Durczewski 1980).

Let us now compare these simple results with some experimental data.

(a) *Ferromagnets*

Of the simple ferromagnets, the electrical resistivities of Fe and Ni at the ferromagnetic–paramagnetic transition have been studied in greatest detail. Some typical results are shown in Figure 8.3.

In order to investigate the critical region, $\mathrm{d}\rho/\mathrm{d}T$ may be represented in a manner proposed by Fisher (1967):

$$\frac{\mathrm{d}\rho}{\mathrm{d}T} = \frac{A}{\lambda}(\varepsilon^{-\lambda} - 1) + B, \tag{8.22}$$

where A and B are only weakly temperature dependent. For Fe and Ni, the phonon contribution at temperature $\sim T_c$ gives $\rho_p(T) \sim T$ and so B is a constant. If $\lambda = 0$, the divergence is logarithmic. By taking the second derivative with respect to temperature, the effects of any slow variations with temperature are minimised and one obtains

$$\frac{\mathrm{d}^2\rho}{\mathrm{d}T^2} = -A\varepsilon^{-(\lambda+1)}, \tag{8.23}$$

which may be plotted on a log–log curve to determine λ. Such a plot is shown in Figure 8.4. These results indicate that at temperatures very close to the critical temperature, $\lambda \sim 1.7$ whereas at higher temperatures the singularity becomes logarithmic with $\lambda = 0$, as expected from the

Fig. 8.3. Normalised resistivity of Ni and its temperature derivative as a function of temperature (after Zumsteg & Parks 1970).

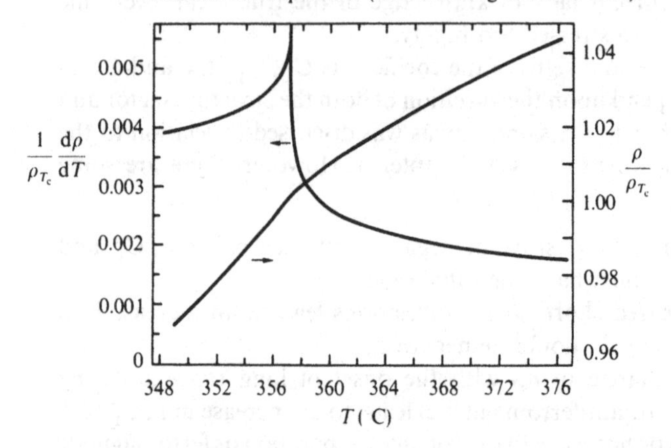

Fig. 8.4. Log–log plot of $d^2\rho/dT$ as a function of ε for Ni with $T > T_c$ (after Craig *et al.* 1967).

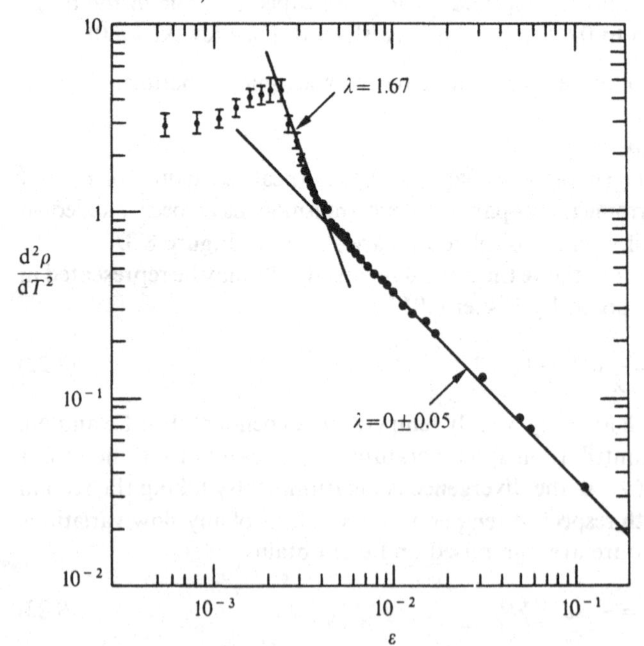

Ornstein–Zernike correlation function. The actual temperature at which the transition in behaviour occurs seems to depend upon the particular specimen studied, but is typically around $\varepsilon \sim 10^{-3}$. Nagy & Pal (1971) have indicated that a slight error in the choice of T_c may lead to the apparent obliteration of any such transition, the divergence then being logarithmic over the whole interval $10^{-4} \leqslant \varepsilon < 10^{-2}$ above or below T_c (see also Sousa *et al.* 1980*a*). It is thus important to obtain some independent means of determining T_c, preferably simultaneously with the resistivity study. The thermoelectric power and magnetic susceptibility are reasonably convenient in this regard (Craig *et al.* 1967; Nagy & Pal 1970, 1971). While it is very difficult to obtain consistent values of λ, the indications are that $\lambda \sim 0.4$ for Fe and ~ 1–1.7 for Ni in the critical (scaling) region, and $\lambda \sim 0$ (logarithmic) outside this region. While the accuracy of these values is a matter of conjecture, it does appear that the sign of $\mathrm{d}\rho/\mathrm{d}T$ remains positive about T_c and that λ is positive, although possibly somewhat larger than the value expected from (8.18a or b) ($\alpha \approx \alpha' \approx 0.125$).

The resistivity of some ferromagnetic alloys around the critical region has also been investigated with similar results. The alloys studied include Ni–Cu (Jackson & Saunders 1968; Sousa *et al.* 1972), Ni–Cu–Si, Ni–Cu–Zn (Lupsa & Burzo 1983) and FePd (Longworth & Tsuei 1968). However, in such cases there is the additional complication of composition inhomogeneities. These can lead to a spatially varying T_c which will effectively smear out the critical behaviour (see e.g. Nieuwenhuys & Boerstoel 1970). Many studies have indicated a lot of 'noise' in $\mathrm{d}\rho/\mathrm{d}T$ around the critical point. Whether this is due to such spatial fluctuations or a magnetic domain phenomena, as suggested by Sousa *et al.* (1980*b*), is not clear. Some further discussion about these effects has been given by Fisher & Ferninand (1967). It might also be noted that structural changes associated with the magnetic transition can also lead to anomalous results (Sousa *et al.* 1979, 1980; Viard & Gavoille 1979). Finally, the behaviour observed in anisotropic materials such as Gd will depend upon the direction of current flow in relation to the crystal axes (see e.g. Zumsteg *et al.* (1970)). This aspect of the problem has been discussed by Parks (1972).

(b) *Antiferromagnets*

As evident from (8.13) and (8.14) the nature of the resistivity anomaly in an antiferromagnet at T_N will depend upon the periodicity and dimensionality of the magnetic structure, and whether $\mathbf{q}$ is parallel or perpendicular to $\mathbf{Q}$ (see e.g. Suezaki & Mori 1968; Richard & Geldart 1977). Thus the sign of $\mathrm{d}\rho/\mathrm{d}T$ about T_N and the dominant critical power law determined by the magnitudes of the coefficients in (8.18a and b) may vary from system to system. It is thus not surprising that antiferromagnetic materials exhibit a variety of $\rho(T)$ and $d\rho(T)/dT$ behaviours. For example Ce–Al and Au–Ho exhibit a smooth decrease in ρ as T decreases through T_N, giving a positive anomaly in $d\rho/dT$ like those discussed above (Fote *et al.* 1970). An example of this behaviour is shown in Figure 8.5. By way of contrast, Cr and its alloys and many of the rare-earth based antiferromagnets exhibit a broad hump in ρ just below T_N (cf. Figure 5.12(*a*)), with $d\rho/dT$ negative around T_N (see e.g. Meaden 1966). A typical result is shown in Figure 8.6.

There have been attempts to deduce the critical exponents in some cases, but the same difficulties as described above are encountered. Another problem which often besets these studies is that the energy gaps introduced by the superlattice Brillouin zone boundaries often lead to a first order phase transition, excluding much of the critical region. It is thus advisable in an experimental study to compare results obtained in increasing and decreasing temperature, and hysteresis indicating a first order nature of the transition.

(c) *Atomic order–disorder*

The order–disorder systems β–CuZn, Fe_3Al and FeCo undergo second order phase transitions at the ordering temperature T_c: β–CuZn (Simons & Salamon 1971) and FeCo (Seehra & Silinsky 1976) exhibit a monotonically decreasing ρ with decreasing temperature through T_c and a corresponding singularity of the same form as Fe or Ni. On the other

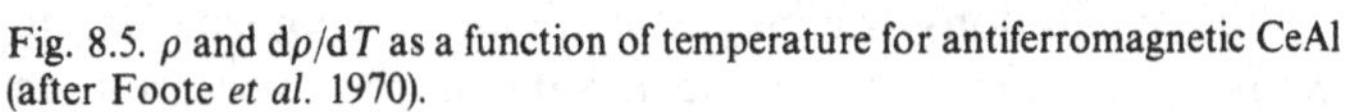

Fig. 8.5. ρ and $d\rho/dT$ as a function of temperature for antiferromagnetic CeAl (after Foote *et al.* 1970).

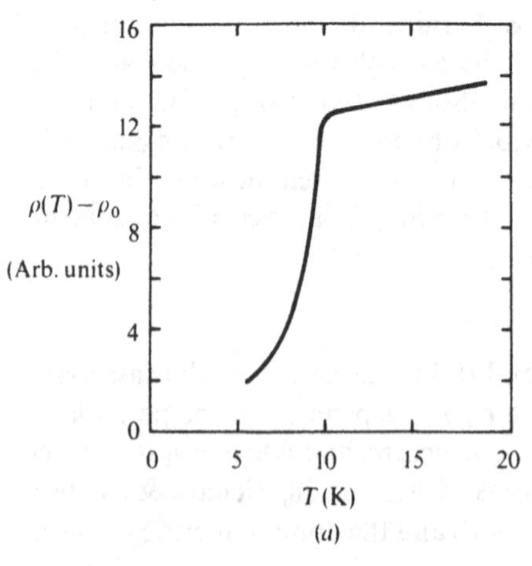

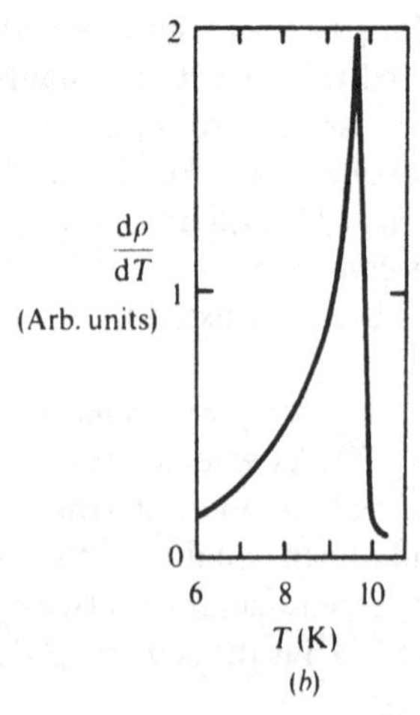

hand Fe_3Al (Thomas *et al.* 1973) has a broad maximum in ρ just below T_c with $d\rho/dT$ negative through T_c. This behaviour is directly analogous to that found in antiferromagnetic systems, as described above, and some representative results are shown in Figures 8.7 and 8.8. The behaviour of these systems at the critical point has been discussed in detail elsewhere (Rossiter 1980*b*) and the behaviour generally found to be in accord with (8.18a) and (8.18b) and the known effects of short range atomic correlations on τ and n_{eff}. For example, increasing short range order leads to a decrease in the resistivity of β–CuZn and Fe_3Al but an increase in FeCo, and so C_2^+ and C_2^- should be positive for the former and negative for the latter. The change in n_{eff} with the onset of long range order gives an increase in the resistivity of ρ in Fe_3Al but a decrease in FeCo (see Section 5.7.4). In CuZn this effect is expected to be small since any new superlattice Brillouin zone boundaries are not expected to significantly affect the Fermi surface (see e.g. Richard & Geldart 1977). Thus C_1^+ and C_1^- should be negative for Fe_3Al, small for β–CuZn and positive for FeCo. The values of the coefficients chosen to give a rough fit to the experimental data with the assumptions $\alpha = \alpha' = \frac{1}{8}$, $\beta = \beta' = \frac{1}{3}$ are given in Table 8.2 and are in general agreement with these predictions. Salamon & Lederman (1974), Balberg & Maman (1979) and Sobotta (1985) give some more detailed comments about curve-fitting to divergent functions. Note also that C_1^+/C_1^- and C_2^+/C_2^- lie in the range 0.3–0.5 (except C_1^+/C_1^- for β–CuZn, but these coefficients are an order of magnitude less than C_2^+/C_2^- and so are prone to large errors in the curve-fitting procedure). The fitted temperature coefficients of resistivity are

Fig. 8.6. ρ and $d\rho/dT$ as a function of temperature for antiferromagnetic Cr–0.06%Al (after Sousa *et al.* 1979).

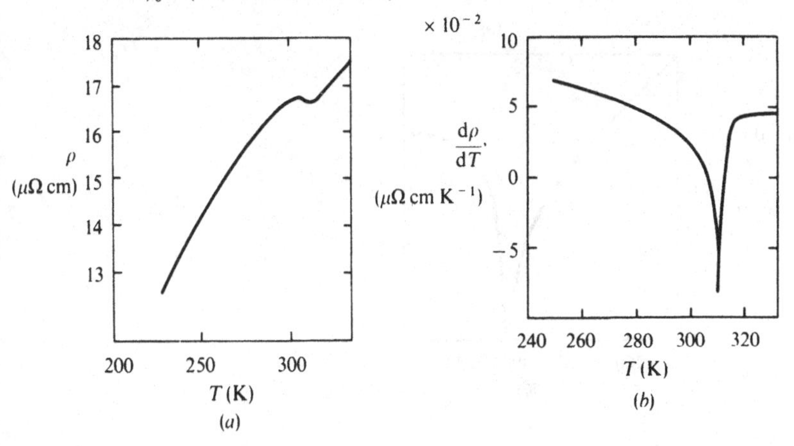

shown in Figures 8.9 and 8.10. While it is admitted that this sort of curve fitting cannot give any detailed information about the critical exponents, it does serve to illustrate the general form of behaviour near the critical point and the constraints on the coefficients.

(d) *Miscibility gap*

A second order phase transition should occur at some critical composition and temperature in an alloy system which contains a miscibility gap. However, studies of the critical behaviour in such systems are virtually non-existent, probably because of the problems associated with metastability, although some liquid metal alloys have been investigated (Schürman & Parks 1971).

Fig. 8.7. ρ and $d\rho/dT$ as a function of temperature for Fe_3Al (after Thomas *et al.* 1973).

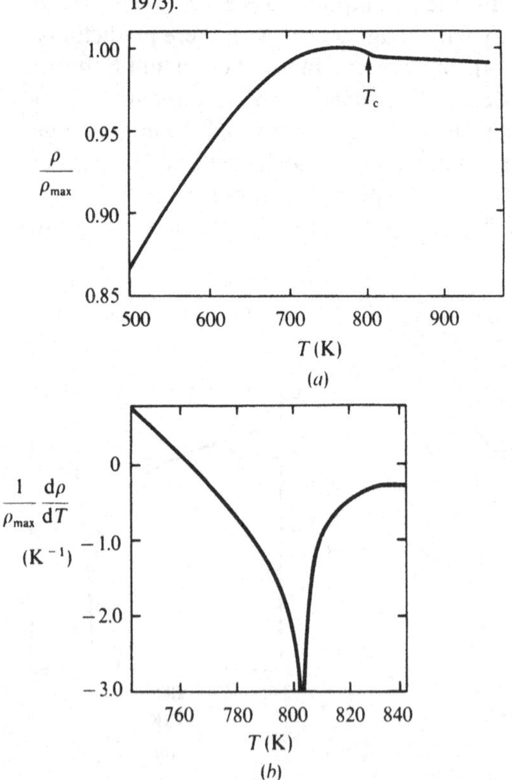

Table 8.2. *Values of the amplitudes C_1^+, C_1^-, C_2^+ and C_2^- chosen to fit the resistivities of the alloys indicated in the vicinity of T_c*

		n_{eff}		τ		
Alloy	T_c(K)	C_1^+	C_1^-	C_2^+	C_2^-	Units
Fe_3Al	803.2	−0.41	−1.18	0.78	2.62	K^{-1}
β–CuZn	466	1.4×10^{-4}	1.1×10^{-4}	1.2×10^{-3}	3.7×10^{-3}	K^{-1}
FeCo	1006	0.012	0.04	−0.017	−0.033	$\mu\Omega$ cm K^{-1}

Fig. 8.8. ρ and $d\rho/dT$ as a function of temperature for FeCo (after Seehra & Silinsky 1976).

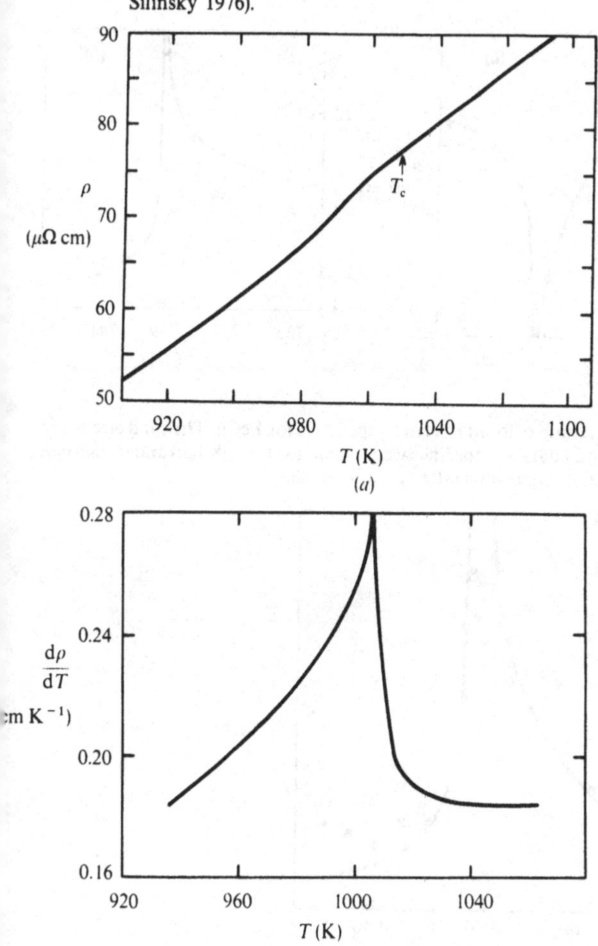

8.1.3 *Related phenomena*

A striking similarity between $d\rho/dT$ and some other physical property such as the specific heat or thermopower has often been observed, as shown in Figures 8.11 and 8.12. In many cases this similarity is to be expected over a limited temperature range because various properties are determined by the same fluctuations at the critical point

Fig. 8.9. (*a*) Variation of $(1/\rho_{max})(d\rho/dT)$ with temperature for Fe_3Al, where ρ_{max} is the peak value of the resistivity. (*b*) Variation of $(1/\rho_{25})(d\rho/dT)$ with temperature for β–CuZn, where ρ_{25} is the electrical resistivity at 25 °C. The solid circles are the experimental data and the lines were calculated from (8.18*a*) and (8.18*b*) (see text for values of constants) (after Rossiter 1980*b*).

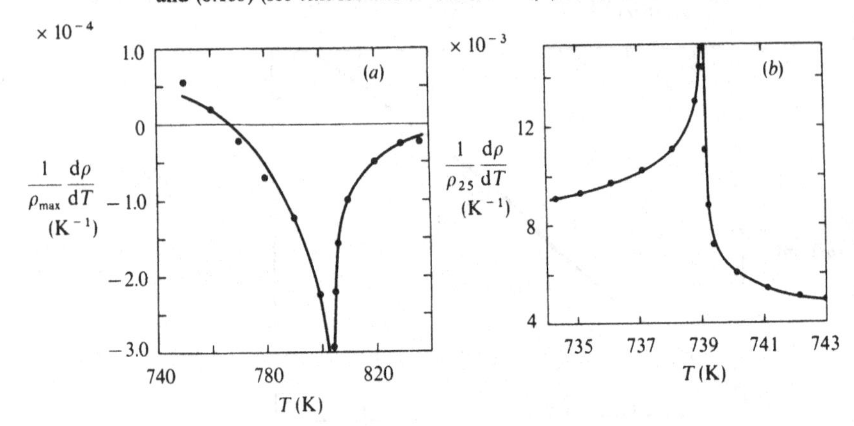

Fig. 8.10. Variation of $(d\rho/dT)$ with temperature for FeCo. The solid circles are the experimental data and the lines were calculated from (8.18*a*) and (8.18*b*) (see text for values of constants) (after Rossiter 1980*b*).

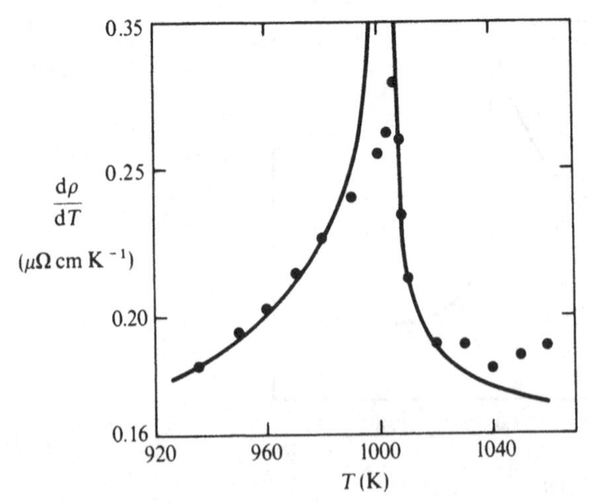

Fig. 8.11. (*a*) Specific heat C_p/R and temperature derivative of resistivity as a function of temperature for β-CuZn. (*b*) Correlation between temperature derivative of resistivity and specific heat from the results shown in (*a*) (after Simmons & Salamon 1971).

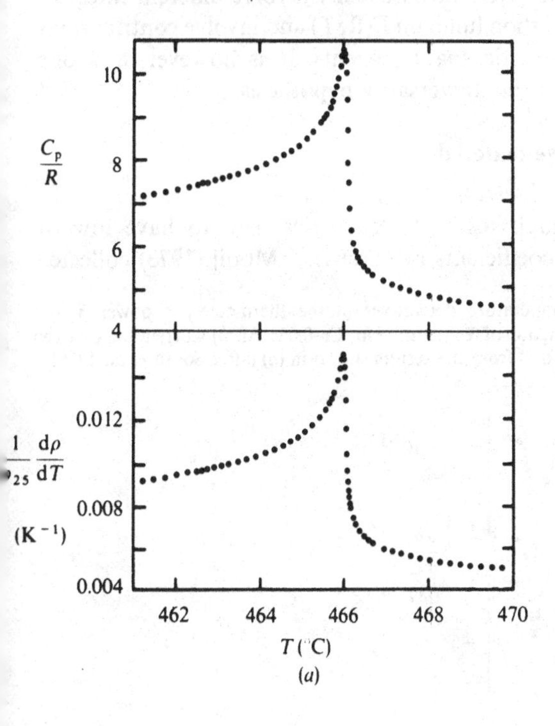

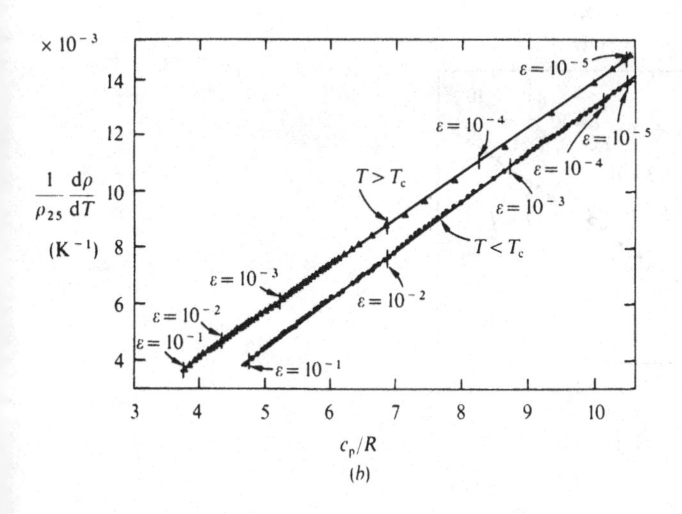

(see e.g. Fisher & Langer 1968; Mannari 1968; Nabutovskii & Patashinskii 1969; Ausloos 1977*a*; Zinov'yev *et al.* 1980). An exact correspondence over a wider temperature range is somewhat surprising as the different physical phenomena usually involve different integrals over the dynamic correlation function $\Gamma(\mathbf{R}, T)$ and involve contributions from different characteristic spatial extents. It is however, a strong argument in support of the universality hypothesis.

8.2 Highly resistive materials

8.2.1 *Some general observations*

Alloys with high residual resistivities tend to have low or negative temperature coefficients of resistivity. Mooij (1973) collected

Fig. 8.12. (*a*) Temperature derivatives of the thermoelectric power *S* and resistivity as a function of temperature for Cr–0.06%Al. (*b*) Correlation between ($\mathrm{d}S/\mathrm{d}T$) and ($\mathrm{d}\rho/\mathrm{d}T$) from the results shown in (*a*) (after Sousa *et al.* 1979).

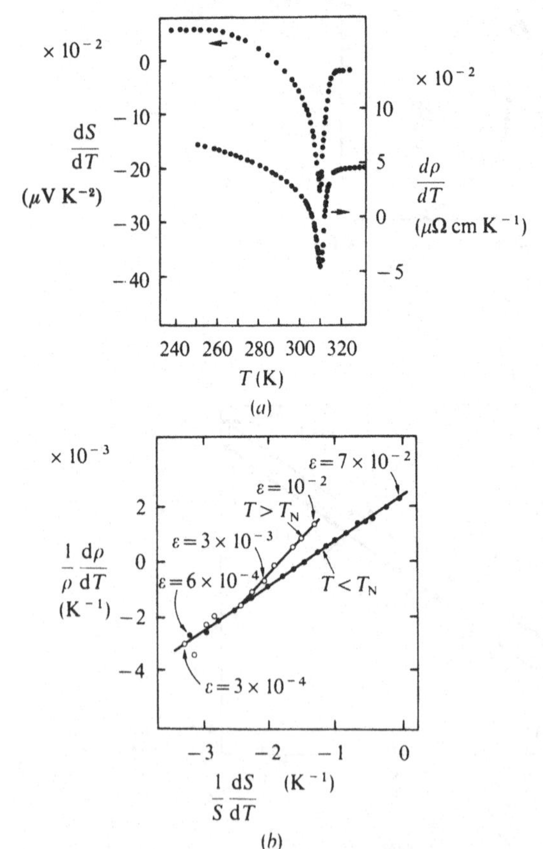

some data for bulk alloys, thin films and amorphous alloys and found that all exhibited a similar correlation between $\mathrm{d}\rho/\mathrm{d}T$ and ρ_0, as shown in Figure 8.13. It does not appear to matter whether the disorder giving rise to a high ρ_0 is due to static structural disorder or temperature effects, and the transition in behaviour occurs in the residual resistivity range $\rho_0 \sim 100$–$160\mu\Omega$ cm over which the mean free path Λ becomes comparable to the lattice spacing a_0.

However, inclusion of more recent data on amorphous metals (see next section) effectively destroys this correlation when $\rho_0 \leqslant 130\mu\Omega$ cm (i.e. when $\Lambda < a_0$). Nevertheless, the fact remains that there appears to be no alloys having $\rho_0 \geqslant 160\mu\Omega$ cm which display a positive $\mathrm{d}\rho/\mathrm{d}T$. Another observation associated with high resistivity alloys which requires explanation is that increasing thermal or compositional disorder ultimately leads to a saturation of the resistivity at about 150–200$\mu\Omega$ cm, i.e. there appears to be a minimum conductivity that can be achieved in any alloy system. Some typical behaviour is shown in Figure 8.14. This saturation value again corresponds roughly to the condition $\Lambda \sim a_0$ and Mooij has suggested that the two effects have a common origin. Note that both of these observations generally relate to non-magnetic, conducting alloys which tend to be concentrated and contain transition metal or rare-earth components. We are not concerned here with phenomena relating to the magnetic or nearly magnetic behaviour that was discussed in Chapter 7.

Fig. 8.13. Temperature coefficient of resistivity as a function of residual resistivity for bulk alloys (■); thin films (▲); and amorphous alloys (□) (after Mooij 1973).

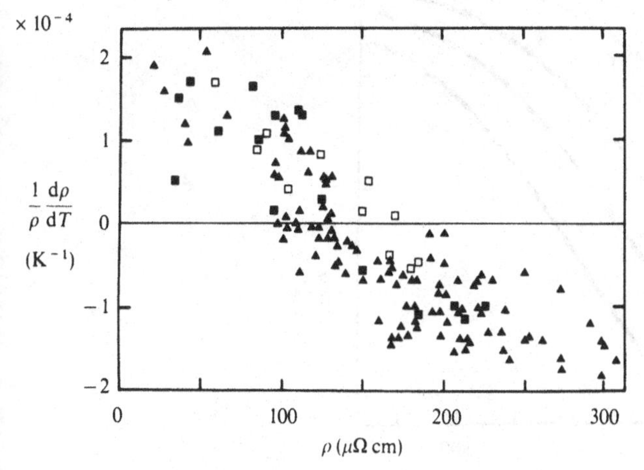

A number of theories have been proposed to explain these two phenomena. (For reviews see Allen 1980, 1981; van Daal 1980; Belitz & Schirmacher 1983). It seems that some of these have been tempted by the apparent Mooij correlation into explaining the behaviour of $d\rho/dT$ in alloys over a wide range of ρ_0. Some are based on the assumption of extended (itinerant) electron states, with the electrons having a well-defined wavenumber **k**. The resistivity is then determined by the scattering mechanisms and density of states in the various bands as discussed in Chapters 5 and 6. Other models are based on the concept of localised electron states, with quite different transport mechanisms. The basic physics of the problem argue against either of these extreme models being applied over the whole range $\Lambda \sim a_0$ to $\Lambda \gg a_0$. If $\Lambda \gg a_0$ the alloys under consideration definitely form extended electron states with well-characterised band structures and associated physical properties, and a localised model is inappropriate. The theoretical principles presented in Chapters 5 and 6 should remain valid, although the complex nature of many of the alloys concerned would suggest that the nearly free electron model was a poor approximation. Over this regime $d\rho/dT$ in most crystalline alloys is positive for reasons which will be discussed shortly, and for amorphous (and liquid) alloys it may be positive *or* negative (see next section). However when Λ becomes comparable to a_0 the electron

Fig. 8.14. Resistivity as a function of temperature for Ti and Ti–Al alloys (compositions as indicated) (after Mooij 1973).

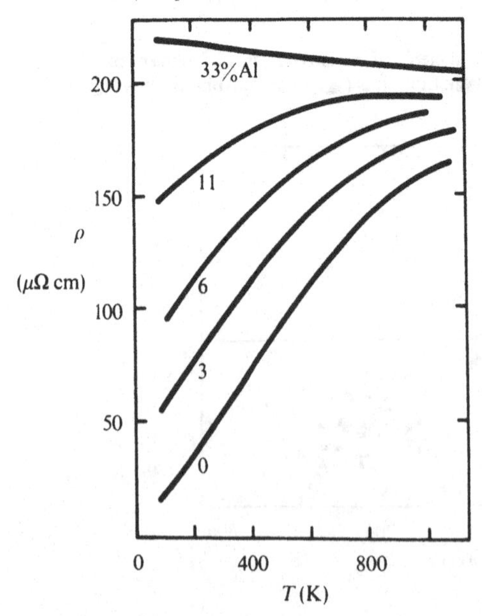

gas analogy implicit in the Boltzmann theory is invalid since the electron wavepackets and associated wavevectors **k** are then no longer clearly defined (see Section 1.8.1). In accordance with the intuitive picture of Kittel (1948) and the more recent arguments of Ioffe & Regel (1960), the mean free path can only be reduced to some minimum value corresponding either to the electron wavelength or the lattice constant. At this stage (or even approaching close to it) a diffraction model or one more generally based on itinerant electron states is then no longer applicable and one would expect a localised electron state model to be more realistic. More recently, a number of generalised transport equations having a similar structure to the Boltzmann equation but derived from two-particle Green's functions or other distribution functions have been proposed to describe the transport properties over the whole range from the long mean free path regime ($\Lambda \gg a_0$) to the localised regime ($\Lambda \approx a_0$). In the latter case the electron scattering is peaked in the backward direction with the maximum momentum change of $2k_F$. In order to clarify these points let us now review some of the proposed mechanisms according to their expected range of validity.

8.2.2 $\Lambda > a_0$

(a) *Diffraction models and the Debye–Waller factor*

The basic approach to the determination of resistivity outlined in this book is based on conduction electron scattering from atomic (or magnetic) disorder. The scattering is determined by the nature of the scattering potential and its spatial distribution described by the structure factor (see Chapter 4). Either a static or dynamic structure factor may be employed and increased disorder can lead to an increase or a decrease in resistivity depending upon the position of the major peaks in the structure factor (see Sections 5.7.3 and 8.3.2). However, as noted in Section 1.4 and discussed at length in Chapter 5, there can be no coherent scattering of the electron by correlations over a length greater than $\sim\Lambda$ and so the structure factor in real space should contain a damping term $\exp(-|\mathbf{R}|/\Lambda)$. If $\Lambda \sim a_0$, virtually all peaks will be removed from this structure factor. The resistivity is then determined by the $R_{ij}=0$ term, i.e. by the random disorder only. If, for the sake of simplicity we assume a nearly free electron model, the discussion in Sections 5.6.3 and 5.6.5 indicate that a general expression for the resistivity in the presence of both atomic and thermal disorder is then

$$\rho(T) = \rho_p(T) + C\int_0^{2k_F} |\bar{w}(\mathbf{q})|^2(1-\exp[-2(M'+M)])q^3\,dq$$
$$+ Cc_Ac_B\int_0^{2k_F} |w^d(\mathbf{q})|^2\exp(-2M)q^3\,dq, \quad (8.24)$$

where the pseudo Debye–Waller factor M' results from static atomic displacements, M is the temperature-dependent Debye–Waller factor and $\rho_p(T)$ the phonon scattering term that is assumed to contain all anharmonic and multiphonon processes. The terms M' and M involve the factor $1/q^2$ but we will assume that the integrals are dominated by the behaviour at $q=2k_F$ (because of the q^3 weighting factor) and so write (8.24) simply

$$\rho(T)=\rho_p(T)+\rho_0 \exp(-2M). \tag{8.25}$$

An alternative approach often adopted is to average the Debye–Waller factor over q and take this average outside the integral leading to the same result. Let us now turn to the phonon term $\rho_p(T)$. As indicated in Section 5.6.2, a very small mean free path effectively excludes long-wavelength phonons from contributing to the scattering of conduction electrons. This means that the integration over ω in (5.148) should be restricted to frequencies corresponding to phonon wavelengths $>\Lambda$. Cote & Meisel (1978) have evaluated the effects of such a cut-off using the Debye model, in which case the integration is limited to phonons with wavenumbers between $2\pi/\Lambda$ and q_D.They found that for $T>\Theta_D/2$, the resistivity due to one-phonon processes was reduced by a factor $1-\gamma$:

$$\rho_p(T)=\rho_{ip}(T)(1-\gamma), \tag{8.26}$$

where ρ_{ip} is the ideal ($\Lambda\to\infty$) phonon contribution and $\gamma=2\pi/\Lambda q_D$. Equations (8.25) and (8.26) then give

$$\rho(T)=(1-\gamma)\rho_{ip}(T)+\rho_0 \exp(-2M). \tag{8.27}$$

The overall temperature dependence is thus determined by a balance between the phonon term reduced by $(1-\gamma)$ and the residual resistivity term reduced by the Debye–Waller factor. To evaluate this expression we note that τ depends upon $1/\Lambda$ and so can be written in terms of ρ:

$$\gamma=\rho/\rho^D. \tag{8.28}$$

The term ρ^D is the resistivity corresponding to $\Lambda=2\pi/q_D$, i.e. the saturation mean free path in the Debye model. For a typical free electron monovalent metal this is $\sim 200\mu\Omega$ cm. In transition metals the large density of states in the d-band $N_d(E_F)$ may increase this value somewhat. The temperature dependence of the resistivity may finally be determined from (8.27) and (8.28). This is shown plotted in Figure 8.15 for various values of ρ_0 and the assumptions $M=0.01T/\Theta_D$ (see Section 5.7.3), $\theta_D=300$ K, $\rho^D=200\mu\Omega$ cm and $\rho_{ip}=\alpha T$ with $\alpha=0.2\mu\Omega$ cm K^{-1}. This result indicates a saturating behaviour as T increases and that a negative $d\rho/dT$ is expected for all temperatures if $\rho_0 \geqslant \rho^D/2$. Note that this behaviour stems mainly from the cutoff in the phonon contribution. If this is omitted, the Debye–Waller factor alone simply leads to a reduction in the overall temperature coefficient of resistivity by a small amount (~5%) unless very small values of α are assumed (as done by Markowitz 1977, for example). Brouers *et al.* (1978) have shown that the form of this nearly free electron result is retained in a system containing narrow bands provided that M is multiplied by a factor related to the density of states and so can be quite large if E_F lies within a peak in the density of states. However, α and ρ^D will also be affected by the band structure and so it is not clear what the overall result may be. Morton *et al.* (1978) have derived a result similar to the above by incorporating an expression due to Pippard (1955, 1960) to take into account the reduction in the electron–phonon interaction for $Q \leqslant 2\pi/\Lambda$. We will discuss this method further in relation to amorphous metals in Section 8.3.2(b). While a comparison of Figures 8.14 and 8.15 shows that the calculated result is very similar to that found experimentally, the theoretical foundation of the analysis is weak if Λ approaches a_0 for the reasons stated above.

(b) *CPA, interband and other band-based calculations*

As discussed in Section 6.7.2, a more rigorous calculation of the resistivity of a disordered alloy with narrow bands shows that $d\rho/dT$ may be reduced or even negative in strongly scattering alloys. The actual mechanism is again the mixing of electron–phonon and potential

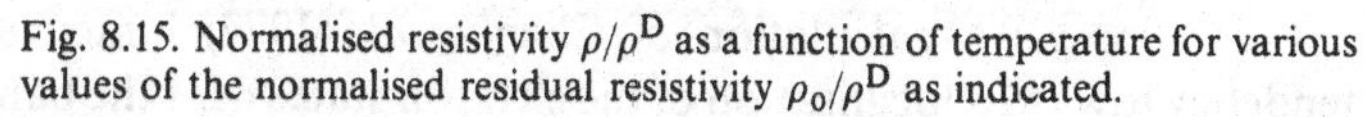

Fig. 8.15. Normalised resistivity ρ/ρ^D as a function of temperature for various values of the normalised residual resistivity ρ_0/ρ^D as indicated.

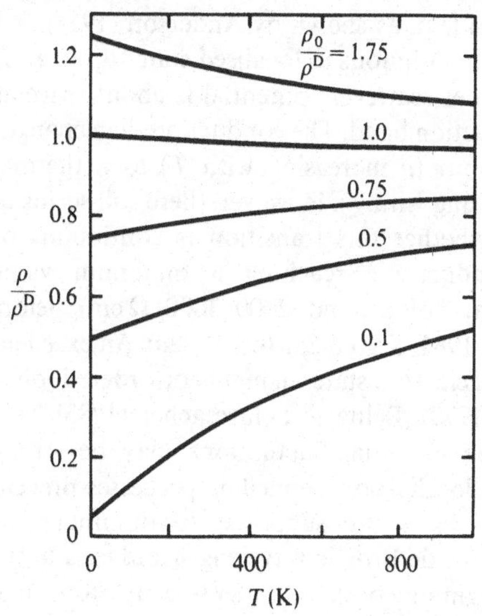

scattering leading to a smearing of the density of states. Christoph (1979) has approached the problem within the force–force correlation formalism using a single-band model which included multiple scattering, and again found that a negative temperature coefficient is expected if the scattering is strong. Mott (1972) has suggested that interband transitions may also lead to a decrease in the resistivity with increasing temperature. This model has been developed formally by Allen & Chakraborty (1981), Chakraborty & Allen (1979), and gives rise to non-classical terms which modify the semiclassical Boltzmann behaviour (see also Guinea 1983). Conceptually the model may be represented as a shunt 'resistor' ρ_{max}, representing the new interband conduction channel and limiting the total resistivity. Such a representation is known to give a good description of much of the available experimental data (Wiesmann *et al.* 1977; see also Allen 1980, 1981) and is in accord with the intuitive picture of Kittel described above. It might be noted that the results of the phonon–cut-off may also be presented in this form (Morton *et al.* 1978). However, all of these theories are based on the notion of energy bands which are themselves a product of coherent diffraction processes and so are of unknown validity if $\Lambda \sim a_0$, although they may well be of importance in explaining the diversity of results observed in alloys having larger Λ ($\rho_0 < 150\mu\Omega$ cm).

8.2.3 $\Lambda \sim a_0$

The CPA calculation of Chen *et al.* (1972) hinted at some tendency toward a localisation of the electron states near the band edges in a strong scattering alloy. In fact, as shown by Anderson (1958) and Mott (1974), a transition from continuous to localised states up to E_F is expected to occur if the disorder scattering potential is about twice as large as the width of the conduction band. The conduction then changes from normal metallic behaviour (ρ increasing with T) to a thermal hopping mechanism (ρ decreasing with T). However, there still seems to be some uncertainty as to whether this transition is continuous or discontinuous with the conductivity reaching a minimum value (corresponding to a resistivity of around 2000–3000$\mu\Omega$ cm) before jumping to zero (see e.g. Mott 1981; Kaveh & Mott 1982*a*). An excellent introduction to the problem of electron states in highly disordered solids is given by Mott & Davis (1979). Belitz & Schirmacher (1983) have argued that even though the potential fluctuations may be strong enough to produce Anderson localisation, tunnelling processes prevent its establishment. These processes are supposed to dominate the conductivity if Λ approaches the de Broglie wavelength and lead to the behaviour described at the beginning of this chapter (see also Jonson &

Girvin 1979). Imry (1980) and Kaveh & Mott (1982*b*) have also suggested that the observed behaviour may in fact be understood as a precursor to the Anderson localisation, the resistivity then being proportional to the mean free path due to inelastic phonon scattering. In the more general theories briefly mentioned at the end of Section 8.2.1, the scattering is proportional to $|\mathbf{k}-\mathbf{k}'|^{-2}$ and so becomes singular when $\mathbf{k}=-\mathbf{k}'$. The momentum transfer is then $2k_F$ (for this reason such scattering is sometimes described as '$2k_F$ scattering') and the resistivity should diverge in the localised limit (see e.g. Morgan *et al.* 1985*a,b*; Morgan & Hickey 1985; and references therein).

8.2.4 *Some general comments*

Can we explain the behaviour described in Section 8.2.1? With regard to resistivity saturation, it appears that in a diffraction model the Debye–Waller factor alone is not strong enough to explain the behaviour. However, if the effects of finite electron mean free path are included (as they must) then a reasonable representation of the experimentally observed behaviour is possible. Similarly, models based on a changing density of states, interband transitions, incipient localisation or tunnelling are also in qualitative agreement with some of the available experimental data, although it is still a matter of conjecture whether the present localised state models can adequately explain why the resistivity appears to be limited to values below some maximum resistivity $\rho_{max} \sim 300$–$400 \mu\Omega$ cm, and why it remains reasonably constant with decreasing temperature down to the lowest temperatures studied.

Auerbach & Allen (1984) have noted that a similar saturation effect exists in the thermal conductivity of insulators. This would appear to indicate that the phenomenon is a general property of waves propagating in a disordered media and rule out any electron- or phonon-specific theories (such as Fermi surface smearing or e–e interations). The ubiquitous nature of the behaviour would also appear to rule out any mechanisms based on particular details of the band structure, although resistivity saturation is usually observed in compounds which have strong electron–phonon coupling and narrow bands near the Fermi level (Ho *et al.* 1978). In such materials the nearly free electron theory will be a poor approximation and the scattering should be determined using other methods, provided that they include the effects of a finite mean free path.

With regard to the negative temperature-coefficient of resistance in high ρ_0 materials, again existing scattering and/or band models give a reasonable description and are quite suitable if $\Lambda > a_0$. Whether the negative $d\rho/dT$ found in materials with $\Lambda \sim a_0$ is explained by the same mechanism or whether this is due to incipient localisation effects is not clear. The fact that it is immaterial whether the disorder is of thermal or static origin, that ρ does not appear to have any unusual (i.e. divergent) behaviour as $T \rightarrow 0$ K, and that roughly the same maximum resistivity is observed in most metallic systems, would seem to argue against any current explanations based on localisation. However this leaves us with the question of why a scattering model should work when $\Lambda \sim a_0$. A partial explanation may lie in the ideas of Mott & Davis (1979) in which the electron wavefunctions in a disordered solid are constructed from basis functions which are locally plane waves but randomly out of phase with each other over distances $\sim \Lambda$. For the reasons described above, the use of the Boltzmann equation is still not justified if $\Lambda \sim a_0$, but the Kubo–Greenwood formula then leads to an expression for the resistivity which is formally quite similar to that obtained from nearly free electron theory (see e.g. Mott & Davis 1979, ch. 2; Ziman 1967). Thus, while the actual parameters used in a nearly free electron model such as k_F and Λ are likely to differ from the true values if $\Lambda \sim a_0$, it is quite possible that the qualitative predictions of the model are in fact valid. More recently, Weger *et al.* (1984); and Weger & Mott (1985) have argued that the decrease in $d\rho/dT$ with increasing temperature is not a saturation effect and that it occurs even when the electron mean free path is ~ 10 times the interatomic spacing. They proposed a mechanism based on a two-band model, suggesting that above a certain temperature the electron–phonon scattering energies may exceed the hybridisation interaction thereby decoupling the s- and d-states. Conduction then occurs mainly by s electrons with a reduced rate of increase of phonon scattering with increasing temperature. What is clear is that more information about the energy states of electrons (and phonons) in such disordered materials is required in order to resolve some of these problems.

8.3 Amorphous metals

8.3.1 *General observations*

Metallic glasses exhibit a wide range of properties and have been classified according to a number of schemes. Here we will classify them as 'simple' if they exhibit nearly free electron characteristics, with no anomalous behaviour associated with localised magnetic moments, exchange enhancement, magnetic ordering or other band structure effects, in accordance with the description of crystalline materials adopted in previous chapters. Metallic glasses which contain significant concentrations of transition metals will generally not fit into this group and so will be classified as 'non-simple'.

The electrical resistivity of these materials exhibits several interesting features. At temperatures $\sim\theta_D$ the resistivity of simple glasses varies roughly linearly with temperature, the slope being either positive or negative. At lower temperatures it varies as T^2 (omitting now from consideration specimens with a very high ρ_0) and may finally pass through a minimum as $T \to 0$.

Some representative results are shown in Figure 8.16. If a magnetic or nearly magnetic (i.e. exchange enhanced or Kondo) state is formed, this behaviour will be modified along the lines discussed in Chapter 7. A low temperature minimum has been observed in a range of amorphous metals with properties varying from strongly ferromagnetic to weakly paramagnetic. An excellent review of the electronic structure of metallic glasses has been given by Mizutani (1983). This review includes a compilation of much experimental data, including the resistivity of a wide range of amorphous metals. Naugle (1984) has also given a good review of electron transport in these materials.

Fig. 8.16. Temperature dependence of the electrical resistivity of amorphous $(Ag_{0.5}Cu_{0.5})_{100-x}Mg_x$ alloys (after Mizutani & Yoshino 1984).

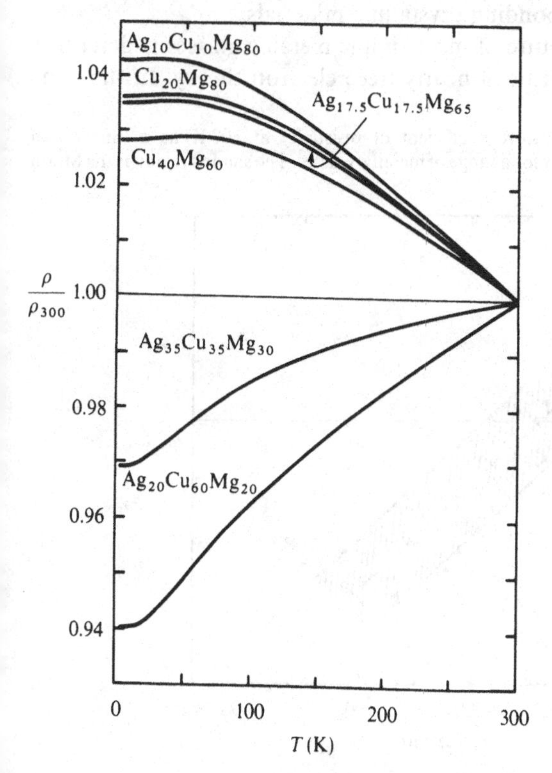

8.3.2 *Resistivity in non-magnetic glasses*

(a) $T \geqslant \Theta_D$

The temperature coefficient of resistivity of a range of simple metallic glasses is plotted as a function of their residual resistivities in Figure 8.17. The data is taken largely from Mizutani (1983) but has been supplemented by some later results. From this diagram it is clear that the Mooij correlation described in Section 8.2.1 and indicated by the shaded region is not followed. In fact it appears that below a residual resistivity of $\sim 120\mu\Omega$ cm, $d\rho/dT$ may be either positive or negative. However, in agreement with Mooij there do not appear to be any positive $d\rho/dT$ results in glasses having a residual resistivity greater than $\sim 160\mu\Omega$ cm. Some results for non-magnetic, non-simple metallic glasses are shown in Figure 8.18 and indicate a very similar trend.

In this section we will restrict consideration to those materials having a residual resistivity $\rho_0 \leqslant 120\mu\Omega$ cm which are expected to be reasonably well-characterised in terms of extended electron states and diffraction effects. Those with $\rho_0 > 160\mu\Omega$ cm fall into the situation where $\Lambda \sim a_0$ and are subject to the same comments made in the previous section in relation to the corresponding crystalline materials.

The electronic structure of most simple metallic glasses is described quite adequately in terms of nearly freee electron theory. In the non-

Fig. 8.17. Temperature coefficient of resistivity at 300 K as a function of residual resistivity for a range of metallic glasses. The shaded region is the Mooij correlation.

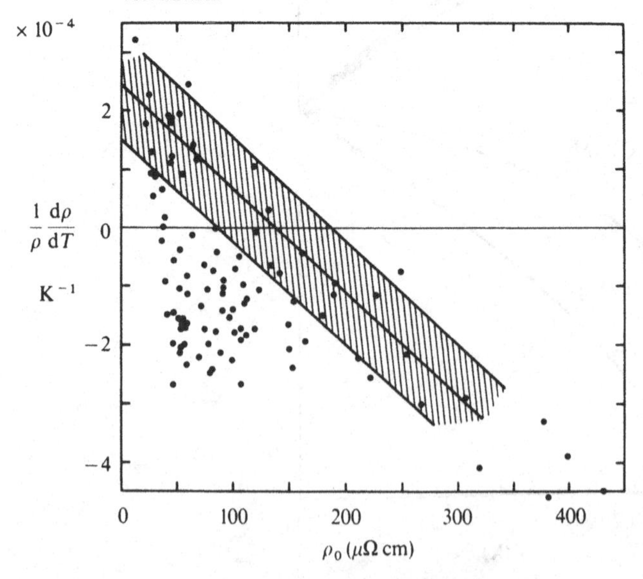

simple glasses, the nearly free electron approximation is on a much weaker footing and details of the band structure should be taken into account (see e.g. Mizutani 1983). At the very least, the scattering potential should be replaced by the appropriate t-matrix, as discussed in Section 6.4. Nevertheless, much of the observed resistivity behaviour can be understood simply on the basis of scattering effects. In general terms the electrical resistivity of a simple metallic glass, for example, can be calculated from equation (5.16) if a static structure factor is used, or (5.148) if a dynamic structure factor is available. However, as discussed in Section 2.8, in an amorphous alloy the correlations between atom types as well as atom positions need to be taken into account, leading to the existence of three partial structure factors $S_{NN}(\mathbf{q})$, $S_{CC}(\mathbf{q})$ and $S_{NC}(\mathbf{q})$. By analogy with (5.160) and (5.167), the appropriate scattering potentials are then $|\bar{w}(\mathbf{q})|^2$, $|w^d(\mathbf{q})|^2$ and $\bar{w}(\mathbf{q})w^d(\mathbf{q})$ respectively. This is made explicit in (5.191) and (5.194). Note that one could retain the alloy pseudopotential in the form (4.64) leading to the definition of the partial structure factors

Fig. 8.18. Temperature coefficient of resistivity as a function of resistivity at 300 K for non-simple, non-magnetic metallic glasses (after Mizutani 1983).

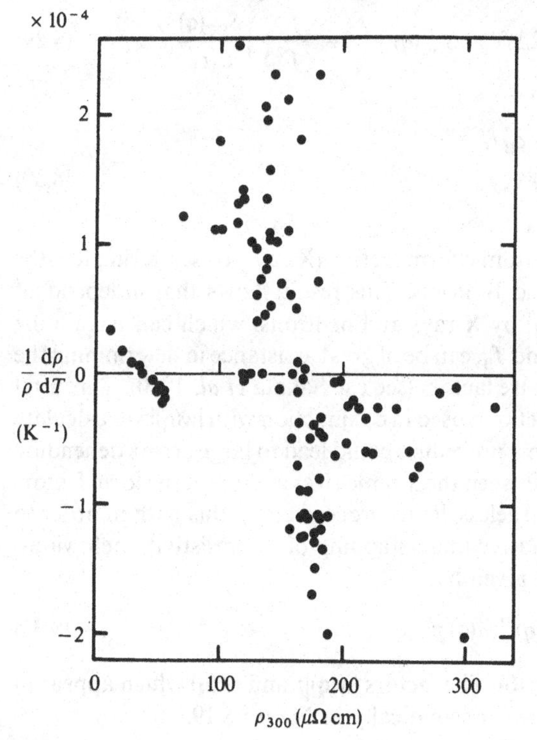

$S_{AA}(\mathbf{q})$, $S_{AB}(\mathbf{q})$ and $S_{BB}(\mathbf{q})$ for AA, AB and BB atom pairs (including the effects of displacements) rather than separate position and atomic correlation partial structure factors. This was the method adopted by Faber & Ziman (1964) for the study of liquid alloys and its application to amorphous materials is usually described as the 'generalised Faber–Ziman theory'. These two approaches must of course inevitably lead to the same results, although the former allows for a clear distinction between the effects of atomic disorder due to position and due to atomic potential.

While such sets of partial structure factors have been determined for some metallic glasses (see e.g. Chieux & Ruppersberg 1980; Wagner & Lee 1980; Sakata *et al.* 1980), they are generally not available over a range of temperatures, although recent calculations by Hafner (1983) suggest that they could be calculated quite successfully from pseudopotential theory (see also Bath & Blétry 1985; Pasturel *et al.* 1985 for similar liquid alloy calculations). Often the only experimental result available is the total interference function or structure factor, which is a mixture of the partial structure factors (Bhatia & Thornton 1970; Wagner & Lee 1980):

$$\mathcal{S}(\mathbf{q}) = \frac{\langle f\rangle^2}{\langle f^2\rangle} S_{NN}(\mathbf{q}) + 2\Delta f\langle f\rangle S_{NC}(\mathbf{q}) + \left(1 - \frac{\langle f\rangle^2}{\langle f^2\rangle}\right)\frac{S_{CC}(\mathbf{q})}{c_A c_B}, \tag{8.29}$$

where, for a binary alloy

$$\begin{aligned} \langle f\rangle &= c_A f_A + c_B f_B \\ \langle f^2\rangle &= c_A f_A^2 + c_B f_B^2 \\ \Delta f &= f_B - f_A, \end{aligned} \tag{8.30}$$

and f_A and f_B are the atomic form factors (X-rays) or scattering lengths (neutrons) of the A and B atoms. This result shows that independent measurements of $\mathcal{S}(\mathbf{q})$ by X-rays and neutrons, which can have quite different values of f_A and f_B, can be of great assistance in determining the separate partial structure factors (see e.g. Sakata *et al.* 1980). This total structure factor is sometimes used in conjunction with $|\bar{w}(\mathbf{q})|^2$ to calculate the resistivity, but such a procedure could lead to large errors depending upon the differences between the atomic sizes and upon the form factors $w_A(\mathbf{q})$ and $w_B(\mathbf{q})$. Nevertheless, let us proceed along this path in order to obtain a simple qualitative understanding of the resistivity behaviour. The resistivity is then given by

$$\rho = C\int_0^{2k_F} |\bar{w}(\mathbf{q})|^2 \mathcal{S}(\mathbf{q}) q^3\, dq, \tag{8.31}$$

where C is given by (5.10). The factors $|\bar{w}(\mathbf{q})|^2$ and $\mathcal{S}(\mathbf{q})$ which appear in the integrand are shown schematically in Figure 8.19.

The temperature dependence of the resistivity enters through the change in the structure factor with temperature. At higher temperatures the correlations become weaker and so $\mathscr{S}(\mathbf{q})$ becomes broader and less intense, as indicated in Figure 8.19. The resistivity is given an integral over the product of these two terms heavily weighted by q^3 in favour of the $q=2k_F$ region. Thus, for a system with $2k_F(1)$, indicated on Figure 8.19, the resistivity will increase with increasing temperature, whereas with $2k_F(2)$ a negative $d\rho/dT$ will be found. From this simple argument it is expected that the behaviour is determined by the relative positions of $2k_F$ and the value of q corresponding to the peak in the structure factor at q_p. Such a correlation is usually observed in practice, as shown in Figure 8.20. Note however that the detailed behaviour can only be determined by resorting to the partial structure factors and appropriate scattering potentials, with the result that the actual value of $2k_F/q_p$ at which $d\rho/dT$ changes sign (if at all) or is maximal is expected to vary from system to system. This simple argument can provide a basis for a correlation between ρ_0 and $d\rho/dT$ *within any particular system* since $d\rho/dT$ should be largest when $2k_F/q_p \approx 1$ and this will also correspond to a maximum ρ_0. Correlations of this kind are common in many metallic glasses, an example being given in Figure 8.21.

In non-simple metallic glasses, the general behaviour described above could be considerably altered by band structure effects (see e.g. Schulz *et*

Fig. 8.19. Schematic representation of the factors $|\bar{w}(\mathbf{q})|^2$ and $\mathscr{S}(\mathbf{q})$ that appear in the integrand of the resistivity integral. The structure factor $\mathscr{S}(\mathbf{q})$ is represented by the full line at the temperature T_1, and the dashed line at $T_2 > T_1$.

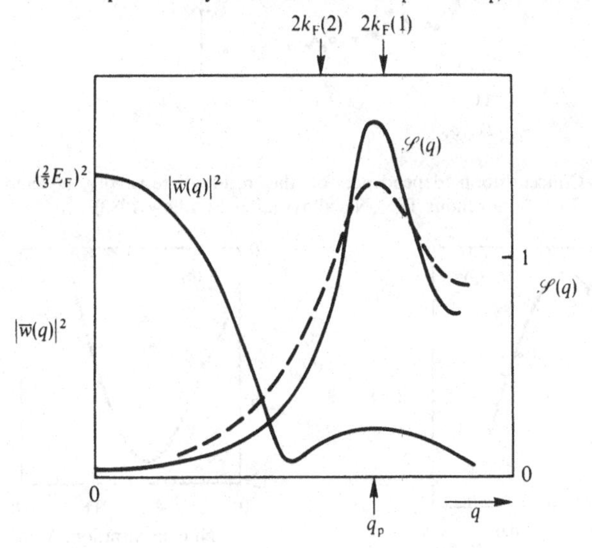

al. 1985). In such cases it may be necessary to employ some of the more advanced formalisms, as discussed in Chapter 6 (see e.g. Mookerjee *et al.* 1985).

(b) $\Theta_D > T > 0$

At temperatures below Θ_D the static structure factor $\mathscr{S}(\mathbf{q})$ must be replaced by the dynamical structure factor $\mathscr{S}(\mathbf{q}, \omega)$ and solution of the Boltzmann equation proceeds along the lines discussed in Section 5.6.2. However, we may again get a qualitative understanding of the behaviour by defining a suitable average structure factor along the lines proposed by Meisel & Cote (1977; Cote & Meisel 1977). They show that the structure factor of metallic glasses having a Debye phonon spectrum can be expanded as

$$\mathscr{S}(\mathbf{q}) = \mathscr{S}^0_\Lambda(\mathbf{q}) + \mathscr{S}^1_\Lambda(\mathbf{q}) + \cdots, \tag{8.32}$$

where $\mathscr{S}^0_\Lambda(\mathbf{q})$ is the static disorder structure factor and $\mathscr{S}^1_\Lambda(\mathbf{q})$ represents the one-phonon processes. The subscript Λ is used to indicate that a long

Fig. 8.20. $d\rho/dT$ as a function of $2k_F/q_p$ for a number of simple metallic glasses: ○, Ag–Cu–Mg; △, Ag–Cu–Al; ■, Ag–Cu–Ge; ●, Mg–Zn; □, Mg–Cu (after Mizutani 1983).

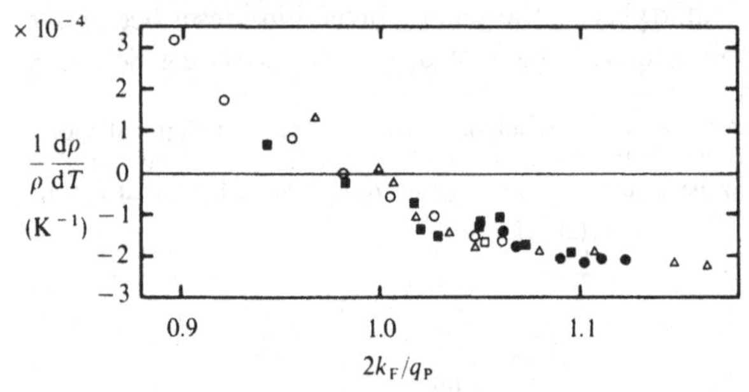

Fig. 8.21. Concentration dependence of the residual resistivity (*a*) and $1/\rho\, d\rho/dT$ (*b*) of amorphous $Ti_{1-x}Ni_x$ alloys (after Buschow 1983).

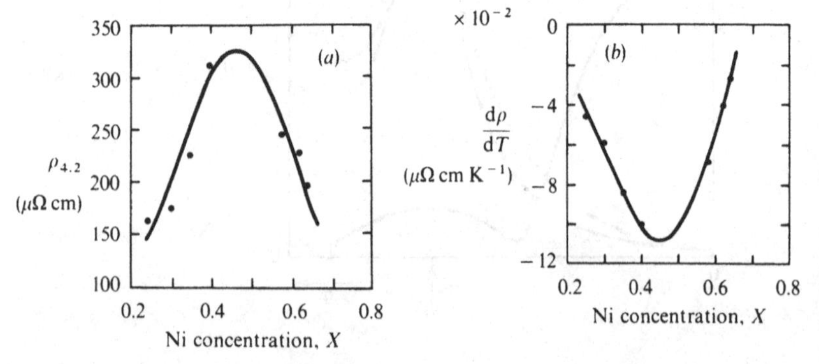

wavelength phonon cutoff is being invoked (see Section 8.2.2). The static structure factor is given by

$$\mathscr{S}^0_\Lambda(\mathbf{q}) = a(\mathbf{q}) \exp(-2M), \tag{8.33}$$

where $a(\mathbf{q})$ is the 'geometrical' structure factor

$$a(\mathbf{q}) = \frac{1}{N} \sum_i \sum_j \exp[\mathrm{i}\mathbf{q}\cdot(\mathbf{r}_i - \mathbf{r}_j)]. \tag{8.34}$$

The one-phonon scattering structure factor is

$$\mathscr{S}^1_\Lambda(\mathbf{q}) \approx \frac{3\hbar^2 q^2 T^2}{(2\pi)^2 M k_\mathrm{B} \Theta_\mathrm{D}^3} \exp(-2M)(1-\gamma) \int \frac{\zeta^2 \, \mathrm{d}\zeta}{\{\exp\zeta - 1\}\{1 - \exp(-\zeta)\}} \times \int \frac{\mathrm{d}\Omega}{4\pi} a(\mathbf{Q}+\mathbf{q}), \tag{8.35}$$

where, as before, $\zeta = \hbar\omega/k_\mathrm{B}T$ and the factor $(1-\gamma)$ is identical to that discussed in 8.2.2(a). At low temperatures the integral over ζ in (8.35) tends to the constant value of 3.29 and so the resistivity due to the phonon scattering goes as $+T^2$. At high temperatures the integral tends to ζ ($\propto 1/T$) and so a linear behaviour is predicted.

From (8.32)–(8.35) the overall temperature dependence is again given by an equation of the form (8.27) and is determined by a balance between the intrinsic phonon temperature dependence and the change in ρ_0 introduced by the Debye–Waller factor M. At low temperatures $M \propto T^2$ and so there is a direct competition between the $+T^2$ phonon term and the $-T^2$ Debye–Waller term. If ρ_0 is small, the phonon term is dominant and gives a $+T^2$ dependence but if ρ_0 is large a negative T^2 dependence will prevail. At high temperatures both the crystalline and amorphous structure factors predict $\rho \propto T$ and so the same behaviour as described in Section 8.2.2 is expected. Note that if ρ_0 is large enough, $\gamma \to 1$ and a negative $\mathrm{d}\rho/\mathrm{d}T$ is expected regardless of the position of the dominant peak in the structure factor. (This peak will also be rendered less important by a damping term – see (5.26).) More detailed calculations along these lines have been given by Meisel & Cote (1983) for amorphous Mg–Zn alloys.

These results indicate that inclusion of the phonon cut-off terms can produce a significant effect even in moderately low resistivity ($\rho_0 \sim 50\mu\Omega$ cm) metallic glasses. A more detailed consideration of this point has been given by Meisel & Cote (1984) (see also Mizutani & Matsuda 1984). Other examples of the application of simple scattering theory to metallic glasses are given by Froböse & Jäckle (1977), Jäckle & Froböse (1979), Wochner & Jäckle (1981), Hafner & Philipp (1984) and references cited therein. In glasses containing a high concentration of transition metals there is again the possibility of additional non-free electron band structure-related temperature effects (see e.g. Cochrane 1978).

8.3.3 *Resistivity of metallic glasses containing magnetic components*

(a) *Ferromagnetic behaviour*

The magnetic resistivity of Fe- and Ni-based ferromagnetic metallic glasses tends to vary as $T^{3/2}$ at temperatures less than about $T_c/2$ and then as T^2 up to T_c (see e.g. Babic *et al.* 1980; Bohnke & Rosenberg 1980). A typical example is shown in Figure 8.22. This is as one would expect on the basis of the mechanisms discussed in Chapter 7. At low temperatures the high degree of disorder in the structure makes coherent spin-wave scattering impossible. The incoherent spin-wave scattering then leads to a $T^{3/2}$ variation. At high temperatures ($T > 0.5T_c$) the spin disorder is more of the kind of random spin reversals (see Chapter 3) which leads to a resistivity contribution which varies as T^2. However, an entirely different behaviour is observed in metallic glasses that contain a large concentration of rare earth elements, particularly if this is in conjunction with Ni. These have low Curie points (~ 15 K) and often exhibit a dramatic increase in the resistivity as the temperature drops through T_c, as shown in Figure 8.23. This behaviour is very sensitive to the application of a magnetic field, as also shown in Figure 8.23, and should be distinguished from the resistivity minima to be discussed in Section 8.3.4. It has been attributed to the particular spin–

Fig. 8.22. The temperature dependence of the resistivity of some $Fe_xNi_{80-x}B_{19}Si_1$ metallic glasses: solid line, $x = 10$; dashed line, $x = 13$; dot-dash line, $x = 16$; dotted line, $x = 20$. The Curie points are marked by an arrow (after Bohnke & Rosenberg 1980).

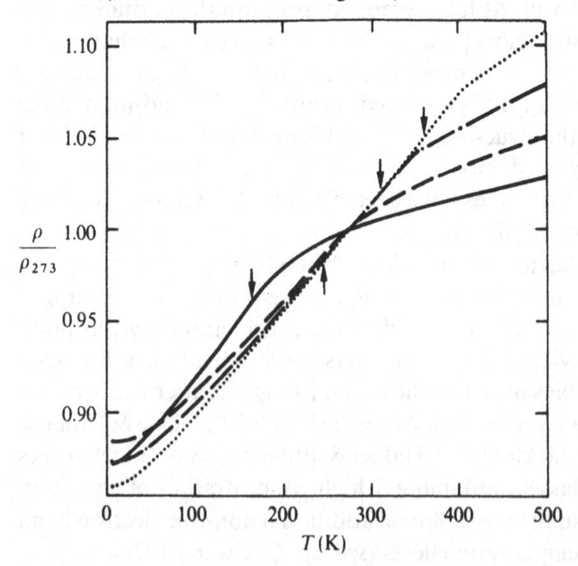

spin correlation function which, in an amorphous structure, is only spatially coherent over short distances. Thus an increase in the degree of magnetic order may give an increase or a decrease in ρ, depending upon the nature of the correlation function in exactly the same way as discussed in Chapters 5 and 7 in relation to atomic or magnetic short range correlations. Detailed calculations of pairwise spin–spin correlations by Bhattacharjee & Coqblin (1978) support this interpretation for both ferromagnetic and antiferromagnetic spin couplings. Similar behaviour has also been found in amorphous Fe–Zn alloys which have a much higher T_c ($\sim$250 K) (Fukamichi *et al.* 1982), Mg–Zn–Gd (Poon *et al.* 1982), and Ag–rare-earth metallic glasses (Ousset *et al.* 1980).

(b) *Spin glasses*

The electrical resistivity due to spin glass formation is quite different from that in crystalline alloys (see Section 7.4). In amorphous

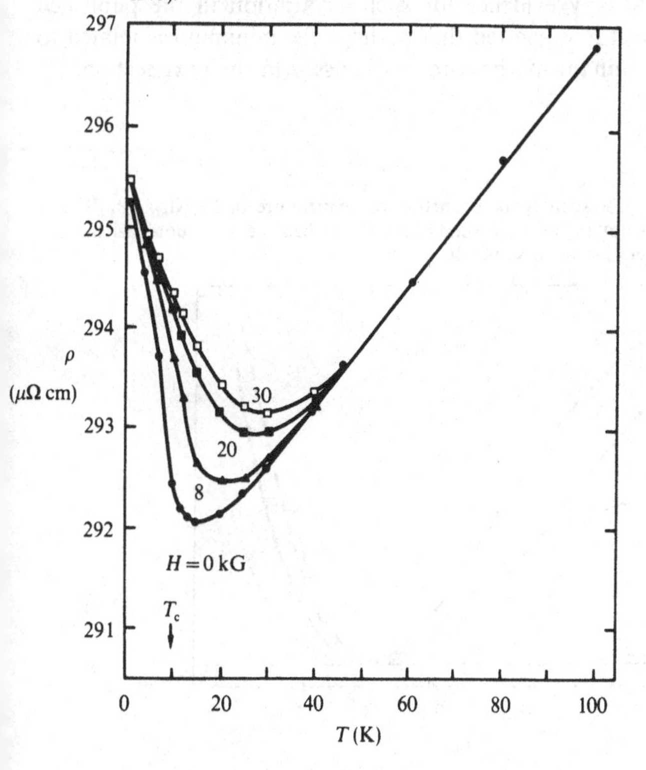

Fig. 8.23. Resistivity of amorphous $Ni_{45}Dy_{25}$ as a function of temperature in several applied magnetic fields as indicated (fields in kG) (after Fert & Asomoza 1979).

spin glass alloys the resistivity exhibits a simple minimum, even at temperatures two orders of magnitude below the spin glass freezing temperature T_0. Some typical results are shown in Figure 8.24. However, this minimum persists with generally the same character over a whole range of compositions which embrace ferromagnetic, mictomagnetic, spin glass and paramagnetic behaviour (Babic *et al.* 1978*a*,*b*, 1980; Gudmundsson *et al.* 1980). It is thus not clear that the minimum is associated with the spin glass state, although its depth and characteristic temperature T_{min} are often most extreme in the spin glass regime. If there is a spin glass contribution to this minimum we might ask why there is no broad maximum in resistivity at some temperature above T_0. However, as we saw in the analysis of the spin glass resistivity of crystalline materials, if the disorder scattering is sufficiently strong the positive $T^{3/2}$ or T^2 behaviour may change sign. This is indeed likely to be the case in amorphous spin glasses, giving a possible explanation for the absence of such a maximum. If this is so, the high scattering rate would also be likely to suppress the long wavelength diffuse magnon modes leaving an expected $-T^2$ behaviour at temperatures far below T_{min}. As there does not seem to be any evidence for such a variation in the published resistivity data it is suggested that perhaps the minimum is related to some other dominant mechanism as discussed in the next section.

Fig. 8.24. Resistivity as a function of temperature of $Fe_xNi_{80-x}P_{14}B_6$ spin glasses: dotted line, $x=0$; solid line, $x=1$; dashed line, $x=5$; dot–dash line, $x=13$ (after Gudmondsson *et al.* 1980).

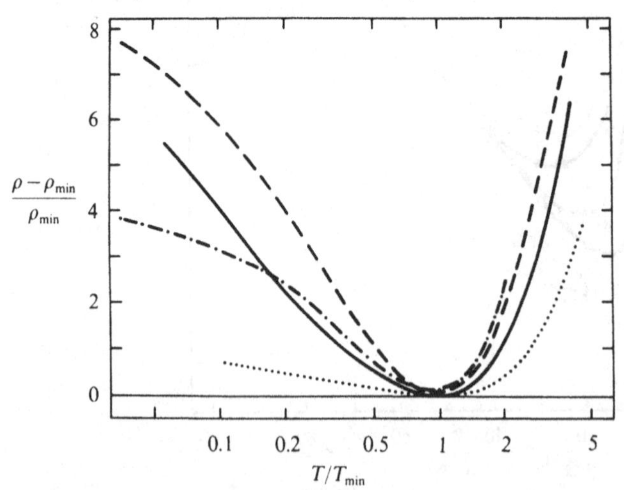

8.3.4 *Resistivity minima*

As indicated in Sections 8.3.1 and 8.3.3(b), many amorphous metals exhibit resistivity minima at low temperatures. This phenomenon does not seem to depend upon whether the glass is paramagnetic, nearly ferromagnetic or strongly ferromagnetic. Furthermore, apart from the special cases described in Section 8.3.3, this minimum is not particularly sensitive to the application of a magnetic field, in marked contrast to the behaviour expected from a Kondo or spin fluctuation system (see Section 7.3). Some typical results for a number of different metallic glasses are shown in Figure 8.25. This low temperature behaviour is characterised by a gradual decrease in resistivity with increasing temperature, followed by a more rapid drop through the liquid He temperature range which is proportional to $-\ln T$. 'Normal' behaviour resumes at about 10–20 K. The ubiquitous nature of this minimum and its apparent insensitivity to alloy composition strongly suggests that it is a fundamental property of the disordered lattice rather

Fig. 8.25. Normalised resistivity $[\rho(T)-\rho(4.2)]/\rho(4.2)$ as a function of temperature. The lower three curves have been offset vertically for clarity. The dashed line is given by equation (8.36) (after Cochrane 1978).

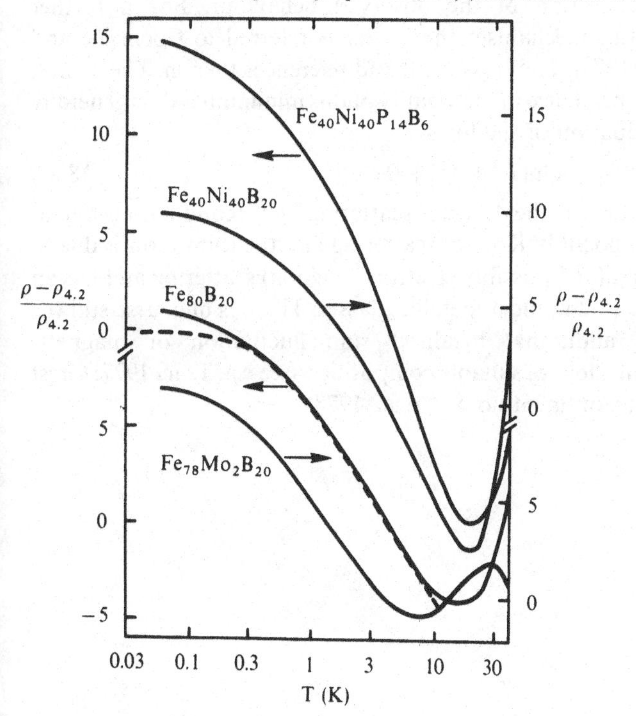

than a magnetic or spin fluctuation effect. Reviews of this behaviour have been given by Tsuei (1977) and Mizutami (1983). Cochrane *et al.* (1975) have suggested a mechanism based on the 'two-level' model introduced by Anderson *et al.* (1972) and Phillips (1972) to explain the linear specific heat of glasses at low temperatures. This model assumes that (by analogy to the Kondo effect) some configurations exist in amorphous metals which are separated from each other by low energy barriers such that tunnelling may occur between them. This leads to a finite number of very low energy excitations and to the observed low-temperature specific heat behaviour. The conduction electrons are scattered from these atomic displacements and, provided that there is some degree of orthogonality of the conduction electron wavefunctions with the atom in each of its alternative positions, the low temperature resistivity is given by

$$\rho = -A \ln(T^2 + \Delta^2), \tag{8.36}$$

where Δ is the energy difference between the two atomic tunnelling states and A is a constant which depends upon the number of contributory sites and strength of the electron scattering, but is independent of magnetic field.

As shown by the dashed line in Figure 8.25, this expression gives a reasonable description of the observed behaviour. For a further discussion of this mechanism the reader is referred to Cochrane and Strom–Olsen (1977), Thomas (1984) and references therein. The overall temperature dependence in the vicinity of this minimum is thus generally given by an equation of the form

$$\rho(T) = \rho_0 - A \ln(T^2 + \Delta^2) + BT^n, \tag{8.37}$$

where the first term is the disorder scattering, the second term is due to two-band (and possibly Kondo) scattering and the third term is due to electron phonon (and possibly electron–magnon) scattering and has an exponent close to 2 in most metallic glasses. There is of course still the possibility of additional localised spin fluctuation or magnetic contributions in alloys of suitable composition (see e.g. Tsuei 1977; Grest & Nagel 1979; Continentino & Rivier 1978).

Appendix A
Units

The equations given in the text assume SI (or rationalised MKSA) units. If these units are used, the constant C multiplying the resistivity integral (equation (5.10)) is given by

$$C=\frac{3\pi m^2\Omega_0}{4\hbar^3 e^2 k_F^6}$$

$$=6.58\times 10^{79}\frac{\Omega_0}{k_F^6}. \quad \text{(A.1)}$$

With a suitable scattering potential in Joules and distances in direct and reciprocal space in m and m^{-1} respectively, this will give a resistivity in Ω m. However, it is often more convenient to employ units of Å and $Å^{-1}$ for these distances in which case the constant is

$$C=6.58\times 10^{29}\frac{\Omega_0}{k_F^6}. \quad \text{(A.2)}$$

Furthermore, most atomic potential studies use atomic units (in which $e=m=\hbar=1$) but often give energies in units of Rydbergs (Ry). If the form factor is used in such units, together with the other parameters in Å, the constant becomes

$$C=3.13\times 10^{-6}\frac{\Omega_0}{k_F^6}. \quad \text{(A.3)}$$

The resistivity of metals is often required in units of μΩ cm, in which case A.3 becomes

$$C=3.13\times 10^{2}\frac{\Omega_0}{k_F^6}. \quad \text{(A.4)}$$

Some useful constants and relationships are:

$e=1.602\times 10^{-19}$ C,

$m=9.110\times 10^{-31}$ kg,

$\hbar=1.055\times 10^{-34}$ J s,

$k_B=1.381\times 10^{-23}$ J K^{-1},

$\varepsilon_0=8.854\times 10^{-12}$ F m^{-1},

1 atomic unit of length = 0.529 Å,

1 atomic unit of energy = 1.09×10^{-18} J,

1 Ry = 2.18×10^{-18} J

(i.e. 1 Ry = 2 atomic units of energy),

1 Å = 10^{-10} m,

1μΩ cm = 10^{-8} Ωm.

Appendix B
Integrations over dk, dS, dE and dΩ

An integration over $\mathbf{dk}$ (often written as d^3k) is an integration over an element of volume in $\mathbf{k}$-space $dk_x\,dk_y\,dk_z$. This element may be regarded as lying between surfaces of energy E and $E+\delta E$ as shown in Figure A.1. These energy surfaces will be separated by a distance δk_n where the subscript n indicates that δk is normal to the surfaces. The element of volume in $\mathbf{k}$-space is then

$$\mathbf{dk}=\delta k_n\,dS, \tag{A.5}$$

where dS is the element of area on the energy surfaces (and so for a spherical Fermi surface $\mathbf{dk}=4\pi k^2\,dk$). However,

$$\delta k_n=\frac{\partial k_n}{\partial E}\,\delta E. \tag{A.6}$$

From (1.18) the gradient of E with respect to k is

$$\frac{\partial E}{\partial k_n}=\hbar v(k), \tag{A.7}$$

where $v(k)$ is the group velocity of the wavepacket. The element of volume then becomes

$$\mathbf{dk}=\frac{\delta E}{\hbar}\frac{dS}{v(k)}. \tag{A.8}$$

In terms of the polar coordinates defined in Figure 1.2

$$\mathbf{dk}=k^2\,dk\,\sin\theta\,d\theta\,d\phi. \tag{A.9}$$

As the results are usually independent of ϕ, this integration can be carried out giving

$$\mathbf{dk}(=\mathbf{d^3k}=dk_x\,dk_y\,dk_z)=2\pi k^2\sin\theta\,d\theta\,dk. \tag{A.10}$$

Fig. A.1. An element of volume in $\mathbf{k}$-space bounded by the energy surfaces E and $E+\delta E$. δS is the element of surface area on the energy surface.

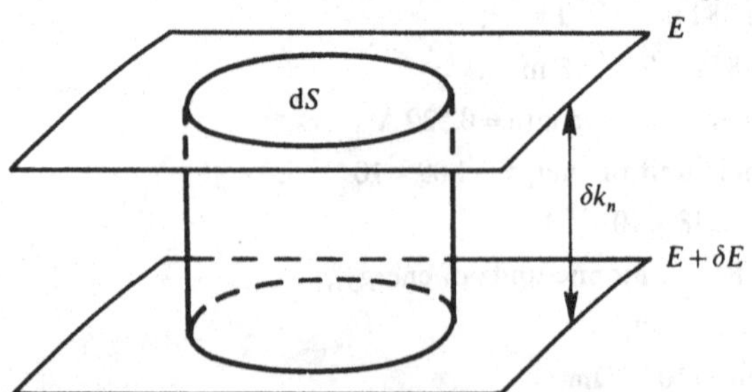

Since

$$\mathrm{d}\mathbf{k} = \mathrm{d}S\,\mathrm{d}k_n \tag{A.11}$$

(see Figure A.1),

$$\mathrm{d}S = 2\pi k^2 \sin\theta\,\mathrm{d}\theta. \tag{A.12}$$

Sometimes the integration is over an element of solid angle $\mathrm{d}\Omega$ (not to be confused with the volume Ω), where

$$\mathrm{d}\Omega = 2\pi \sin\theta\,\mathrm{d}\theta, \tag{A.13}$$

i.e.

$$\mathrm{d}S = k^2\,\mathrm{d}\Omega. \tag{A.14}$$

This last result also follows immediately from the equivalence of the ratios

$$\frac{\mathrm{d}\Omega}{4\pi} = \frac{\mathrm{d}S}{4\pi k^2}, \tag{A.15}$$

Finally, it may be noted that for a spherical Fermi surface

$$\frac{\mathrm{d}E}{\mathrm{d}k_n} = \frac{\hbar^2 k}{m}, \tag{A.16}$$

and so an integral over $k\,\mathrm{d}k$ is often transformed into an integral over energy and vice-versa.

Appendix C
The average $\langle \exp(ix) \rangle$

We want to consider the average $\langle \exp(ix) \rangle$. If x is small the exponential can be expanded as a power series

$$\langle \exp(ix) \rangle = \langle 1 + ix - \tfrac{1}{2}x^2 - \tfrac{1}{6}ix^3 + \tfrac{1}{12}x^4 + \cdots \rangle. \tag{A.17}$$

If x has an equal probability of being positive or negative (or if it follows a Gaussian distribution; see Warren 1969, p. 36), the odd power averages are zero and so

$$\begin{aligned} \langle \exp(ix) \rangle &= 1 - \tfrac{1}{2}\langle x^2 \rangle + \tfrac{1}{12}\langle x^4 \rangle + \cdots \\ &= \exp(-\tfrac{1}{2}\langle x^2 \rangle). \end{aligned} \tag{A.18}$$

Appendix D
High and low temperature limits of $\rho_p(T)$

From (5.9a) and (5.138) or (5.139):

$$\rho = CN \int_0^{2k_F} |\langle \mathbf{k}+\mathbf{q}|\bar{w}(\mathbf{r})|\mathbf{k}\rangle|^2 |\mathbf{q}\cdot\mathbf{u}_{\mathbf{Q},j}|^2 q^3\, dq. \tag{A.19}$$

If we assume that $\bar{w}(\mathbf{q})$ $(=\langle \mathbf{k}+\mathbf{q}|\bar{w}(\mathbf{r})|\mathbf{k}\rangle)$ is a slowly varying function of $\mathbf{q}$, to a first approximation it may be averaged and taken outside the integral. Furthermore,

$$\zeta = \frac{\hbar\omega_{\mathbf{Q}}}{k_B T} = \frac{q\Theta_D}{q_D T}. \tag{A.20}$$

From (5.141), (A.19) and (A.20), the resistivity integral may thus be written in terms of ζ:

$$\rho = C\frac{2\hbar^2 q_D^6 T^5}{M k_B \Theta_D^6}\,\overline{w(\mathbf{q})^2} \int_0^{k_F\Theta_D/q_D T} \frac{\zeta^4\, d\zeta}{\exp\zeta - 1}. \tag{A.21}$$

At low temperatures the integral tends to a constant value of (124.4) and so $\rho \propto T^5$. At high temperatures the integral tends to $(k_F\Theta_D/q_D T)^4$ and so $\rho \propto T$.

Appendix E
Determination of $2k_F R_i$ in a nearly free electron solid

Consider a cube of edge L containing n electrons per unit volume. The electronic wavefunction must be periodic in real space with a period L and so the allowed values of the wavevector are

$$k=\left(\frac{2\pi}{L}\right)M, \tag{A.22}$$

where $M=i+j+k$ and i, j, k are integers. The number of states within the Fermi level (i.e. $M<M_F$) is $2(4\pi/3)M_F$ since for every M there are two allowed spin orientations. At absolute zero all states are occupied and so

$$\left(\frac{8\pi}{3}\right)M_F^3=nL^3. \tag{A.23}$$

From (A.22) and (A.23)

$$k_F^3=3\pi^2 n. \tag{A.24}$$

If there are η atoms per unit cell and N electrons per atom

$$n=\frac{N\eta}{a_0^3}. \tag{A.25}$$

Equations (A.24) and (A.25) give

$$(k_F a_0)^3=3\pi^2 N\eta, \tag{A.26}$$

which leads immediately to the result

$$2k_F R_i=(3\pi^2 N\eta)^{1/3}(l^2+m^2+n^2)^{1/2}, \tag{A.27}$$

and the integers l, m, n are all even or all odd for a bcc lattice and $l+m+n$ is even for a fcc lattice.

From (1.10) and (A.22) the energy of an electron may be written as

$$E_M=\frac{\hbar^2}{2m}\left(\frac{2\pi}{L}\right)^2 M^2, \tag{A.28}$$

and so, from (A.22) and (A.28), the electron velocity at the Fermi surface is

$$v_F=\frac{(3\pi^2 n)^{1/3}\hbar}{m}. \tag{A.29}$$

References

Aaronson, H. I., Laughlin, D. E., Sekerka, R. F. & Wayman, C. M. (1982) *Solid St. Phase Trans.*, Met. Soc., AIME.

Adamowicz, L. (1977) *J. Phys. F: Met. Phys.*, **7**, 2401.

Ahmad, H. M. & Greig, D. (1974) *J. de Phys. Colloque*, **35**, C4-223.

Aitken, R. G., Cheung, T. D. & Kouvel, J. S. (1981) *Phys. Rev. B*, **24**, 1219.

Akai, H. (1977) *Physica*, **86–8**, 539.

Aldred, A. T., Rainford, B. D. & Stringfellow, M. W. (1970) *Phys. Rev. Lett.*, **24**, 897.

Allen, P. B. (1978) *Phys. Rev. B*, **17**, 3725.

Allen, P. B. (1980) *Superconductivity in d- and f-band Metals*, eds. Suhl, H. & Maple, M. B., Academic Press, NY, p. 291.

Allen, P. B. & Chakraborty, B. (1981) *Phys. Rev. B*, **23**, 4815.

Allen, P. B. (1981) Inst. Phys. Conf. Ser. No. 55, *Physics of Transition Metals 1980*, p. 425.

Allen, S. M. & Cahn, J. W. (1979) *Acta Met.*, **27**, 1085.

Amamou, A., Gautier, F. & Loegel, B. (1974) *J. de Phys. Colloque*, **35**, C4–217.

Amamou, A., Gautier, F. & Loegel, B. (1975) *J. Phys. F: Met. Phys.*, **5**, 1342.

Anderson, P. W. (1958) *Phys. Rev.*, **109**, 1492.

Anderson, P. W. (1961) *Phys. Rev.*, **124**, 41.

Anderson, P. W., Halperin,.B. I. & Varma, C. M. (1972) *Phil. Mag.*, **25**, 1.

Animalu, A. E. O. (1966) *Proc. Roy. Soc.*, **294**, 376.

Animalu, A. E. O. (1973) *Phys. Rev. B*, **8**, 3542.

Animalu, A. E. O. & Heine, V. (1965) *Phil. Mag.*, **12**, 1249.

Anquetil, M. C. (1962) *J. de Phys. et Rad.*, **23**, 113.

Appapillai, M. & Williams, A. (1973) *J. Phys. F: Met. Phys.*, **3**, 759.

Appelbaum, J. A. & Penn, D. R. (1971) *Phys. Rev. B*, **3**, 942.

Arajs, S. & Colvin, R. V. (1964) *Phys. Stat. Sol.*, **6**, 797.

Arajs, S., Rao, K. V., Yao, Y. D. & Teoh, W. (1977) *Phys. Rev. B*, **15**, 2429.

Arko, A. J., Brodsky, M. B. & Nellis, W. J. (1972) *Phys. Rev. B*, **5**, 4564.

Asch, A. E. & Hall, G. L. (1963) *Phys. Rev.*, **132**, 1047.

Ashcroft, N. W. (1966) *Phys. Lett.*, **23**, 48.

Ashcroft, N. W. (1968) *J. Phys. C (Proc. Phys. Soc.)*, **1**, 232.

Ashcroft, N. W. & Langreth, D. C. (1967) *Phys. Rev.*, **159**, 500.

Ashcroft, N. W. & Mermin, N. D. (1976) *Solid State Physics*, Holt, Rinehart & Winston, NY.

Aubauer, H. P. (1978) *Phys. Stat. Sol. (b)*, **90**, 345.

Aubauer, H. P. & Rossiter, P. L. (1981) *Phys. Stat. Sol. (b)*, **104**, 685.

Auerbach, A. & Allen, P. B. (1984) *Phys. Rev. B*, **29**, 2884.

Ausloos, M. (1976) *J. Phys. F: Met. Phys.*, **6**, 1723.

Ausloos, M. (1977*a*) *Solid St. Comm.*, **21**, 373.

Ausloos, M. (1977*b*) *Physica*, **86–8B**, 338.

Ausloos, M. & Durczewski, K. (1980) *Phys. Rev. B*, **22**, 2439.

Azbel, M. Ta. & Gurzhi, R. N. (1962) *J. Exp. Theor. Phys.*, **15**, 1133.

Babic, E., Krsnik, R., Leontic, B., Ocko, M., Vucic, Z., Zoric, I. & Girst, E. (1972) *Solid St. Comm.*, **10**, 691.
Babic, E., Krsnik, R. & Rizzuto, C. (1973) *Solid St. Comm.*, **13**, 1027.
Babic, E., Marohnic, Z., Cooper, J. R., Hamzic, A., Saub, K. & Pivac, B. (1978*a*) *J. de Phys. Colloque*, **39**, C6–946.
Babic, E., Marohnic, Z. & Ivkov, J. (1978*b*) *Solid St. Comm.*, **27**, 441.
Babic, E., Ocko, M., Marohnic, Z., Schaafsma, A. S. & Vincze, I. (1980) *J. de Phys. Colloque*, **41**, C8–473.
Bailyn, M. (1960*a*) *Phys. Rev.*, **52**, 381.
Bailyn, M. (1960*b*) *Phil. Mag.*, **5**, 1059.
Balberg, I. & Binenbaum, N. (1983) *Phys. Rev. B*, **28**, 3799.
Balberg, I. & Maman, A. (1979) *Physica*, **96B**, 54.
Ball, M. A. & Islam, Md. M. (1980) *J. Phys. F: Met. Phys.*, **10**, 1943.
Ballentine, L. E. & Heaney, W. J. (1974) *J. Phys. C: Solid St. Phys.*, **7**, 1985.
Bardeen, J. (1937) *Phys. Rev.*, **52**, 688.
Bartels, A., Kemkes, Chr. & Lücke, K. (1985) *Acta Met.*, **33**, 1887.
Bass, J. (1972) *Adv. in Phys.*, **21**, 431.
Bass, J. (1984) *Landolt–Bornstein Numerical Data and Functional Relationships in Science and Technology*, ed. Hellwege, K.-H., Springer-Verlag (Berlin), Vol. 15a, p. 1.
Bath, A. & Blétry, J. (1985) *J. Phys. F: Met. Phys.*, **15**, 1703.
Batirev, I. G., Katsnelson, A. A., Kertesz, L. & Szasz, A. (1980) *Phys. Stat. Sol.* (*b*), **101**, 163.
Baym, G. (1964) *Phys. Rev.*, **135**, A1691.
Beck, P. A. (1972) *J. Less-Common Metals*, **28**, 193.
Beeby, J. L. & Edwards, S. F. (1963) *Proc. R. Soc. A*, **274**, 395
Behari, J. (1973) *J. Phys. F: Met. Phys.*, **3**, 959.
Belitz, A. D. & Schirmacher, W. (1983) *J. Phys. C: Solid St. Phys.*, **16**, 913.
Benedek, R. & Baratoff, A. (1971) *J. Phys. Chem. Sol.*, **32**, 1015.
Benjamin, J. D., Adkins, C. J. & van Cleve, J. E. (1984) *J. Phys. C: Solid St. Phys.*, **17**, 559
Benneman, K. H. & Garland, J. W. (1971) Report AD-716031, Univ. of Rochester, USA (NTIS, Springfield, Va. 22151).
Bennett, L. H., Swartzendruber, L. J. & Watson, R. E. (1969) *Phys. Rev. Lett.*, **23**, 1171.
Bergman, D. J. (1978) *Phys. Reports*, **43**, 377.
Bernasconi, J. & Wiesmann, H. J. (1976) *Phys. Rev. B*, **13**, 1131.
Best, P. R. & Lloyd, P. (1975) *J. Phys. C: Solid St. Phys.*, **8**, 2219.
Betts, D. D., Bhatia, A. B. & Wyman, M. (1956) Phys. Rev., **104**, 37.
Bhatia, A. B. & Thornton, D. E. (1970) *Phys. Rev. B*, **2**, 3004.
Bhatia, A. B. & Thornton, D. E. (1971) *Phys. Rev. B*, **4**, 2325.
Bhattacharjee, A. K. & Coqblin, B. (1978) *J. Phys. F: Met. Phys.*, **8**, L221.
Bhattacharya, A. K., Brauwers, M., Brouers, F. & Joshi, S. K. (1976) *Phys. Rev.*, **B13**, 5214.
Binder, K. & Stauffer, D. (1976) *Z. Phys. B*, **24**, 407.
Black, J. E. (1974) *Can. J. Phys.*, **52**, 345.
Blaker, J. W. & Harris, R. (1971) *J. Phys. C: Solid St. Phys.*, **4**, 569.
Blandin, A. & Friedel, J. (1959) *J. de Phys. Rad.*, **20**, 160.
Blatt, F. J. (1957*a*) *Phys. Rev.*, **108**, 285.
Blatt, F. J. (1957*b*) *Phys. Rev.*, **108**, 1204.
Blatt, F. J. (1968) *Physics of Electronic Conduction in Solids*, McGraw-Hill, Maidenhead.

Blatt, F. J. & Satz, H. G. (1960) *Helv. Phys. Acta*, **33**, 1007.
Bloch, F. (1928) *Z. Phys.*, **52**, 555.
Boerrigter, P. M., Lodder, A. & Molenaar, J. (1983) *Phys. Stat. Sol. (b)*, **119**, K91.
Bohnke, G. & Rosenberg, M. (1980) *J. de Phys. Colloque*, **41**, C8–481.
Boltan, A. (1960) *Trans. Met. Soc. AIME*, **218**, 812.
Borchi, G. & de Gennaro, S. (1970) *Phys. Lett.*, **32A**, 301.
Borie, B. (1957) *Acta Cryst.*, **10**, 89.
Born, M. & Huang, K. (1954) *The Dynamical Theory of Crystal Lattices*, Clarendon Press, Oxford.
Born, M. & Misra, R. D. (1940) *Proc. Camb. Phil. Soc.*, **36**, 466.
Born, M. & Sarginson, K. (1941) *Proc. R. Soc. A*, **179**, 69.
Borodachev, Y. A., Volkov, V. A. & Masharov, S. I. (1976) *Phys. Met. Metall.*, **42** (No. 6), 19.
Boucai, E., Lecoanet, B., Pilon, J., Thoulence, J. L. & Tournier, R. (1971) *Phys. Rev. B*, **3**, 3834.
Bourquart, A., Daniel, E. & Fert, A. (1968) *Phys. Letts.*, **26A**, 260.
Boyer, L. L. & Hardy, J. R. (1971) *Phys. Rev. B*, **4**, 1079.
Bozorth, R. M. (1951) *Ferromagnetism*, Van Nostrand, New Jersey.
Braeter, H. & Priezzhev, V. B. (1975) *Phys. Stat. Sol. (b)*, **72**, 353.
Bragg, W. L. & Williams, E. J. (1935) *Proc. R. Soc. A*, **151**, 540.
Brandli, G. & Cotti, P. (1965) *Helv. Phys. Acta*, **38**, 801.
Braspenning, P. J. & Lodder, A. (1981) *J. Phys. F: Met. Phys.*, **11**, 79.
Braspenning, P. J., Zeller, R., Lodder, A. & Dederichs, P. H. (1984) *Phys. Rev. B*, **29**, 703.
Brauwers, M., Giner, J., Van der Rest, J. & Brouers, F. (1975) *Solid St. Comm.*, **17**, 229.
Brenner, S. S., Camus, P. P., Miller, M. K. & Soffa, W. A. (1984) *see* Taskalakos, T. p. 557.
Brockhouse, B. N., Hallman, E. D. & Ng, S. C. (1967) Met. Soc. Conf., *Magnetic and Inelastic Scattering of Neutrons by Metals*, **43**, 161.
Brockhouse, B. N. & Watanabe, H. (1963) *IAEA Symposium*, Chalk River, IAEA, Vienna, 297.
Bronsveld, P. M. & Radelaar, S. (1974) *J. de Phys. Suppl.*, **35**, C4–19.
Brouers, F. & Brauwers, M. (1975) *J. de Phys. (Letts.)*, **36**, L–17.
Brouers, F. & Vedyayev, A. V. (1972) *Phys. Rev. B*, **5**, 348.
Brouers, F., Vedyayev, A. V. & Giorgino, M. (1973) *Phys. Rev. B*, **7**, 380.
Brouers, F., Giner, J. & van der Rest, J. (1974*a*) *J. Phys. F: Met. Phys.*, **4**, 214.
Brouers, F., Ducastelle, F., Gautier, F. & van der Rest, J. (1974*b*) *J. de Phys. Colloque*, **35**, C4–89.
Brouers, F., Brauwers, M. & Rivory, J. (1974*c*) *J. Phys. F: Met. Phys.*, **4**, 928.
Brouers, F., Kumar, N. & Litt, C. (1976) *J. de Phys.*, **37**, 245.
Brouers, F., Gomes, A. A., de Menezes, O. L. T. & Troper, A. (1978) *Solid St. Comm.*, **27**, 931.
Brouers, F. & Brauwers, M. (1978) *Solid St. Comm.*, **25**, 785.
Brown, H. R. & Morgan, G. J. (1971) *J. Phys. F: Met Phys.*, **1**, 132.
Brown, R. A. (1977) *J. Phys. F: Metal Phys.*, **7**, 1477.
Brown, R. A. (1982) *Can. J. Phys.*, **60**, 766.
Brown, W. F. Jr. (1955) *J. Chem. Phys.*, **23**, 1514.
Buchmann, R., Falke, H. P., Jablonski, H. P. & Wassermann, E. F. (1977) *Physica*, **86B**, 835.
Bullett, D. W. (1980) *Solid St. Phys.*, **35**, 129.

Burns, F. P. & Quimby, S. L. (1955) *Phys. Rev.*, **97**, 1567.
Buschow, K. H. J. (1983) *J. Phys. F: Met. Phys.*, **13**, 563.
Butler, W. H. (1981) *Physics of Transition Metals 1980*, Inst. of Physics, London, p. 505.
Butler, W. H. (1982) *Can. J. Phys.*, **60**, 735.
Butler, W. H. & Stocks, G. M. (1984) *Phys. Rev. B*, **29**, 4217.
Cable, J. W. (1977) *Phys. Rev. B*, **15**, 3477.
Cable, J. W. & Wollan, E. O. (1973) *Phys. Rev. B*, **7**, 2005.
Cabrera, G. G. (1977) *J. Phys. F: Met. Phys.*, **7**, 827.
Cahn, J. W. (1961) *Acta Met.*, **9**, 795.
Cahn, J. W. (1962) *Acta Met.*, **10**, 179.
Cahn, J. W. (1968) *Trans. Met. Soc. AIME*, **242**, 166.
Cahn, R. W. (1982) *J. de Phys. Colloque*, **43**, C9–55.
Cahn, R. W. & Feder, R. (1960) *Phil. Mag.*, **5**, 451.
Callaway, J. & March, N. H. (1984) *Solid St. Phys.*, **33**, 135.
Campbell, I. A., Fert, A. & Jaoul, O. (1970) *J. Phys. C: Metal Phys. Suppl.*, **1**, S95.
Campbell, I. A., Ford, P. J. & Hamzic, A. (1982) *Phys. Rev. B*, **26**, 5195.
Capellmann, H. (1981) in *Physics of Transition Metals*, ed. Rhodes, P., Inst. of Phys., London, p. 205.
Caplin, A. D. & Rizzuto, C. (1968) *Phys. Rev. Lett.*, **21**, 746.
Carcia, P. F. & Suna, A. (1983) *J. Appl. Phys.*, **54**, 2000.
Cargill, G. S. (1975) *Solid St. Phys.*, **30**, 227.
Chakraborty, B. & Allen, P. B. (1979) *Phys. Rev. Lett.*, **42**, 736.
Chambers, R. G. (1968) *Solid St. Phys.*, **1**, 314.
Chambers, R. G. (1969) in Ziman, 1969, p. 175.
Chardon, J. Th. & Radelaar, S. (1969) *Phys. Stat. Sol.*, **33**, 439.
Charsley, P. & Shimmin, A. F. (1979) *Mater. Sci. Eng.*, **38**, 23.
Chen, A-B. (1973) *Phys. Rev. B*, **7**, 2230.
Chen, A-B., Weisz, G. & Sher, A. (1972) *Phys. Rev. B*, **5**, 2897.
Chen, H. S. (1980) *Rep. Prog. Phys.*, **43**, 351.
Chen, H. & Cohen, J. B. (1979) *Acta Met.*, **27**, 603.
Cheung, T. D. & Kouvel, J. S. (1983) *Phys. Rev. B*, **28**, 3831.
Chieux, P. & Ruppersberg, H. (1980) *J. de Phys. Colloque*, **41**, C8–145.
Chikazumi, S. (1964) *Physics of Magnetism*, Wiley, NY.
Chipman, D. R. (1956) *J. Appl. Phys.*, **27**, 739.
Chouteau, G., Tournier, R. & Mollard, P. (1974) *J. de Phys. Colloque*, **35**, C4–185.
Christoph, V., Vojta, G. & Ropke, G. (1974) *Phys. Stat. Sol.* (*b*), **66**, K101.
Christoph, V. (1979) *Phys. Stat. Sol.* (*b*), **91**, 593.
Christoph, V. & Richter, J. (1980) *J. Phys. F: Met. Phys.*, **10**, L215.
Christoph, V., Richter, J. & Schiller, W. (1980) *Phys. Stat. Sol.* (*b*), **100**, 595.
Clapp, P. C. (1971) *Phys. Rev. B*, **4**, 255.
Clapp, P. C. (1981) *Met. Trans. A*, **12A**, 589.
Clapp, P. C. & Moss, S. C. (1968) *Phys. Rev.*, **171**, 754.
Clark, J. A. & Dawber, P. G. (1972) *J. Phys. F: Met. Phys.*, **2**, 930.
Claus, H., Sinha, A. K. & Beck, P. A. (1967) *Phys. Lett.*, **26A**, 38.
Cochran, W. (1973) *The Dynamics of Atoms in Crystals*, Arnold, London.
Cochrane, R. W. (1978) *J. de Phys. Colloque*, **39**, 6–1540.
Cochrane, R. W. & Strom-Olsen, J. O. (1977) *J. Phys. F: Met. Phys.*, **7**, 1799.
Cochrane, R. W., Harris, R., Strom-Olson, J. O. & Zuckermann, M. J. (1975) *Phys. Rev. Lett.*, **35**, 676.
Cohen, J. B. (1970) *Phase Transformations*, ASM, Ohio, p. 561.

Cohen, J. B. & Georgopoulos, P. (1984), *Phase Transformations in Solids*, ed. Tsakalakos, T., Mats. Res. Soc., North Holland, NY.
Cohen, M. L. (1984) *Phys. Reports*, **110**, 293.
Cohen, M. L. & Heine, V. (1970) *Sol. St. Phys.*, **24**, 37.
Coleridge, P. T. (1972) *J. Phys. F: Met. Phys.*, **2**, 1016.
Coleridge, P. T., Holzwarth, N. A. W. & Lee, M. J. G. (1974) *Phys. Rev. B*, **10**, 1213.
Coles, B. R. (1974) *J. de Phys. Colloque*, **35**, C4–203.
Coles, B. R. & Taylor, J. C. (1962) *Proc. R. Soc. A*, **267**, 139.
Coles, B. R. (1960) *Physica*, **26**, 143.
Coles, B. R. & Caplin, A. D. (1976) *The Electronic Structures of Solids*, Arnold, London.
Coles, B. R., Sarkissian, B. V. B. & Taylor, R. H. (1978) *Phil. Mag. B*, **37**, 489.
Comstock, R. J., Cohen, J. B. & Harrison, H. R. (1985) *Acta Met.*, **33**, 423.
Continentino, M. A. & Rivier, N. (1978) *J. Phys. F: Met. Phys.*, **8**, 1187.
Cook, H. E. (1973) *Acta Met.*, **21**, 1431.
Cook, H. E. (1975) *Acta Met.*, **23**, 1027.
Corey, C. L. & Lisowsky, B. (1967) *Trans. Met. Soc. AIME*, **239**, 239.
Cornut, B., Perrier, J. P., Tissier, B. & Tournier, R. (1971) *J. de Phys. Colloque*, **32**, C1–746.
Cote, P. J. & Meisel, L. V. (1977) *Phys. Rev. Lett.*, **39**, 102.
Cote, P. J. & Meisel, L. V. (1978) *Phys. Rev. Lett.*, **40**, 1586.
Cowley, J. M. (1960) *Phys. Rev.*, **120**, 1648.
Cowley, J. M. (1965) *Phys. Rev.*, **138**, A1384.
Cowley, J. M. (1968) *Acta Cryst.*, **A24**, 557.
Cowley, J. M. (1975) *Diffraction Physics*, North Holland, Amsterdam.
Cowley, J. M., Cohen, J. B., Salamon, M. B. & Wuensch, B. J. (1979) AIP Conf. Proc. No. 53, *Modulated Structures – 1979*, AIP, NY.
Cowley, R., Woods, A. D. B. & Dolling, G. (1966) *Phys. Rev.*, **150**, 487.
Craig, P. P., Goldburg, W. I., Kitchens, T. A. & Budnick, J. I. (1967) *Phys. Rev. Lett.*, **19**, 1334.
Crangle, J. (1977) *The Magnetic Properties of Solids*, Arnold, London.
Crangle, J. & Butcher, P. J. L. (1970) *Phys. Lett.*, **32A**, 80.
Csanak, G. Y., Taylor, H. S. & Yaris, R. (1971) *Adv. in Atomic and Mol. Phys.*, **7**, 287.
Cullity, B. D. (1972) *Introduction to Magnetic Materials*, Addison-Wesley, London.
Czycholl, G. (1978) *Z. Phys. B*, **30**, 383.
Czycholl, G. & Zittartz, J. (1978) *Z. Phys. B*, **30**, 375.
Dagens, L. (1976) *J. Phys. F: Met. Phys.*, **6**, 1801.
Damask, A. C. (1956*a*) *J. Phys. Chem. Sol.*, **1**, 23.
Damask, A. C. (1956*b*) *J. Appl. Phys.*, **27**, 610.
Darby, J. K. & March, N. H. (1964) *Proc. Phys. Soc.*, **84**, 591.
Davidson, A. & Tinkham, M. (1976) *Phys. Rev. B*, **13**, 3261.
Davies, R. G. (1963) *J. Phys. Chem. Sol.*, **24**, 985.
Daybell, M. D. & Steyert, W. A. (1968) *Rev. Mod. Phys.*, **40**, 380.
de Faget de Casteljau, P. & Friedel, J. (1956) *J. de Phys. et Rad.*, **17**, 27.
de Fontaine, D. (1970) *Acta Met.*, **18**, 275.
de Fontaine, D. (1979) *Solid St. Phys.*, **34**, 73.
de Fontaine, D. & Kulik, J. (1985) *Acta Met.*, **33**, 145.
De Gennaro, S. & Rettori, A. (1985) *J. Phys. F: Met. Phys.*, **15**, 2177.
de Gennes, P. G. (1962) *J. de Phys. et Rad.*, **23**, 630.
de Gennes, P. G. & Friedel, J. (1958) *J. Phys. Chem. Sol.*, **4**, 71.
de Hoff, R. T. & Rhines, F. N. (1968) *Quantitative Microscopy*, McGraw-Hill, NY.

Deegan, R. A. & Twose, W. D. (1967) *Phys. Rev.*, **164**, 993.
Deutz, J., Dederichs, P. H. & Zeller, R. (1981) *J. Phys. F: Met. Phys.*, **11**, 1787.
Diamond, J. B. (1972) *Intern. J. Mag.*, **2**, 241.
Dickey, J. M., Meyer, A. & Young, W. H. (1967) *Phys. Rev.*, **160**, 490.
Digges, T. G. & Tauber, R. N. (1971) *Met. Trans.*, **2**, 1683.
Dimmich, R. (1985) *J. Phys. F: Met. Phys.*, **15**, 2477.
Dimmock, J. O. (1971) *Solid St. Phys.*, **26**.
Dingle, R. B. (1950) *Proc. R. Soc. A*, **201**, 546.
Ditchek, B. & Schwartz, L. H. (1980) *Acta Met.*, **28**, 807.
Doniach, S. & Sondheimer, E. H. (1974) *Green's Functions for Solid State Physics*, Benjamin, Reading, Mass.
Dorleijn, J. W. F. (1976) *Philips Res. Rep.*, **31**, 287.
Dorleijn, J. W. F. & Miedema, A. R. (1975*a*) *J. Phys. F: Met. Phys.*, **5**, 487.
Dorleijn, J. W. F. & Miedema, A. R. (1975*b*) *J. Phys. F: Met. Phys.*, **5**, 1543.
Doyama, M. & Cotterill, R. M. J. (1967) *Lattice Defects and their Interactions*, ed. Hasiguti, R. R., Gordon & Breach, NY, p. 79.
Drchal, V. & Kudrnovský, J. (1976) *J. Phys. F: Met. Phys.*, **6**, 2247.
Dreirach, O. (1973) *J. Phys. F: Met. Phys.*, **3**, 577.
DuCharme, A. R. & Edwards, L. R. (1970) *Phys. Rev. B*, **2**, 2940.
Dubinin, S. F., Izyumov, Yu. A., Sidorov, S. K., SyromYatnikov, V. N. & Teplouchov, S. F. (1981) *Phys. Stat. Sol. (a)*, **67**, 75.
Dugdale, J. S. (1977) *The Electrical Properties of Metals and Alloys*, Arnold, London.
Dugdale, J. S. & Guénault, A. M. (1966) *Phil. Mag.*, **13**, 503.
Dunleavy, H. N. & Jones, W. (1978) *J. Phys. F: Met. Phys.*, **8**, 1477.
Durand, J. & Gautier, F. (1970) *J. Phys. Chem. Solids*, **31**, 2773.
Dworin, L. & Narath, A. (1970) *Phys. Rev. Lett.*, **25**, 1287.
Dynes, R. C. & Carbotte, J. P. (1968) *Phys. Rev.*, **175**, 913.
Ebert, H., Weinberger, P. & Voitländer, J. (1985) *Phys. Rev. B*, **31**, 7557.
Edwards, J. T. & Hillel, A. J. (1977) *Phil. Mag.*, **35**, 1221.
Edwards, L. R., Chen, C. W. & Legvold, S. (1970) *Solid St. Comm.*, **8**, 1403.
Edwards, S. F. (1965) *Proc. Phys. Soc.*, **86**, 977.
Edwards, S. F. & Anderson, P. W. (1975) *J. Phys. F: Met. Phys.*, **5**, 965.
Eguchi, T., Kinoshita, C. & Tomokiyo, Y. (1978) *Trans. Japan Inst. Met.*, **19**, 198.
Ehrenreich, H. & Schwartz, L. M. (1976) *Solid St. Phys.*, **31**, 149.
Elk, K., Richter, J. & Christoph, V. (1978) *Phys. Stat. Sol. (b)*, **89**, 537.
Elk, K., Richter, J. & Christoph, V. (1979) *J. Phys. F: Met. Phys.*, **9**, 307.
Elliott, R. J. & Wedgwood, F. A. (1963) *Proc. Phys. Soc.*, **81**, 846.
Elliott, R. J., Krumhausl, J. A. & Leath, P. L. (1974) *Rev. Mod. Phys.*, **46**, 465.
Engquist, H-L. (1980) *Phys. Rev. B*, **21**, 2067.
Eshelby, J. A. (1956) *Solid St. Phys.*, **3**, 79.
Eshelby, J. D. (1954) *J. Appl. Phys.*, **25**, 255.
Eshelby, J. D. (1956) *Solid St. Phys.*, **3**, 79.
Essam, J. W. (1972) *Phase Transitions and Critical Phenomena*, eds. Domb, C. & Green, M. S., Academic Press, NY.
Essam, J. W. (1980) *Rep. Prog. Phys.*, **43**, 833.
Evans, R., Gaspari, G. D. & Gyorffy, B. L. (1973*a*) *J. Phys. F: Met. Phys.*, **3**, 39.
Evans, R., Ratti, V. K. & Gyorffy, B. L. (1973*b*) *J. Phys. F: Met. Phys.*, **3**, L199.
Evseev, A. M. & Dergachev, V. M. (1976) *Sov. Phys. Solid State*, **18**, 1323.
Faber, T. E. (1969) in Ziman, 1969, p. 282.
Faber, T. E. & Ziman, J. M. (1964) *Phil. Mag.* **11**, 153.
Falicov, L. M. & Heine, V. (1961) *Adv. Phys.*, **10**, 57.

Farrell, T. & Greig, D. (1968) *J. Phys. C (Proc. Phys. Soc.)*, **1**, 1359.
Faulkner, J. S. (1977) *J. Phys. C: Solid St. Phys.*, **10**, 4661.
Faulkner, J. S., Painter, G. S., Butler, W. & Coghlan, W. A. (1974) *J. de Phys. Colloque*, **35**, C4–85.
Fert, A. (1969) *J. Phys. C: Solid St. Phys.*, **2**, 1784.
Fert, A. & Asomoza, R. (1979) *J. Appl. Phys.*, **50**, 1886.
Fert, A. & Campbell, I. A. (1971) *J. de Phys. Colloque*, **32**, C1–46.
Fert, A. & Campbell, I. A. (1976) *J. Phys. F: Met. Phys.*, **6**, 849.
Fischer, K. H. (1970) *Springer Tracts in Modern Physics*, Vol. 54, p. 1.
Fischer, K. H. (1974) *J. Low Temp. Phys.*, **17**, 87.
Fischer, K. H. (1978) *Phys. Reports*, **47**, 226.
Fischer, K. H. (1979) *Z. Physik B*, **34**, 45.
Fischer, K. H. (1980) *Z. Physik B*, **39**, 37.
Fischer, K. H. (1981) *Z. Physik B*, **42**, 27.
Fischer, K. H. (1984) *Landolt–Bornstein Numerical Data and Functional Relationships in Science and Technology*, ed. Hellwege, K.-H., Springer-Verlag, Berlin, Vol. 15a, p. 289.
Fisher, M. E. (1964) *J. Math. Phys.*, **5**, 944.
Fisher, M. E. (1967) *Rep. Prog. Phys.*, **30**, 615.
Fisher, M. E. (1974) *Rev. Mod. Phys.*, **46**, 597.
Fisher, M. E. (1982) *Lecture Notes in Physics No. 186: Critical Phenomena*, ed. Hahne, F. J. W., Springer-Verlag, p. 1.
Fisher, M. E. & Burford, R. J. (1967) *Phys. Rev.*, **156**, 583.
Fisher, M. E. & Ferdinand, A. E. (1967) *Phys. Rev. Lett.*, **19**, 169.
Fisher, M. E. & Langer, J. S. (1968) *Phys. Rev. Lett.*, **20**, 665
Fisher, M. E. & Selke, W. (1980) *Phys. Rev. Lett.*, **44**, 1502.
Fisher, M. E. & Selke, W. (1981) *Phil. Trans. R. Soc.*, **302**, 1.
Flocken, J. W. & Hardy, J. R. (1969) *Phys. Rev.*, **177**, 1054.
Flynn, C. P. (1962) *J. de Phys. et Rad.*, **23**, 654.
Foiles, C. L. (1973) *Phil. Mag.*, **27**, 757.
Ford, P. J. & Mydosh, J. A. (1974) *J. de Phys. Colloque*, **35**, C4–241.
Fote, A., Lutz, H., Mihalisin, T. & Crow, J. (1970) *Phys. Lett.*, **33A**, 416.
Franck, J. P., Manchester, F. D. & Martin, D. L. (1961) *Proc. R. Soc. A*, **263**, 494.
Freeman, S., Jr. (1965) *J. Phys. Chem. Sol.*, **26**, 473.
Fricke, H. (1924) *Phys. Rev.*, **24**, 577.
Fridel, J. (1958) *Nuovo Cimento Suppl.*, **7**, 287.
Friedel, J. (1969) in Ziman, 1969, p. 340.
Friedel, J. (1973) *J. Phys. F: Met Phys.*, **3**, 785.
Froböse, K. & Jäckle, J. (1977) *J. Phys. F: Met. Phys.*, **7**, 2331.
Frommeyer, G. & Brion, H. G. (1981) *Phys. Stat. Sol. (a)*, **66**, 559.
Fuchs, K. (1938) *Proc. Camb. Phil. Soc.*, **34**, 100.
Fukai, Y. (1968) *Phys. Lett.*, **27A**, 416.
Fukai, Y. (1969) *Phys. Rev.*, **186**, 697.
Fukamichi, K., Gambino, R. J. & McGuire, T. R. (1982) *J. Appl. Phys.*, **53**, 2310.
Gaidukov, Yu P. & Kadletsova, Ta (1970*a*) *Phys. Stat. Sol. (a)*, **2**, 407.
Gaidukov, Yu P. & Kadletsova, Ya (1970*b*) *Sov. Phys. JETP*, **30**, 637.
Gaskell, P. H. (1979) *J. Phys. C: Solid St. Phys.*, **12**, 4337.
Gautier, F., Brouers, F. & van der Rest, J. (1974) *J. de Phys.*, **35**, C4–207.
Gehlen, P. C. & Cohen, J. B. (1965) *Phys. Rev.*, **139**, A844.
Gersch, H. A., Schull, C. G. & Wilkinson, M. K. (1956) *Phys. Rev.*, **103**, 525.
Ghosh, D. K. & Bhattacharaya, P. (1975) *Phys. Rev. B*, **11**, 2642.

Gibson, J. B. (1956) *J. Phys. Chem. Sol.*, **1**, 27.
Gilat, G. (1970) *Solid St. Comm.*, **8**, 2053.
Gilat, G. & Dolling, G. (1965) *Phys. Rev.*, **138**, A1053.
Glässer, U. H. (1984) *Phys. Stat. Sol. (b)*, **124**, 423.
Glässer, U. H. & Löffler, H. (1984) *Phys. Stat. Sol. (b)*, **126**, 741.
Goedsche, F., Richter, R. & Richter, A. (1975) *Phys. Stat. Sol. (b)*, **69**, 213.
Goedsche, F., Richter, R. & Vojta, G. (1978) *Phys. Stat. Sol. (b)*, **867**, 177.
Goedsche, F., Mobius, A. & Richter, A. (1979) *Phys. Stat. Sol. (b)*, **96**, 279.
Gohel, V. B., Achary, C. K. & Jani, A. R. (1984) *Phys. Stat. Sol. (b)*, **124**, 137.
Gonis, A. & Freeman, A. J. (1984) *Phys. Rev. B*, **29**, 4277.
Gonis, A., Stocks, G. M., Butler, W. H. & Winter, H. (1984) *Phys. Rev. B*, **29**, 555.
Gordon, B. E. A., Temmerman, W. E. & Gyorffy, B. L. (1981) *J. Phys. F: Met. Phys.*, **11**, 821.
Gragg, J. E., Bardhan, P. & Cohen, J. B. (1971) in *Critical Phenomena in Alloys, Magnets & Superconductors*, eds. Mills, Ascher & Jaffee, McGraw-Hill, NY.
Granqvist, C. G. & Hunderi, O. (1978) *Phys. Rev. B*, **18**, 1554.
Gratz, E. (1983) *CRC Handbook of Electrical Resistivity of Binary Metallic Alloys*, ed. Schröder, CRC Press.
Gratz, E. & Zuckermann, M. J. (1982) *Journal of Magnetism and Magnetic Materials*, **29**, 181.
Grayevskaya, Ya. I., Iveronova, V. I., Katsnel'son, A. A. & Popova, I. I. (1975) *Fiz. Met. Metall.*, **40**, 195.
Greene, M. P. & Kohn, W. (1965) *Phys. Rev.*, **137**, A513.
Greene, R. F. (1964) *Surface Sci.*, **2**, 101.
Greene, R. F. & O'Donnell, R. W. (1966) *Phys. Rev.*, **147**, 599.
Greenwood, D. A. (1958) *Proc. Phys. Soc.*, **71**, 585.
Greenwood, D. A. (1966) *Proc. Phys. Soc.*, **87**, 775.
Greig, D. & Rowlands, J. A. (1974) *J. Phys. F: Met. Phys.*, **4**, 536.
Grest, G. S. & Nagel, S. R. (1979) *Phys. Rev. B*, **19**, 3571.
Grier, B. H. (1980) Ph.D. Thesis, Rochester, USA.
Grimes, H. H. & Rice, J. H. (1968) *J. Phys. Chem. Sol.*, **29**, 1481.
Gudmundsson, H., Rao, K. V., Egami, Anderson, A. C. & Astrom, H. U. (1980) *Phys. Rev. B*, **22**, 3374.
Guinea, F. (1983) *Phys. Rev. B*, **28**, 1148.
Guinier, A. & Fournet, G. (1955) *Small Angle Scattering of X-rays*, Wiley, NY.
Gunnarson, O. & Lundqvist, B. I. (1976) *Phys. Rev. B*, **10**, 4274.
Gupta, O. P. (1968) *Phys. Rev.*, **174**, 668.
Gupta, R. P. & Benedek, R. (1979) *Phys. Rev. B*, **19**, 583.
Guyot, P. (1970) *Phys. Stat. Sol.*, **38**, 409.
Guyot, P. & Simon, J. P. (1977) *Scripta Met.*, **11**, 751.
Haas, H. & Lücke, K. (1972) *Scripta Met.*, **6**, 715.
Hafner, J. (1973) *Phys. Stat. Sol (b)*, **56**, 579.
Hafner, J. (1975*a*) *J. Phys. F: Met. Phys.*, **5**, 1439.
Hafner, J. (1975*b*) *Z. Phys. B*, **22**, 351.
Hafner, J. (1976) *J. Phys. F: Met. Phys.*, **6**, 1243.
Hafner, J. (1980) *Liquid and Amorphous Metals*, NATO-ASI series E, 36, eds. Luscher, E. & Coufal, H., Alpen van den Rijn, Netherlands: Sijlhoff and Noordhoff, p. 199.
Hafner, J. (1983) *Phys. Rev. B*, **27**, 678.
Hafner, J. & Philipp, A. (1984) *J. Phys. F: Met. Phys.*, **14**, 1685.
Hall, G. L. & Christy, D. O. (1966) *Amer. J. Phys.*, **34**, 526.

Haller, H., Korn, D. F. & Zibold, G. (1981) *Phys. Stat. Sol.* (*a*), **68**, 135.
Hallman, E. D. & Brockhouse, B. N. (1969) *Can. J. Phys.*, **47**, 1117.
Harris, R. (1973) *J. Phys. F: Met. Phys.*, **3**, 89.
Harrison, W. A. (1958) *Phys. Rev.*, **110**, 14.
Harrison, W. A. (1963) *Phys. Rev.*, **131**, 2433.
Harrison, W. A. (1966) *Pseudopotentials in the Theory of Metals*, Benjamin, NY.
Harrison, W. A. (1969) *Phys. Rev.*, **181**, 1036.
Harrison, W. A. (1970) *Solid State Theory*, McGraw-Hill, NY.
Hartmann, W. M. (1968) *Phys. Rev.*, **172**, 677.
Hasegawa, H. (1983*a*) *J. Phys. F: Met. Phys.*, **13**, 1915.
Hasegawa, H. (1983*b*) *J. Phys. F: Met. Phys.*, **13**, 2655.
Hasegawa, H. (1984) J. Phys. F: Met. Phys., **14**, 1235.
Hasegawa, H. & Kanamori, J. (1971) *J. Phys. Soc. Japan*, **31**, 382.
Hashin, Z. & Shtrikman, S. (1962) *J. Appl. Phys.*, **33**, 3125.
Hautojärvi, P., Heinio, J., Manninen, M. & Nieminen, R. (1977) *Phil. Mag.*, **35**, 973.
Hayman, B. & Carbotte, J. P. (1971) *Can. J. Phys.*, **49**, 1952.
Hayman, B. & Carbotte, J. P. (1972) *J. Phys. F: Met. Phys.*, **2**, 915.
Heeger, A. J. (1969) *Solid St. Phys.*, **23**, 283.
Heeger, A. J., Klein, A. P. & Tu, P. (1966) *Phys. Rev. Lett.*, **17**, 803.
Heidsiek, H., Scheffel, R. & Lucke, K. (1977) *J. de Phys. Colloque*, **38**, C7–174.
Heine, V. (1969) in Ziman (1969), p. 1.
Heine, V. (1980) *Solid St. Phys.*, **35**, 1.
Heine, V. & Abarenkov, I. (1964) *Phil. Mag.*, **9**, 451.
Heine, V. & Samson, J. H. (1981) in *Physics of Transition Metals*, ed. Rhodes, P., Inst. of Phys., London, p. 221.
Heine, V. & Weaire, D. (1970) *Solid St. Phys.*, **24**, 249.
Heller, P. (1967) *Rep. Prog. Phys.*, **30**, 731.
Henger, U. & Korn, D. (1981) *J. Phys. F: Met. Phys.*, **11**, 2575.
Herbstein, F. H., Borie, B. S. Jr. & Averbach, B. L. (1956) *Acta Cryst.*, **9**, 466.
Herring, C. (1960) *J. Appl. Phys.*, **31**, 1939.
Hicks, T. J., Holden, T. M. & Low, G. G. (1968) J. Phys. C: Solid St. Phys., **1**, 528.
Hicks, T. J., Rainford, B., Kouvel, J. S., Low, G. G. & Comly, J. B. (1969) *Phys. Rev.*, **22**, 531.
Hillel, A. J. (1970) *Acta Met.*, **18**, 253.
Hillel, A. J. (1983) *Phil. Mag. B*, **48**, 237.
Hillel, A. J. & Edwards, J. T. (1977) *Phil. Mag.*, **35**, 1231.
Hillel, A. J. & Rossiter, P. L. (1981) *Phil. Mag. B*, **44**, 383.
Hillel, A. J., Edwards, J. T. & Wilkes, P. (1975) *Phil. Mag.*, **32**, 189.
Ho, K. M., Cohen, M. L. & Picket, W. E. (1978) *Phys. Rev. Lett.*, **41**, 815.
Ho, P. S. (1971) *Phys. Rev. B*, **3**, 4035.
Hoare, F. E., Matthews, J. C. & Walling, J. C. (1953) *Proc. R. Soc. A*, **216**, 502.
Hoffman, D. W. & Cohen, M. (1973) *Acta Met.*, **21**, 1215.
Hogan, L. M. (1968) *Advances in Materials Research*, ed. Herman, H., Vol. 5, p. 84.
Hohenberg, P. & Kohn, W. (1964) *Phys. Rev.*, **136**, B864.
Holwech, I. & Jeppesen, J. (1967) *Phil. Mag.*, **15**, 217.
Honeycombe, R. W. K. & Boas, W. (1948) *Nature*, **161**, 612.
Hoshino, K. & Watabe, M. (1975) *J. Phys. C: Solid St. Phys.*, **8**, 4193.
Houghton, M. E. & Rossiter, P. L. (1978) *Phys. Stat. Sol.* (*a*), **48**, 71.
Houghton, R. W., Sarachik, M. P. & Kouvel, J. S. (1970*a*) *Phys. Rev. Lett.*, **25**, 238.
Houghton, R. W., Sarachik, M. P. & Kouvel, J. S. (1970*b*) *Solid St. Comm.*, **8**, 943.
House, D., Gyorffy, B. L. & Stocks, G. M. (1974) *J. de Phys. Colloque*, **35**, C4–81.

Houston, W. V. (1929) *Phys. Rev.*, **34**, 279.
Howie, A. (1960) *Phil. Mag.*, **5**, 251.
Huang, K. (1947) *Proc. R. Soc.*, **A190**, 102.
Huang, K. (1948) *Proc. Phys. Soc. Lond.*, **60**, 161.
Hubbard, J. (1957*a*) *Proc. R. Soc.*, **A240**, 359.
Hubbard, J. (1957*b*) *Proc. R. Soc.*, **A243**, 236.
Hurd, C. M. (1974) *Adv. in Phys.*, **23**, 315.
Hurd, C. M. & McAllister, S. P. (1980) *Phil. Mag. B*, **42**, 221.
Hurd, C. M., McAlister, S. P. & Shiozaki, I. (1981) *J. Phys. F: Met. Phys.*, **11**, 457.
Hwang, C. M., Meichle, M., Salamon, M. B. & Wayman, C. M. (1983*a*) *Phil. Mag. A*, **47**, 9.
Hwang, C. M., Meichle, M., Salamon, M. B. & Wayman, C. M. (1983*b*) *Phil. Mag. A*, **47**, 31.
Ikeda, K. & Tanosaki, K. (1984) *Trans. Japan Inst. Met.*, **25**, 447.
Imry, Y. (1980) *Phys. Rev. Lett.*, **44**, 469.
Ioffe, A. F. & Regel, A. R. (1960) *Prog. Semicond.*, **4**, 237.
Ivchenko, V. A. & Syutkin, N. N. (1983) *Sov. Phys. Solid State*, **25**, 1759.
Jaccarino, V. & Walker, L. R. (1965) *Phys. Rev. Lett.*, **15**, 258.
Jäckle, J. & Froböse, K. (1979) *J. Phys. F: Met. Phys.*, **9**, 967.
Jackson, P. J. & Saunders, N. H. (1968) *Phys. Lett.*, **28A**, 19.
Jan, J. P. (1957) *Solid St. Phys.*, **5**, 1.
Jaoul, O. & Campbell, I. A. (1975) *J. Phys. F: Met. Phys.*, **5**, L69.
Jenkins, T. A. & Arajs, S. (1983) *Phys. Stat. Sol.* (*a*), **75**, 73.
Jernot, J. L., Chermant, J. L. & Coster, M. (1982) *Phys. Stat. Sol.* (*a*), **74**, 475.
Jo, T. (1976) *J. Phys. Soc. Japan*, **40**, 715.
Johansson, C. H. & Linde, J. O. (1936) *Ann. Phys.*, **25**, 1.
Johnson, K. H. (1973) *Advances in Quantum Chemistry*, ed. Lowden, P. O., Academic, NY, **7**, 143.
Johnson, K. H., Vvedensky, D. D. & Messmer, R. P. (1979) *Phys. Rev. B*, **19**, 1519.
Johnson, M. D., Hutchinson, P. & March, N. H. (1964) *Proc. R. Soc. A*, **282**, 283.
Jones, F. W. & Sykes, C. (1938) *Proc. R. Soc. A*, **166**, 376.
Jones, H. (1956) *Handb. Phys.*, **19**, 269.
Jones, R. D., Rowlands, D. & Rossiter, P. L. (1971) *Scripta Met.*, **5**, 915.
Jones, W. (1974) *J. Phys. C: Solid St. Phys.*, **7**, 1974.
Jonson, M. & Girvin, S. M. (1979) *Phys. Rev. Lett.*, **43**, 1447.
Joynt, R. (1984) *J. Phys. F: Met. Phys.*, **14**, 2363.
Julianus, J. A. & de Chatel, P. F. (1984) *J. Phys. F: Met. Phys.*, **14**, 149.
Julianus, J. A., Bekker, F. F. & de Chatel, P. F. (1984) *J. Phys. F: Met. Phys.*, **14**, 2061.
Jullien, R., Beal-Monod, M. T. & Coqblin, B. (1974) *Phys. Rev. B*, **9**, 1441.
Kaiser, A. B. & Doniach, S. (1970) *Int. J. Mag.*, **1**, 11.
Kanadoff, L. P. *et al.* (1967) *Rev. Mod. Phys.*, **39**, 395.
Kanzaki, H. (1957) *J. Phys. Chem. Sol.*, **2**, 24.
Kasuya, T. (1956) *Prog. Theor. Phys.*, **16**, 58.
Katsnel'son, A. A. & Alimov, A. A. (1967). *Phys. Met. Metall.*, **24** (No. 6), 129.
Katsnel'son, A. A., Alimov, Sh. A., Dazhayev, P. Sh., Silonov, V. M. & Stupina, N. N. (1968) *Phys. Met. Metall.*, **26** (No. 6), 26.
Katsnel'son, A. A., Silonov, V. M. & Skorobogatova, T. V. (1978) *Phys. Met. Metall.*, **44** (No. 2), 193.
Kaul, S. N. (1977) *J. Phys. F: Met. Phys.*, **7**, 2091.
Kaveh, M. & Mott, N. F. (1982*a*) *J. Phys. C: Solid St. Phys.*, **15**, L697.

Kaveh, M. & Mott, N. F., (1982*b*) *J. Phys. C: Solid St. Phys.*, **15**, L707.
Kedves, F. J., Hordos, F. J. & Schuszter, F. (1976) *Phys. Stat. Sol.* (*a*), **38**, K123.
Keller, J. & Smith, P. V. (1972) *J. Phys. C: Solid St. Phys.*, **5**, 1109.
Kenny, P. N., Trott, A. J. & Heald, P. T. (1973) *J. Phys. F: Met. Phys.*, **3**, 513.
Khachaturyan, A. G. (1979) *Prog. Mats. Sci.*, **22**, 1.
Khanna, S. N. & Jain A. (1974) *J. Phys. F: Met. Phys.*, **4**, 1982.
Khanna, S. N. & Jain, A. (1977) *J. Phys. Chem. Sol.*, **38**, 447.
Khannanov, Sh. Kh. (1977) *Phys. Met. Metall.*, **43** (No. 2), 178.
Kharoo, H. L., Gupta, O. P. & Hemkar, M. P. (1978) *Phys. Rev. B*, **18**, 5419.
Kim, D-J. (1964) *Prog. Theor. Phys.*, **31**, 921.
Kim, D. J. (1970) *Phys. Rev. B*, **1**, 3725.
Kim, M. J. & Flanagan, W. F. (1967) *Acta Met.*, **15**, 747.
King, W. F. & Cutler, P. H. (1970) *Phys. Lett.*, **31A**, 150.
Kinzel, W. & Fischer, K. H. (1977) *J. Phys. F: Met. Phys.*, **7**, 2163.
Kirkpatrick, S. (1973) *Rev. Mod. Phys.*, **45**, 574.
Kittel, C. (1948) *Phys. Rev.*, **75**, 972.
Kittel, C. (1963) *Quantum Theory of Solids*, Wiley, NY.
Kittel, C. (1976) *Introduction to Solid State Physics*, 5th ed., Wiley, NY.
Klein, A. P. & Heeger, A. J. (1966) *Phys. Rev.*, **144**, 458.
Kleinman, L. (1967) *Phys. Rev.*, **160**, 585.
Kleinman, L. (1968) *Phys. Rev.*, **172**, 383.
Kneller, E. (1958) *Z. Phys.*, **152**, 574.
Knowles, K. M. & Goodhew, P. J. (1983*a*) *Phil. Mag. A*, **48**, 527.
Knowles, K. M. & Goodhew, P. J. (1983*b*) *Phil. Mag. A*, **48**, 555.
Koehler, W. C. (1972) *Magnetic Properties of Rare Earth Metals*, ed. Elliott, R. J., Plenum Press, NY, p. 81.
Kohl, W., Scheffel, R., Heidsiek, H. & Lucke, K. (1983) *Acta Met.*, **31**, 1895.
Köhler, M. (1938) *Ann. Phys.*, **32**, 211.
Köhler, M. (1948) *Z. Phys.*, **124**, 772.
Kohn, W. & Luttinger, J. M. (1957) *Phys. Rev.*, **108**, 590.
Kohn, W. & Olsen, C. E. (1972) *J. de Phys. Colloque*, **33**, C3–135.
Kondo, J. (1969) *Solid State Physics*, **23**, 183.
Koon, N. C., Schlinder, A. I. & Mills, D. L. (1972) *Phys. Rev. B*, **6**, 4241.
Korevaar, B. M. (1961) *Acta Met.*, **9**, 297.
Kornilov, I. I. (1974) *Order–disorder Transformation in Alloys*, ed. Warliamont, Springer-Verlag, p. 132.
Köster, W. & Schüle, W. (1957) *Z. Metallkde.*, **48**, 592.
Köster, W. & Rave, H-P. (1961) *Z. Metallkde.*, **52**, 161.
Kouvel, J. S. (1969) in *Magnetism and Metallurgy*, eds. Berkowitz, A. & Kneller, E., Academic Press, NY, Vol. 2, p. 523.
Kovacs, I. & Szenes, G. (1971) *Phys. Stat. Sol.* (*a*), **5**, 231.
Kovacs-Csetenyi, E., Kedves, F. J. & Gergely, L. (1973) *Phys. Stat. Sol.* (*a*), **15**, K57.
Kozlov, E. V., Dementryev, V. M., Emelyanov, V. N., Kormin, N. M., Taylashev, A. S. & Stern, D. M. (1974) in Warliamont (1974), p. 58.
Krivoglaz, M. A. (1969) *Theory of X-ray and Thermal Neutron Scattering by Real Crystals*, Plenum Press, NY.
Kubo, R. (1957) *J. Phys. Soc., Japan*, **12**, 570.
Kudrnovský, J. & Velický, B. (1977) *Czech. J. Phys.*, **27**, 71.
Kumar, S. (1975) *Indian J. Phys.*, **49**, 615.
Kunitomi, N., Tsunoda, Y. & Hirai, Y. (1973) *Solid St. Comm.*, **13**, 495.
Kuranov, A. A. (1982) *Phys. Met. Metall.*, **54** (No. 3), 65.

Kus, F. W. (1978) *J. Phys. F: Met. Phys.*, **8**, 1483.
Landau, L. D. & Lifschitz, E. M. (1958) *Quantum Mechanics*, Pergamon, Oxford.
Landauer, R. (1952) *J. Appl. Phys.*, **23**, 779.
Landauer, R. (1978) *AIP Conference Proceedings No. 40*, eds. Garland, J. C. & Tanner, D. B., p. 2.
Langer, J. S. (1973) *Acta Met.*, **21**, 1649.
Langer, J. S., Bar-on, M. & Miller, H. D. (1975) *Phys. Rev. A*, **11**, 1417.
Langreth, D. (1969) *Phys. Rev.*, **181**, 753.
Larsen, U. (1976) *Phys. Rev. B*, **14**, 4356.
Larsen, U. (1977) *Solid St. Comm.*, **22**, 311.
Larsen, U. (1978) *Phys. Rev. B*, **18**, 5014.
Larsson, L. E. (1967) *Acta Met.*, **15**, 35.
Lasseter, R. H. & Soven, P. (1973) *Phys. Rev. B*, **8**, 2476.
Lawley, A. & Cahn, R. W. (1961) *J. Phys. Chem. Sol.*, **20**, 204.
Le Doug Khoi, Viellet, P. & Campbell, I. A. (1976) *J. Phys. F: Met. Phys.*, **6**, L197.
Leavens, C. R. (1977) *J. Phys. F: Met. Phys.*, **7**, 163.
Leavens, C. R. & Laubitz, M. J. (1976) *J. Phys. F: Met. Phys.*, **6**, 1851.
Lederer, P. & Mills, D. L. (1968) *Phys. Rev.*, **165**, 837.
Legvold, S., Peterson, D. T., Burgardt, P., Hofer, R. J., Lundell, B. & Vryostek, T. A. (1974) *Phys. Rev. B*, **9**, 2386.
Lekner, J. (1970) *Phil. Mag.*, **22**, 663.
Leung, K. M. (1984) *Phys. Rev. B*, **30**, 647.
Levin, K. & Mills, D. L. (1974) *Phys. Rev. B*, **9**, 2354.
Levin, K., Velicky, B. & Ehrenreich, H. (1970) *Phys. Rev. B*, **2**, 1771.
Levin, K., Bass, R. & Bennemann, K. H. (1971) *Phys. Rev. Lett.*, **27**, 589.
Levin, K., Bass, R. & Bennemann, K. H. (1972*a*) *Phys. Rev. B*, **5**, 3770.
Levin, K., Bass, R. & Bennemann, K. H. (1972*b*) *Phys. Rev. B*, **6**, 1865.
Levin, K., Soukoulis, C. M. & Grest, G. S. (1979) *J. Appl. Phys.*, **50**, 1695.
Lewis, H. W. (1958) *Solid St. Phys.*, **7**.
Liebmann, W. K. & Miller, E. A. (1963) *J. Appl. Phys.*, **34**, 2653.
Lin, L. S., Goldstein, J. I. & Williams, D. B. (1977) *Geochimica et Cosmochimica Acta*, **41**, 1861.
Linde, J. O. (1932) *Ann. Phys.*, **15**, 219.
Liu, K. S., Kai-Shue Lam & George Th. F. (1983) *Phys. Stat. Sol.* (*b*), **120**, 695.
Lloyd, P. (1965) *Proc. Phil. Soc.*, **86**, 825.
Lodder, A. (1976) *J. Phys. F: Met. Phys.*, **6**, 1885.
Lodder, A. (1977) *J. Phys. F: Met. Phys.* **7**, 139.
Lodder, A. & Braspenning, P. J. (1980) *J. Phys. F: Met. Phys.*, **10**, 2259.
Lodder, A., Boerrigter, P. M. & Braspenning, P. J. (1982) *Phys. Stat. Sol.* (*b*), **114**, 405.
Loegel, B. & Gautier, F. (1971) *J. Phys. Chem. Sol.*, **32**, 2723.
Löffler, H. & Radomsky, C. (1981) *Phys. Stat. Sol.* (*a*), **65**, K127.
Long, P. D. & Turner, R. E. (1970) *J. Phys. C: Solid St. Phys. Suppl.*, **2**, S127.
Longworth, G. & Tsuei, C. C. (1968) *Phys. Lett.*, **27A**, 258.
Loram, J. W. & Mirza, K. A. (1985) *J. Phys. F: Met. Phys.*, **15**, 2213.
Lormand, G. (1979) *Scripta Met.*, **13**, 27.
Lormand, G. (1982) *J. de Phys. Colloque*, **33**, C6–283.
Loucks, T. L. (1967) *Augmented Plane Wave Method*, Benjamin, NY.
Low, G. G. (1969) *Adv. in Phys.*, **18**, 371.
Low, G. G. & Collins, M. F. (1963) *J. Appl. Phys.*, **34**, 1195.
Low, G. G. & Holden, T. M. (1966) *Proc. Phys. Soc.*, **89**, 19.
Lowde, R. D., Moon, R. M., Pagonis, B., Perry, C. H., Sokoloff, J. B. & Vaughan-

Watkins (1983) *J. Phys. F: Met. Phys.*, **13**, 249.
Luborsky, F. E. (1983) *Amorphous Metallic Alloys, Butterworths Monographs in Metals*, Butterworths.
Lücke, K. & Haas, H. (1973) *Scripta Met.*, **7**, 781.
Lücke, K., Haas, H. & Schulze, H. A. (1976) *J. Phys. Chem.*, **37**, 979.
Luiggi, N., Simon, J. P. & Guyot, P. (1980) *J. Phys. F: Met. Phys.*, **10**, 865.
Luiggi, N. J. (1984) *J. Phys. F: Met. Phys.*, **14**, 2601.
Lupsa, I. & Burzo, E. (1983) *Phys. Stat. Sol. (a)*, **78**, K155.
Luttinger, J. M. & Kohn W. (1958) *Phys. Rev.*, **109**, 1892.
Ma, S. (1973) *Rev. Mod. Phys.*, **45**, 589.
Mahan, G. D. (1984) *Physics Reports*, **110**, 321.
Maksymowicz, A. Z. (1976) *J. Phys. F: Met. Phys.*, **6**, 1705.
Maletta, H. (1981) *J. Mag. Magnetic Materials*, **24**, 179.
Malinowska-Adamska, C. & Wojtczak, L. (1979) *Phys. Stat. Sol. (b)*, **96**, 697.
Malmström, G. & Geldart, D. J. W. (1980) *Phys. Rev. B*, **21**, 1133.
Mannari, I. (1968) *Phys. Letts.*, **26A**, 134.
Maradudin, A. A. (1966) *Solid State Phys.*, **18**, 273.
Maradudin, A. A. (1969) *Lattice Dynamics*, Benjamin, NY.
Maradudin, A. A., Montroll E. W. & Weiss, G. H. (1963) *Theory of Lattice Dynamics in the Harmonic Approximation*, Academic Press, NY.
March, J.-F. (1978) *Z. Metallkde.*, **69**, 377.
Markowitz, D. (1977) *Phys. Rev. B*, **15**, 3617.
Marro, J., Bortz, A. B., Kalos, M. H. & Lebowitz, J. L. (1975) *Phys. Rev. B*, **12**, 2000.
Marsch, E. (1976) *J. Phys. C: Solid St. Phys.*, **9**, L117.
Martin, D. L. (1972) *Phys. Lett.*, **39A**, 320.
Martin, D. L. (1976) *Phys. Rev. B*, **14**, 369.
Martin, J. W. (1968) *Precipitation Hardening*, Pergamon, Oxford.
Martin, J. W. (1972) *J. Phys. F: Met. Phys.*, **2**, 842.
Martin, J. W. & Doherty, R. D. (1976) *Stability of Microstructure in Metallic Systems*, Cambridge University Press.
Martin, J. W. & Paetsch, R. (1973) *J. Phys. F: Met. Phys.*, **3**, 907.
Masharov, S. I. (1966) *Phys. Met. Metall.*, **21** (No. 4), 1.
Matsubara, E. & Cohen, J. B. (1983) *Acta Metall.*, **31**, 2129.
Mattheiss, L. F. (1964) *Phys. Rev.*, **133**, A1399.
Mayadas, A. F. & Shatzkes, M. (1970) *Phys. Rev. B*, **1**, 1382.
Mazin, I. I., Savitskii, E. M. & Upenskii, Yu. A. (1984) *J. Phys. F: Met. Phys.*, **14**, 167.
Meaden, G. T. (1965) *Electrical Resistance of Metals*, Heywood Books, London
Medina, R. & Parra, R. E. (1982) *Phys. Rev. B*, **26**, 5187.
Meisel, L. V. & Cote, P. J. (1977) *Phys. Rev. B*, **16**, 2978.
Meisel, L. V. & Cote, P. J. (1983) *Phys. Rev. B*, **27**, 4617.
Meisel, L. V. & Cote, P. J. (1984) *Phys. Rev. B*, **30**, 1743.
Meisterle, P. & Pfeiler, W. (1983*a*) *Acta Met.*, **31**, 1543.
Meisterle, P. & Pfeiler, W. (1983*b*) *Phys. Stat. Sol. (a)*, **80**, K37.
Merlin, J. & Vigier, G. (1980) *Phys. Stat. Sol. (a)*, **58**, 571.
Mertig, I., Mrosan, E. & Schopke, R. (1982) *J. Phys. F: Met. Phys.*, **12**, 1689.
Messiah, A. (1961) *Quantum Mechanics*, Wiley, NY.
Meyer, A., Young, W. H. & Heyey, T. M. (1971) *Phil. Mag.*, **23**, 977.
Mills, D. L. & Lederer, P. (1966) *J. Phys. Chem. Sol.*, **27**, 1805.
Mills, D. L., Fert, A. & Campbell, I. A. (1971) *Phys. Rev. B*, **4**, 196.
Mimault, J., Delafond, J., Junqua, A., Naudon, A. & Grilhe, J. (1978) *Phil. Mag. B*, **38**, 255.

Mimault, J., Delafond, J. & Grilhe, J. (1981) *Phys. Stat. Sol.* (*a*), **65**, K121.
Miwa, H. (1962) *Prog. Theor. Phys.*, **28**, 209.
Miwa, H. (1963) *Prog. Theor. Phys.*, **29**, 477.
Mizutani, U. (1983) *Prog. Mats. Sci.*, **28**, 97.
Mizutani, U. & Yoshino, K. (1984) *J. Phys. F: Met. Phys.*, **14**, 1179.
Mohri, T., Sanchez, J. M. & de Fontaine, D. (1985) *Acta Met.*, **33**, 1171.
Monod, P. (1968) PhD. Thesis, Orsay, France.
Montgomery, H., Pells, G. P. & Wray, E. M. (1967) *Proc. R. Soc.*, **A301**, 261.
Mooij, J. H. (1973) *Phys. Stat. Sol.* (*a*), **17**, 521.
Mookerjee, A., Thakur, P. K. & Yussouff, M. (1985) *J. Phys. C: Solid St. Phys.*, **18**, 4677.
Moorjani, K., Tanaka, T., Sokolski, M. M. & Bose, S. M. (1974) *J. Phys. C: Solid St. Phys.*, **7**, 1098.
Morgan, G. J. (1966) *Proc. Phys. Soc.*, **89**, 365.
Morgan, G. J. & Hickey, B. J. (1985) *J. Phys. F: Met. Phys.*, **15**, 2473.
Morgan, G. J., Howson, M. A. & Saub, K. (1985*a*) *J. Phys. F: Met. Phys.*, **15**, 2157.
Morgan, G. J., Gilbert, R. & Hickey, B. J. (1985*b*) *J. Phys. F: Met. Phys.*, **15**, 2171.
Moriarty, J. A. (1970) *Phys. Rev. B*, **1**, 1363.
Moriarty, J. A. (1972) *Phys. Rev. B*, **6**, 1239.
Moriarty, J. A. (1974) *Phys. Rev. B*, **10**, 3075.
Morinaga, H. (1970) *Phys. Lett.*, **32A**, 75.
Morita, T., Horiguchi, T. & Chen, C. C. (1975) *J. Phys. Soc.* (*Japan*), **38**, 981.
Moriya, T. (1985) *Spin Fluctuations in Itinerant Electron Magnetism*, *Springer Series in Solid-State Sciences*, Vol. 56, Springer-Verlag, Berlin.
Morton, N., James, B. W. & Wostenholm, G. H. (1978) *Cryogenics*, **18**, 131.
Moruzzi, V. L., Williams, A. R. & Janak, J. F. (1977) *Phys. Rev. B*, **15**, 2854.
Moss, S. C. (1964) *J. Appl. Phys.*, **35**, 3547.
Moss, S. C. & Clapp, P. C. (1968) *Phys. Rev.*, **171**, 764.
Mott, N. F. (1935) *Proc. Phys. Soc.*, **47**, 571.
Mott, N. F. (1937) *J. Inst. Metals*, **60**, 267.
Mott, N. F. (1964) *Adv. Phys.*, **13**, 325.
Mott, N. F. (1972) *Phil. Mag.*, **26**, 1249.
Mott, N. F. (1974) *Metal Insulator Transitions*, Taylor and Francis, London.
Mott, N. F. (1981) *Phil. Mag. B*, **44**, 265.
Mott, N. F. & Davis, E. A. (1979) *Electronic Processes in non-Crystalline Materials*, Oxford, Clarendon.
Mott, N. F. & Jones, H. (1936) *The Theory of the Properties of Metals and Alloys*, Oxford.
Mouritsen, O. G. (1984) *Computer Studies of Phase Transitions and Critical Phenomena*, Springer-Verlag.
Mrosan, E. & Lehmann, G. (1976*a*) *Phys. Stat. Sol.* (*b*), **77**, K161.
Mrosan, E. & Lehmann, G. (1976*b*) *Phys. Stat. Sol.* (*b*), **78**, 159.
Müller, E. W. & Tien Tzou Tsong (1969) *Field Ion Microscopy, Principles and Applications*, Elsevier.
Murakami, T. (1953) *J. Phys. Soc.* (*Japan*), **8**, 458.
Murani, A. P. (1974) *Phys. Rev. Lett.*, **33**, 91.
Muth, P. (1983*a*) *Phys. Stat. Sol.* (*b*), **118**, K117.
Muth, P. (1983*b*) *Phys. Stat. Sol.* (*b*), **118**, K137.
Muth, P. (1985) *Phys. Stat. Sol.* (*b*), **127**, K151.
Muth, P. & Christoph, V. (1981) *J. Phys. F: Met. Phys.*, **11**, 2119.
Muto, T. (1936) *Inst. of Phys. and Chem. Res.* (*Tokyo*), *Scientific Papers*, **30**, 99.

Muto, T. (1938) *Inst. Phys. Chem. Res. (Tokyo), Scientific Papers*, **34**, 377.
Muto, T. & Takagi, Y. (1955) *Solid St. Phys.*, **1**, 193.
Mydosh, J. A. (1978) *J. Mag. Magn. Mat.*, 7, 237.
Myers, H. P., Wallden, L. & Karlsson, A. (1968) *Phil. Mag.*, **18**, 725.
Nabutovskii, V. M. & Patashinskii, A. Z. (1969) *Sov. Phys. Solid St.*, **10**, 2462.
Nagy, E. & Nagy, I. (1962) *J. Phys. Chem. Sol.*, **23**, 1605.
Nagy, I. & Pal, L. (1970) *Phys. Rev. Lett.*, **24**, 894.
Nagy, I. & Pal, L. (1971) *J. de Phys. Colloque*, **32**, C1–531.
Nakagawa, Y. G. & Weatherly, G. C. (1972) *Mater. Sci. Eng.*, **10**, 223.
Nakamura, M. (1984) *Phys. Rev. B*, **29**, 3691.
Narath, A. & Gossard, A. C. (1969) *J. Phys. Rev.*, **183**, 391.
Nascimento, L. C. S. & Caberra, G. G. (1981) *J. Low Temp. Phys.*, **45**, 481.
Naugle, D. G. (1984) *J. Phys. Chem. Sol.*, **45**, 367.
Nieminen, R. M. & Puska, M. (1980) *J. Phys. F: Met. Phys.*, **10**, L123.
Nieuwenhuys, G. J. & Boerstoel, B. M. (1970) *Phys. Lett.*, **33A**, 281.
Nix, F. C. & Shockley, W. (1938) *Rev. Mod. Phys.*, **10**, 1.
Nordheim, L. (1931) *Ann.* Phys., **9**, 607.
Norris, C. & Myers, H. P. (1971) *J. Phys. F: Met. Phys.*, **1**, 62.
Norris, C. & Nilsson, P. O. (1968) *Solid St. Comm.*, 7, 649.
Ododo, J. C. (1985) *J. Phys. F: Met. Phys.*, **15**, 941.
Ogawa, S. (1974) in *Order–Disorder Transformations in Alloys*, ed. Warliamont, Springer-Verlag, 1974, p. 240.
Okamura, K., Iwasaki, H. & Ogawa, S. (1968) *J. Phys. Soc. Japan*, **24**, 569.
Olsen, C. E. & Elliott, R. O. (1965) *Phys. Rev.*, **139**, A437.
Olsen, J. L. (1958) *Helv. Phys. Acta*, **31**, 713.
Oomi, G. & Woods, S. B. (1985) *Solid St. Comm.*, **53**, 223.
Osamura, K., Otsuka, N. & Murakami, Y. (1982) *Phil. Mag. B*, **45**, 583.
Osamura, K. O., Hiraoka, Y. & Murakami, V. (1973) *Phil. Mag.*, **28**, 809.
Otsuka, K., Kubo, H. & Wayman, C. M. (1981) *Met. Trans. A*, **12A**, 595.
Ott, H. (1935) *Phys. Zeits.*, **36**, 51.
Ousset, J. C., Ulmet, J. P., Asomoza, R., Bieri, J. B., Askenazy, S. & Fert, A. (1980) *J. de Phys. Colloque*, **41**, C8–470.
Paja, A. (1976) *J. Phys. C: Solid St. Phys.*, **9**, 1445.
Paja, A. (1981) *Physics of Transition Metals 1980*, Inst. of Phys., London, p. 471.
Pal, S. (1973) *J. Phys. F: Met. Phys.*, **3**, 1296.
Panseri, C. & Federighi, T. (1960) *Acta Met.*, **8**, 217.
Parent, L. G., Ueba, H. & Davidson, S. G. (1982) *Phys. Rev. B*, **26**, 753.
Parks, R. D. (1972) A.I.P. Conference Proc., *Magnetism and Magnetic Materials – 1971*, AIP, NY, **5**, 630.
Parrott, J. E. (1965) *Proc. Phys. Soc.*, **85**, 1143.
Pasturel, A., Hafner, J. & Hicter, P. (1985) *Phys. Rev. B*, **32**, 5009.
Pawlek, F. & Rogalla, D. (1966) *Cryogenics*, **6**, 14.
Perrier, J. P., Tissier, B. & Tournier, R. (1970) *Phys. Rev. Lett.*, **24**, 313.
Perrin, J. & Rossiter, P. L. (1977) *Phil. Mag.*, **36**, 109.
Pfeiler, W., Meisterle, P. & Zehetbauer, M. (1984) *Acta Met.*, **32**, 1053.
Pfeiler, W., Reihsner, R. & Trattner, D. (1985) *Scripta Met.*, **19**, 199.
Phillips, W. A. (1972) *J. Low Temp. Phys.*, **7**, 351.
Pichard, C. R., Tellier, C. R. & Tosser, A. J. (1979) *Thin Solid Films*, **62**, 189.
Pichard, C. R., Tellier, C. R. & Tosser, A. J. (1980) *Phys. Stat. Sol. (b)*, **99**, 353.
Pichard, C. R., Tellier, C. T. & Tosser, A. J. (1981) *Phys. Stat. Sol. (a)*, **65**, 327.
Pindor, A. J., Temmerman, W. M., Gyorffy, B. L. & Stocks, G. M. (1980) *J. Phys. F: Met. Phys.*, **10**, 2617.

Pines, D. (1955) *Solid St. Phys.*, **1**, 1.
Pippard, A. B. (1955) *Phil. Mag.*, **46**, 1104.
Pippard, A. B. (1960) *Proc. R. Soc. A*, **247**, 165.
Plischke, M. & Mattis, D. (1973) *Phys. Rev. B*, **7**, 2430.
Podloucky, R., Zeller, R. & Dederichs, P. H. (1980) *Phys. Rev. B*, **22**, 5777.
Poerschke, R., Theis, U. & Wollenberger, H. (1980) *J. Phys. F: Met. Phys.*, **10**, 67.
Poerschke, R. & Wollenberger, H. (1976) *J. Phys. F: Met. Phys.*, **6**, 27.
Poerschke, R. & Wollenberger, H. (1980) *Radiation Effects*, **49**, 225.
Poon, S. J., Dunn, P. L. & Smith, L. M. (1982) *J. Phys. F: Met. Phys.*, **12**, L101.
Popovic, Z., Carbotte, J. P. & Piercy, G. R. (1973) *J. Phys. F: Met. Phys.*, **3**, 1008.
Poquette, G. E. & Mikkola, D. E. (1969) *Trans. Met. Soc. AIME*, **245**, 743.
Porter, D. A. & Easterling, K. E. (1981) *Phase Transformations in Metals and Alloys*, Van Nostrand Reinhold, NY.
Poulsen, U. K., Kollar, J. & Andersen, O. K. (1976) *J. Phys. F: Met. Phys.*, **6**, L241.
Price, D. L., Singwi, D. S. & Tosi, M. P. (1970) *Phys. Rev. B*, **2**, 2983.
Punz, G. & Hafner, J. (1985) *Z. Phys. B*, **61**, 231.
Pureur, P., Kunzler, J. V., Schreiner, W. H. & Brandao, D. E. (1982) *Phys. Stat. Sol. (a)*, **70**, 11.
Radomsky, C. & Löffler, H. (1978) *Phys. Stat. Sol. (a)*, **50**, 123.
Radomsky, C. & Löffler, H. (1979*a*) *Phys. Stat. Sol. (a)*, **54**, 615.
Radomsky, C. & Löffler, H. (1979*b*) *Phys. Stat. Sol. (a)*, **56**, 711.
Ramanan, V. R. V. & Berger, L. (1981) *J. Appl. Phys.*, **52**, 2211.
Lord Rayleigh (1892) *Phil. Mag.*, **34**, 481.
Raynor, D. & Silcock, J. M. (1968) *Scripta Met.*, **2**, 399.
Raynor, D. & Silcock, J. M. (1970) *J. Mats. Sci.*, **4**, 121.
Reihsner, R. & Pfeiler, W. (1985) *J. Phys. Chem. Sol.*, **46**, 1431.
Reitz, J. R. (1955) *Solid St. Phys.*, **1**, 1.
Reynolds, J. A. & Hough, J. M. (1957) *Proc. Phys. Soc.*, **70**, 769.
Rice, M. J. & Bunce, O. (1970) *Phys. Rev. B*, **2**, 3833.
Richard, T. G. & Geldart, D. J. W. (1977) *Phys. Rev. B*, **15**, 1502.
Richter, F., Benedick, W. & Pepperhoff, W. (1974) *Z. Metallkde.*, **65**, 42.
Richter, J. & Schiller, W. (1979) *Phys. Stat. Sol.*, **92**, 511.
Richter, J. & Schubert, G. (1983) *Phys. Stat. Sol. (b)*, **116**, 597.
Ricker, Von T. & Pfluger, E. (1966) *Z. Metallkde.*, **57**, 39.
Riseborough, P. S. (1984) *Phys. Rev. B*, **29**, 4134.
Rivier, N. (1968) PhD. Thesis, University of Cambridge.
Rivier, N. (1974) *J. Phys. F: Met. Phys.*, **4**, L249.
Rivier, N. & Adkins, K. (1975) *J. Phys. F: Met. Phys.*, **5**, 1745.
Rivier, N. & Zlatic, V. (1972*a*) *J. Phys. F: Met. Phys.*, **2**, L87.
Rivier, N. & Zlatic, V. (1972*b*) *J. Phys. F: Met. Phys.*, **2**, L99.
Rivier, N. & Zuckermann, M. J. (1968) *Phys. Rev. Letts.*, **21**, 904.
Rivier, N., Sunjic, M. & Zuckermann, M. J. (1969) *Phys. Lett.*, **28A**, 492.
Rivier, N. & Zitkova, J. (1971) *Adv. Phys.*, **20**, 143.
Rizzuto, C., Babic, E. & Stewart, A. M. (1973) *J. Phys. F: Met. Phys.*, **3**, 825.
Robbins, C. G., Claus, H. & Beck, P. A. (1969) *Phys. Rev. Lett.*, **22**, 1307.
Robinson, J. E. & Dow, J. D. (1968) *Phys. Rev.*, **171**, 815.
Rodberg, L. S. & Thaler, R. M. (1967) *Introduction to the Quantum Theory of Scattering*, Academic Press, NY.
Ropke, G. & Christoph, V. (1975) *J. Phys. C: Solid St. Phys.*, **8**, 3615.
Rossiter, P. L. (1977) *J. Phys. F: Met. Phys.*, **7**, 407.
Rossiter, P. L. (1979) *J. Phys. F: Met. Phys.*, **9**, 891.

Rossiter, P. L. (1980*a*) *J. Phys. F: Met. Phys.*, **10**, 1459.
Rossiter, P. L. (1980*b*) *J. Phys. F: Met. Phys.*, **10**, 1787.
Rossiter, P. L. (1981*a*) *J. Phys. F: Met. Phys.*, **11**, 615.
Rossiter, P. L. (1981*b*) *J. Phys. F: Met. Phys.*, **11**, 2105.
Rossiter, P. L. & Bykovec, B. (1978) *Phil. Mag. B*, **38**, 555.
Rossiter, P. L. & Houghton, M. E. (1978) *Phys. Stat. Sol.* (*a*), **47**, 597.
Rossiter, P. L. & Wells, P. (1971) *J. Phys. C: Solid St. Phys.*, **4**, 354.
Rowland, T., Cusack, N. E. & Ross, R. G. (1974) *J. Phys. F: Met. Phys.*, **4**, 2189.
Rowlands, J. A., Greig, D. & Blood, P. (1971) *J. Phys. F: Met. Phys.*, **1**, L29.
Rowlands, J. A., Stackhouse, B. J. & Woods, S. B. (1977) *J. Phys. F: Met. Phys.*, **7**, L301.
Ruderman, M. A. & Kittel, C. (1954) *Phys. Rev.*, **96**, 99.
Rundman, K. B. & Hilliard, J. E. (1967) *Acta Met.*, **15**, 1025.
Rusby, R. L. (1974) *J. Phys. F: Met. Phys.*, **4**, 1265.
Sakata, M., Cowlam, N. & Davies, H. A. (1980) *J. de Phys.*, **41**, C8–19.
Salamon, M. B. & Lederman, F. L. (1974) *Phys. Rev. B*, **10**, 4492.
Sambles, J. R. & Elsom, K. C. (1980) *J. Phys. F: Met. Phys.*, **10**, 1487.
Sambles, J. R. & Preist, T. W. (1982) *J. Phys. F: Met. Phys.*, **12**, 1971.
Sambles, J. R., Elsom, K. C. & Priest, T. W. (1982) *J. Phys. F: Met. Phys.*, **12**, 1169.
Sanchez, J. M. & de Fontaine, D. (1978*a*) *Phys. Rev. B*, **17**, 2926.
Sanchez, J. M. & de Fontaine, D. (1978*b*) *Acta Met.*, **26**, 1083.
Sanchez, J. M. & de Fontaine, D. (1980) *Phys. Rev. B*, **21**, 216.
Sanchez, J. M. & de Fontaine, D. (1982) *Phys. Rev. B*, **25**, 1759.
Sarachick, M. P. (1968) *Phys. Rev.*, **170**, 679.
Sarkissian, B. V. B. (1981) *J. Phys. F: Met. Phys.*, **11**, 2191.
Sarkissian, B. V. B. & Taylor, R. H. (1974) *J. Phys. F: Met. Phys.*, **4**, L243.
Sato, H. & Toth, R. S. (1965) in *Alloying Behaviour and Effects in Concentrated Solid Solutions*, Gordon and Breach, NY, p. 295.
Schaafsma, A. S., Vincze, I., Van der Woude, F., Kemeny, T. & Lovas, A. (1980) *J. de Phys. Colloque*, **41**, C8–246.
Schiff, L. I. (1968) *Quantum Mechanics* (3rd ed.), McGraw-Hill, NY.
Schilling, J. S., Ford, P. J., Larsen, U. & Mydosh, J. A. (1976) *Phys. Rev. B*, **14**, 4368.
Schindler, A. I. & Rice, M. J. (1967) *Phys. Rev.*, **164**, 759.
Schlawne, F. (1983) *J. Phys. F: Met. Phys.*, **13**, 2437.
Schoenberg, D. (1969) in Ziman, 1969, p. 62.
Schöpke, R. & Mrosan, E. (1978) *Phys. Stat. Sol.* (*b*), **90**, K95.
Schotte, K. D. (1978) *Physics Reports*, **46**, 93.
Schreiner, W. H., Brandao, D. E., Ogiba, F. & Kunzler, J. V. (1982) *J. Phys. Chem. Sol.*, **43**, 777.
Schröder, K. (1983) *CRC Handbook of Electrical Resistivity of Binary Metallic Alloys*, CRC Press, p. 1.
Schryvers, D., van Tendeloo, G. & Amelinckx, S. (1985) *Phys. Stat. Sol.* (*a*), **87**, 401.
Schubert, G. & Löser, W. (1985) *Phys. Stat. Sol.* (*b*), **130**, 775.
Schubert, G. & Richter, J. (1984) *Phys. Stat. Sol.* (*b*), **125**, 839.
Schüle, W. & Kehrer, H-P. (1961) *Z. Metallkde.*, **52**, 168.
Schulz, R., Mehra, M. & Johnson, W. L. (1985) *J. Phys. F: Met. Phys.*, **15**, 2109.
Schürmann, H. K. & Parks, R. D. (1971) *Phys. Rev. Lett.*, **26**, 367, 835.
Schwarz, K., Mohn, P. & Kubler, J. (1984) *J. Phys. F: Met. Phys.* **14**, 2659.
Schwerer, F. C. (1971) *Int. J. Mag.*, **2**, 381.
Seeger, A. & Schottky, G. (1959) *Acta Metal.*, **7**, 945.
Seehra, M. S. & Silinsky, P. (1976) *Phys. Rev. B*, **13**, 5183.
Segall, B. (1961) *Phys. Rev.*, **124**, 1797.

Segall, B. & Ham, F. S. (1968) *Methods in Computational Physics*, eds. Adler *et al.*, Academic Press, NY, p. 8.
Segenreich, B., Pfeiler, W. & Trieb, L. (1978) *Scripta Met.*, **12**, 1131.
Servi, I. W. & Turnbull, D. (1966) *Acta Met.*, **14**, 161.
Sham, L. J. (1965) *Proc. R. Soc. A*, **283**, 3.
Sham, L. J. & Ziman, J. M. (1963) *Solid St. Phys.*, **15**, 221.
Shante, V. K. S. & Kirkpatrick, S. (1971) *Adv. in Phys.*, **20**, 325.
Shapiro, S. M. (1981) *Mat. Trans. A*, **12A**, 567.
Shatzkes, M., Chaurhari, P., Levi, A. A. & Mayadas, A. F. (1973) *Phys. Rev. B.*, **7**, 5058.
Shaw, R. W. (1968) *Phys. Rev.*, **174**, 769.
Shirane, G., Minkiewicz, V. J. & Nathans, R. (1968) *J. Appl. Phys.*, **39**, 383.
Shukla, R. C. (1980) *Phys. Rev. B*, **22**, 5810.
Shukla, R. C. & Muller, E. R. (1980) *Phys. Rev. B*, **21**, 544.
Shukla, R. C. & Taylor, R. (1976) *J. Phys. F: Met. Phys.*, **6**, 531.
Shukla, R. C. & VanderSchans, M. (1980) *Phys. Rev. B*, **22**, 5818.
Shull, C. G. (1968) *Magnetic and Inelastic Scattering by Neutrons*, eds. Rowland, T. J. & Beck, P. A., Gordon and Breach, NY.
Shyu, W. M., Brust, D. & Fumi, F. G. (1967) *J. Phys. Chem. Sol.*, **28**, 717.
Siegel, S. (1940) *Phys. Rev.*, **57**, 537.
Sikka, S. K., Vohra, Y. K. & Chidambaram, R. (1982) *Prog. Mats. Sci.*, **27**, 245.
Silcock, J. M. (1970*a*) CERL Report No. RD/L/N 160/70.
Silcock, J. M. (1970*b*) *Scripta Met.*, **4**, 25.
Silvertsen, J. M. (1964) in *Resonance and Relaxation in Metals*, ed. Vogel Jr., Plenum Press, p. 383.
Simoneau, R. & Bégin, G. (1973) *J. Appl. Phys.*, **44**, 1461.
Simoneau, R. & Bégin, G. (1974) *J. Appl. Phys.*, **45**, 3828.
Simons, D. S. & Salamon, M. B. (1971) *Phys. Rev. Lett.*, **26**, 750.
Sinclair, R. & Thomas, G. (1975) *J. Appl. Cryst.*, **8**, 206.
Sinclair, R., Schneider, K. & Thomas, G. (1975) *Acta Met.*, **23**, 873.
Singh, N., Singh, J. & Prakash, S. (1977) *Phys. Stat. Sol.* (*b*), **79**, 787.
Singwi, K. S. and Tosi, M. P. (1981) *Solid St. Phys.*, **36**, 177.
Singwi, K. S., Tosi, M. P., Sjolander, A. & Land, R. H. (1968) *Phys. Rev.*, **176**, 589.
Singwi, K. S., Sjolander, A., Tosi, M. P. & Land, R. H. (1970) *Phys. Rev. B*, **1**, 1044.
Slater, J. C. (1936) *Phys. Rev.*, **49**, 537.
Slater, J. C. (1951) *Phys. Rev.*, **84**, 179.
Slater, J. C. (1972) *Adv. Quantum. Chem.*, **6**, 1.
Slater, J. C. & Johnson, K. H. (1972) *Phys. Rev. B*, **5**, 844.
Smart, J. S. (1966) *Effective Field Theories of Magnetism*, Saunders, Pa.
Smit, J. (1951) *Physica*, **17**, 612.
Sobotta, G. (1985) *J. Phys. C: Solid St. Phys.*, **18**, 2065.
Soffer, S. B. (1967) *J. Appl. Phys.*, **38**, 1710.
Sondheimer, E. H. (1950) *Phys. Rev.*, **80**, 401.
Sondheimer, E. H. (1952) *Adv. Phys.*, **1**, 1.
Sondheimer, E. H. (1962) *Proc. R. Soc.*, **A268**, 100.
Sorbello, R. S. (1973) *Solid St. Comm.*, **12**, 287.
Sorbello, R. S. (1974*a*) *J. Phys. F: Met. Phys.*, **4**, 503.
Sorbello, R. S. (1974*b*) *J. Phys. F: Met. Phys.*, **4**, 1665.
Sorbello, R. S. (1977) *Phys. Rev. B*, **15**, 3045.
Sorbello, R. S. (1980) *Phys. Stat. Sol.* (*b*), **100**, 347.
Sorbello, R. S. (1981) *Phys. Stat. Sol.* (*b*), **107**, 233.
Sousa, J. B. *et al.* (1980*a*) *J. Phys. F: Met. Phys.*, **10**, 933.

Sousa, J. B. *et al.* (1980*b*) *J. Mag. Magn. Mats.*, **15–18**, 892.
Sousa, J. B., Chaves, M. R., Pinto, R. S. & Pinheiro, M. F. (1972) *J. Phys. F: Met. Phys.*, **2**, L83.
Sousa, J. B., Pinto, R. S., Amado, M. M., Moreira, J. M., Braga, M. E., Ausloos, M. & Balberg, I. (1979) *Solid St. Comm.*, **31**, 209.
Springer, G. S. & Tsai, S. W. (1967) *J. Composite Materials*, **1**, 166.
Springford, M. (1971) *Adv. Phys.*, **20**, 493.
Squires, G. L. (1963) in *Inelastic Scattering of Neutrons in Solids and Liquids*, IAEA, Vienna, Vol. 2, p. 55.
Srivastava, C. M., Sheikh, A. W. & Chandra, G. (1981*a*) *Physica*, **108B**, 861.
Srivastava, C. M., Shiekh, A. W. & Chandra, G. (1981*b*) *J. Mag. Magn. Mat.*, **25**, 147.
Stanley, H. E. (1971) *Introduction to Phase Transitions and Critical Phenomena*, Oxford, Clarendon.
Star, W. M. (1972) *Physica*, **58**, 623.
Star, W. M., Basters, F. B., Nap, G. M., de Vroede, E. & van Baarle, C. (1972*a*) *Physica*, **58**, 585.
Star, W. M., de Vroede, E. & van Baarle, C. (1972*b*) *Physica*, **59**, 128.
Starke, E. A. Jr., Gerold, V. & Guy, A. G. (1965) *Acta Met.*, **13**, 957.
Stauffer, D. (1977) *Amorphous Magnetism*, eds. Levy, R. A. & Hassegawa, B., Plenum Press, NY, Vol. 2.
Stauffer, D. (1979) *Phys. Reports*, **54**, 1.
Stern, E. A. (1966) *Phys. Rev.*, **144**, 545.
Stern, E. A. (1967) *Phys. Rev.*, **157**, 544.
Stern, E. A. (1968) *Energy Bands in Metals and Alloys*, Gordon and Breach, NY, p. 151.
Stern, E. A. (1970) *Phys. Rev. B*, **1**, 1518.
Stern, E. A. (1971) *Phys. Rev. Lett.*, **26**, 1630.
Stocks, G. M. & Butler, W. H. (1981) *Physics of Transition Metals 1980*, Inst. of Phys., London, p. 467.
Stocks, G. M., Williams, R. W. & Faulkner, J. S. (1973) *J. Phys. F: Met. Phys.*, **3**, 1688.
Stoner, E. C. (1946) *Rept. Prog. Phys.*, **9**, 43.
Stoner, E. C. (1947) *Rept. Prog. Phys.*, **11**, 43.
Stoner, E. C. (1951) *J. de Phys. Rad.*, **12**, 372.
Stratton, J. A. (1941) *Electromagnetic Theory*, McGraw-Hill, NY.
Stroud, D. (1980) *Phys. Rev. Lett.*, **44**, 1708.
Suck, J. B., Rudin, H., Guntherhodt, H. J., Thomanek, D., Beck, H. & Markel, C. (1980) *J. de Phys.*, **41**, C8–17.
Seuzaki, Y. & Mori, H. (1968) *Phys. Lett.*, **28A**, 70.
Svensson, E. C., Brockhouse, B. N. & Rowe, J. M. (1967) *Phys. Rev.*, **155**, 619.
Sykes, C. & Evans, H. (1936) *J. Inst. Metals*, **58**, 255.
Synecek, V. (1962) *J. de Phys. et Rad.*, **23**, 787.
Szenes, G., Kovacs, I. & Nagy, E. (1971) *Phys. Stat. Sol.* (*a*), **7**, 251.
Takada, S. (1971) *Prog. Theor. Phys.*, **46**, 15.
Takai, O., Fukusako, T., Yamamoto, R. & Doyama, M. (1972) *J. Phys. F: Met. Phys.*, **2**, L80.
Takai, O., Yamamoto, R. & Doyama, M. (1974) *J. Phys. Chem. Sol.*, **35**, 1257.
Takayama, S. (1976) *J. Mats. Sci.*, **11**, 164.
Tanigawa, S. & Doyama, M. (1973) *J. Phys. F: Met. Phys.*, **3**, 977.
Taylor, P. L. (1963) *Proc. R. Soc.*, **A275**, 200.
Tebble, R. S. & Craik, D. J. (1969) *Magnetic Materials*, Wiley, NY.

Tellier, C. R., Pichard, C. R. & Tosser, A. J. (1979) *Thin Solid Films*, **61**, 349.
Temmerman, W. M., Gyorffy, B. L. & Stocks, G. M. (1978) *J. Phys. F: Met. Phys.*, **8**, 2461.
Tholence, J. L. & Tournier, R. (1970) *Phys. Rev. Lett.*, **25**, 867.
Thomas, G. A., Giray, A. B. & Parks, R. D. (1973) *Phys. Rev. Lett.*, **31**, 241.
Thomas, H. (1951) *Z. Phys.*, **129**, 219.
Thomas, N. (1984) *J. Phys. C: Solid St. Phys.*, **17**, L59.
Thompson, J. C. (1974) *J. de Phys. Colloque*, **35**, C4–367.
Tibballs, J. (1974) PhD. Thesis, Melbourne University, Australia.
Torrens, H. M. & Gerl, M. (1968) *Phys. Rev.*, **187**, 912.
Tournier, R. & Blandin, A. (1970) *Phys. Rev. Lett.*, **8**, 397.
Towers, G. R. (1972) PhD. Thesis, Melbourne University, Australia.
Trattner, D. & Pfeiler, W. (1983*a*) *J. Phys. F: Met. Phys.*, **13**, 739.
Trattner, D. & Pfeiler, W. (1983*b*) *Scripta Met.*, **17**, 909.
Trieb, L. & Veith, G. (1978) *Acta Met.*, **26**, 185.
Tripp, J. H. & Farrell, D. E. (1973) *Phys. Rev. B*, **7**, 571.
Troper, A. & Gomes, A. A. (1975) *Phys. Stat. Sol.* (*b*), **68**, 99.
Tsakalakos, T. (1984) *Mat. Res. Soc. Symp. Proc.*, **21**, North-Holland, NY.
Tsuei, C. C. (1977) *Amorphous Magnetism II*, eds. Levy, R. A. & Hasegawa, R., Plenum Press, NY, p. 187.
Tsukada, M. (1972) *J. Phys. Soc. Japan*, **32**, 1475.
Ueda, K. & Moriya, T. (1975) *J. Phys. Soc. Japan*, **39**, 605.
Underwood, E. E. (1970) *Quantitative Stereology*, Addison-Wesley, Reading, Mass.
Van Daal, H. J. (1981) *Physics of Transition Metals 1980*, eds. Rhodes, P., IOP, London, p. 435.
van Dam, J. E. & van den Berg, G. J. (1970) *Phys. Stat. Sol.* (*a*), **3**, 11.
van der Rest, J. (1977) *J. Phys. F: Met. Phys.*, **7**, 1051.
van der Voort, E. & Guyot, P. (1971) *Phys. Stat. Sol.* (*b*), **47**, 465.
van Dijk, C. & Bergsma, J. (1968) *Neutron Inelastic Scattering*, IAEA, Vienna, **1**, 233.
van Hove, L. (1954) *Phys. Rev.*, **95**, 249.
van Royen, E. W., Brandsma, G., Molder, A. L. & Radelaar, S. (1973) *Scripta Met.*, **7**, 1125.
van Tendeloo, G. & Amelinckx, S. (1978) *Phys. Stat. Sol.* (*a*), **50**, 53.
van Tendeloo, G. & Amelinckx, S. (1984) *Phys. Stat. Sol.* (*a*), **84**, 185.
van Vucht, R. J. M., van Kempen, H. & Wyder, P. (1985) *Rep. Prog. Phys.*, **48**, 853.
Velický, B. & Levin, K. (1970) *Phys. Rev. B*, **2**, 938.
Velický, B., Kirkpatrick, S. & Ehrenreich, H. (1968) *Phys. Rev.*, **175**, 747.
Viard, M. & Gavoille, G. (1979) *J. Appl. Phys.*, **50**, 1828.
Vigier, G. & Merlin, J. (1983) *Phil. Mag. B*, **47**, 299.
Vigier, G., Pelletier, J. M. & Merlin, J. (1983) *J. Phys. F: Met. Phys.*, **13**, 1677.
von Bassewitz, A. & Mitchell, E. N. (1969) *Phys. Rev.*, **182**, 712.
von Heimendahl, L. (1979) *J. Phys. F: Met. Phys.*, **9**, 161.
Vrijen, J. (1977) *Netherlands Energy Research Foundation*, ECN Petten Report, ECN-31.
Vuillemin, J. J. & Priestley, M. G. (1965) *Phys. Rev. Lett.*, **14**, 307.
Wagner, C. N. J. (1978) *J. Non-Crystalline Solids*, **31**, 1.
Wagner, C. N. J. & Lee, D. (1980) *J. de Phys.*, **41**, C8–24.
Wagner, W., Poerschke, R., Axmann, A. & Schwann, D. (1980) *Phys. Rev. B*, **21**, 3087.
Wagner, W., Poerschke, R. & Wollenberger, H. (1981) *Phil. Mag. B*, **43**, 345.
Wagner, W., Poerschke, R. & Wollenberger, H. (1982) *J. Phys. F: Met. Phys.*, **12**, 405.

Wagner, W., Piller, J., Poerschke, R. & Wollenberger, H. (1984) *Microstructural Characterisation of Materials by non-Microscopical Techniques*, eds. Anderson, N. H. *et al.*, Risø Nat. Lab., Roskilde.
Walker, C. B. & Keating, D. T. (1961) *Acta Cryst.*, **14**, 1170.
Wallace, D. J. & Zia, R. K. P. (1978) *Rep. Prog. Phys.*, **41**, 1.
Wang, K. P. (1970) *Phys. Lett.*, **32A**, 282.
Wang, K. P. (1973) *Can. J. Phys.*, **51**, 1678.
Wang, K. P. & Amar, H. (1970) *Phys. Rev. B*, **1**, 582.
Warliamont, H. (1974) *Order–Disorder Transformations in Alloys*, Springer-Verlag.
Warren, B. E. (1969) *X-ray Diffraction*, Addison-Wesley, NY.
Warren, B. E., Averbach, B. L. & Roberts, B. W. (1951) *J. Appl. Phys.*, **22**, 1493.
Waseda, W. & Suzuki, K. (1971) *Phys. Lett.*, **34A**, 69.
Waseda, Y. & Suzuki, K. (1972) *Phys. Stat. Sol. (b)*, **49**, 339.
Watanabe, D. & Terasaki, O. (1984) *Phase Transformations in Solids*, ed. Tsakalakos, T., Mats. Res. Soc. Symp. Proc., **21**, 231.
Watson, B. P. & Leath, P. L. (1974) *Phys. Rev. B*, **9**, 4893.
Watson, G. N. (1944) *Theory of Bessel Functions*, rev. ed., Macmillan, NY, p. 128.
Watson, W. G., Hahn, W. C. & Kraft, R. W. (1975) *Met. Trans. A*, **6A**, 151.
Weger, M. & Mott, N. F. (1985) *J. Phys. C: Solid St. Phys.*, **18**, L201.
Weger, M., de Groot, R. A., Mueller, F. M. & Kaveh, M. (1984) *J. Phys. F: Met. Phys.*, **14**, L207.
Weiss, R. J. & Marotta, A. S. (1959) *J. Phys. Chem. Sol.*, **9**, 302.
Wells, P. & Rossiter, P. L. (1971) *Phys. Stat. Sol. (a)*, **4**, 151.
Wendrock, G. (1983) *Phys. Stat. Sol. (a)*, **78**, 497.
Werner, K., Schmatz, W., Bauer, G. S., Seitz, E., Fenzl, H. J. & Baratoff, A. (1978) *J. Phys. F: Met. Phys.*, **8**, L207.
White, G. K. & Woods, S. B. (1959) *Phil. Trans. R. Soc.*, **A251**, 273.
Wiesmann, H., Gurvitch, M., Lutz, H., Ghosh, A., Schwarz, B., Strongin, M., Allen, P. B. & Halley, J. W. (1977) *Phys. Rev. Lett.*, **38**, 782.
Wilkes, P. (1973) *Solid State Theory in Metallurgy*, Cambridge University Press.
Williams, R. O. (1960) *Trans. ASM*, **52**, 530.
Williams, R. O. (1978) *Acta Cryst.*, **A34**, 65.
Wilson, F. G. & Pickering, F. B. (1968) *Scripta Met.*, **2**, 471.
Wilson, K. G. (1972) *Phys. Rev. Lett.*, **28**, 548.
Wilson, K. G. (1974) *Nobel Symposia – Medicine and Natural Sciences*, Academic Press, NY, p. 24.
Wilson, K. G. & Fisher, M. E. (1972) *Phys. Rev. Lett.*, **28**, 240.
Wilson, K. G. & Kogut, J. (1974) *Phys. Rep.* (*Phys. Lett. C* (Netherlands)), **12C**, 75.
Winter, H. & Stocks, G. M. (1983) *Phys. Rev. B*, **27**, 882.
Wiser, N. (1982) *Physica Scripta*, **T1**, 118.
Wiser, N. (1966) *Phys. Rev.*, **143**, 393.
Wochner, P. & Jäckle, J. (1981) *Z. Phys. B*, **44**, 293.
Wolff, P. A. (1962) *Phys. Rev.*, **124**, 1030.
Wong, J. (1981) *Glassy Metals – 1*, eds. Guntherodt, H. J. & Beck, H., Springer, *Topics in Appl. Phys.*, **46**, 45.
Woodruff, T. O. (1957) *Solid St. Phys.*, **4**, 367.
Woolley, R. G. & Mattuck, R. D. (1973) *J. Phys. F: Met. Phys.*, **3**, 75.
Wright, P. & Thomas, K. (1958) *Br. J. Appl. Phys.*, **9**, 330.
Wu, T. B. (1984) *Mat. Res. Soc. Symp. Proc.*, **21**, 19.
Wu, T. B. & Cohen, J. B. (1983) *Acta Metall.*, **31**, 1929.

Wynblatt, P. & Gjostein, N. A. (1967) *J. Phys. Chem. Sol.*, **28**, 2108.
Yamamoto, R., Doyama, M., Takai, O. & Fukusako, T. (1973) *J. Phys. F: Met. Phys.*, **3**, 1134.
Yarnell, J. L., Warren, J. L. & Koenig, S. H. (1965) *Lattice Dynamics*, ed. Wallis, R. F., Pergamon Press, NY, p. 57.
Yim, W. M. & Stofko, E. J. (1967) *J. Appl. Phys.*, **38**, 5211.
Yonemitsu, K. & Matsuda, T. (1976) *Phys. Stat. Sol.* (*a*), **36**, 791.
Yonezawa, F. & Morigaki, K. (1973) *Prog. Theor. Phys. Suppl.*, **53**, 1.
Yosida, K. (1957) *Phys. Rev.*, **106**, 893.
Yosida, K., Okiji, A. & Chikazumi, S. (1965) *Progr. Theoret. Phys.* (*Kyoto*), **33**, 559.
Yu, Z. Z., Haerle, M., Zwart, J. W., Bass, J., Pratt, W. P. Jr. & Schroeder, P. A. (1984) *Phys. Rev. Lett.*, **52**, 368.
Zeller, R. (1981) in *Physics of Transition Metals*, ed. Rhodes, P., Inst. of Phys., London, p. 265.
Zeller, R., Podloucky, R. & Dederichs, P. H. (1980) *Z. Physik B*, **38**, 165.
Zibold, G. (1979) *J. Phys. F: Met. Phys.*, **9**, 917.
Ziman, J. M. (1960) *Electrons and Phonons*, Cambridge University Press
Ziman, J. M. (1961*a*) *Phys. Rev.*, **121**, 1320.
Ziman, J. M. (1961*b*) *Adv. Phys.*, **10**, 1.
Ziman, J. M. (1964) *Adv. in Phys.*, **13**, 89.
Ziman, J. M. (1967) *Adv. in Phys.*, **16**, 551.
Ziman, J. M. (1969) *The Physics of Metals. 1 Electrons*, Cambridge University Press.
Ziman, J. M. (1971) *Solid St. Phys.*, **26**, 1.
Ziman, J. M. (1972) *Principles of the Theory of Solids*, Cambridge University Press.
Zin, A. & Stern, E. A. (1985) *Phys. Rev. B*, **31**, 4954.
Zinov'yev, V. Ye., Petrova, L. N., Dik, Ye. G. & Ivliyev, A. D. (1980) *Phys. Met. Metall.*, **50** (No. 4), 193.
Zumsteg, F. C. & Parks, R. D. (1970) *Phys. Rev. Lett.*, **24**, 520.
Zumsteg, F. C., Cadieu, F. J., Marcelja, S. & Parks, R. D. (1970) *Phys. Rev. Lett.*, **15**, 1204.

Index

adiabatic approximation, 61, 135
Ag, resistivity, 218–19, 289, (T) 290, 291, (T) 292
Ag–Al, resistivity, (T) 240, 252
Ag–Au, resistivity, (T) 224, (T) 228
Ag–Bi, resistivity, 262
Ag–Ca, resistivity (T), 270
Ag–Cu–Mg, resistivity, 381
Ag–Mg, antiphase domain structure, 53
Ag–Mn
  resistivity, 352, 354
  spin glass, 85
Ag–Pd
  effects of alloying on band structure, 301–2
  resistivity, 315
Ag–rare earth metallic glasses, resistivity, 389
Ag–Sn, resistivity (T), 288
Ag–Zn, resistivity, (T) 240, 253
Al
  band structure, 11
  phonon dispersion curves, 63, 66
  resistivity of transition metal impurities, 293–4
Al–Ag
  decomposition sequence (T), 47
  resistivity, 224, (T) 228, (T) 231, 257–8
Al–Au, resistivity (T), 228
Al–Ca, resistivity (T), 226
Al–Cd, resistivity (T), 231
Al–Co, resistivity, 262, (T) 270
Al–Cu
  decomposition sequence, 45–6, (T) 47
  GP zone sites (T), 46
  resistivity, (T) 228, (T) 231, 257–60
Al–Cu–Mg, decomposition sequence (T), 47
Al–Fe, resistivity, 262, 268–9, (T) 270, (T) 231
Al–Ga, resistivity (T), 231
Al–Ge, resistivity (T), 231
Al–Mg, resistivity, 224, (T) 226, (T) 231
Al–Mg–Si, decomposition sequence (T), 47
Al–Mn, resistivity, 342
Al–Ni, resistivity, 262, 266, (T) 270
Al–Si, resistivity (T), 231
Al–Zn
  local mean free path (T), 158
  resistivity, 224, (T) 226, (T) 231
  resistivity of GP zones, 147, 156–7, 257–9
Al–Zn–Mg decomposition sequence (T), 47
alkali metals, resistivity of various alkali impurities (T), 230
amorphous alloys (*see also* resistivity of amorphous alloys)
  atomic structure, 71ff
  dynamic radial distribution function, 75
  frequency spectrum, 75
  pair correlation function, 71
  radial distribution function, 71
  short range atomic correlations, 72
  short range structural correlations, 72
  spin–spin correlation function, 388
  structure, 69ff
  structural models, 72
  tunnelling between states, 392
Anderson and Wolff alloys (T), 346
Anderson model, 82, 345
anisotropic scattering and relaxation time, 7, 153 (*see also* anisotropic relaxation time, resistivity calculations, scattering theory)
  Ni, Fe and Al impurities in Cu, 297
antiferromagnet, *see* magnetic structure
antiphase domains, 50–1 (*see also* atomic LRO, resistivity due to surfaces, resistivity of ordering alloys)
  data for different alloys (T), 53
APW, *see* augmented plane wave method
atomic clustering (*see also* atomic phase separation, resistivity of phase separating alloys)
  atomic distribution, 43
  composition waves, 39
  computer modelling, 45
  diffuse neutron scattering from, 43
  GP zones, 45, 143ff
  spinodal decomposition, 39
atomic displacements, 57, 242–6 (*see also* resistivity due to atomic

atomic displacements, *continued*
displacements, resistivity due to thermal effects)
associated with antiphase domain boundaries (T), 61
Debye model, 65
deformable-ion, 134
diffuse X-ray scattering due to, 59
displacements due to point defects, 195ff, 227–9
displacive phase transitions, 67, 207–8
due to different atomic sizes, 58
due to long range atomic order, 60, 207
dynamic, 61ff (*see also* lattice vibrations)
effects on screening, 135
Einstein model, 62
lattice relaxations around vacancies, 233
pseudopotential associated with, 134
rigid-ion model, 134
rigid muffin-tim potential, 291
static displacements in concentrated alloys, 203ff
thermally induced displacements, 198ff
atomic long range order (LRO), 8, 51ff (*see also* antiphase domains, long period superlattice, long range order, resistivity of ordering alloys)
antiphase domains, 167–8
bandstructure effects, 164
Bragg–Williams approximation, 32, 253
composition waves, 39–40
conduction electron scattering effects, 160ff
Cowley parameters, 36
inhomogeneous order, 167–9
spinodal ordering, 41
variation with composition, 162
atomic ordering (*see* atomic long range order, atomic short range order)
atomic phase separation, 169ff (*see also* atomic clustering, resistivity of phase separating alloys)
models of two phase alloys, 170ff
phase separation, 169ff
atomic short range order (SRO), 8, 36, 49ff
atomic configuration, 41ff
conduction electron scattering effects, 139
effect on resistivity, 139ff (*see also* resistivity of ordering alloys)
in conjunction with inhomogeneous LRO, 167–9
shell model, 36
type I homogeneous (statistical), 49ff
type II(a) heterogeneous (microdomain), 50
type II(b) heterogeneous (antiphase domains), 50
atomic structure
see atomic LRO, atomic SRO, atomic phase separation, atomic clustering, atomic displacements)
atomic volume of various metals (T), 223
Au
resistivity, 218–19, 289, (T) 290
Au–Ag, resistivity, 252–3, (T) 240, (T) 288
Au–Co
Curie–Weiss temperature, 101
resistivity, 348
Au–Cr
resistivity, 352, 354
spin glass, 85
Au–Cu, resistivity, 253
Au–Fe
giant moments and atomic SRO, 106
inhomogeneous approach to magnetic LRO, 89
magnetic phase diagram, 89–90
resistivity, 354
spin glass, 85
Au–Ga, resistivity (T), 288
Au–Ho, resistivity at T_N, 366
Au–Mn
antiphase domain structure, 53
effect of atomic LRO on magnetic structure, 103
giant moments and atomic SRO, 106
resistivity, 352, 354
spin glass, 85
Au–Pd, resistivity, 252
Au–V
resistivity, 342
spin fluctuation temperature, 101
Au–Zn
antiphase domain structure, 53
resistivity (T), 288
augmented plane wave method (APW), 111ff
average t-matrix approximation (ATA), 306ff

B_2 structure, 32
backscattering, 288 (*see also* scattering theory, resistivity calculations)
band structure
Al, 11
exchange splitting, 76

basis states, 113
Bi–Ag
  resistivity (T), 270
  temperature dependence of resistivity anisotropy, 184
Bi–Cu, resistivity, 261–2
Bi–Mn
  resistivity (T), 270
  temperature dependence of resistivity anisotropy, 184
Bi–Sn, resistivity, 262
Bi–Tb, resistivity, 262
binary alloys, electrical properties, 1
Bloch wavefunctions, 107
Boltzmann equation, 17ff, 144 (*see also* distribution function, scattering theory, anisotropic scattering, resistivity calculations)
  anisotropic relaxation time solutions, 24, 285
  in two sub-band model, 323–4
  iterative solution, 257
  linearised, 21ff
  relaxation time approximation, 22
  relaxation to other formalisms, 28–9
Born approximation, 286
  first, 139, 128
  phase shifts, 282
  second, 128
  t-matrix, 285
  T-matrix, 284
Born expansion, 284
Bragg scattering, 5, 16, 140
  from zones, 148–9
  in nearly free electron theory, 126
  reduced by Debye–Waller term, 200
  reduced by pseudo Debye–Waller term, 205, 244
  superlattice, 163
  superlattice, bandstructure effects, 164
Bragg–Williams approximation, 32, 253 (*see also* atomic LRO)
  order parameter, 32–4
  resistivity, 162
Brillouin zone
  bcc, 10
  fcc, 10, 13
  special points, 41

Cd–Ag, resistivity (T), 228
Cd–Au, resistivity (T), 228
Cd–Cu, resistivity (T), 228
Cd–Pb, resistivity, 262
Ce–Al, resistivity at T_N, 366
Ce–La–Th, resistivity, 348
charge density, about many electron atom, 109
Co, resistivity (T), 275
Co–Pt, Debye–Waller and pseudo Debye–Waller factors (T), 245
coherent potential approximation (CPA), 306ff, 315
  cluster extensions, 316
  in resistivity studies, 310ff
  localised moment in alloys, 83
  self consistent, 315–16
collective electron model
  see magnetic structure
composition waves, 39ff
  in resistivity expression, 142
concentration waves
  see composition waves
conduction electrons (*see also* free electrons, resistivity calculations)
  mass, 4
  mean free path, 7 (*see also* local mean free path)
  mean free time, 7–8 (*see also* relaxation time)
  number per unit volume, 4
  scattering, 1ff (*see also* scattering theory, anisotropic scattering, anisotropic relaxation time, resistivity calculations)
  spin state, 16
conductivity tensor, 23, 26
  in Kubo formalism, 28
conductivity
  see resistivity
conservation of momentum, 15ff
core radii
  alkali metals (T), 214
  Cu and various impurities (T), 223
correlation length, near T_c, 358
correlation parameters, 30ff
  fluctuations near critical point, 356ff
  Ornstein–Zernike, 359–60
  pair, 35
  scaling, 359–60
  variation with distance, 36–7
  Warren–Cowley, 35–6
correlations, short and long range, 8 (*see also* atomic LRO, atomic SRO, atomic clustering, magnetic LRO, magnetic SRO, magnetic structure)
Coulomb interaction
  between electrons and nucleus, 108
  between electrons in virtual bound state, 81
  effect of local environment, 101
  origin of magnetic moment, 101

Coulomb interaction, *continued*
  screening of, 121
Cr
  magnetic structure, 78
  resistivity at T_N, 366
Cr–Al, temperature derivative of
    resistivity and thermopower, 372
critical point (*see also* resistivity at
    critical point)
  antiferromagnetic, 365–6
  atomic order–disorder, 366–8
  exponents, 357ff, (T) 357
  ferromagnetic, 363–5
  miscibility gap, 368
  relation of resistivity to specific heat
    and thermopower, 370–1
  renormalisation group, 358
  scaling, 357–8
  universality hypothesis, 356
crystal momentum, 15, 18
Cs, resistivity, 213
Cu
  core radius and atomic volume (T), 223
  Debye frequency spectrum, 65
  resistivity, 218–19, 289, (T) 290–1
Cu–Ag, resistivity (T), 224, 228
Cu–Al, resistivity, (T) 240, 252–3, 288
Cu–Au
  antiphase domain structure, 53
  atomic displacements due to LRO, 60
  Debye–Waller and pseudo Debye–
    Waller factors (T), 245
  diffuse X-ray scattering, 49
  pairwise correlation parameters (T), 38
  resistivity, 224, 228, 236–7, 239–40,
    (T) 248, 253–5, 256–7, 288
  resistivity due to SRO, SE and PDW
    (T), 247
Cu–Be
  decomposition sequence (T), 47
  resistivity, 257
Cu–Co
  decomposition sequence (T), 47
  resistivity, 348
  spin fluctuation temperature, 101
Cu–Fe
  resistivity, 262, (T) 288, 342–3
  spin fluctuation temperature, 101
Cu–Ge, resistivity (T), 288
Cu–Mn
  giant magnetic moments and atomic
    SRO, 106
  resistivity, 252
  spin glass, 85
Cu–Ni
  diffuse neutron scattering, 43
  effect of alloying on bandstructure, 300
  effect of atomic SRO on magnetic
    structure, 104–5
  inhomogeneous approach to magnetic
    LRO, 89
  pairwise correlation parameters (T), 44
  resistivity, 250, 252, (T) 288, 306, 350
  resistivity of magnetic clusters, 336–9
  spin fluctuation temperature, 101
Cu–Ni–Fe
  small angle neutron scattering, 48
  spin fluctuation temperature, 101
Cu–P, microstructure, 54
Cu–Pd, antiphase domain structure, 53
Cu–Pt, antiphase domain structure, 53
Cu–Sb, resistivity, 262
Cu–Ti, diffuse X-ray scattering of
    amorphous, 70
Cu–Zn
  effect of alloying on bandstructure, 300
  elastic constants, 67
  microstructure, 54
  phonon dispersion curves, 66–7
  resistivity, (T) 240, 252, at T_c, 366–8
  temperature derivative of resistivity
    and specific heat, 371

Debye model
  atomic displacements, 65
  cutoff frequency, 64
  frequency spectrum, 64–5
  frequency spectrum for Cu, 65
  in alkali metals, 213
  long wavelength cutoff in resistivity
    expression, 202, 375
  phonon energy upper limit, 15
Debye temperature, 15
Debye wave number, 16
Debye–Waller factor, 143, 209, 248
  for various alloys (T), 245
  in highly resistive alloys, 375ff
  in resistivity of amorphous metals, 387
deformable-ion model, 134
deformation potential, 134
density of states, 12
  in s- and d-bands, 14
  local, 82
deviation (difference) potential
  see deviation lattice
deviation (difference) lattice, 137, 139
  atomic displacements, 195ff
  in long range ordered lattices, 160–1
  potential, 130ff
dielectric constant, screening, 120–1
diffuse scattering
  from zones, 149

diffuse scattering, *continued*
of X-rays and neutrons, 142
dilute alloys, 30 (*see also* Kondo effect, localised spin fluctuations, spin fluctuations)
effect of magnetic near neighbours, 80
magnetic configuration, 80
magnetic moment, 80, (T) 83
resistivity, 221ff
Dirac notation, 125
discontinuous growth
see nucleation and growth, 42
dislocations, 185
displacement waves
dynamic, *see* Debye model
static, 68
displacive transitions, 67ff
distribution function, 18, 138 (*see also* Boltzmann equation)
and surface scattering, 187
Drude formula, 4, 24
Dy, magnetic structure (T), 91
dynamical matrix, 63, 227
in structure factor, 202

effective medium theory
in percolation problem, 183–4
resistivity of two phase alloy, 173–4
see also coherent potential approximation
Einstein model, 61ff
Einstein temperature, 62
elastic constants
effect of LRO, 67
effect of precipitation, 67
related to phonon dispersion curves, 66
electron waves, fundamental concepts, 1
electron wavefunction, free electron, 13 (*see also* conduction electron)
electron–electron interactions, 119ff
electronic structure
at order–disorder interface, 314
determination by APW method, 111
determination by cluster methods, 114
determination by KKR–Green's function method, 113
determination by OPW method, 111–12
determination by plane wave method, 111
determination by tight binding method, 110
empty lattice, 116
energy gaps, due to ordering at T_c, 362
Er, magnetic structure (T), 91
eutectic alloys, resistivity, 261ff

exchange and correlation, 122ff, (T) 123
Hubbard–Sham, 123
Kleinman, 123
Singwi, 123
$X\alpha$ local statistical approximation, 123
exchange interaction, 16
exchange-enhanced alloys, resistivity, 341

Faber–Ziman theory, 384
Fe
magnetic correlation (SRO) parameters, (T) 96, (T) 321
resistivity, (T) 275, (T) 290, (T) 321, 325–6
resistivity at T_c, 363–5
spin wave spectra, 98
temperature dependence of magnetisation, 93
Fe–Al
antiphase domain structure, 52
effects of LRO on phonon dispersion curves, 66
resistivity, 255–6
resistivity at T_c, 366–8
Fe–C
decomposition sequence (T), 47
magnetisation of two phase alloy, 104
Fe–Co
effect of atomic LRO on magnetic structure, 103
resistivity, 256
resistivity at T_c, 366–8
spin density, 79
Fe–Co–Ni–V, microstructure, 54
Fe–Cr, resistivity, 325–6
Fe–Cr–Co
composition profile, 55
effect of phase separation, 103
microstructure, 47
Fe–Mn
magnetic structure in ordered, 77
resistivity, 325–6
Fe–Mo–B, resistivity, 391
Fe–N, decomposition sequence (T), 47
Fe–Ni
concentration gradient at α–γ boundary, 55
effect of atomic LRO on magnetic structure, 103
resistivity, 325–6
Fe–Ni–B–Si, resistivity, 388
Fe–Ni–P–B, resistivity, 390–1
Fe–Pd, resistivity at T_c, 365
Fe–Pt, magnetic structure in ordered, 77
Fe–Rh, magnetic structure in ordered, 77
Fe–Zn, resistivity, 389

Fe_2O_3, spin wave spectra, 98
Fermi energy, in relation to phonon energy, 15
Fermi level, 13 (*see also* Fermi surface and Fermi energy)
Fermi surface, 2, 11
  fcc in reduced zone scheme, 13
  free electron, 11
  reduced zone scheme, 11
  spherical, 4
Fermi–Dirac distribution, 19
ferromagnet
  see magnetic structure
Flinn occupation parameter, 35
force–force correlation functions, 29 (*see also* resistivity calculations)
form factor, 129 (*see also* model potential, pseudopotential)
  Bardeen, 218–19ff
  Borchi & De Gennaro, 219ff
  Moriarty, 219ff, 277
Fourier transforms, discrete and continuous, 41
free electrons (*see also* conduction electrons)
  density of states, 12
  Fermi surface, 11
  theory, 4
  wavefunction, 13
Friedel oscillations, 121
Friedel sum rule, 280–2

Ga–Ag, resistivity (T), 228
Ga–Au, resistivity (T), 228
Ga–Cu, resistivity (T), 228
galvanomagnetic effects, 1
Gd
  magnetic structure (T), 91
  resistivity at T_c, 365
Ge–Ag, resistivity (T), 228
Ge–Au, resistivity (T), 228
Ge–Cu, resistivity (T), 228
giant polarisation clouds, 102
golden rule, 2
GP zone
  see atomic clustering
grain boundaries (*see also* resistivity due to surfaces)
  reflection coefficient, 185
  transmission coefficient, 185, 194
Green's functions, 113ff, 287ff

Hamiltonian, 107
Hartree self-consistent field approximation
  see screening
Hartree–Fock, criterion for local magnetic moment, 82 (*see also* spin fluctuations, localised spin fluctuations)
Heisenberg interaction, 91
Heusler alloy, effect of atomic LRO on magnetic structure, 103
Hg–Ag, resistivity (T), 228
Hg–Au, resistivity (T), 228
Hg–Cu, resistivity (T), 228
highly resistive alloys
  bandstructure effects, 377–8
  electron localisation, 378–9
  resistivity, 372ff (*see also* resistivity of amorphous alloys, resistivity of strongly scattering alloys)
Ho, magnetic structure (T), 91
Hubbard model density of states, 308
Hubbard–Sham, exchange and correlation correction, 123
hybridisation
  in CPA, 308–9
  in transition metals, 275

impurity interstitial
  see point defects
impurity level
  see virtual bound state
In–Ag, resistivity (T), 228
In–Au, resistivity (T), 228
In–Cu, resistivity (T), 228
In–Sb, resistivity, 262
integrations over d**k**, d**S**, dE and dΩ, 394
interfaces (*see also* atomic phase separation, antiphase domains, grain boundaries, magnetic domains, magnetic structure, resistivity due to surfaces)
  associated defects, 57
  structure, 54
intermetallic compounds
  electrical properties, 1
  rare earth, 1
Ir–Fe, resistivity, 344

K, resistivity, 213, 215, 217
KKR–Green's function method, 113ff, 287ff
  localised moment formation, 83
Kleinman, exchange and correlation correction, 123 (*see also* screening)
Kondo effect, 339–42 (*see also* localised spin fluctuations, spin fluctuations, spin fluctuation temperature, resistivity calculations)
  Kondo–Nagoaka theory, 102

Kondo effect, *continued*
spin glasses, 353
magnetic clusters, 335
Kubo formalism, 28

$L1_0$ structure, 33
$L1_2$ structure, 33
lattice distortions (*see also* atomic displacements, resistivity due to atomic displacements)
effective valency, 283
lattice potential (*see also* form factor, pseudopotential, model potential)
in alloys, 130
periodicity, 107
lattice vibrations (*see also* atomic displacements, Debye model, Einstein model, phonons, resistivity due to thermal effects)
effects of alloying, 65
effects of LRO, 66
investigated by neutron scattering, 65
lattice waves
see lattice vibrations
LCAO method
see tight binding approximation
Li, resistivity, 213
Li–Mg
Debye–Waller and pseudo Debye–Waller factor (T), 245
resistivity, 271
Linde–Norbury rule, 222
Lindhard dielectric function
see random phase approximation (RPA)
local and non-local pseudopotentials, 115
local charge density approximation, 123
local mean free path, 157–8, 187, 335, 359
determination of, 154ff
in precipitating systems, 257–61
with inhomogeneous LRO, 176ff
with phase separation, 143ff
local spin density approximation, 123
localised energy levels
see virtual bound state
localised moment model
see magnetic structure
localised moments, Hartree–Fock criterion (T), 83
localised spin fluctuations, 83, 98ff (*see also* Kondo effect, spin fluctuations)
effect of atomic interactions, 345–8
resistivity, 339ff
long period superlattice, 51 (*see also* atomic LRO)
long range order (LRO), 8, 50ff (*see also* atomic long range order, magnetic long range order)
single site average, 32
sublattices, 32–3
LRO
see long range order

magnetic anisotropy, effect on superparamagnet, 85
magnetic clusters, 87ff (*see also* magnetic structure)
magnetic domains (*see also* magnetic structure, resistivity of magnetic alloys)
resistivity, 322
wall thickness, 94
magnetic long range order, 89ff (*see also* magnetic structure)
inhomogeneous approach to, 89
molecular field approximation, 91–2, 319
resistivity, 319–20 (*see also* resistivity of magnetic alloys)
spin–spin correlation function, 92
temperature dependence, 94
magnetic short range order (*see also* magnetic structure)
above T_c, 79, 94
pairwise correlation parameter, 93
resistivity, 320 (*see also* resistivity of magnetic alloys)
temperature dependence, 94
magnetic structure (*see also* magnetic LRO, magnetic SRO)
band or itinerant model, 76
collective electron model, 76
domains, 94
effect of atomic order, 77, 103–4
effect of atomic phase separation, 103
effect of structural transformation, 103
giant polarisation clouds, 102
localised moment model, 76
magnetic moment density in Ni_3Fe, 77
of two phase alloy, 103
spin density, 76
spin polarisation wave, 93
spin waves (magnons), 95
variation of moment with composition, 79
magnons
see magnetic structure, spin waves
martensitic transformation, discontinuous transformation, 69
Matthiessen's rule, 9ff
as an inequality, 28
deviations from, 1, 9, 259, 267–8, 325, 328

mean free path (*see also* conduction electron)
comparable to lattice spacing, 373
local, 144, 150–1 (*see also* local mean free path)
metallic glasses
see amorphous alloys
Mg–Al, resistivity (T), 226
Mg–Cd, resistivity (T), 224
Mg–Li, resistivity (T), 226
Mg–Pb, resistivity, 262
Mg–Sn, resistivity, 262
Mg–Zn, resistivity, 262
Mg–Zn–Cd, resistivity, 389
microdomains, 50
microscopic reversibility, 22
mictomagnet
see magnetic clusters, magnetic structure
Mn, resistivity (T), 275
Mn–As, magnetic structure in ordered, 77
Mn–Au
effect of atomic LRO on magnetic structure, 103
magnetic structure in ordered, 77
Mn–Bi, effect of atomic LRO on magnetic structure, 103
Mn–Cr, magnetic structure in disordered, 77
Mn–Cu, magnetic structure in disordered, 77
Mn–Ni, effect of atomic LRO on magnetic structure, 103
Mn–Se, magnetic structure in ordered, 77
Mn–Te, magnetic structure in ordered, 77
Mo, resistivity, (T) 275, (T) 290, (T) 292
model potential, 115ff (*see also* form factor, pseudopotential)
as a weak potential, 117
Ashcroft (*see* empty-core)
Dagens resonant, 278
δ-function (*see* point-ion)
empty-core, 118ff, 213ff
Heine–Abarenkov–Animalu (HAA), 118ff, 213ff
muffin tin, 111, 289ff
point-ion, 117ff, 224ff, (T) 238
resonant, 277–9
rigid muffin-tin, 291
Shaw optimised, 118ff, 215ff
transition metal, 277–9, (T) 279, 291
molecular field approximation
see magnetic LRO
Mooij correlation, 312, 372–3, 382
Mott's rule, 224
muffin-tin potential, 111, 289ff
multiple OPW's, 249 (*see also* relaxation time anisotropy, resistivity calculations)
multiple scattering, 285, 299 (*see also* resistivity calculations)

N-scattering
see normal scattering, resistivity due to thermal effects)
Na, resistivity, 213, 215
Nb, resistivity, (T) 290, (T) 292
nearly free electron theory, 124ff
Ni
resistivity, (T) 275, 289, (T) 290
resistivity at T_c, 363–5
temperature dependence of magnetisation, 93
Ni alloys
resistivity (T), 325
resistivity of nearly magnetic alloys, 350
Ni–Al, resistivity, 256
Ni–Au
Debye–Waller and pseudo Debye–Waller factors (T), 245
variation of magnetic moment with composition, 81
Ni–Co, resistivity, 334
Ni–Cr, resistivity, 252, 334
Ni–Cr–Ti–Al, decomposition sequence (T), 47
Ni–Cu
resistivity, 332–3
resistivity at T_c, 365
Ni–Cu–Sn, resistivity at T_c, 365
Ni–Cu–Zn, resistivity at T_c, 365
Ni–Dy, resistivity, 389
Ni–Fe
effects of LRO on phonon dispersion curves, 66
magnetic moment density, 77
resistivity, 332–3
Ni–Mn
magnetic structure in ordered, 77
resistivity, 334
Ni–Mo, resistivity, 334
Ni–Ru, resistivity, 334
Ni–Ti, partial radial distribution functions, 73
Ni–V, resistivity, 334
Ni–W, resistivity, 334
noble metal
form factors, 275–7

noble metal, *continued*
noble metal alloy resistivities, 218ff, (T) 220, (T) 228, (T) 280, (T) 299
resistivity of various impurities (T), 232–3
non-simple metal, 10, 13
resistivity, 272ff
Nordheim's rule, 222
normal scattering, 16 (*see also* resistivity due to thermal effects)
Np, resistivity, 348
nucleation and growth, 42 (*see also* atomic clustering, atomic LRO, atomic SRO)

optical theorem, 29
OPW
see orthogonalised plane wave method
ordering wave, 39
orthogonalised plane wave method (OPW), 111ff

paramagnons
see spin fluctuations
partial structure factor, 211–12
in amorphous metals, 383–4
in inhomogeneous alloys, 152
of a zone, 148
Pb–Cd, resistivity (T), 231
Pb–In, resistivity (T), 231
Pb–Mg, resistivity (T), 231
Pb–Sb, resistivity, (T) 231, 262
Pd
exchange enhancement, 101
resistivity, (T) 275, (T) 290, (T) 292, 348
Pd–Ag, resistivity, 303–5, 349
Pd–Au, resistivity, 303–5, 349
Pd–Cr
giant moments and atomic SRO, 106
spin fluctuation temperature, 101
Pd–Fe, giant polarisation clouds, 102
Pd–Mn
antiphase domain structure, 53
effect of atomic LRO on magnetic structure, 103
resistivity, 354
Pd–Ni, Curie–Weiss temperature, 101
Pd–Ni–Fe, spin fluctuation temperature, 101
Pd–Pt, resistivity (T), 224
Pd–Rh, resistivity, 306
percolation
in magnetic system, 86
resistivity, 181–4 (*see also* resistivity in phase separating alloys)
perturbation expansion, 124
perturbation characteristic, 121
perturbation theory, 124ff
use with pseudopotentials, 116
phase shifts, 279ff (*see also* resistivity calculation)
with *t*-matrix, 285
with pseudopotential, 282–3
phase transformations, first and second order, 34
phonons (*see also* Debye model, lattice vibrations, atomic displacements)
creation and destruction, 16
dispersion curves, 63
drag, 16
Planck distribution, 62, 64
plane wave method, 111
point defects, 196ff (*see also* atomic displacements, resistivity due to structural disorder)
point ion potential (*see also* model potential)
values of β for various alloys (T), 238
potential
see form factor, model potential, pseudopotential
precipitation (*see also* atomic clustering, atomic phase separation)
decomposition sequences (T), 47
resultant morphology, 45ff
projection operators, 31
pseudo Debye–Waller factors (PDW), 204–5, 243–6
for various alloys (T), 245
pseudo-Schrödinger (pseudopotential) equation, 115
pseudoatoms, 135
pseudopotential, 115ff
as a wek potential, 116
average, 131
dependence upon atomic environment, 133
difference, 131ff
form factor, 130 (*see also* form factor)
in alloys, 130ff
in deformed lattice, 134ff
local and non-local, 115
not unique, 155
relation to phase shifts, 282–3
transition metal, 274–7
use with perturbation theory, 116
Pt, resistivity (T), 275
Pt–Au, resistivity, 349
Pt–Cr
resistivity, 342
spin fluctuation temperature, 101

Pt–Pd, resistivity, 349
Pu, resistivity, 348
pure metals, electrical properties, 1

radial distribution function, 71
  chemical, 73
  partial, 73
  reduced, 71
random phase approximation (RPA), dielectric function, 121 (*see also* screening)
rare earth
  intermetallic compounds, 1
  localised spins, 79
  magnetic structures (T), 91
Rb, resistivity, 213–15
Rb–Na, atomic displacements (T), 229
reciprocal space, composition waves in, 41
reciprocal lattice vector, 16
reflection coefficient
  see grain boundaries, resistivity due to surfaces
relaxation time, 3
  averaging, 27–8, 148–9, 153–4
relaxation time approximation, 22
  general resistivity expression, 138
  resistivity in, 22ff
relaxation time anisotropy, 7, 24, 153–5 (*see also* relaxation time)
  in polyvalent metals, 229–36
  in alkali metals, 213, 216–17
  in precipitating systems, 257–61
  Ni, Fe and Al impurities in Cu, 297
residual resistivity, 9
resistivity calculations, 1–2, 138, 249–50
  backscattering, 284, 375
  bandstructure effects, 273ff, 377–8
  cluster method, 289, 296–7, 299
  Cu–Au with various form factors (T), 239
  CPA techniques, 310ff
  electron–electron interactions, 274
  Faber–Ziman theory, 384
  force–force correlation method, 317
  homogeneous atomic correlations, 141ff
  in Bragg–Williams model, 162ff
  inhomogeneous atomic correlations, 143ff
  KKR–Green's function method, 294–5, 299
  Kubo–Greenwood formula, 380
  multiple OPW's, 229–36
  multiple scattering, 285, 299
  phase shift method, 279ff, 294
  polyvalent metals, 229–36
  relativistic effects, 299
  relaxation time anisotropy, 257–61
  simple metals, 137ff
  upper and lower bounds for two phase alloys, 170–2
  using pseudopotentials recalculated for different environments, 226
  using different pseudopotentials, exchange and screening corrections, 224, (T) 225, (T) 239
  using X-ray scattering data, 259
  virtual bound states, 291ff
resistivity at the critical point, 356ff
  temperature coefficient near T_c, 358ff
resistivity due to structural disorder, 195ff, 203ff
  displacive transitions, 207–8, 271
  lattice distortions around point defects, 195–8, 227–36
  microvoids, 236
  SRO, atomic displacement and pseudo Debye–Waller factor contributions (T), 247
  vacancies in various metals (T), 230
  vacancy clusters, 236
resistivity due to surfaces
  antiphase domain boundaries, 256–7
  external surfaces, 187–9
  foils and wires, (T) 192–3, 187–95
  grain boundaries, 185–6
  multilayer films, 195
  stacking faults, 185
  twin boundaries, 185
resistivity due to thermal effects, 198ff
  alkali metals, 213–18, (T) 213, (T) 215, (T) 217
  anharmonic process in noble metals, 219
  Debye–Waller terms in amorphous alloys, 387
  Debye–Waller terms in highly resistive alloys, 375–7
  effect of measuring temperature, 155ff
  effect of phonon cutoff, 387
  effect of phonon induced changes in $N(E_F)$, 310–13
  multiphonon and Debye–Waller terms, 218
  noble metals, 218ff, (T) 220
  phonon scattering, 212ff
  rigid-ion model, 134
  rigid muffin-tin model, 291
  temperature coefficient, 158, 358
  temperature dependence, 165, 199–203, 259–60, 273–4, 305

resistivity of amorphous alloys, 380ff (*see also* amorphous alloys)
Debye–Waller term, 387
effect of phonon cutoff, 387
incoherent spinwave scattering, 388–9
minima, 381, 391–2
particular spin–spin correlations, 388–9
spin glass alloys, 389–90
tunnelling between states, 392
resistivity of disordered alloys
Al alloys, (T) 225, (T) 226
alkali impurities in alkali hosts (T), 230
impurities in polyvalent hosts, 294
monovalent impurities in monovalent hosts, 299
noble metal alloys, (T) 288, (T) 299
noble metal impurities in various solutes (T), 228
point defects, 195–8
random solid solutions, 220ff
transition metal alloys, 303ff
transition metal impurities in Ni, Fe or Co, 329–32
transition metal impurities in noble metals, 295, 300
transition metal impurities in Pd, Pt or Rh, 341
various impurities in noble metal hosts, (T) 232–3
various solutes in Al alloys (T), 231
resistivity of highly scattering alloys, 213, 372ff (*see also* amorphous alloys)
saturation, 348
resistivity of ordering alloys
antiphase domain in boundaries, 256–7
coexisting LRO and SRO, 162
combined effect of atomic correlations and displacements, 208ff
combined SRO, atomic displacement and pseudo Debye–Waller factor contributions (T), 247
composition waves, 142–3
effect of atomic LRO, 253ff
effect of atomic SRO in simple alloys, 237ff
effect of atomic SRO in various non-simple alloys, (T) 250
order dependence in CPA, 313–14
order dependence in force–force correlation method, 317
studies of atomic diffusion, 251–3
resistivity of magnetic alloys, 318ff, (T) 319 (*see also* amorphous alloys)
clusters, 334ff
domain walls, 322
spin-flip, 324ff
spin glasses, 351ff, 389–90
spin waves, 274, 322, 326–7
resistivity of nearly magnetic alloys
exchange enhanced alloys, 341ff
local spin fluctuations, 339ff
nearly magnetic Ni alloys, 350
transition from Kondo to exchange enhanced behaviour, 345
resistivity of phase separating alloys, 261ff
anisotropy in eutectic, 184
effect of particle shape in two phase alloy, 177ff
effects of precipitation, 257ff
electrical analog of two phase alloy, 178ff
eutectics, 261ff
parallel and transverse to fibres, 179–81
percolation in two phase alloy, 181–4
resistivity maximum, 147
spinodal decomposition, 142–3, 261
two phase alloy (T), 175–6
resistivity relationships
Linde–Norbury rule, 222
Matthiessen's rule, 9ff
Mooij correlation, 312, 372–3
Mott's rule, 224
Nordheim's rule, 222, 236
resistivity size effect, 264–71
foils and wires, 187–95, (T) 192–3
scale of phases $\sim\Lambda$, 184ff
Rh, resistivity (T), 292
Rh–Fe, resistivity, 344, 354
rigid band approximation, 300ff
rigid-ion model, 134
rigid muffin-tin model, 291
RKKY interaction, 84, 144
asymptotic form, 85
between magnetic clusters, 87–8
damping by mean free path effects, 85
effect on magnetic ground states, 91
effect on spin fluctuations, 87
in spin glasses, 353–4
Rudermann–Kittel oscillations, 121

saturation magnetisation, 91 (*see also* magnetic structure)
scattering theory (*see also* anisotropic scattering, form factor, pseudo-potential, resistivity entries, structure factor)
anisotropy, 7, 24, 153–5
amplitude, 3

scattering theory, *continued*
 coherence, 8
 conduction electrons, 1ff
 elastic and inelastic, 15ff
 geometry, 3–4
 potential, 2 (*see also* deviation lattice)
 probability, 2–3, 20
 scattering matrix, 127, 129ff, 145
 wave vector, 2
scattering by defects, 138 (*see also* resistivity entries, scattering theory)
 antiphase domains, 167–8
 cutoff at small phonon wavelengths, 376–7
 external surfaces, 187–9
 grain boundaries, 185–6
 interfacial, 184ff
 interference between atomic and magnetic, 322
 magnetic, 10, 16
 normal scattering, 16
 phonons, 9
 s–d, 273ff, 303–5
 spin-flip, 17
 Umklapp scattering, 16
scattering matrix (*see* form factor, model potential, pseudopotential, resistivity entries, scattering theory)
Schrödinger equation
 nearly free electron theory, 124–6
 phase shift solution, 280
 time independent, 107
screened Coulomb potential, 121
screening, 108ff
 adiabatic approximation, 135
 Hartree self consistent field approximation, 108
 in an alloy, 133
 in metals, 120
 Lindhard (RPA), 121
 potentials in transition metals, 275
 Thomas–Fermi, 120, 122
 with atomic displacements, 135
self-interstitials
 see point defects
short range order (SRO), 8 (*see also* atomic SRO, magnetic SRO)
 correlation parameter, 35
simple metals, 10, 13
 resistivity of, 137ff
single site average, 31
Singwi exchange and correlation correction, 123
site occupation parameters, 30
size effect, 144, 167, 184 (*see also* atomic displacements, local mean free path, resistivity due to surfaces)
 foils and wires, 187–95, (T) 192–3
Slater–Pauling curve, 78
Sn–Ag, resistivity (T), 228
Sn–Au, resistivity (T), 228
Sn–Cu, resistivity (T), 228
Sn–Te, resistivity, 262
specularity parameter, 187–9, 194, 265, 267–8
 and surface roughness, 189
 apparent temperature dependence, 187–9
spin fluctuations, 98ff, 336 (*see also* Kondo effect, localised spin fluctuations, resistivity of nearly magnetic alloys)
 effect of Coulomb interaction, 99
 effects of local environment, 100–1
 enhanced susceptibility, 99
 in collective electron model, 78
 in spin glasses, 87
 mean lifetime, 99
 of magnetic clusters, 87
 rate compared to experimental probe characteristic time, 83
spin fluctuation temperature, 85, 99ff (*see also* localised spin fluctuations, spin fluctuations)
 characteristic values in dilute alloys (T), 10
 effect of atomic interactions, 345–8
 in spin glasses, 87, 353
spin glasses, 84ff
 competition between RKKY and Kondo interactions, 353–4
 freezing as cooperative effect, 98
 freezing temperature T_0, 85
 magnetic susceptibility, 85
 relaxation times in, 85–6
 specific heat, 85
spin waves (*see also* magnetic structure)
 diffuse, in spin glasses, 351–3
 in amorphous metals, 388
 spectra, 97
spin-flip resistivity, 324ff (*see also* resistivity of magnetic alloys)
spinodal decomposition, 41 (*see also* atomic clustering)
 morphology resulting from, 46ff
spinodal ordering, 41 (*see also* atomic LRO)
SRO
 see short range order
Stoner enhancement factor, 99 (*see also* localised spin fluctuations, spin fluctuations, spin fluctuation temperature)

structure factor, 129 (*see also* partial structure factor, resistivity entries, scattering theory, scattering by defects)
  boundaries, 150
  coexisting LRO and SRO, 163–4
  composition wave, 142–3
  dynamical, 201–2, 210
  general, 140–1
  GP zone, 149ff
  homogeneous atomic correlations, 141
  partial, 130
  self-interstitial, 198
  sublattice (homogeneous LRO), 161–2
  substitutional impurity, 197
  thermal displacements, 199, 201–2
  vacancy, 195–6
  zones, 149ff
sublattices, 32ff (*see also* atomic LRO)
substitutional impurity
  see point defects
superlattice
  see atomic LRO
superparamagnet, 87 (*see also* magnetic clusters)
  relaxation time as a function of energy (T), 89
  blocking temperature, 85

T-matrix, 128, 284ff
  in optical theorem, 29
t-matrix, 285ff
Ta, resistivity (T), 290
Tb, magnetic structure (T), 91
temperature, distinction between quench and measuring, 9
Thomas Fermi screening, 120 (*see also* screening)
Ti–Al, resistivity, 374
tight binding approximation, 110ff
Tm, magnetic structure (T), 91
transition matrix
  see *T*-matrix
transition metals (*see also* dilute alloys, pseudopotentials, model potentials, resistivity calculation)
  screening and hybridisation, 275
transmission coefficient
  see grain boundaries
two sub-band model, 323ff
two-phase mixtures (*see also* atomic phase separation, resistivity due to surfaces, resistivity of phase separating alloys)
  geometrical configuration, 57
  microstructure, 54

U-scattering
  see Umklapp scattering
Umklapp scattering, 16
units, 1, 393

vacancies (*see also* point defects, atomic displacements)
  resistivity in various metals, 195, (T) 234–5, 252
    (*see also* resistivity due to structural disorder)
  lattice relaxation around, 233
van Hove singularities, 12
virtual bound states, 80ff
  as a scattering problem, 81–2
  in transition metals, 292
  lifetime, 99
  magnetic moment of, 81
virtual crystal approximation, 130, 139, 300ff

W, resistivity (T), 290
ω-transformation, 68
  as a continuous transformation, 69
Warren–Cowley SRO parameter (*see also* atomic SRO, correlation parameters)
  in resistivity expression, 142ff
wavefunctions
  atomic-like, 109
  model, 117
  one electron, 108
  periodicity, 107
  pseudo-, 115
wavepackets, 17ff
  classical limit, 18
  in highly resistive alloys, 375
  group velocity, 17
wavevector
  scattering, 115, 139
  Thomas Fermi screening, 120
Wolff model, 345

$X\alpha$ local statistical exchange approximation, 123

Y, resistivity (T), 275
Y–Ce, resistivity, 342
Yb, magnetic structure (T), 91
Young's modulus
  see elastic constants

Ziman approximation, 27
  in *t*-matrix, 286
Ziman equation, 139
Zn, core radius and atomic volume (T), 223
Zn–Ag, resistivity (T), 228

Zn–Au, resistivity (T), 228
Zn–Cu, resistivity (T), 228
Zn–Fe, resistivity, 342
Zn–Mn, resistivity, 342
zones (*see also* atomic clustering, GP zones, resistivity of phase separating alloys, scattering theory, scattering due to defects)
  effect of measuring temperature on resistivity, 155ff
zones, *continued*
  three dimensional, structure factor, 153–4
  two dimensional, structure factor, 153–4
zone boundaries
  composition profile, 151
  scattering from, 150–1
Zr, resistivity (T), 275

Printed in the United States
By Bookmasters